LONDON MATHEMATICAL SOCIETY LECTURE NOTE SERIES

Managing Editor:
Professor N.J. Hitchin,
Mathematical Institute, 24–29 St. Giles, Oxford OX1 3DP, UK

All the titles listed below can be obtained from good booksellers or from
Cambridge University Press. For a complete series listing visit
http://publishing.cambridge.org/stm/mathematics/lmsn/

London Mathematical Society Lecture Note Series: 213

General Theory of
Lie Groupoids and Lie Algebroids

Kirill C. H. Mackenzie
University of Sheffield

CAMBRIDGE UNIVERSITY PRESS
Cambridge, New York, Melbourne, Madrid, Cape Town, Singapore, São Paulo, Delhi

Cambridge University Press
The Edinburgh Building, Cambridge CB2 8RU, UK

Published in the United States of America by Cambridge University Press, New York

www.cambridge.org
Information on this title: www.cambridge.org/9780521499286

First published 2005

A catalogue record for this publication is available from the British Library

ISBN 978-0-521-49928-6 paperback

Transferred to digital printing 2009

To Margaret

Contents

Prologue

Groupoids possess many of the features which give groups their power and importance, but apply in situations which lack the symmetry which is characteristic of group theory and its applications. Though only developed since the mid 20th century, the modern concept of Lie groupoid is as much entitled as is the familiar concept of Lie group to be regarded as the rigorous formulation of the 19th century notion which went under the then vague term 'continuous group of local transformations'; a case could be made that the modern concept of Lie group has been a transitional stage in the evolution of the notion of Lie groupoid.

Groups arise primarily, though not exclusively, in connection with symmetry; that is, as sets of automorphisms of geometric or other mathematical structures. From this viewpoint, groupoids are the natural formulation of a symmetry system for objects which have a bundle structure. The most immediate illustration from geometry is to think of a tangent bundle: with each tangent space there is at first associated a general linear group, and the presence of a geometric structure on the manifold — such as a metric, or a complex structure — is reflected in the replacement of this group by a subgroup — such as the orthogonal or complex linear group. But a tangent space is a linear approximation to the manifold only near a single point and any geometrical study will involve moving from point to point within the manifold. This being so, it is necessary to consider the isomorphisms between different tangent spaces, and one is thus led to the groupoid of all linear isomorphisms between all tangent spaces, or to a subgroupoid of isomorphisms which preserve a given additional structure.

A distinct but equally important source of groupoid theory is the fundamental groupoid of a space. Here one has a similar relationship between fundamental group and fundamental groupoid: the fundamental group at a specific point consists of (homotopy classes of) loops at the chosen point, whereas in the fundamental groupoid one considers (homotopy classes of) paths between arbitrary points. The fundamental groupoid acts upon the system of fundamental groups (and on the systems of higher homotopy groups).

Both these cases are, in groupoid terms, somewhat special. They are locally trivial in a sense which is similar to that for fibre bundles, and because of this there is a compromise position, which yields a principal bundle. In the first case, rather than considering the system of all isomorphisms between the tangent spaces of a manifold M, one fixes a reference point m_0 and considers isomorphisms from $T_{m_0}(M)$ to all other tangent spaces; these can be readily identified with isomorphisms from a standard reference space $\mathbb{R}^n$, and one arrives at the full frame bundle or a reduction of it. In the second case, fixing a reference point and considering (homotopy classes of) paths from it to arbitrary points within the manifold yields the universal cover, with its structure as a principal bundle over the given space. For a general Lie groupoid, however, not locally trivial, choosing a point of the base manifold and considering the arrows which radiate from it, yields a structure which is not equivalent to the original groupoid.

It was at one time very common to find groupoid theory criticised as an artificial generalization of group theory. Personally, I find generalization an always useful technique for understanding basic concepts; certainly as useful as the dual study of examples. A well–chosen generalization supplies for a theory the effect of a frame on a painting, and enables it to be looked at 'from the outside' as it were. Nonetheless, the case for Lie groupoids and Lie algebroids does not rely on such considerations. I outline here four of the most important.

(I) The Lie theory of Lie groups and Lie algebras is one of the cornerstones of 20th century mathematics, and it is hard to disentangle the importance of Lie groups and Lie algebras as encoders of symmetry from the importance of the classification results for Lie algebras. There is no comparable classification of Lie algebroids — and I personally do not believe that such a classification is possible — but the basic processes of Lie theory carry over to Lie groupoids and Lie algebroids and provide a unified approach to many of the fundamental constructions in first–order differential geometry. Differential geometry is, after all, the study of geometry by means of differentiation — or linearization — and the process of taking the Lie algebroid of a Lie groupoid demonstrates that this basic construction has no necessary relationship with the case of groups and all the familiar symmetry which is present in that case.

The three classical results of Lie theory — the integrability of morphisms, the integrability of subobjects, and the integrability of abstract Lie algebras — all generalize to meaningful questions in the context of Lie groupoids and Lie algebroids. The answers are of course more complex, but what is important is that they embody results of intrinsic geometric interest, both known and new.

In Part II I give a detailed exposition of this aspect in the case of locally trivial Lie groupoids and transitive Lie algebroids. In this case the integrability of morphisms embodies the triviality of a bundle which has a flat connection and a simply–connected base; the integrability of Lie subalgebroids embodies the Ambrose–Singer and Reduction Theorems; and the integrability question for abstract transitive Lie algebroids gives criteria for the existence of connections when curvature data is prescribed. Expressed in terms of principal bundles these results fit into no clear framework.

(II) The relationship between the results of connection theory mentioned in **(I)** and the classical Lie theory of Lie groups and Lie algebras is concealed — very effectively — by the awkward nature of the algebra of principal bundles. Indeed, there hardly exists a recognizable algebraic theory of principal bundles, perhaps in part because there has been no clear model on which to build it: principal bundles do not behave very much like modules, nor like vector bundles. Nonetheless, it is possible to develop an algebraic theory of groupoids (set–theoretic and Lie) to an extent which may be surprising at first. This algebraic theory begins by being modelled on standard group theory, but diverges from it in some important respects; it is possible, for example, to characterize many algebraic constructions by classes of morphisms in a way which is impossible for ordinary groups. These constructions have analogues for Lie algebroids and together they form a technique of great value in treating geometric questions involving linearization and globalization. This process is described in detail in Chapters 2 and 4.

All frame groupoids and all fundamental groupoids are locally trivial. Lie groupoids which are locally trivial are symmetry structures of vector bundles — or of more general fibre bundles — and it is precisely the class of locally trivial groupoids for which the concept of principal bundle provides an alternative. It is intrinsic to the nature of general Lie groupoids that they are not determined by their vertex structure in the way that is true in the locally trivial case.

It is sometimes suggested that the notion of symmetry should be extended so that general Lie groupoids may be regarded as symmetry structures. This seems to me debatable; the concept of symmetry as it passes from ordinary usage into mathematics retains features which link it firmly to group symmetry. To extend the word to encompass arbitrary groupoids — in effect, to treat every equivalence relation as if it were the orbit relation of a group action — robs the word of much of its significance. However, one could distinguish groupoids from groups by the possibility of a continuous variation of symmetry.[1]

(III) Large families of Lie groupoids which are not locally trivial arise naturally: as the symplectic realizations of Poisson manifolds, in the study of non–transitive Lie group actions, and in foliation theory, for example. Whereas many basic constructions have been known for general Lie groupoids for a number of years, it is only very recently that strong results have been obtained in the general case. The fact that practical results and working techniques now exist for general Lie groupoids is a compelling argument for abandoning the concentration on the principal bundle approach in the locally trivial case.

Most of Part I, and most of Part III, applies to general Lie groupoids and Lie algebroids. However, it has not been possible to give a complete account of all recent developments, and some topics are only described in outline in the Appendix.

(IV) Where future developments are concerned, the most important distinction between groupoids and groups lies in the existence of higher–dimensional forms of the concept. It is widely appreciated that iterating the group concept by considering a group object in the category of groups leads to nothing new: a group object in the category of groups is merely an abelian group, a result which is the abstract form of the fact that the higher–order homotopy groups are all abelian. However, a groupoid object in the category of groupoids is a genuinely new and different object. These double groupoids arise naturally in Poisson geometry and may be regarded as a semi–classical form of the use of multiple category theory in quantization.

This theory lies largely beyond the present book, though some aspects of double structures are treated in Part III. Further references are given in the Appendix.

---

[1] Plotinus (A.D. 205–269/70) defined beauty as symmetry irradiated by life: 'There must be symmetry, achieved by the perfect realisation of geometrical possibilities. There must be a feeling of movement, for movement meant life.' [Runciman, 2004].

Introduction

As with many books, this one is best read piecewise backwards. In describing the contents, I accordingly begin with Part III.

Groupoid theory was transformed in the mid 1980s by the introduction of the notion of symplectic groupoid and the methods of Poisson geometry. The announcement by Weinstein [1987] and the seminar notes of Coste, Dazord, and Weinstein [1987] on symplectic groupoids and Poisson geometry became available about late 1986. In fact a similar approach to the use of groupoid structures in Poisson geometry had been given by Karasëv [1989] in papers deposited in VINITI in Moscow in 1981 but not generally available until much later. The two papers of Zakrzewski [1990a,b], gave a third and independent treatment. In most of the discussion in this Introduction I will treat these three very different approaches as if they were a single body of work.

The work of these authors transformed both the subject and the applications of Lie groupoid and Lie algebroid theory. Until that time only the case of locally trivial Lie groupoids and transitive Lie algebroids was well–understood. Despite the very general programme and results announced by Pradines in four short notes [1966], [1967a], [1967b], [1968], and some isolated work on specific aspects of general Lie groupoids and Lie algebroids, there seemed to be little compelling reason to understand the very difficult general theory.

The reciprocal influence — the importance of groupoid theory in Poisson geometry — is based on two fundamental observations. Firstly, that the Poisson bracket of 1–forms on a Poisson manifold P makes the cotangent bundle T^*P a Lie algebroid — this fact in itself was found by a number of authors; the survey of Huebschmann [1990] gives a detailed account of the history. Secondly, on the relationship between realizations of Poisson manifolds and groupoid structures. A realization[1] of a Poisson manifold P is a surjective submersion $S \to P$ which is a Poisson map from a symplectic manifold S to P. The simplest interesting example is that the linear Poisson structure on the dual $\mathfrak{g}^*$ of a Lie algebra $\mathfrak{g}$ has a realization $\mathscr{R}: T^*G \to \mathfrak{g}^*$ where G is any Lie group integrating $\mathfrak{g}$ and $\mathscr{R}$ is right–translation. That this map, and the corresponding left–translation $\mathscr{L}$, give symplectic realizations had been known for a considerable time: the new and crucial observation in the 1980s was that $\mathscr{R}$ and $\mathscr{L}$ are the source and target maps for a natural groupoid structure on T^*G defined by the coadjoint action, and that the canonical symplectic structure on T^*G is compatible with this groupoid structure in a natural way.

The concept of symplectic groupoid extends this example and links the two fundamental observations tightly together. For any symplectic groupoid Σ with base a Poisson manifold P, the target map is a symplectic realization of P and the source map is a symplectic realization of the opposite structure. Thus Σ with its symplectic structure may be regarded as a desingularization of P with its Poisson structure. Most remarkably, the Lie algebroid $A\Sigma$ of the Lie groupoid structure and the cotangent Lie algebroid T^*P of the Poisson manifold P are canonically isomorphic. Thus the realization problem for Poisson manifolds has been reduced to an aspect of a generalized Lie theory. (It is, furthermore, true that for P a Poisson manifold, any Lie groupoid which integrates T^*P and which is suitably connected, has a canonical symplectic structure making it a symplectic groupoid with base P.) In particular, integrating a Lie algebra $\mathfrak{g}$ and finding a symplectic realization of $\mathfrak{g}^*$ are equivalent problems.

Thus the Lie theory of Lie groupoids and Lie algebroids embodies the classical Lie theory of Lie groups and Lie algebras not only in its standard form, but also in the dual form which is a special case of the relationship between symplectic groupoids and Poisson manifolds.

Symplectic groupoids are only rarely locally trivial, and their behaviour is very far removed from the features of the locally trivial case. The existence of a symplectic groupoid structure is a very strong constraint on a groupoid. While symplectic groupoids provided a definite reason for studying Lie theory without the restriction of local triviality, it has until recently proved difficult to construct large families of examples of symplectic groupoids.

---

[1] This is actually a *full* realization in the terminology of Weinstein [1982]; I will not consider realizations in which the map is not a surjective submersion.

The importance of the work of Karasev, Weinstein and Zakrzewski for Lie groupoid and Lie algebroid theory themselves rests primarily on two further developments.

Firstly, one consequence is a duality between Lie algebroids and vector bundles with a Poisson structure which respects the linear and bundle structures: this book calls these simply *Poisson vector bundles*. This duality, given in detail by Courant [1990], extends the classical duality between Lie algebras and linear Poisson structures, and also gives a clear meaning to the statement that the canonical symplectic structure on a cotangent bundle is the dual of the bracket of vector fields.

This duality is nontrivial in the specific sense that it takes place outside each of the categories with which it is concerned: one must consider sections of the Lie algebroid and functions (or 1–forms) on the dual bundle. Taken together with the process which associates the cotangent Lie algebroid to any Poisson manifold, this means that there are two processes which pass between the Poisson category and the Lie algebroid category, and although these are far from giving an equivalence, they frequently allow problems on one side to be usefully transformed into problems in the other. Because these processes are non–trivial and not genuinely inverse, they often deliver a substantial benefit.

Secondly, Weinstein [1988] introduced a concept of Poisson groupoid which unites the two extreme cases of symplectic groupoids and Poisson Lie groups. That such a unification would be possible does not seem obvious even in retrospect: symplectic groupoids were introduced as global realizations of Poisson manifolds, whereas the concept of Poisson Lie group was introduced by Drinfel'd [1983] as a semi–classical form of the notion of quantum group, and was then seen to also provide a valuable tool in work on complete integrability. Nonetheless the concept of Poisson groupoid provides a continuum of structures linking symplectic groupoids to Poisson Lie groups. Poisson groupoids have been shown to provide an appropriate general framework in which to study the classical dynamical Yang–Baxter equation [Etingof and Varchenko, 1998]. Their infinitesimal form, the concept of Lie bialgebroid defined by myself and Ping Xu [1994] is part of a family of algebraic concepts under development in mathematical physics [Xu, 1999]. Poisson groupoids and Lie bialgebroids have turned out to be crucially important in the study of double Lie groupoids and double Lie algebroids [Mackenzie, 1998].

The account of Poisson groupoids which I give here is a theoretical one — there has been no space to deal with examples beyond the most fundamental. The treatment is new and appears in print here for the first time, though I have had the good fortune to be able to set it out in series of lectures at Utrecht and Amsterdam in 2000, and at Queen Mary, London in 2002.

The crux of this approach is the observation that the compatibility condition between a Poisson structure and a Lie groupoid structure on a manifold G is equivalent to a compatibility condition between the Lie groupoid structure on T^*G induced by the groupoid structure on G, and the anchor of the Lie algebroid structure induced on T^*G by the Poisson structure on G. This condition is of categorical or diagrammatic type — it is of the same nature as the compatibility conditions on double structures in the categorical sense: that the structure maps of one structure be morphisms with respect to the other. In fact the groupoid structure maps in T^*G, for G a Poisson groupoid, are Lie algebroid morphisms, so that T^*G is an $\mathscr{LA}$–groupoid in the terminology of [Mackenzie, 1992]. General $\mathscr{LA}$–groupoids occupy an intermediate place between double Lie groupoids and double Lie algebroids. This double Lie theory is not treated in the present book, but the treatment of Poisson groupoids given here provides much of the background necessary for it.

A more immediate consequence of this treatment of Poisson groupoids is that the basics of symplectic groupoid theory may be developed without any use of genuine symplectic geometry. I deduce the basic properties of symplectic groupoids in §11.5 as an immediate corollary of the general Poisson groupoid theory, merely by imposing the nondegeneracy condition of the Poisson anchor. The first accounts of symplectic groupoid theory made extensive use of nontrivial and genuinely symplectic results, but the approach given here avoids this entirely. This is not, of course, because of any lack of appreciation of the power and importance of symplectic geometry, but to demonstrate that those aspects of symplectic groupoid theory which often surprise people new to the subject, are not in fact consequences of symplectic geometry as such. This procedure extends beyond the aspects treated in this book, and may be applied, for example, to deduce results on symplectic groupoid actions from general Lie theory.

One might call the principle underlying this approach a 'cotangent philosophy' — that where general constructions with Poisson structures are concerned, it is conceptually simpler to work with the cotangent Lie algebroid rather than with the actual manifold. Readers not already acquainted with this approach may at first question the word 'simpler' — there is, for instance, a good deal of apparatus in §11.5. However, the calculus of canonical isomorphisms which I develop here behaves in a direct, categorical fashion, and applies with minor changes to much more general situations.

It was not possible, in a book of reasonable length, to provide a treatment of work on double Lie structures while also covering most of the basic theory and constructions of Lie groupoids and Lie algebroids. What I have been able to do — I trust — is to show in how completely natural a way work with Poisson groupoids (and hence Poisson Lie groups and symplectic groupoids) is clarified by the consideration of double structures. One sees this already when considering Poisson structures on a vector bundle: these can be very easily handled through the associated double vector bundles.

Part III therefore begins in **Chapter 9** with an account of double vector bundles. The notion of double vector bundle has been around for some time — the general notion and an extensive account of much general theory was given by Pradines [1974a], and the double tangent bundle of a manifold and the tangent of a general vector bundle were sometimes used in accounts of connection theory in the 1960s and 1970s [Dieudonné, 1972], [Besse, 1978], [Yano and Ishihara, 1973]. Iterated tangent and cotangent bundles have a well–established place in some treatments of classical mechanics [Tulczyjew, 1989]. The chapter starts with an account of the general concept and progresses immediately to recent results on the duals of a double vector bundle, due independently to myself [Mackenzie, 1999] and to Konieczna and Urbański [1999]. Briefly stated, a double vector bundle may be dualized either along its horizontal structure or its vertical structure, and these two duals are themselves dual. Thus successive dualizations of a double vector bundle return to the original structure not at the second dual, but at the third. This phenomenon is likely to be of great significance in further work on multiple structures.

Applying this result to the tangent double vector bundle of a vector bundle, Chapter 9 recovers the canonical isomorphisms between the cotangents of a vector bundle and of its duals, introduced by myself and Ping Xu [1994, 1998] and the canonical pairing between the tangents of a vector bundle and of its dual. These will be used repeatedly in the subsequent chapters.

Chapter 10 gives the crucial relationships between Poisson structures and Lie algebroids. The basic definitions and properties of Poisson structures are included, but a reader entirely new to Poisson geometry

will need to supplement this with other sources. I give the construction
of the cotangent Lie algebroid of a Poisson manifold and in §10.3 treat
the duality between Poisson vector bundles and Lie algebroids. In §10.2
I briefly consider Poisson cohomology as an example of Lie algebroid
cohomology, and describe the associated Batalin–Vilkovisky structures.
§10.4 gives the correspondences for subobjects and for morphisms between Poisson structures and Lie algebroids; in principle all concepts of
Poisson geometry may be translated to Lie algebroids, and reciprocally.

Chapter 11 treats Poisson groupoids, beginning with a brief resumé
of Poisson Lie groups in terms of the Lie groupoid and Lie algebroid
structures on T^*G. The techniques normally used for a Poisson group
G exploit the fact that the tangent group TG is trivializable as a bundle
and is thus a semi–direct product, with respect to the adjoint representation, as a group, Thus one commonly defines a Poisson structure and a
Lie group structure to be compatible if the map $G \to \mathfrak{g} \wedge \mathfrak{g}$ derived from
the Poisson tensor π is an Ad–cocycle. This formulation fits into no pre-
existing framework and it can take some time to build up a feel for how
to proceed. Alternatively, one may define a Poisson structure and a Lie
group structure to be compatible if the multiplication map is Poisson.
This appears to be a compatibility condition of a standard type, but the
'backward flipping' or contravariant nature of Poisson structures means
that this intuition can mislead: group inversion is not a Poisson map,
but anti–Poisson, and right– and left– translations are neither Poisson
nor anti–Poisson; the analogy with more familiar compatibility conditions breaks down. I demonstrate in §11.1 that naturality is restored by
working with the structures on the cotangent bundle.

For a general Lie groupoid $G \rightrightarrows P$ there is no simple version of the
adjoint and coadjoint representations. The original treatment of Weinstein [1988] was in terms of the coisotropic calculus introduced in the
same paper. Here I make systematic use of the structures on the cotangent bundle. Both the tangent bundle TG and the cotangent bundle
T^*G are 'double' objects in a categorical sense: they possess a vector
bundle structure and a Lie groupoid structure, and the structure maps
of each structure are morphisms with respect to the other structure.
This kind of compatibility condition has been used in category theory
since at least the 1960s, when Ehresmann made what at the time must
have seemed to be a complete change of direction from differential geometry to multiple category theory. This use of the word 'double' should
be carefully distinguished from its use in work on the Drinfel'd dou-

ble, though in fact the two usages are related in a non–obvious fashion [Mackenzie, 1998].

The double structure — more precisely the $\mathscr{VB}$–groupoid structure as defined by Pradines [1988] — of TG and T^*G supplies a replacement for the actions which characterize the structure of a tangent or cotangent group. In §11.2 I define the abstract notion of $\mathscr{VB}$–groupoid, a groupoid object in the category of vector bundles, and give the duality which exists for such structures. This extends the duality for double vector bundles which was treated in Chapter 9. The brief §11.3 deduces explicitly the properties of the cotangent groupoid and §11.4 then gives the basic properties of Poisson groupoids.

In §11.5 I first deduce the basic properties of symplectic groupoids as a special case of Poisson groupoids. The main part of the section proves that $T^*G \rightrightarrows A^*G$, for G a general Lie groupoid, is a symplectic groupoid with respect to the canonical symplectic structure on T^*G. The proof proceeds by showing that the canonical isomorphisms R, I and Θ associated with iterated tangent and cotangent bundles respect the relevant groupoid structures. This 'calculus of canonical isomorphisms' can be readily extended to more general situations.

Chapter 12 is a short account of the basic theory of Lie bialgebroids. I take as definition the Schouten calculus formula used by myself and Ping Xu in the original paper [Mackenzie and Xu, 1994] and prove its self–duality by a mixture of the proof given in [Mackenzie and Xu, 1994] and the much improved formulation of Kosmann–Schwarzbach [1995]. §12.2 gives the proof that this definition is equivalent to a morphism criterion which is parallel to the definition of Poisson groupoid. The following §12.3 gives an alternative and often preferable construction of the Poisson structure on the Lie algebroid of a Poisson groupoid. Finally in §12.4 I consider Poisson actions and moment maps for Poisson Lie groups; this is an introduction to a large further subject.

Part II

Part II of the book is devoted to the theory of transitive Lie algebroids and locally trivial Lie groupoids. This case is inextricably bound up with connection theory.

Connection theory was a very large and important area of differential geometry for most of the 20th century; there now seems to be some danger of it falling out of common knowledge. From the introduction of Christoffel symbols and the identities of Bianchi around 1870 to the explosive development of gauge theory about a hundred years later, connection theory was understood to be a fundamental concept in geometry and mathematical physics. In the late 19th century a connection was a structure associated with a surface or higher–dimensional Riemannian manifold; in modern terminology this is the Levi–Civita connection of a Riemannian manifold. The concept was then separated from a metric structure by Weyl in an early attempt at a unified field theory, giving rise to what we now think of as an affine connection in a manifold. Some 20 to 30 years later, in the late 1950s, Ehresmann formulated a very general notion of connection in a principal bundle or locally trivial Lie groupoid. It was this notion, in its principal bundle form, which Kobayashi and Nomizu [1963] put at the start of their classic treatment of the foundations of differential geometry. In the 1970s the treatment of [Kobayashi and Nomizu, 1963, Chap 2] was widely regarded as mysterious and difficult; some 20 or 30 years of gauge theory has lessened that attitude somewhat, but it is still I think the case that the treatment of connection theory in principal bundles is regarded as a less easily assimilated piece of mathematics than one expects of such a fundamental concept. Indeed many treatments still focus almost exclusively on Koszul connections in vector bundles, a concept which is much more easily absorbed. However, to dismiss principal bundles as an optional elaboration of vector bundles is a modern version of the old misconception that the only Lie groups which matter are matrix groups.

The formulation of connection theory in terms of Lie groupoids and Lie algebroids makes clear that connection theory follows the pattern of the Lie theory of Lie groups and Lie algebras. Indeed connection theory *is*, in a sense which is made precise below, the Lie theory of locally trivial Lie groupoids and Lie algebroids. To justify this statement quickly, recall first of all the basic three theorems[1] of Lie:

[**Lie–1**] Let G and H be Lie groups with G simply–connected and with Lie algebras $\mathfrak{g}$ and $\mathfrak{h}$. Let $\varphi\colon \mathfrak{g} \to \mathfrak{h}$ be a morphism of Lie algebras. Then there is a unique morphism of Lie groups $F\colon G \to H$ which induces φ.

[**Lie–2**] Let G be a Lie group with Lie algebra $\mathfrak{g}$ and let $\mathfrak{h}$ be a Lie subalgebra of $\mathfrak{g}$. Then there is a unique connected Lie subgroup H of G for which the Lie algebra is $\mathfrak{h}$.

---

[1] These are, of course, modern formulations and do not exactly correspond to the three theorems as stated in, for example, expository books of the early 1900s. Perhaps they should be called the *three integrability theorems of Lie*.

[**Lie–3**] Given a Lie algebra $\mathfrak{g}$, there is a Lie group G with Lie algebra isomorphic to $\mathfrak{g}$.

Throughout the book, I take all Lie groups and Lie algebras to be finite–dimensional and real, except where explicitly indicated otherwise. Cognate with these theorems are the following important riders:

[**Lie–4**] Let G be a Lie group. Then the connected component of the identity element, G_0, is a Lie group and the inclusion $G_0 \to G$ induces the identity map of Lie algebras $\mathfrak{g} \to \mathfrak{g}$.

[**Lie–5**] Let G be a connected Lie group. Then the universal cover $\widetilde{G}$ is a Lie group and the covering projection $\widetilde{G} \to G$ induces the identity map of Lie algebras $\mathfrak{g} \to \mathfrak{g}$.

I show in Part II that each of these results, [**Lie–1**] to [**Lie–5**], has an extension to locally trivial Lie groupoids and transitive Lie algebroids, which embodies substantial results for principal bundles or their connection theory. The extension of [**Lie–1**] to locally trivial Lie groupoids (6.2.11) has the same form and implies the following well–known result for principal bundles:

[**CT–1**] Let $P(M,G)$ be a principal bundle with M simply–connected. If P admits a flat connection, then it is trivializable.

Likewise, the extension of [**Lie–2**] to locally trivial Lie groupoids (6.2.1) has the same form and implies the following result, which combines the Ambrose–Singer and Reduction Theorems for principal bundles:

[**CT–2**] Let $P(M,G)$ be a principal bundle and let γ be a connection in $P(M,G)$. There is a least Lie subalgebroid of the Lie algebroid constructed from $P(M,G)$ which contains the values of γ and the values of its curvature form, and this Lie subalgebroid is the Lie algebroid corresponding to any holonomy bundle of γ.

The extension of [**Lie–3**] to transitive Lie algebroids is a more complex matter, treated in Chapter 8. For a real valued closed 2–form ω on a manifold M, a classical lemma of Weil states that ω is the curvature of a connection in a circle bundle over M if and only if the Čech class corresponding to the de Rham class of ω is integral. The extension of [**Lie–3**] to transitive Lie algebroids gives a comprehensive generalization of this to 2–forms which take values in an arbitrary Lie algebra bundle. See 8.3.9 for a precise statement.

The two riders extend to locally trivial Lie groupoids in a straight-forward fashion. The corresponding results for principal bundles are as follows:

[**CT–4**] Let $P(M, G)$ be a principal bundle and let P_0 be any connected component of P. Then there is an open subgroup G_1 of G such that $P_0(M, G_1)$ is a principal subbundle of $P(M, G)$, and the inclusion $P_0 \to P$ induces an isomorphism of the corresponding Lie algebroids.

[**CT–5**] Let $P(M, G)$ be a principal bundle with P connected. Then the universal cover $\widetilde{P}$ has a principal bundle structure $\widetilde{P}(M, H)$ for a certain Lie group H, which is an extension of G, such that the covering projection $\widetilde{P} \to P$ is a morphism of principal bundles over M and induces an isomorphism of the corresponding Lie algebroids.

Thus the basic Lie theorems for locally trivial Lie groupoids and transitive Lie algebroids give results for connection theory. The proofs are not, in themselves, fundamentally different to the standard proofs given by [Kobayashi and Nomizu, 1963]. Rather, the Lie groupoid/Lie algebroid formulation provides a conceptual framework in which the results arise naturally and inevitably. This formulation is scarcely possible if one works solely with principal bundles.

Perhaps more surprisingly, one may prove [**CT–1**] and [**CT–2**] directly and then deduce from them the locally trivial Lie groupoid/Lie algebroid versions of [**Lie–1**] and [**Lie–2**]. Thus connection theory is actually coextensive with the Lie theory of locally trivial Lie groupoids and Lie algebroids.

From a pedagogical point of view, the demonstration that connection theory is a generalization of Lie theory fits it firmly into the broad framework of contemporary mathematics: Lie theory is the study of a functor, and its fundamental results concern the faithfulness and fullness of this functor.

Nonetheless this has not been the place to give a full expository treatment of connection theory: considerations of space and balance mean that Part II contains no significant examples or motivation outside the theory itself.

Looking at connection theory in a broad sense, there have always been three distinct approaches. One may treat it 'globally' in terms of differential forms and exterior calculus; one may treat it 'locally', either by a localization of the differential form approach, or by some form of tensor analysis; finally, one may treat it in terms of path lifting and holonomy. These approaches correspond respectively to the three great cohomology theories which concern differential geometry: the de Rham cohomology, Čech cohomology, and singular cohomology.

The close relationships between these approaches sometimes obscures the fact that they may be treated independently: I demonstrate in the course of Part II that both the global and the local approaches are intrinsic to transitive Lie algebroids and may be developed entirely independently of any underlying Lie groupoid. The place of classical de Rham cohomology is taken by Lie algebroid cohomology, which emerges from a straightforward extension of the calculus of differential forms. On the other hand, considerations involving path–lifting and holonomy require an underlying locally trivial Lie groupoid. The relationships between these three approaches can now be seen as instances of the Lie theory of Lie groupoids and Lie algebroids.

In the writing of Part II, I have had in mind a reader who has met the standard theory of connections in vector bundles, and who does not need to be given geometric motivation for the notion. Nonetheless, all basic definitions are included.

Here now is a brief description of the contents of the individual chapters. **Chapter 5** is an account of the infinitesimal part of the theory of connections in transitive Lie algebroids, both the global calculus and the local description. For the latter, I anticipate Chapter 8 by using a temporary concept of *locally trivial Lie algebroid*; Theorem 8.2.1 shows that all transitive Lie algebroids are locally trivial.

Chapter 6 is primarily concerned with path connections and holonomy. I begin however with proofs of [**Lie–1**], [**Lie–2**] and [**Lie–5**] for locally trivial Lie groupoids, with [**Lie–5**] in §6.1 and [**Lie–1**] and [**Lie–2**] in §6.2. The proofs (given for the first time in [Mackenzie, 1987a]) depend heavily on local triviality, but are for that reason more immediately accessible. Path connections and their holonomy are treated in §6.3. The important §6.4 deduces forms of the Ambrose–Singer and Reduction Theorems of connection theory from [**Lie–1**] and [**Lie–2**]. Using the results of §6.4, §6.5 establishes several 'local constancy' results for transitive Lie algebroids: the kernel of the anchor map is a Lie algebra bundle; morphisms of transitive Lie algebroids are of locally constant rank, and so on.

Chapter 7 is an extended account of the cohomology of Lie algebroids. On a formal level this cohomology is a simple generalization of Chevalley–Eilenberg cohomology of Lie algebras and de Rham cohomology, and has been studied on many occasions; however, to make use of this formalism for smooth structures, it is necessary to ensure at all stages that the constructions respect the underlying vector bundle structure, and this requires the results of §6.5. The account of non–abelian extension theory in §7.3 is (I hope the reader will agree) an elegant demonstration of the geometric content of a purely algebraic formalism: for example the (second) Bianchi identity emerges as the condition that the obstruction class of the coupling (*abstract kernel* in the terminology of Mac Lane [1995]) associated to a transitive Lie algebroid is zero. The results of this section are the foundation for Chapter 8.

§7.4 treats the spectral sequence of a transitive Lie algebroid; this unites the Hochschild–Serre spectral sequence of an extension of Lie algebras with the Leray–Serre spectral sequence of a principal bundle.

Chapter 8 is concerned with the integrability problem for transitive Lie algebroids; that is, with [**Lie–3**]. The main result, Theorem 8.3.9, produces a Čech class associated to any transitive Lie algebroid, the *(cohomological) integrability obstruction*. If $\widetilde{M}$ is the universal cover of the base of the Lie algebroid A, and $Z\widetilde{G}$ is the centre of the universal covering Lie group of the fibre type of the adjoint bundle of A, then the integrability obstruction is in $\check{H}^2(\widetilde{M}, Z\widetilde{G})$. For the Lie algebroid to be integrable, it is not necessary that this class be zero, but that it lie within a discrete subgroup of $Z\widetilde{G}$; equivalently, that it can be sent to zero by the map on cohomology induced by a covering map $\widetilde{G} \to G$, for some G. This class is thus a non–abelian version of the first Chern class of a circle bundle and §8.1 is spent in recalling the various aspects of the first Chern class which are relevant here. The detailed construction of the class is in §8.3. The construction of the integrability obstruction depends crucially on the result 8.2.1 that a transitive Lie algebroid on a contractible base admits a flat connection. This result also provides, in §8.2, a local description of transitive Lie algebroids by families of local Maurer–Cartan forms; this is an infinitesimal analogue of the description of principal bundles by transition functions.

I have not considered the construction of non–integrable transitive Lie algebroids at any length; this, of course, requires techniques outside of Lie groupoid theory. The most natural source is the theory of transversally complete foliations developed by Molino [1988]; the original examples of non–integrable Lie algebroids of Almeida and Molino

[1985] are of this type. A thorough account of the Molino theory and its relation to integrability has been given by Moerdijk and Mrčun [2003].

The extension of [**Lie–1**] to [**Lie–5**] to general Lie groupoids and Lie algebroids is a far more complex matter, and has been the subject of much recent work. Some account of the problems involved and recent achievements are given in the Appendix.

Part I

Lie groupoids (*groupoïdes différentiables*) were introduced by Charles Ehresmann in the 1950s as an alternative to the principal bundle language which he had developed at about the same time as Steenrod. (A detailed history is given in [Ehresmann, 1984, Vol. I].) Ehresmann appears to have been most concerned with higher–order prolongation processes and the notion of a connection of arbitrary order, but it is evident that from the outset he was also concerned with Lie groupoids as holonomy structures in foliation theory.

Lie algebroids were defined by Pradines in the second of a series of short papers [1966, 1967a, 1967b, 1968] outlining a Lie theory for Lie groupoids. The cognate, purely algebraic concept of a Lie pseudoalgebra had already been discussed by a number of authors, and forms of this concept have continued to be introduced since; Pradines' concept of Lie algebroid is unique in positing an underlying structure of smooth vector bundle, and it is this which makes possible the Lie theory of Lie groupoids. Pradines' four short papers were mainly concerned with setting out the theory, which in the case of general Lie groupoids is intimately bound up with foliation theory. This theory lay largely undeveloped for several years.

In [Mackenzie, 1987a] I set out to establish Pradines' Lie theory in rigorous detail for the locally trivial case and, at the same time, to demonstrate the close links with connection theory which have been described above. As well as the restriction to the locally trivial case, [Mackenzie, 1987a] also largely restricted itself to morphisms which are base–preserving; the two restrictions are equally natural when connection theory is the main application.

Arbitrary morphisms of general Lie algebroids are — rather surprisingly — hard to handle; a detailed treatment was given by Philip Higgins and myself [Higgins and Mackenzie, 1990a]. This paper showed that Lie algebroids possess an infinitesimal version of the properties characteristic of general groupoids and introduced a notion of quotient, or descent, for Lie algebroids which had hitherto been lacking.

Part I is concerned with the differentiation processes which relate Lie groupoids and Lie algebroids. **Chapter 1** is concerned with the most basic constructions and fundamental examples for Lie groupoids. Local triviality is treated in full, including the relationship with principal bundles: the fact that this is not a true equivalence has as a consequence that the automorphism groups of a locally trivial Lie groupoid and of a corresponding principal bundle do not correspond. Most of this chapter deals with constructions which resemble processes familiar from Lie group theory, with [**Lie–4**] in §1.5. One exception is the notion of bisection treated in §1.4. Bisections — or 'generalized elements' — correspond to left– and right– translations on a groupoid and are needed later for the adjoint formulas and exponential map. I include the striking description, due to Xu [1995], of the multiplication in a tangent groupoid in terms of bisections.

The algebraic properties of groupoids differ fundamentally from those of groups in that several basic processes may be characterized in terms of morphisms. The first of these is the characterization of actions given in §1.6. From a groupoid action may be constructed an *action groupoid* and a morphism from it to the acting groupoid; these *action morphisms* may be characterized intrinsically and are in bijective correspondence with the actions. This construction enables groupoid actions, and their infinitesimal forms, to be subsumed under the study of the Lie functor for morphisms. This construction goes back to [Ehresmann, 1957]. It was considerably developed by Ronnie Brown who gave a detailed demonstration that, in particular, it provides an algebraic model of covering spaces [Brown, 1988].

The final §1.7 presents the theory of frame groupoids and linear actions of Lie groupoids on vector bundles. Theorem 1.6.23, which shows that stabilizer subgroupoids are, under simple conditions, locally trivial Lie subgroupoids, is applied to tensor structures on vector bundles. This material is central to the connection theory of Part II.

Chapter 2 is chiefly concerned with quotienting processes for Lie groupoids. I begin with the case of vector bundles, which are a simple special case of both Lie groupoids and of Lie algebroids. Here it is well known that the kernel of a surjective morphism $F\colon E \to E'$ over a surjective submersion $f\colon M \to M'$ does not determine F except when the base map is a diffeomorphism; one may say that the 'First

Isomorphism Theorem' breaks down. To restore it, one clearly needs to use the kernel pair $R(f) = \{(y, x) \mid f(y) = f(x)\} \subseteq M \times M$ of f, but it is also necessary to have $R(f)$, regarded as a Lie groupoid on M, acting on E/K, where K is the usual concept of kernel. With the concept of kernel enlarged to incorporate $R(f)$ and the action, one recovers a First Isomorphism Theorem for vector bundles [Higgins and Mackenzie, 1990a].

The class of vector bundle morphisms which is thus characterized by the enlarged concept of kernel, is that where both the map of the total spaces and of the base spaces are surjective submersions. For Lie groupoids the corresponding class of morphisms is the *fibrations*; these arose in the context of set–groupoids as models of Hurewicz fibrations in topology [Brown, 1988], and elsewhere in category theory, and are the broadest simply–defined class of Lie groupoid morphisms for which a First Isomorphism Theorem can be expected. This is the subject of §2.4.

Fibrations also arise in connection with semi–direct products. The general concept of action appropriate to groupoids leads in §2.5 to a very general concept of semi–direct product. It turns out that the natural projection from a general semi–direct product to the acting groupoid is a fibration which is *split* in a specific sense, and conversely, every split fibration defines a semi–direct product.

The long **Chapter 3** contains the basic theory of Lie algebroids and the properties of the Lie functor. For the convenience of readers who know the theory of principal bundles well, and who wish to start directly with Lie algebroids, §3.1 and §3.2 give the construction of the Atiyah sequence of a principal bundle independently of Chapters 1 and 2. The basic definitions and examples of abstract Lie algebroids are in §3.3.

In order to prepare for the Lie algebroids of frame groupoids, §3.4 treats linear vector fields on vector bundles in some detail. A linear vector field on a vector bundle (E, q, M) corresponds to a *derivation* in E — that is, to a first–order differential operator with scalar symbol (named a 'covariant differential operator' in [Mackenzie, 1987a]) — and there are several different formulations, in part developed by myself and Ping Xu [1998], which will be needed throughout the rest of the book. For any Koszul connection ∇ in E, the operators ∇_X, $X \in \mathfrak{X}(M)$, are derivations and some of the results of this section are intrinsic forms of results well known for connections.

The construction of the Lie algebroid of a Lie groupoid and the basic examples are given in §3.5. The two sections which follow, §3.6 and §3.7, treat the exponential map and the adjoint formulas for a general Lie groupoid. The exponential map of a Lie groupoid takes values in (local) bisections rather than in elements and I have endeavoured to make clear how to work in practice with these formulas. §3.6 also contains the calculation of the Lie algebroids of the frame groupoids of a vector bundle with a tensor structure; these results are basic to Part II.

Chapter 4 is concerned with infinitesimal versions of the results of Chapter 2. Whereas most constructions for Lie groupoids differentiate readily enough to Lie algebroids, if can be unexpectedly difficult to give an abstract form of these results for arbitrary Lie algebroids, independently of any integrability assumptions.

Although the crucial difficulty is with the concept of morphism, I begin with the case of actions of Lie algebroids. A natural abstract definition presents itself (4.1.1) in this case. Given an action of a Lie algebroid A with base M on a smooth map $f\colon M' \to M$, there is an *action Lie algebroid $A \lessdot f$* on M' and the natural projection from $A \lessdot f$ to A enjoys an infinitesimal form of the property characteristic of action morphisms of Lie groupoids. I take this infinitesimal property as defining an *action morphism of Lie algebroids* and establish an equivalence between action morphisms of Lie algebroids and Lie algebroid actions, which commutes with the Lie functor and the corresponding equivalence on the groupoid level.

The case of actions provides a paradigm for the other results of this chapter: (i) characterize a groupoid construction in terms of a class of groupoid morphisms or maps; (ii) apply the Lie functor to this class and characterize the resulting Lie algebroid morphisms abstractly, without reference to the differentiation process by which they were obtained; (iii) prove that this class of Lie algebroid morphisms corresponds to a Lie algebroid construction analogous to the original groupoid construction. This method, developed in [Higgins and Mackenzie, 1990a], avoids a considerable amount of unilluminating computation.

The general concept of morphism for Lie algebroids is dealt with in §4.3, following necessary preliminaries on direct products and pullbacks in §4.2. The Lie functor set out in §3.5 is straightforward; what is at issue in §4.3 is the abstract definition. The definition given here includes maps arising from the Lie functor, base–preserving morphisms, pullback projections, and action morphisms. It has been shown [Mackenzie and Xu, 2000] to pass the final test, that such a morphism between the Lie algebroids of Lie groupoids integrates suitably to a morphism of Lie groupoids.

Using this concept of morphism and the paradigm described above, I give concepts of fibration and quotient (§4.4) and general semi–direct product and split fibration (§4.5) for Lie algebroids. This, like most of the material of this Chapter, is based upon work of Philip Higgins and myself [1990a].

The general concepts of morphism for Lie groupoids and Lie algebroids have been crucial for much subsequent work: the criterion 12.2.1 for a Lie bialgebroid requires a general concept of Lie algebroid morphism, as does the development of a Lie theory for double groupoids.

Preface

This book was originally intended to be largely disjoint from my earlier book *Lie groupoids and Lie algebroids in differential geometry*, [Mackenzie, 1987a], written in the period from 1980 to mid 1985 and published in the London Mathematical Society Lecture Note series in 1987. However I have included, in Part II of the present book, the central chapters of the earlier book on transitive Lie algebroids, their cohomology and connection theory and their integrability. Despite the dramatic results since 2000 on the integrability problem for general Lie algebroids, and work on more sophisticated cohomology theories, it seems to me well worth while to continue to treat the case of locally trivial Lie groupoids and transitive Lie algebroids independently of the general theory. As I have noted already, the systematic use of Lie groupoids and abstract Lie algebroids provides a thoroughgoing reformulation of standard connection theory, and is likely to retain its own character independent of the more general results. This material has in some cases been rewritten and in others left almost unchanged, though typos and obscurities have, I hope, always been caught.

The earlier book was intended to be readable without a detailed prior knowledge of connection theory, and certainly without any acquaintance with groupoids or Lie algebroids, and was consequently leisurely in pace. I feel it is no longer necessary to argue throughout the book for the importance of groupoids in differential geometry, and I have now also assumed that readers have a basic knowledge of connection theory and principal bundles, as well as the standard processes of manifolds, vector bundles and Lie groups.

Readers familiar with the earlier book may note the following broad changes:

- Almost all of the material on topological groupoids in Chapter II of [Mackenzie, 1987a] has been omitted. That part which is relevant to the theory of Lie groupoids has been retained, but is now given only in terms of smooth structures. Most of the remaining material dealt with phenomena which cannot occur for Lie groupoids, and perhaps it will be of interest in some future study of topological groupoids. However, I do not pursue that possibility here.

- The material on locally trivial Lie groupoids and transitive Lie algebroids has been thoroughly revised and developed, though without altering the nature of the approach in any fundamental way. I have added a brief expository section, §5.2, on the connection theory of vector bundles. The aim has been to make the account of transitive Lie algebroids technically self–contained, but §5.2 is not a complete introduction to connections in vector bundles.

  I have expanded the treatment of the integrability obstruction so as to give a more complete account of this subject, and elsewhere I have compressed the earlier account, where it seemed too detailed.

- Historical material is now placed in Notes at the end of each chapter, together with comments on other approaches, omitted material, and so on. These are not comprehensive historical surveys, but are as complete as I could make them in a reasonable time.

*   *   *   *   *

This book contains several subbooks, which may be read independently of the rest of the text:

  • Readers interested in the locally trivial/transitive theory of Part II need read only Chapter 1 and Chapter 3 as preparation.

  • Those interested in Lie algebroids but not the underlying groupoid theory may concentrate on §2.1, §3.1 to §3.4, and Chapter 4. For a reader who already knows the standard theory of connections in vector bundles, Chapter 5 and Chapter 7 will then be accessible.

  • Those wanting an introduction to double vector bundles may read §3.4 and then sections 9.1 to 9.6.

• Those interested mainly in Part III may omit Part II. Much of Part III requires only Chapter 1 and Chapter 3, but reference back to Chapter 2 and Chapter 4 will be needed for some results.

I hope that this book will be used both as a reference, and as a source for those learning the subject. With the first aim in view, topics have been placed, with one or two short exceptions, in their logical position, rather than where they are first needed, and there is extensive cross–referencing. On the other hand, I have gone to some lengths to explain the general approach and most individual transitions, and I hope that readers new to the subject will not lack for signposts.

*   *   *   *   *

This book is by no means an account of everything that is known on the subject of Lie groupoids and Lie algebroids; topics which are not covered include the general theory of holonomy and monodromy, the several recent very general integrability results, and multiple Lie theory. Some comments and references on topics which have been omitted are collected in the Appendix.

Apart from the universal constraints of space and availability under which everyone works, my defence for omissions is that I have set out to present three main strands of the theory — with which I have been most involved — in a philosophically coherent and motivated way. Certainly the subject is rich enough to be developed from other perspectives than those adopted here.

Listed below are monographs and survey articles that contain substantial alternative treatments of aspects of groupoid theory or of Lie algebroids. Full references are given in the bibliography.

I. Moerdijk and J. Mrčun, *Introduction to Foliations and Lie groupoids.* Cambridge University Press, 2003.

A. Cannas da Silva and A. Weinstein, *Geometric Models for Noncommutative Algebras*, American Mathematical Society, 1999.

A. Paterson, *Groupoids, Inverse Semigroups, and their Operator Algebras*, Birkhäuser Boston Inc. 1999.

N. P. Landsman, *Mathematical Topics between Classical and Quantum Mechanics*, Springer–Verlag, 1998.

A. Weinstein, 'Groupoids: Unifying internal and external symmetry', *Notices of the Amer. Math. Soc.*, 1996.

K. C. H. Mackenzie, "Lie algebroids and Lie pseudoalgebras", *Bull. London Math. Soc.*, 1995.

A. Connes, *Noncommutative Geometry*, Academic Press, 1994.

I. Vaisman, *Lectures on the Geometry of Poisson Manifolds*, Birkhäuser, 1994.

M. V. Karasëv and V. P. Maslov, *Nonlinear Poisson brackets, Geometry and Quantization*, American Mathematical Society, 1993.

R. Brown, *Topology: a Geometric Account of General Topology, Homotopy Types and the Fundamental Groupoid*, Ellis Horwood, 1988.

R. Brown, "From groups to groupoids: a brief survey", *Bull. London Math. Soc.*, 1987.

Feedback will be welcomed and may be sent to me via the web page
http://www.shef.ac.uk/~pm1kchm/gt.html

TERMINOLOGY AND NOTATION

★ All manifolds are pure, Hausdorff, finite–dimensional and second–countable, except where explicitly stated otherwise.

★ The expression 'Lie groupoid' in this book does not include a local triviality condition. We take Lie groupoids in the same sense (except as regards concepts of manifold; see the previous point) as the differentiable groupoids of [Pradines, 1966] and [Mackenzie, 1987a], and the smooth groupoids of [Connes, 1994].

★ For a vector bundle E, following [Kosmann–Schwarzbach and Mackenzie, 2002], I have replaced the term 'covariant differential operator on E' by 'derivation on E' or 'derivative endomorphism of ΓE' and replaced the notation $\mathrm{CDO}(E)$ by $\mathfrak{D}(E)$.

★ The semi–direct product notation $\ltimes$ has in the past been frequently used in three distinct senses, both for Lie groupoids and Lie algebroids: for action (or transformation) structures, for semi–direct products in which the base manifold remains fixed, and for 'general semi–direct products', combining the first two constructions. This could be seriously confusing: for example one was led to write $T^*G \cong G \ltimes \mathfrak{g}^*$ and $TG \cong G \ltimes \mathfrak{g}$, for a Lie group G, where in the first case $G \ltimes \mathfrak{g}^*$ was the action Lie groupoid and in the second $G \ltimes \mathfrak{g}$ was the usual semi–direct product group. I denote these three senses by, respectively, the symbols $\vartriangleleft$, $\ltimes$, $\sqsubset$.

★ The exterior powers of a vector bundle A are denoted $\Lambda^k(A)$ and those of its dual $\Lambda^k(A^*)$. Since the symbol Ω is used for a locally trivial Lie groupoid, the p–forms on a manifold M are denoted not by $\Omega^p(M)$ but by $\Omega^p(M)$. The graded ring of all differential forms is $\Omega^\bullet(M)$. The module of vector fields is $\mathfrak{X}(M)$ and the Schouten algebra of all multivector fields is $\mathfrak{X}^\bullet(M)$.

ACKNOWLEDGEMENTS

This book has evolved over a considerable period of time and I have reason to be grateful to many people.

My greatest acknowledgements are to Phillip Higgins and to Xu Ping, extended collaborations with each of whom underpin much of the work presented here. Alan Weinstein's introduction of symplectic groupoids and of Poisson groupoids in the mid 1980s transformed and revitalized the subject of Lie groupoids and Lie algebroids for me, as undoubtedly for many others. Over the years since that time I have benefitted repeatedly from conversations with him on the subjects treated here, and I am very grateful to him.

I have been announcing the account of Poisson and symplectic groupoids given in Part Three since the mid 1990s and have had the good fortune to be able to give series of lectures upon it on two occasions; at Utrecht and Amsterdam thanks to Ieke Moerdijk and Klaas Landsman, and at Queen Mary, London, thanks to Shahn Majid. It has been a very considerable help to have been able to present this material to these audiences.

Beyond these lectures, I am indebted to Ieke Moerdijk for providing much moral support and on numerous occasions a conducive environment for work, for a project whose length was not then apparent.

I am very grateful to Yvette Kosmann–Schwarzbach who corrected me on various matters concerning differential operators, and for guidance and instruction on graded structures.

I owe a particular debt of gratitude to Cambridge University Press, most especially to David Tranah, but also to Roger Astley and Jonathon Walthoe, for having continued to support this project through a long period of preparation. I am especially grateful to David Tranah for his care and attention to the minutiae in the final stages of preparation.

Sheffield, 2004

ACKNOWLEDGMENT

PART ONE:
THE GENERAL THEORY

1

Lie groupoids:
Fundamental theory

The first two sections of this chapter are concerned with the most basic definitions and examples. In §1.3 we describe the class of locally trivial Lie groupoids: those Lie groupoids which are — subject to an important proviso — equivalent to principal bundles. In a sense which will be clear after §1.5, locally trivial groupoids are the simplest interesting class of Lie groupoids.

In §1.4 we deal with the basic properties of bisections; these may be thought of as 'generalized elements' of a groupoid. In Theorem 1.4.14 we give the important description, due to Xu, of the tangent groupoid structure in terms of bisections.

The first part of §1.5 extends to Lie groupoids the elementary properties of the identity component of a Lie group. We then consider the partition of the base manifold of a Lie groupoid by transitivity components, or *orbits*, and prove the fundamental result of Pradines that the restriction of a Lie groupoid to an orbit is locally trivial.

The final two sections deal with actions of Lie groupoids on general maps and in particular with linear representations on vector bundles. For groupoids there is a diagrammatic characterization of actions in terms of the notion of action groupoid and action morphism. This is a feature of groupoid theory not available for groups. Whereas a group action can be described diagrammatically only by using the action map $G \times M' \to M'$, which is not a morphism, or the associated $G \to \mathrm{Diff}(M')$, which can raise problems of differentiability, a Lie groupoid action can be described in terms of a single morphism of Lie groupoids. This enables one to give conceptually simple proofs of many basic results, especially those which involve passing from a groupoid action to the corresponding Lie algebroid action.

3

1.1 Groupoids and Lie groupoids

To introduce the concept of groupoid, take as example the set, denoted $\Phi(E)$, of all linear isomorphisms between the various fibres of a vector bundle (E, q, M). Each such isomorphism $\xi \colon E_x \to E_y$ has associated with it two points of M, namely the points x and y which label the fibres which are its domain and range; we denote x by $\alpha(\xi)$ and y by $\beta(\xi)$ and call $\alpha, \beta \colon \Phi(E) \to M$ the source and target projections of $\Phi(E)$; the isomorphism ξ can be composed with an isomorphism $\eta \colon E_{y'} \to E_z$ if and only if $y' = y$, that is, if and only if $\alpha(\eta) = \beta(\xi)$. Thus composition is a partial multiplication on $\Phi(E)$ with domain the set $\Phi(E) * \Phi(E) = \{(\eta, \xi) \in \Phi(E) \times \Phi(E) \mid \alpha(\eta) = \beta(\xi)\}$. Note that when the composition $\eta\xi$ is defined, we have $\alpha(\eta\xi) = \alpha(\xi)$ and $\beta(\eta\xi) = \beta(\eta)$. This partial multiplication has properties which resemble the properties of a group multiplication as closely as is possible: each point $x \in M$ has associated with it the identity isomorphism id_{E_x}, here denoted 1_x, and the elements 1_x, $x \in M$, act as unities for any multiplication in which they can take part; each isomorphism $\xi \colon E_x \to E_y$ has an inverse isomorphism $\xi^{-1} \colon E_y \to E_x$ and $\xi\xi^{-1}$ and $\xi^{-1}\xi$ are the unities $1_{\beta(\xi)}$ and $1_{\alpha(\xi)}$ respectively. These properties are abstracted into the following definition.

Definition 1.1.1 A *groupoid* consists of two sets G and M, called respectively the *groupoid* and the *base*, together with two maps α and β from G to M, called respectively the *source projection* and *target projection*, a map $1 \colon x \mapsto 1_x$, $M \to G$ called the *object inclusion map*, and a partial multiplication $(h, g) \mapsto hg$ in G defined on the set $G * G = \{(h, g) \in G \times G \mid \alpha(h) = \beta(g)\}$, all subject to the following conditions:

   (i) $\alpha(hg) = \alpha(g)$ and $\beta(hg) = \beta(h)$ for all $(h, g) \in G * G$;
   (ii) $j(hg) = (jh)g$ for all $j, h, g \in G$ such that $\alpha(j) = \beta(h)$ and $\alpha(h) = \beta(g)$;
   (iii) $\alpha(1_x) = \beta(1_x) = x$ for all $x \in M$;
   (iv) $g1_{\alpha(g)} = g$ and $1_{\beta(g)}g = g$ for all $g \in G$;
   (v) each $g \in G$ has a two-sided inverse g^{-1} such that $\alpha(g^{-1}) = \beta(g)$, $\beta(g^{-1}) = \alpha(g)$ and $g^{-1}g = 1_{\alpha(g)}$, $gg^{-1} = 1_{\beta g}$.

$\square$

An element of M may be called an *object* of the groupoid G and an element of G may be called an *arrow*. The arrow 1_x corresponding to an object $x \in M$ may also be called the *unity* or *identity* corresponding to x. To justify this terminology and to prove that the inverse in (v) is unique, we have the following proposition, whose proof is a simple exercise.

Proposition 1.1.2 *Let G be a groupoid with base M, and consider $g \in G$ with $\alpha(g) = x$ and $\beta(g) = y$.*

(i) *If $h \in G$ has $\alpha(h) = y$ and $hg = g$, then $h = 1_y$.*
If $j \in G$ has $\beta(j) = x$ and $gj = g$, then $j = 1_x$.
(ii) *If $h \in G$ has $\alpha(h) = y$ and $hg = 1_x$, then $h = g^{-1}$.*
If $j \in G$ has $\beta(j) = x$ and $gj = 1_y$, then $j = g^{-1}$.

In place of the phrase 'a groupoid with base M', we often write 'a groupoid on M' and we often indicate a groupoid and its base by $G \rightrightarrows M$. For a groupoid G on M and $x, y \in M$ we write G_x for $\alpha^{-1}(x)$, G^y for $\beta^{-1}(y)$ and G_x^y for $G_x \cap G^y$. The set G_x is called the *α–fibre* over x and G^y the *β–fibre* over y. The set G_x^x, obviously a group under the restriction of the partial multiplication in G, is called the *vertex group* at x. For any subsets $U, V \subseteq M$ we likewise write G_U, G^V and G_U^V for $\alpha^{-1}(U)$, $\beta^{-1}(V)$ and $G_U \cap G^V$ respectively. The set of identity elements $\{1_m \mid m \in M\}$ is denoted 1_M.

Definition 1.1.3 A *Lie groupoid* is a groupoid G on base M together with smooth structures on G and M such that the maps $\alpha, \beta \colon G \to M$ are surjective submersions, the object inclusion map $x \mapsto 1_x$, $M \to G$ is smooth, and the partial multiplication $G * G \to G$ is smooth.   □

Here $G * G = (\alpha \times \beta)^{-1}(\Delta_M)$ is a closed embedded submanifold of $G \times G$, since α and β are submersions.

When we wish to refer to a groupoid without manifold structure, as in 1.1.1, we call it a *set–groupoid*. As a preliminary exercise, we show that the inversion map is smooth. As in the case of Lie groups, this is a consequence of the inverse function theorem. The following notation will be used throughout the book.

Definition 1.1.4 Let G be a Lie groupoid on M, and take $g \in G$ with $\alpha g = x$, $\beta g = y$.

The *left–translation* corresponding to g is $L_g \colon G^x \to G^y$, $h \mapsto gh$; the *right–translation* corresponding to g is $R_g \colon G_y \to G_x$, $h \mapsto hg$.   □

The α–fibres and β–fibres of a Lie groupoid G are closed embedded submanifolds of G. A Lie groupoid is α–*connected* if each of its α–fibres is connected; likewise for any property P, it is α–P if each α–fibre has property P. If $x, y \in M$ and there exists $g \in G_x^y$, then G_x and G_y are diffeomorphic, and G^x and G^y likewise are diffeomorphic.

Proposition 1.1.5 *Let G be a Lie groupoid on base M. The inversion map $\iota\colon g \mapsto g^{-1}$ is a diffeomorphism.*

Proof The tangent bundle to $G * G$ is

$$TG * TG = \{(Y, X) \in T(G) \times T(G) \mid T(\alpha)(Y) = T(\beta)(X)\}.$$

Denote the multiplication in G by κ. Take $g, h \in G$ with $\alpha h = \beta g$ and suppose that $Y \in T_h(G_{\alpha h})$ and $X \in T_g(G^{\beta g})$. Then apply the Leibniz rule in the usual way to the map $G_{\alpha h} \times G^{\beta g} \to G$ and we have

$$T(\kappa)(Y, X) = T\big(R_g\big)_h(Y) + T(L_h)_g(X). \tag{1}$$

Now define $\theta\colon G * G \to G \times_\beta G$ by $(h, g) \mapsto (h, hg)$. Then θ is a bijection, by the algebraic properties of G. To see that θ is an immersion, take $(Y, X) \in T(G * G)_{(h,g)}$ and suppose $T(\theta)(Y, X) = (0, 0)$. Since $\pi_2 \circ \theta = \pi_2$, it follows that $Y = 0$. Hence $X \in T(G^{\beta g})_g$ and (1) can be applied to show that $X = 0$. Since α and β are both submersions, $G * G$ and $G \times_\beta G$ have the same dimension and so θ is étale everywhere on $G * G$. Hence it is a diffeomorphism and the composite of $G \to G \times_\beta G$, $h \mapsto (h, 1_{\beta h})$, followed by θ^{-1}, followed by $\pi_2\colon G * G \to G$, is smooth; this is ι. Inversion is its own inverse and is therefore a diffeomorphism. $\qquad\square$

Note further that the object inclusion map $1\colon M \to G$ in a Lie groupoid is an immersion, since either projection is left–inverse to it, and is easily seen to be a homeomorphism onto 1_M. Thus 1_M is a closed embedded submanifold of G.

The following examples are of basic importance.

Example 1.1.6 Any manifold M may be regarded as a Lie groupoid on itself with $\alpha = \beta = \mathrm{id}_M$ and every element a unity. A groupoid in which every element is a unity will be called a *base groupoid*. $\qquad\boxtimes$

Example 1.1.7 Let M be a manifold and G a Lie group. We give $M \times G \times M$ the structure of a Lie groupoid on M in the following way: α is the projection onto the third factor of $M \times G \times M$ and β is the projection onto the first factor; the object inclusion map is $x \mapsto 1_x =$

$(x, 1, x)$ and the partial multiplication is $(z, h, y')(y, g, x) = (z, hg, x)$, defined if and only if $y' = y$. The inverse of (y, g, x) is (x, g^{-1}, y). This is usually called the *trivial groupoid* on M with group G.

In particular, any Lie group may be considered to be a Lie groupoid on a singleton manifold, and any cartesian square $M \times M$ is a groupoid on M. This groupoid $M \times M$ is called the *pair groupoid* on M. ⊠

Example 1.1.8 Let $q: M \to Q$ be a surjective submersion. Then

$$R(q) = M \times_q M = \{(x, y) \in M \times M \mid q(x) = q(y)\}$$

is a Lie groupoid on M with respect to the restriction of the pair groupoid structure. Each α–fibre $R(q)_x$, $x \in M$, may be naturally identified with the fibre of q through x; thus the α–fibres of a Lie groupoid need not be diffeomorphic. The vertex groups of $R(q)$ are trivial.

The point here is that the source projection $R(q) \to M$, $(y, x) \mapsto x$, must be a surjective submersion. By Godement's criterion, an equivalence relation on a manifold M, the graph of which is a closed embedded submanifold of $M \times M$, satisfies this condition if and only if it is $R(q)$ for some surjective submersion. Thus in terms of Definition 1.2.11 these examples $R(q)$ are the only closed embedded wide Lie subgroupoids of the pair groupoid $M \times M$. ⊠

Example 1.1.9 Let $G \times M \to M$ be a smooth action of a Lie group G on a manifold M. Give the product manifold $G \times M$ the structure of a Lie groupoid on M in the following way: α is the projection onto the second factor of $G \times M$ and β is the action $G \times M \to M$ itself; the object inclusion map is $x \mapsto 1_x = (x, 1)$ and the partial multiplication is $(g_2, y)(g_1, x) = (g_2 g_1, x)$, defined if and only if $y = g_1 x$. The inverse of (g, x) is (g^{-1}, gx). With this structure we denote $G \times M$ by $G \ltimes M$ and call it the *action groupoid* of $G \times M \to M$.

The α–fibre $(G \ltimes M)_x$ is $G \times \{x\}$, and the β–fibres can also be identified with the group G. The vertex group $(G \ltimes M)_x^x$ is naturally isomorphic to the stabilizer group G_x. ⊠

Example 1.1.10 Applying the construction of 1.1.9 to the action $\mathbb{R} \times \mathbb{S}^1 \to \mathbb{S}^1$, $(t, z) \mapsto e^{2\pi i t} z$ gives a groupoid structure on the cylinder $\mathbb{R} \times \mathbb{S}^1$. The base may be identified with the circle $t = 0$, the α–fibres are straight lines orthogonal to $t = 0$, the β–fibres are the helices which make an angle of $\pi/4$ with the circles $t = constant$, and the vertex groups are the $\mathbb{Z} \times \{z\}$ for $z \in \mathbb{S}^1$.

The reader may construct similarly visualizable examples on the torus, using the actions $\mathbb{S}^1 \times \mathbb{S}^1 \to \mathbb{S}^1$, $(w, z) \mapsto w^n z$, for given $n \in \mathbb{Z}$. ⊠

Example 1.1.11 Let M be a manifold. Then the set $\Pi(M)$ of homotopy classes $\langle \gamma \rangle$ relative endpoints of smooth paths $\gamma \colon [0, 1] \to M$ is a groupoid on M with respect to the following structure: the source and target projections are $\alpha(\langle \gamma \rangle) = \gamma(0)$ and $\beta(\langle \gamma \rangle) = \gamma(1)$, the object inclusion map is $x \mapsto 1_x = \langle \kappa_x \rangle$, where κ_x is the path constant at x, and the partial multiplication is $\langle \delta \rangle \langle \gamma \rangle = \langle \delta \gamma \rangle$ where $\delta \gamma$ is the standard concatenation of γ followed by δ, namely $(\delta \gamma)(t) = \gamma(2t)$ for $0 \leqslant t \leqslant \frac{1}{2}$, $(\delta \gamma)(t) = \delta(2t - 1)$ for $\frac{1}{2} \leqslant t \leqslant 1$. The inverse of $\langle \gamma \rangle$ is $\langle \gamma^{\leftarrow} \rangle$ where $\gamma^{\leftarrow}$ is the reverse of the path γ, namely $\gamma^{\leftarrow}(t) = \gamma(1 - t)$.

With this structure $\Pi(M)$ is the *fundamental groupoid* of M; its vertex groups are the fundamental groups $\pi_1(M, x)$, $x \in M$, and if M is connected, then its α–fibres are the universal covering spaces of M.

⊠

Example 1.1.12 Let (E, q, M) be a vector bundle. Let $\Phi(E)$ denote the set of all vector space isomorphisms $\xi \colon E_x \to E_y$ for $x, y \in M$. Then $\Phi(E)$ is a Lie groupoid on M with respect to the following structure: for $\xi \colon E_x \to E_y$, $\alpha(\xi)$ is x and $\beta(\xi)$ is y; the object inclusion map is $x \mapsto 1_x = \mathrm{id}_{E_x}$, and the partial multiplication is the composition of maps. The inverse of $\xi \in \Phi(E)$ is its inverse as a map. With this structure $\Phi(E)$ is called the *frame groupoid* or *linear frame groupoid* of (E, q, M).

The smooth structure on $\Phi(E)$ is induced from that of E as follows. Let $\{ \psi_i \colon U_i \times V \to E_{U_i} \}$ be an atlas for E. For each i and j, define

$$\overline{\psi}_i^j \colon U_j \times GL(V) \times U_i \to \Phi(E)_{U_i}^{U_j}, \quad (y, A, x) \mapsto \psi_{j,y} \circ A \circ \psi_{i,x}^{-1}.$$

Clearly each $\overline{\psi}_i^j$ is a bijection and any $(\overline{\psi}_k^l)^{-1} \circ \overline{\psi}_i^j$ which has a non-void domain is a diffeomorphism. Hence there is a well-defined smooth structure on $\Phi(E)$ for which each $\overline{\psi}_i^j$ is a diffeomorphism.

That $\Phi(E)$ is now a Lie groupoid is straightforward: one works locally and the details are similar to those for a trivial groupoid. ⊠

Example 1.1.13 Given a local diffeomorphism $\varphi \colon U \to V$ between open sets of a manifold M, and given $x \in U$, let $j_x^1 \varphi$ be the 1–jet of φ at x. Then $J^1(M, M)$, the set of all such 1–jets, has a natural groupoid structure given by $\alpha(j_x^1 \varphi) = x$, $\beta(j_x^1 \varphi) = \varphi(x)$, and

$(j^1_{\varphi(x)}\psi)(j^1_x\varphi) = j^1_x(\psi \circ \varphi)$; it is a Lie groupoid with respect to the natural smooth structure and is naturally isomorphic to $\Phi(TM)$ under $j^1_x\varphi \mapsto T_x(\varphi)$. $\boxtimes$

For the next example we recall the definition of principal bundle.

Definition 1.1.14 A *principal bundle* $P(M,G,\pi)$ consists of a manifold P, a Lie group G, and a free right action of G on P denoted $(u,g) \mapsto ug$, such that the orbits of the action coincide with the fibres of the surjective submersion $\pi\colon P \to M$, and such that M is covered by the domains of local sections $\sigma\colon U \to P$, $U \subseteq M$, of π. $\square$

Example 1.1.15 Let $P(M,G,\pi)$ be a principal bundle. Let G act on $P \times P$ to the right by $(u_2, u_1)g = (u_2g, u_1g)$; denote the orbit of (u_2, u_1) by $\langle u_2, u_1 \rangle$ and the set of orbits by $\frac{P \times P}{G}$. Then $\frac{P \times P}{G}$ is a groupoid on M with respect to the following structure: the source and target projections are $\alpha(\langle u_2, u_1 \rangle) = \pi(u_1)$, $\beta(\langle u_2, u_1 \rangle) = \pi(u_2)$; the object inclusion map is $x \mapsto 1_x = \langle u, u \rangle$, where u is any element of $\pi^{-1}(x)$ and the partial multiplication is defined by

$$\langle u_3, u_2' \rangle \langle u_2, u_1 \rangle = \langle u_3, u_1 \delta(u_2'.u_2) \rangle.$$

Here $\delta\colon P \times_\pi P \to G$ is the map $(ug, u) \mapsto g$. That $(u_2', u_2) \in P \times_\pi P$ is ensured by the condition $\alpha(\langle u_3, u_2' \rangle) = \beta(\langle u_2, u_1 \rangle)$. Note that one can always choose representatives so that $u_2' = u_2$ and the multiplication is then simply

$$\langle u_3, u_2 \rangle \langle u_2, u_1 \rangle = \langle u_3, u_1 \rangle.$$

The inverse of $\langle u_2, u_1 \rangle$ is $\langle u_1, u_2 \rangle$. This groupoid is the *gauge groupoid associated to* $P(M,G,\pi)$.

Denote $\frac{P \times P}{G}$ by Ω. That Ω is a quotient manifold of $P \times P$ is a standard exercise. Denote the projection $P \times P \to \Omega$, $(u_2, u_1) \mapsto \langle u_2, u_1 \rangle$ by p. Now the source projection $\alpha\colon \Omega \to M$ composed with p is equal to $\pi \circ \mathrm{pr}$; since $\pi \circ \mathrm{pr}$ is a surjective submersion, it follows that α also is.

We prove that the groupoid multiplication is smooth. Since p is a surjective submersion, its square $p \times p\colon P^2 \times P^2 \to \Omega \times \Omega$ is also. Now $(p \times p)^{-1}(\Omega * \Omega) = P \times (P \times_\pi P) \times P$ and hence the restriction of $p \times p$ to $P \times (P \times_\pi P) \times P \to \Omega * \Omega$ is a surjective submersion. Evidently

$$P \times (P \times_\pi P) \times P \xrightarrow{(u_4,u_3,u_2,u_1) \mapsto (u_4,u_1\delta(u_3,u_2))} P \times P$$

$$p \times p \downarrow \qquad\qquad\qquad\qquad\qquad \downarrow p$$

$$\Omega * \Omega \xrightarrow{\quad\text{groupoid multiplication}\quad} \Omega$$

commutes and so the smoothness of the groupoid multiplication follows from that of the top map. It is clear that the object inclusion map is smooth. Thus the gauge groupoid is a Lie groupoid.

If P is a frame bundle for a vector bundle E on M and one thinks of an element $u_1 \in P$ as an isomorphism to E_{x_1}, where $x_1 = \pi(u_1)$, from some standard fibre type V, then $\langle u_2, u_1 \rangle$ should be thought of as the composite isomorphism $u_2 \circ u_1^{-1}$ from E_{x_1} to E_{x_2}. This is a useful picture to keep in mind even in the abstract case. The construction of a frame bundle for a vector bundle is noncanonical inasmuch as a choice of reference point, or of standard fibre type, is required; the corresponding frame groupoid is intrinsically defined. We deal in detail with frame groupoids in §1.7.   ⊠

Example 1.1.16 Let G be a Lie groupoid on base M. Applying the tangent functor to each of the maps defining G yields a Lie groupoid structure on TG with base TM, source $T(\alpha)$ and target $T(\beta)$, and multiplication $T(\kappa)\colon TG * TG \to TG$, where κ is the multiplication in G. The identity at $X \in TM$ is $T(1)(X)$. This is the *tangent prolongation groupoid* of $G \Longrightarrow M$; we usually omit the word *prolongation*.

An explicit formula for the multiplication in TG is given in §1.4.   ⊠

Example 1.1.17 Let G be a Lie group with Lie algebra $\mathfrak{g}$. Define a groupoid structure on T^*G with base $\mathfrak{g}^*$ as follows. Given $\theta \in T_g^*G$, the source and the target are

$$\beta(\theta) = \theta \circ T(R_g), \qquad \alpha(\theta) = \theta \circ T(L_g),$$

where R_g and L_g are the right and left translations for G. The multiplication is now determined by the groupoid axioms; for θ as above and $\varphi \in T_h^*G$ it is

$$\varphi \bullet \theta = \varphi \circ T(R_{g^{-1}}) = \theta \circ T(L_{h^{-1}}).$$

The check that $T^*G \Longrightarrow \mathfrak{g}^*$ is a Lie groupoid is straightforward.   ⊠

Example 1.1.18 For the principal bundle $SU(2)(SO(3), \mathbb{Z}_2, \pi)$, applying 1.1.15 yields a groupoid which, though it has dimension 6, is perhaps somewhat visually accessible. Here the action of $\mathbb{Z}_2 = \{I, -I\} \subseteq SU(2)$

on $SU(2)$ is by matrix multiplication and π is essentially the adjoint representation. The groupoid $(SU(2) \times SU(2))/\mathbb{Z}_2$ can be naturally identified with $SO(4)$: regarding $SU(2)$ as the unit sphere in the space of quaternions $\mathbb{H}$, each pair $(p, q) \in SU(2) \times SU(2)$ defines a map $\mathbb{H} \to \mathbb{H}$, $x \mapsto pxq^{-1}$, which, as a map $\mathbb{R}^4 \to \mathbb{R}^4$, is a proper rotation. As is well known, this map $SU(2) \times SU(2) \to SO(4)$ is a surjective morphism of Lie groups with kernel $\{(I, I), (-I, -I)\}$.

Thus we obtain a groupoid structure on $SO(4)$ with base $\mathbb{R}P^3$, α– and β–fibres which are 3–spheres, and vertex groups which are $\mathbb{Z}_2$s. However it seems that this groupoid multiplication has no clear geometrical significance. $\boxtimes$

From the example of the pair groupoids $M \times M$ it is clear that a Lie groupoid need not have a unique compatible analytic structure, and that a topological groupoid may have several nondiffeomorphic compatible structures of Lie groupoid.

The final class of examples in this section will be needed constantly in what follows.

Definition 1.1.19 A *Lie group bundle*, or *LGB*, is a smooth fibre bundle (K, q, M) in which each fibre $K_m = q^{-1}(m)$, and the fibre type G, has a Lie group structure, and for which there is an atlas $\{\psi_i \colon U_i \times G \to K_{U_i}\}$ such that each $\psi_{i,m} \colon G \to K_m$, $m \in U_i$, is an isomorphism of Lie groups.

A *morphism of LGBs from* (K, q, M) *to* (K', q', M') is a morphism (F, f) of fibre bundles such that each $F_m \colon K_m \to K'_{f(m)}$ is a morphism of Lie groups. $\square$

Given any principal bundle $P(M, G)$, the group G acts on itself by inner automorphisms and the resulting associated fibre bundle (see 1.3.8), sometimes called the gauge group bundle, is an LGB. We will call it the *inner group bundle* or *inner LGB*.

Any LGB may be considered a Lie groupoid, with the source and target projections both equal to the bundle projection. In particular, a vector bundle may be regarded as a Lie groupoid.

1.2 Morphisms and subgroupoids

The treatment of morphisms in this section is very brief. A detailed study is postponed to Chapter 2.

Definition 1.2.1 Let G and G' be groupoids on M and M' respectively. A *morphism* $G \to G'$ is a pair of maps $F\colon G \to G'$, $f\colon M \to M'$ such that $\alpha' \circ F = f \circ \alpha$, $\beta' \circ F = f \circ \beta$ and $F(hg) = F(h)F(g)$, for all $(h,g) \in G * G$. We also say that F is a *morphism over* f. If $M = M'$ and $f = \mathrm{id}_M$ we say that F is a *morphism over* M, or that F is a *base–preserving morphism*.

If G and G' are Lie groupoids, then (F,f) is a *morphism of Lie groupoids* if both F and f are smooth. □

The conditions $\alpha' \circ F = f \circ \alpha$ and $\beta' \circ F = f \circ \beta$ ensure that $F(h)F(g)$ is defined whenever hg is.

Note that for Lie groupoids the smoothness of f follows from that of F. We sometimes write $F_0 = f$ for the base map of a morphism F. The proof of the following result is immediate.

Proposition 1.2.2 *Let $F\colon G \to G'$, $f\colon M \to M'$ be a groupoid morphism. Then*

(i) $F(1_m) = 1_{f(m)}$ *for all $m \in M$,*
(ii) $F(g^{-1}) = F(g)^{-1}$ *for all $g \in G$.*

For $m, n \in M$ we denote the various restrictions of F, to $G_m \to G'_{f(m)}$, to $G^n \to G'^{f(n)}$ and to $G^n_m \to G'^{f(n)}_{f(m)}$, by F_m, F^n and F^n_m, respectively.

Definition 1.2.3 A morphism $F\colon G \to G'$, $f\colon M \to M'$ is an *isomorphism of Lie groupoids* if F and (hence) f are diffeomorphisms.

□

The inverse of an isomorphism of Lie groupoids is itself an isomorphism. The following are some very basic examples of morphisms.

Example 1.2.4 For any groupoid G on base M, the map

$$\chi = (\beta, \alpha)\colon G \to M \times M, \ g \mapsto (\beta(g), \alpha(g))$$

is a morphism over M from G to the pair groupoid $M \times M$, called the *anchor* of G. ⊠

Example 1.2.5 A morphism of trivial groupoids $F\colon M \times G \times M \to M' \times G' \times M'$ can be written in the form

$$F(y, g, x) = \big(f(y), \theta(y)s(g)\theta(x)^{-1}, f(x)\big)$$

for a Lie group morphism $s \colon G \to G'$ and a smooth map $\theta \colon M \to G'$. The maps θ and s are not unique; for any chosen point $m_0 \in M$, such maps are defined by $F(x, 1, m_0) = \bigl(f(x), \theta(x), f(m_0)\bigr)$ and $F(m_0, g, m_0) = (f(m_0), s(g), f(m_0))$. ☒

From this example it is clear that a continuous morphism of Lie groupoids need not be smooth: take $M \times M \to M \times G \times M$, $(y, x) \mapsto (y, \theta(y)\theta(x)^{-1}, x)$ for a suitable $\theta \colon M \to G$.

Example 1.2.6 If G is a Lie group the division map $\delta(g_2, g_1) = g_2 g_1^{-1}$ is a groupoid morphism $G \times G \to G$ where the domain is the pair groupoid on the manifold G and the range is the group G itself.

More generally, for any Lie groupoid G on M, the pullback manifold $G \times_\alpha G = \{(h, g) \in G \times G \mid \alpha h = \alpha g\}$ is a Lie groupoid with respect to the operations restricted from the pair groupoid $G \times G$, and $\delta \colon G \times_\alpha G \to G$, $(h, g) \mapsto hg^{-1}$ is a groupoid morphism over $\beta \colon G \to M$. ☒

Example 1.2.7 Let G and G' be Lie groups acting on manifolds M and M'. Then if $\varphi \colon G \to G'$ is a smooth morphism and $f \colon M \to M'$ is a map equivariant with respect to φ, then $\varphi \times f \colon G \times M \to G' \times M'$ is a morphism of the action groupoids. ☒

Example 1.2.8 Let G be a Lie group and form the action groupoid $G \ltimes \mathfrak{g}^*$ with respect to the coadjoint action $\mathrm{Ad}^*_g \varphi = \varphi \circ \mathrm{Ad}_{g^{-1}}$. Then left trivialization $G \times \mathfrak{g}^* \to T^*G$, $(g, \varphi) \mapsto \varphi \circ T(L_{g^{-1}})$, is an isomorphism over $\mathfrak{g}^*$ to the Lie groupoid of 1.1.17. ☒

Example 1.2.9 If $F(f, \varphi) \colon P(M, G) \to P'(M', G')$ is a morphism of principal bundles, then $\langle u_2, u_1 \rangle \mapsto \langle F(u_2), F(u_1) \rangle$ is a morphism of the gauge groupoids. ☒

Example 1.2.10 For any Lie groupoid G on base M the tangent bundle projection $p_G \colon TG \to G$ is a morphism of Lie groupoids over $p_M \colon TM \to M$. ☒

We now briefly consider Lie subgroupoids.

Definition 1.2.11 Let G be a Lie groupoid on M. A *Lie subgroupoid* of G is a Lie groupoid G' on base M' together with injective immersions $\iota \colon G' \to G$ and $\iota_\circ \colon M' \to M$, such that $(\iota, \iota_\circ)$ is a morphism of Lie groupoids.

A Lie subgroupoid $G' \rightrightarrows M'$ of $G \rightrightarrows M$ is *embedded* if G' and M' are embedded submanifolds of G and M.

A Lie subgroupoid $G' \rightrightarrows M'$ of $G \rightrightarrows M$ is *wide* if $M' = M$ and $\iota_\circ = \mathrm{id}$.

The *base subgroupoid* or *identity subgroupoid* of G is the subgroupoid $1_M = \{1_x \mid x \in M\}$.

$\square$

If $(\iota, \iota_\circ)$ is a morphism of Lie groupoids and ι is an injective immersion, then $\iota_\circ$ is also. If ι is an embedding, then $\iota_\circ$ is also.

A Lie subgroupoid is strictly speaking an equivalence class of injective immersive morphisms; with a few exceptions however (one of which follows immediately), we regard subgroupoids as subsets of the ambient groupoid.

Example 1.2.12 Let G be a Lie group and H a closed subgroup of G. Then

$$\frac{G \times G}{H} \quad \rightarrow \quad (G/H) \times G \times (G/H)$$

$$\langle g_2, g_1 \rangle \quad \mapsto \quad \left(g_2 H, g_2 g_1^{-1}, g_1 H\right)$$

is a smooth base–preserving morphism, from the gauge groupoid of $G(G/H, H)$ to the trivial groupoid on G/H with group G.

With respect to this map $\frac{G \times G}{H}$ is a closed embedded Lie subgroupoid of $(G/H) \times G \times (G/H)$. $\boxtimes$

If $U \subseteq M$ is an open submanifold, then G_U^U is clearly a Lie subgroupoid of G on base U, the *restriction of G to U*. We deal with the restriction of G to more general submanifolds of the base in §1.5.

One other case that frequently occurs is that of the closed subset

$$\mathscr{I}G = \{g \in G \mid \alpha g = \beta g\}.$$

For a general Lie groupoid $G \rightrightarrows M$ this is algebraically a subgroupoid, but — as the example of action groupoids in 1.1.9 shows — not necessarily with any good smooth structure. We call it the *inner subgroupoid*, being careful to omit the word 'Lie'. In the locally trivial case, $\mathscr{I}G$ is a locally trivial Lie group bundle; see 1.3.9.

1.3 Local triviality

Definition 1.3.1 A groupoid G on base M is *transitive* if for each pair $x, y \in M$, there exists $g \in G$ with $\alpha(g) = x$ and $\beta(g) = y$. $\square$

This definition applies to any set–groupoid, as defined in 1.1.1. In the case of set–groupoids, the transitivity condition forces G to be isomorphic to a trivial groupoid. Namely, choose any $m \in M$ as reference point, and let $\sigma\colon M \to G_m$ be any right–inverse to the restriction $\beta_m\colon G_m \to M$, and define

$$S\colon M \times G_m^m \times M \to G, \qquad (y, \ell, x) \mapsto \sigma(y)\ell\sigma(x)^{-1}.$$

This result is less significant than it may appear. It is comparable to the fact that every finite–dimensional real vector space is isomorphic to some $\mathbb{R}^n$; in the absence of a canonical isomorphism, it is usually preferable to work with the structure as presented. Further, a Lie groupoid may very well be transitive without possessing smooth global right–inverses σ.

The concept of local triviality is the version of transitivity appropriate to Lie groupoids.

Definition 1.3.2 Let Ω be a Lie groupoid on M. Then Ω is *locally trivial* if the anchor $(\beta, \alpha)\colon \Omega \to M \times M$ is a surjective submersion. ☐

In particular, a locally trivial Lie groupoid is transitive. We usually denote locally trivial Lie groupoids by upper–case Greek letters.

Proposition 1.3.3 *Let Ω be a Lie groupoid on M. The following conditions are equivalent:*

(i) *Ω is locally trivial;*

(ii) *$\beta_m\colon \Omega_m \to M$ is a surjective submersion for one, and hence for all, $m \in M$;*

(iii) *δ_m, the restriction to $\Omega_m \times \Omega_m \to \Omega$ of the division map, is a surjective submersion for one, and hence for all, $m \in M$.*

Proof (i) $\Longrightarrow$ (ii). $M \times \{m\}$ is a closed embedded submanifold of $M \times M$, and its preimage under (β, α) is Ω_m. So the restriction of (β, α) to $\Omega_m \to M \times \{m\}$ is a surjective submersion, and this identifies with β_m.

(ii) $\Longrightarrow$ (i). Since $(\beta, \alpha) \circ \delta_m = \beta_m \times \beta_m$, it follows that if β_m is a surjective submersion, then (β, α) is also.

(ii) $\Longrightarrow$ (iii). Form the pullback

$$
\begin{array}{ccc}
\Omega_m \times_M \Omega & \xrightarrow{\ \theta\ } & \Omega \\
\downarrow & & \downarrow \beta \\
\Omega_m & \xrightarrow[\ \beta_m\ ]{} & M;
\end{array}
$$

since β_m is a surjective submersion, θ is also. Now

$$\Omega_m \times \Omega_m \xrightarrow{\psi} \Omega_m \times_M \Omega, \qquad (\eta, \zeta) \mapsto (\eta, \eta\zeta^{-1}),$$
$$\Omega_m \times_M \Omega \to \Omega_m \times \Omega_m, \qquad (\eta, \xi) \mapsto (\eta, \xi^{-1}\eta),$$

are smooth and mutually inverse, so ψ is a diffeomorphism. Since $\delta_m = \theta \circ \psi$, it follows that δ_m is a surjective submersion.

(iii) $\Longrightarrow$ (ii). We have $\beta \circ \delta_m = \beta_m \circ \mathrm{pr}\colon \Omega_m \times \Omega_m \to M$, so if δ_m is a surjective submersion, it follows that $\beta_m \circ \mathrm{pr}$, and therefore β_m, is a surjective submersion. $\qquad\square$

Given a locally trivial Lie groupoid Ω on M, and fixing any $m \in M$, there is an open cover $\{U_i\}$ of M by the domains of local sections $\sigma_i\colon U_i \to \Omega_m$ of β_m. Writing $G = \Omega_m^m$ there are then base–preserving isomorphisms

$$\Sigma_i\colon U_i \times G \times U_i \to \Omega_{U_i}^{U_i}, \qquad (y, g, x) \mapsto \sigma_i(y)g\sigma_i(x)^{-1}, \qquad (2)$$

from a trivial groupoid to the restriction of Ω to U_i. The converse is not quite true: a Lie groupoid may be locally isomorphic to trivial groupoids in this sense, but not transitive.

We refer to σ_i as a *decomposing section* of Ω and to $\{\sigma_i\}$ as a *section atlas* for Ω.

Examples 1.3.4 The frame groupoid $\Phi(E)$ of a vector bundle E, as defined in Example 1.1.12, is locally trivial. This may be seen by identifying it with the gauge groupoid of any frame bundle of E, or directly from the definition of the smooth structure in 1.1.12.

Likewise the fundamental groupoid $\Pi(M)$ of a connected manifold M is locally trivial; it is the gauge groupoid of the principal bundle $\widetilde{M}(M, \pi(M))$, where $\widetilde{M}$ is the universal cover of M.

If $\Omega = G \ltimes M$ is the action groupoid for a Lie group G acting on a manifold M, then β_m, for any fixed $m \in M$, identifies with the evaluation map $g \mapsto gm$. This map is a surjective submersion if and only if the action is transitive. When this is the case, M may be identified with a homogeneous space G/H. In fact, $G \ltimes (G/H)$ is then isomorphic to the gauge groupoid of $G(G/H, H)$, under

$$(g_2, g_1 H) \mapsto \langle g_2 g_1, g_1 \rangle.$$

$$\boxtimes$$

It is an easy exercise, using 1.3.3, to see that the gauge groupoid of any principal bundle is locally trivial. Conversely, consider a locally trivial Lie groupoid $\Omega \rightrightarrows M$. Fix $x \in M$. Since $\beta_x \colon \Omega_x \to M$ is a surjective submersion, Ω_x^x is a closed embedded submanifold of Ω and the restriction of the groupoid multiplication makes Ω_x^x a Lie group. Further, the action $\Omega_x \times \Omega_x^x \to \Omega_x$ is smooth, and is clearly free; lastly, the orbits of the action are the fibres of β_x. Altogether, $\Omega_x(M, \Omega_x^x, \beta_x)$ is a principal bundle, the *vertex bundle at* x.

If $y \in M$ is a second point, and ζ is any element in Ω_x^y, then

$$R_{\zeta^{-1}}(\mathrm{id}_M, I_\zeta) \colon \Omega_x(M, \Omega_x^x) \to \Omega_y(M, \Omega_y^y)$$

is an isomorphism of principal bundles over M. Here $I_\zeta \colon \Omega_x^x \to \Omega_y^y$ is the Lie group isomorphism $\lambda \mapsto \zeta \lambda \zeta^{-1}$; this is sometimes (abusively) called the *inner automorphism induced by* ζ.

These correspondences between principal bundles and locally trivial Lie groupoids are essentially mutually inverse, though not in a canonical way.

Proposition 1.3.5 (i) *Let* $P(M, G, \pi)$ *be a principal bundle. Choose* $u_0 \in P$ *and write* $x_0 = \pi(u_0)$. *Then the map*

$$P \to \left. \frac{P \times P}{G} \right|_{x_0}, \quad u \mapsto \langle u, u_0 \rangle$$

is a diffeomorphism and the map

$$G \to \left. \frac{P \times P}{G} \right|_{x_0}^{x_0}, \quad g \mapsto \langle u_0 g, u_0 \rangle$$

is an isomorphism of Lie groups. Together they form an isomorphism of principal bundles over M.

Let $F(f, \varphi) \colon P(M, G) \to P'(M', G')$ *be a morphism of principal bundles, denote by* $\widetilde{F} \colon \frac{P \times P}{G} \to \frac{P' \times P'}{G'}$ *the induced morphism of groupoids, and choose* $u_0 \in P$, $u_0' \in P'$ *such that* $u_0' = F(u_0)$. *Write*

$x_0 = \pi(u_0)$, $x_0' = \pi'(u_0')$. Then

$$
\begin{array}{ccc}
\dfrac{P \times P}{G}\Big|_{x_0} & \xrightarrow{\;\widetilde{F}|_{x_0}\;} & \dfrac{P' \times P'}{G'}\Big|_{x_0'} \\[2ex]
\Big\uparrow & & \Big\uparrow \\[1ex]
P & \xrightarrow{\quad F \quad} & P'
\end{array}
$$

commutes, where the vertical maps are the isomorphisms corresponding
to u_0 and u_0'.

(ii) Let Ω be a locally trivial Lie groupoid on M, and choose $x \in M$.
Then the map

$$
\frac{\Omega_x \times \Omega_x}{\Omega_x^x} \to \Omega, \quad \langle \eta, \xi \rangle \mapsto \eta \xi^{-1},
$$

to Ω from the gauge groupoid of the vertex bundle at x is an isomorphism of Lie groupoids over M.

Let $\varphi \colon \Omega \to \Omega'$ be a morphism of locally trivial Lie groupoids over
$\varphi_0 \colon M \to M'$ and choose $x \in M$, $x' \in M'$ such that $x' = \varphi_0(x)$. Then

$$
\begin{array}{ccc}
\dfrac{\Omega_x \times \Omega_x}{\Omega_x^x} & \xrightarrow{\;\widetilde{\varphi_x}\;} & \dfrac{\Omega'_{x'} \times \Omega'_{x'}}{\Omega'^{x'}_{x'}} \\[2ex]
\Big\downarrow & & \Big\downarrow \\[1ex]
\Omega & \xrightarrow{\quad \varphi \quad} & \Omega'
\end{array}
$$

commutes.

Proof In both cases the algebraic assertions are easily verified. To
prove the smoothness of the inverse of $P \to \frac{P \times P}{G}\big|_{x_0}$ in (i), consider
Figure 1.1(a), in which the oblique arrow is $(u, u_0 g) \mapsto u g^{-1}$. This map

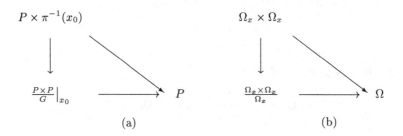

Fig. 1.1.

is smooth since $\delta \colon P \times_\pi P \to G$ is smooth, and the vertical map is a surjective submersion since it is the restriction of a surjective submersion to a preimage.

In (ii) we are in the situation of Figure 1.1(b) and the bottom arrow is a surjective submersion since the other two maps are; since it is also a bijection it is a diffeomorphism. □

The following simple consequence is much used.

Proposition 1.3.6 *Let Ω be a locally trivial Lie groupoid on M, let Ω' be any Lie groupoid on M' and let $\varphi \colon \Omega \to \Omega'$ be a morphism of set groupoids.*

(i) *If any one $\varphi_m \colon \Omega_m \to \Omega'_{\varphi_0(m)}$ is smooth, then φ is smooth.*

(ii) *If φ is smooth on a neighbourhood $\mathcal{U}$ of 1_M in Ω, then it is smooth everywhere on Ω.*

Proof (i) follows directly from 1.3.5. (ii) Choose $m \in M$. Now for every $\xi \in \Omega_m$, $\varphi_{\beta\xi}$ is smooth in a neighbourhood of $1_{\beta\xi}$ and $R_\xi \colon \Omega_{\beta\xi} \to \Omega_m$ maps $1_{\beta\xi}$ to ξ. Since $\varphi_m \circ R_\xi = R_{\varphi(\xi)} \circ \varphi_{\beta\xi}$ it follows that φ_m is smooth at ξ. Now apply (i). □

Proposition 1.3.5 is often paraphrased as the statement that 'locally trivial Lie groupoids are equivalent to principal bundles', and indeed 1.3.5 does enable many concepts familiar from principal bundle theory to be reformulated in terms of locally trivial Lie groupoids.

However the dependence of 1.3.5 on the choice of reference points prevents a correspondence between the automorphism structures of a principal bundle and of its corresponding gauge groupoid. Consider the following, the proof of which is an easy computation.

Proposition 1.3.7 *With the notation of 1.3.5(i), let u'_0 be a second reference point in $\pi^{-1}(x_0)$, say $u'_0 = u_0 h$ where $h \in G$. Then the composite automorphism*

$$P \xrightarrow[(u_0)]{} \left.\frac{P \times P}{G}\right|_{x_0} \xrightarrow[(u'_0)]{} P, \qquad G \xrightarrow[(u_0)]{} \left.\frac{P \times P}{G}\right|^{x_0}_{x_0} \xrightarrow[(u'_0)]{} G$$

is $u \mapsto uh$, $g \mapsto h^{-1}gh$.

Thus G acts as a group of automorphisms of the principal bundle $P(M, G)$ with $h \in G$ acting as $u \mapsto uh^{-1}$, $g \mapsto hgh^{-1}$ or, briefly, as $R_{h^{-1}}(\mathrm{id}_M, I_h)$. All automorphisms of this type correspond to the identity automorphism of the corresponding gauge groupoid.

It follows that there is no straightforward relationship between group actions on a locally trivial Lie groupoid and group actions on a corresponding principal bundle. In many cases — such as the lift of a group action on a manifold to its fundamental groupoid — it is the action on the groupoid which is natural.

Whenever a phenomenon in principal bundle theory is dependent on a reference point, one may be sure that changing the reference point within its fibre will map the phenomenon under an automorphism of this type; one may also be sure that if the phenomenon is formulated in groupoid terms then it will be an intrinsic concept, independent of reference points. The clearest example of this is the replacement of the various mutually conjugate holonomy groups and isomorphic holonomy bundles arising from a connection, by a single holonomy groupoid. (See §6.3.)

Lastly in this section, we need to recall the notion of associated fibre bundle.

Construction 1.3.8 Suppose given a principal bundle $P(M, G, \pi)$ and an action of G on a manifold F. Let G act on the product manifold $P \times F$ by $(u, f)g = (ug, g^{-1}f)$. Then the quotient manifold E exists, thanks to the fact that the action of G is free. With respect to the projection $E \to M$ induced by $(u, f) \mapsto \pi(u)$, E is a fibre bundle over M, the *associated fibre bundle to $P(M, G)$ with respect to the action of G on F*, denoted $\frac{P \times F}{G}$. In most instances of this construction, F has algebraic structure with G acting by automorphisms, and when this is so we denote elements of E by $\lfloor u, f \rfloor$, in order to avoid confusion with the construction in 1.1.15.   ⊺

Associated with any $P(M, G)$ is the *inner LGB* $\frac{P \times G}{G}$ defined by the inner automorphism action of G on itself. Thus elements are $\lfloor u, g \rfloor$ where $\lfloor uh, g \rfloor = \lfloor u, hgh^{-1} \rfloor$ for $h \in G$. The proof of the following is an easy computation.

Proposition 1.3.9 *Let $\Omega \rightrightarrows M$ be a locally trivial Lie groupoid, and choose $m \in M$. Write $P = \Omega_m$, $G = \Omega_m^m$, $\pi = \beta_m$. Then the isomorphism $\frac{P \times P}{G} \to \Omega$ of 1.3.5(ii) carries the inner LGB $\frac{P \times G}{G}$ to the inner subgroupoid $\mathscr{I}\Omega$ of Ω by $\lfloor u, g \rfloor \mapsto \langle u, ug \rangle$.*

Construction 1.3.10 Given a principal bundle $P(M, G, \pi)$ and any morphism $\varphi \colon G \to H$ of Lie groups, let G act on H by $(g, h) \mapsto \varphi(g)h$.

Form the associated fibre bundle $Q = \frac{P \times H}{G}$; in this context, where G is not acting by automorphisms of H, we denote elements of Q by $\langle u, h \rangle$. Thus $\langle ug, h \rangle = \langle u, \varphi(g)h \rangle$. Now Q is itself an H–principal bundle over M with projection $\langle u, h \rangle \mapsto \pi(u)$ and action $\langle u, h \rangle h' = \langle u, hh' \rangle$. We call $Q(M, H)$ the *produced principal bundle of* $P(M, G)$ *along* φ, and denote it $\frac{P \times H}{G}$ in order to avoid confusion with other associated fibre bundles.

Notice that $P \to Q$, $u \mapsto \langle u, 1 \rangle$, is a morphism of principal bundles with respect to $\varphi \colon G \to H$ and id_M; we call it the *produced morphism* and denote it $\widetilde{\varphi}$. $\boxtimes$

In one sense, every base–preserving morphism of principal bundles is a produced morphism. By the same techniques as in 1.3.5, given $F(\mathrm{id}, \varphi) \colon P_1(M, G_1) \to P_2(M, G_2)$, the codomain bundle P_2 is isomorphic to the principal bundle produced from P_1 along φ, by an isomorphism which carries F to $\widetilde{\varphi}$. However, this result is less significant than it might appear: it does not, for example, mean that if $F'(\mathrm{id}, \varphi') \colon P_1(M, G_1) \to P_2(M, G_2)$ is a second base–preserving morphism and $\varphi' = \varphi$, then $F' = F$. These results carry over to locally trivial Lie groupoids via 1.3.5.

Some further results on the nature of morphisms of locally trivial Lie groupoids are given in §2.2.

1.4 Bisections

On a group G, the left translations $L_g \colon x \mapsto gx$ form a group which is isomorphic to G itself under $g \mapsto L_g$ and the right–translations $R_g \colon x \mapsto xg$ likewise form a group with $g \mapsto R_g$ now an anti–isomorphism. In the case of groupoids, an immediate problem is that the translations of 1.1.4 are only defined on the fibres of the source and target projections. These concepts of right– and left–translation are not always sufficient: there are many purposes for which one needs a concept of left– or right–translation which is a map of the whole groupoid.

Consider a Lie groupoid G and a diffeomorphism $G \to G$ which is the union of left–translations $L_g \colon G^{\alpha g} \to G^{\beta g}$; such a map is not characterized by a single element of G, but by a family of elements, one g with $\alpha g = m$ for each $m \in M$. In order that the resulting map be a diffeomorphism, it is necessary that $m \mapsto \beta g$ be a diffeomorphism. This leads to the following concept.

Definition 1.4.1 Let G be a Lie groupoid on M.

A *left–translation* on G is a pair of diffeomorphisms $\varphi\colon G \to G$, $\varphi_\circ\colon M \to M$, such that $\beta \circ \varphi = \varphi_\circ \circ \beta$, $\alpha \circ \varphi = \alpha$, and such that each $\varphi^x\colon G^x \to G^{\varphi_\circ(x)}$ is L_g for some $g \in G_x^{\varphi_\circ(x)}$.

A *bisection* of G is a smooth map $\sigma\colon M \to G$ which is right–inverse to $\alpha\colon G \to M$ and is such that $\beta \circ \sigma\colon M \to M$ is a diffeomorphism. The set of bisections of G is denoted $\mathscr{B}(G)$. $\quad\square$

These bisections may be regarded as generalized elements of the groupoid, since in the generalizations of the adjoint and exponential formulas for Lie groups, they play the rôle which, in the case of groups, is taken by the group elements themselves.

Let $\varphi, \varphi_\circ$ constitute a left–translation on G. For $x \in M$ choose $g \in G^x$ and define $\sigma(x) = \varphi(g)g^{-1}$. If h is another element of G^x then $h = g\ell$ for some $\ell \in G$. Now $\varphi(g) = \theta g$ for some $\theta \in G_x$ and $\varphi(g\ell) = \theta g\ell$ with the same θ. Thus $\varphi(g\ell) = \varphi(g)\ell$ and so $\sigma(x)$ is well-defined. The map σ is smooth since $\beta\colon G \to M$ is a surjective submersion. Clearly σ is a bisection.

Conversely, given a bisection σ, define

$$L_\sigma\colon G \to G, \quad g \mapsto \sigma(\beta g)g.$$

Then L_σ and $\beta \circ \sigma\colon M \to M$ constitute a left–translation on G, with inverse

$$(L_\sigma)^{-1}(h) = \sigma((\beta \circ \sigma)^{-1}(\beta h))^{-1}h.$$

We call L_σ (with $\beta \circ \sigma$ understood) the *left–translation corresponding to σ*.

Clearly the set of left–translations is a group under composition. We transfer its group structure to $\mathscr{B}(G)$. The proof of the following result is straightforward.

Proposition 1.4.2 *Let G be a Lie groupoid on M. Then $\mathscr{B}(G)$ is a group with respect to the multiplication $\star$ defined by*

$$(\sigma \star \tau)(x) = \sigma\big((\beta \circ \tau)(x)\big)\tau(x), \quad x \in M,$$

with identity the object inclusion map $x \mapsto 1_x$, denoted in this context by id, *and inversion*

$$\sigma^{-1}(x) = \sigma((\beta \circ \sigma)^{-1}(x))^{-1}, \quad x \in M.$$

Further, $\sigma \mapsto L_\sigma$ is a group isomorphism; that is $L_{\sigma\star\tau} = L_\sigma \circ L_\tau$.

Note that $\sigma \mapsto \beta \circ \sigma$ is a group morphism from $\mathscr{B}(G)$ to the group of diffeomorphisms of M.

The definition of bisection involves a lack of symmetry which can be avoided. Let σ be a bisection of a Lie groupoid G and consider the image

$$L = \{\sigma(m) \mid m \in M\}.$$

This is a closed embedded submanifold of G and the restrictions of both α and β to $L \to M$ are diffeomorphisms. Conversely, any closed embedded submanifold $L \subseteq G$ such that $\alpha, \beta \colon L \to M$ are both diffeomorphisms is the image of a unique bisection. We will sometimes identify a bisection with its image.

Example 1.4.3 Consider a trivial Lie groupoid $M \times G \times M$. The set $\mathscr{B}(M \times G \times M)$ can be identified with the set of pairs (φ, θ), where $\varphi \colon M \to M$ is a diffeomorphism and $\theta \colon M \to G$ is any smooth map, by identifying (φ, θ) with $x \mapsto (\varphi(x), \theta(x), x)$. The multiplication is then

$$(\varphi_2, \theta_2) \star (\varphi_1, \theta_1) = (\varphi_2 \circ \varphi_1, (\theta_2 \circ \varphi_1)\theta_1)$$

with inversion $(\varphi, \theta)^{-1} = (\varphi^{-1}, \theta^{-1} \circ \varphi^{-1})$; here θ^{-1} refers to the point-wise inverse of a group–valued map and φ^{-1} to the composition–inverse of a diffeomorphism. ⊠

Example 1.4.4 Consider a vector bundle (E, q, M). For $\sigma \in \mathscr{B}\Phi(E)$, define a vector bundle morphism, denoted $\overline{\sigma}$, over $\beta \circ \sigma$, by $u \mapsto \sigma(qu)(u)$. Then $\overline{\sigma \star \tau} = \overline{\sigma} \circ \overline{\tau}$ and $\overline{\mathrm{id}} = \mathrm{id}_E$, and so each $\overline{\sigma} \colon E \to E$ is a vector bundle automorphism. Conversely, given a vector bundle automorphism $\varphi \colon E \to E$, $f \colon M \to M$, the map $\sigma \colon x \mapsto \varphi_x \in \Phi(E)_x^{f(x)}$ is a bisection of $\Phi(E)$ (smoothness is proved by using the local triviality of E) and $\overline{\sigma} = \varphi$.

We will generally use not these vector bundle automorphisms but the maps of sections which they induce. For σ a bisection, $\overline{\sigma} \colon E \to E$, $\beta \circ \sigma \colon M \to M$, induces a map, also denoted $\overline{\sigma}$, from ΓE to ΓE, by

$$\overline{\sigma}(\mu)(x) = \sigma\big((\beta \circ \sigma)^{-1}(x)\big)\mu\big((\beta \circ \sigma)^{-1}(x)\big), \quad x \in M,$$

and a map $C^\infty(M) \to C^\infty(M)$, $u \mapsto u \circ (\beta \circ \sigma)^{-1}$, again denoted $\overline{\sigma}$, and these maps satisfy

$$\overline{\sigma \star \tau}(\mu) = \overline{\sigma}\big(\overline{\tau}(\mu)\big),$$
$$\overline{\sigma}(\mu_1 + \mu_2) = \overline{\sigma}(\mu_1) + \overline{\sigma}(\mu_2),$$
$$\overline{\sigma}(u\mu) = \overline{\sigma}(u)\overline{\sigma}(\mu),$$

as can be easily checked. For future use we also note that

$$\overline{\sigma^{-1}}(\mu)(x) = \sigma(x)^{-1}\mu((\beta \circ \sigma)(x)).$$

This interpretation can be extended to any fibre bundle, dropping the linearity conditions, providing that the frame groupoid can be given a suitable smooth structure.   ⊠

Example 1.4.5 Consider an action groupoid $G \ltimes M$ where G is a Lie group acting on a manifold M. If $\varphi\colon M \to M$ is a diffeomorphism then a bisection σ of $G \ltimes M$ with $\beta \circ \sigma = \varphi$ may be identified with a map $F\colon M \to G$ such that $F(m)m = \varphi(m)$, for all $m \in M$.

If the action is not transitive, then there will exist diffeomorphisms of M for which no such bisection exists.

For the cotangent groupoid $T^*G \rightrightarrows \mathfrak{g}^*$ of a Lie group G, as in 1.1.17, each $g \in G$ defines a bisection σ_g which to $\varphi \in \mathfrak{g}^*$ assigns $\varphi \circ T(L_{g^{-1}})$. The corresponding diffeomorphism $\mathfrak{g}^* \to \mathfrak{g}^*$ is Ad_g^*.   ⊠

In general the map $\sigma \mapsto \beta \circ \sigma$ goes into the subgroup consisting of those diffeomorphisms $M \to M$ which preserve the orbit partition of M (see §1.5).

Definition 1.4.6 Let G be a Lie groupoid on M, and take $\sigma \in \mathscr{B}(G)$. The *right–translation* defined by σ is

$$R_\sigma \colon G \to G, \qquad g \mapsto g\sigma\big((\beta \circ \sigma)^{-1}(\alpha g)\big).$$

The *inner automorphism* defined by σ is

$$I_\sigma \colon G \to G, \qquad g \mapsto \sigma(\beta g)g\sigma(\alpha g)^{-1}.$$

□

Clearly $R_{\sigma \star \tau} = R_\tau \circ R_\sigma$ and $I_{\sigma \star \tau} = I_\sigma \circ I_\tau$. Also, $R_{\sigma^{-1}}(g) = g\sigma(\alpha g)^{-1}$ and $I_\sigma = L_\sigma \circ R_{\sigma^{-1}} = R_{\sigma^{-1}} \circ L_\sigma$. Note that $I_\sigma \colon G \to G$ is an isomorphism of Lie groupoids over $\beta \circ \sigma$. Right–translations can be defined intrinsically as in 1.4.1. Since the Lie algebroid of a Lie groupoid will be defined using right–invariant vector fields and the flows of such fields are local left–translations, we will not use local right–translations very extensively.

Example 1.4.7 Let $P(M,G,\pi)$ be a principal bundle and let $\varphi(\varphi_\circ, \mathrm{id})$ be an automorphism. Thus $\pi \circ \varphi = \varphi_\circ \circ \pi$ and $\varphi(ug) = \varphi(u)g$, for all $u \in P$, $g \in G$. For $x \in M$ choose $u \in \pi^{-1}(x)$ and write $\sigma(x) = \langle \varphi(u), u \rangle$; this is clearly well defined and σ is smooth since π is an surjective submersion. This σ is a bisection of $\frac{P \times P}{G}$ and I_σ is $\langle v, u \rangle \mapsto \langle \varphi(v), \varphi(u) \rangle$, which, in terms of the isomorphism of 1.3.5(i), or in terms of 1.2.9, is the automorphism which corresponds to $\varphi \colon P \to P$.

Conversely consider a locally trivial Lie groupoid Ω on M, and an inner automorphism $I_\sigma \colon \Omega \to \Omega$ over $\beta \circ \sigma \colon M \to M$. At each $x \in M$ there is the vertex principal bundle $\Omega_x(M, \Omega_x^x)$ described in §1.3. Fix $x_0 \in M$ and write $x_1 = \beta(\sigma(x_0))$. Then I_σ restricts to $\Omega_{x_0} \to \Omega_{x_1}$. However we can identify Ω_{x_0} and Ω_{x_1} by right–translation by the single element $\sigma(x_0)^{-1}$, without reference to the rest of σ. If we do this, then the restriction of I_σ is identified with the restriction of L_σ.

Automorphisms of principal bundles of the form $\varphi(f, \mathrm{id})$ might thus legitimately be called either *inner automorphisms* or *left–translations*. The usual name is of course *gauge transformation*. Those for which $f = \mathrm{id}_M$ correspond to those bisections of $\frac{P \times P}{G}$ which take values in the inner group bundle $\frac{P \times G}{G}$.   ⊠

*   *   *   *   *

In most cases where bisections are used, the question is local and it suffices to consider local bisections.

Definition 1.4.8 Let G be a Lie groupoid on M. For $U \subseteq M$ open, a *local bisection of G on U* is a map $\sigma \colon U \to G$ which is right–inverse to α and for which $\beta \circ \sigma \colon U \to (\beta \circ \sigma)(U)$ is a diffeomorphism from U to the open set $(\beta \circ \sigma)(U)$ in M. The set of local bisections of G on U is denoted $\mathscr{B}_U G$.

For $\sigma \in \mathscr{B}_U G$ with $V = (\beta \circ \sigma)(U)$, the *local left–translation induced by σ* is

$$L_\sigma \colon G^U \to G^V, \quad g \mapsto \sigma(\beta g)g;$$

the *local right–translation defined by σ* is

$$R_\sigma \colon G_V \to G_U, \quad g \mapsto g\sigma\big((\beta \circ \sigma)^{-1}(\alpha g)\big);$$

and the *local inner automorphism defined by* σ is

$$I_\sigma \colon G_U^U \to G_V^V, \qquad g \mapsto \sigma(\beta g)g\sigma(\alpha g)^{-1}.$$

$\square$

Proposition 1.4.9 *Let* $G \rightrightarrows M$ *be a Lie groupoid. Given* $g \in G$ *there is a local bisection* σ *with* $\sigma(\alpha g) = g$.

Proof   Note first that, by linear algebra, there is a subspace I of $T_g(G)$ such that $T_g(G) = T_g(G_x) \oplus I$ and $T_g(G) = T_g(G^y) \oplus I$, where $x = \alpha g$, $y = \beta g$. Since α is a surjective submersion, it has a local section $\sigma \colon U \to G$ such that $\sigma(\alpha g) = g$. Use a local α–preserving diffeomorphism in G to arrange that I is the image of $T_x(\sigma)$. Then $\beta \circ \alpha \colon U \to M$ is étale at x and hence σ can be restricted so that $\beta \circ \sigma$ is a diffeomorphism onto its image. $\square$

The following immediate consequences are of great importance in what follows.

Corollary 1.4.10 *Let* $G \rightrightarrows M$ *be a Lie groupoid. For each* $x \in M$, $\beta_x \colon G_x \to M$ *is of constant rank.*

Proof Take g, $h \in G_x$. Then $j = gh^{-1}$ is defined and so there is a local bisection $\sigma \in \mathscr{B}_U G$ with $\beta h \in U$ and $\sigma(\beta h) = j$.

Now $L_\sigma \colon G_x^U \to G_x^V$, where $V = (\beta \circ \sigma)(U)$, maps h to g and $\beta_x \circ L_\sigma = (\beta \circ \sigma) \circ \beta_x$. Hence the ranks of β_x at g and h are equal. $\square$

Corollary 1.4.11 *Let* G *be a Lie groupoid on* M. *Then for all* $x, y \in M$, G_x^y *is a closed embedded submanifold of* G_x, *of* G^y *and of* G. *In particular, each vertex group* G_x^x *is a Lie group.*

Proof Since $G_x^y = \beta_x^{-1}(y)$ is the preimage of a point under a map of constant rank, it is a closed embedded submanifold of G_x and hence of G. A similar argument applies for $G_x^y \subseteq G^y$. It follows that $G_x^x \times G_x^x$ is a closed embedded submanifold of $G * G$ and so the restriction of the multiplication in G is smooth. $\square$

The set of all L_σ for $\sigma \in \mathscr{B}_U G$ and $U \subseteq M$ open is not a pseudogroup on G since it is not closed under restriction. The following result, which will be needed in Chapter 3, shows that this is unimportant.

Proposition 1.4.12 *Let G be a Lie groupoid on M. Let $\varphi\colon \mathcal{U} \to \mathcal{V}$ be a diffeomorphism from $\mathcal{U} \subseteq G$ open to $\mathcal{V} \subseteq G$ open, and let $f\colon U \to V$ be a diffeomorphism from $U = \beta(\mathcal{U}) \subseteq M$ to $V = \beta(\mathcal{V}) \subseteq M$, such that $\alpha \circ \varphi = \alpha$, $\beta \circ \varphi = f \circ \beta$ and $\varphi(gh) = \varphi(g)h$ whenever $(g,h) \in G * G$, $g \in \mathcal{U}$ and $gh \in \mathcal{U}$. Then φ is the restriction to $\mathcal{U}$ of a unique local left–translation $L_\sigma\colon G^U \to G^V$ where $\sigma \in \mathcal{B}_U G$.*

Proof For $x \in U$ choose $g \in \mathcal{U}^x$ and define $\sigma(x) = \varphi(g)g^{-1}$; clearly $\sigma(x)$ is well defined. Since the restriction $\beta\colon \mathcal{U} \to U$ is a submersion, σ is smooth and is therefore a local bisection on U with $\beta \circ \sigma = f$. That $L_\sigma(g) = \varphi(g)$ for $g \in \mathcal{U}$ is clear, as is the uniqueness. $\qquad\square$

Thus any local diffeomorphism $\mathcal{U} \to \mathcal{V}$ which commutes with the $R_h\colon G_{\beta h} \to G_{\alpha h}$ in the sense of 1.4.12 is the restriction of a local left–translation $L_\sigma\colon G^{\beta(\mathcal{U})} \to G^{\beta(\mathcal{V})}$.

In any case, for a general Lie groupoid G we will often treat the set of local bisections as if it were a pseudogroup on M with law of composition $\star$: if $\sigma \in \mathcal{B}_U G$ with $(\beta \circ \sigma)(U) = V$ and $\tau \in \mathcal{B}_{V'} G$ with $(\beta \circ \tau)(V') = W$, then $\tau \star \sigma$ is the local bisection in $(\beta \circ \sigma)^{-1}(V') \cap V$ defined by $(\tau \star \sigma)(x) = \tau\big((\beta \circ \sigma)(x)\big)\sigma(x)$, providing $(\beta \circ \sigma)^{-1}(V') \cap V$ is not void. We will — abusively — refer to the set of all local bisections, together with this composition, as the *pseudogroup of local bisections* of G, and will denote it by $\mathcal{B}^{\mathrm{loc}}(G)$.

We will often need to consider the effect of base–preserving morphisms on bisections.

Definition 1.4.13 Let $\varphi\colon G \to G'$ be a morphism of Lie groupoids over M. Then the maps $\mathcal{B}G \to \mathcal{B}G'$, $\sigma \mapsto \varphi \circ \sigma$, and $\mathcal{B}^{\mathrm{loc}}G \to \mathcal{B}^{\mathrm{loc}}G'$, $\sigma \mapsto \varphi \circ \sigma$, are both denoted by $\tilde{\varphi}$ and called *induced morphisms* (of groups, and — abusively — pseudogroups, respectively). $\qquad\square$

The effect of general, base changing, morphisms on bisections is troublesome, and we will not need it in this book. It can be handled in a way similar to that for general morphisms of Lie algebroids (see Chapter 4).

The following important formula for the multiplication in a tangent groupoid makes essential use of the concept of bisection.

Theorem 1.4.14 *Let* $G \rightrightarrows M$ *be a Lie groupoid, and denote the multiplication by* κ. *Let* $X \in T_g(G)$ *and* $Y \in T_h(G)$ *have* $T(\alpha)(X) = T(\beta)(Y) = w$. *Then*

$$X \bullet Y = T(\kappa)(X, Y) = T(L_\sigma)(Y) + T(R_\tau)(X) - T(L_\sigma)T(R_\tau)(T(1)(w)) \tag{3}$$

where σ, τ *are any (local) bisections of* G *for which* $\sigma(\alpha g) = g$ *and* $\tau(\alpha h) = h$.

Proof First suppose that $\xi = 1_m$. Then $X - T(1)T(\alpha)(X)$ is defined and is annulled by $T(\alpha)$. So $(X - T(1)T(\alpha)(X)) \bullet 0_h$ is defined and we can write

$$X \bullet Y = (X - T(1)T(\alpha)(X)) \bullet 0_h + T(1)T(\beta)(Y) \bullet Y.$$

Using (1) from p. 6, and remembering that $T(1)T(\beta)(Y)$ is an identity in TG, we have

$$X \bullet Y = T(R_h)(X - T(1)T(\alpha)(X)) + Y. \tag{4}$$

Now consider the general case. We have $L_\sigma(1_{\alpha g}) = g$; write $Z = T(L_\sigma^{-1})(X)$. Since L_σ commutes with α, we can apply (4) to get

$$Z \bullet Y = T(R_g)(Z - T(1)T(\alpha)(Z)) + Y.$$

Now R_g is the restriction to $G_{\beta h} \to G_{\alpha h}$ of R_τ. Also, $X \bullet Y = T(L_\sigma)(Z) \bullet Y = T(L_\sigma)(Z \bullet Y)$, since $\kappa \circ (L_\sigma \times \mathrm{id}) = L_\sigma \circ \kappa$. From this equation (3) follows. $\qquad\square$

All told, this gives a complete description of the multiplicative structure of TG. In particular, if $X \in T_g G$ and σ is a (local) bisection with $\sigma(\alpha g) = g$, then

$$T(\iota)(X) = T(R_\sigma^{-1})T(1)T(\alpha)(X) + T(L_\sigma^{-1})T(1)T(\beta)(X)$$
$$- T(R_\sigma^{-1})T(L_\sigma^{-1})(X). \tag{5}$$

1.5 Components and transitivity

We first extend to groupoids two elementary facts about topological groups: the component of the identity is a subgroup and that subgroup is generated by any neighbourhood of the identity.

Proposition 1.5.1 *Let* $G \rightrightarrows M$ *be a Lie groupoid. For each* $x \in M$, *let* C_x *denote the connectedness component of* 1_x *in* G_x. *Then* $C =$

$C(G) = \bigcup_{x \in M} C_x$ *is a wide Lie subgroupoid of* G, *called the* identity–component subgroupoid *of* G.

Proof By definition C contains each 1_x, $x \in M$, so it is certainly wide. Take $g \in C_x^y$ and $h \in C_y^z$ and consider $hg = R_g(h) \in G_x^z$. Because $R_g \colon G_y \to G_z$ is a diffeomorphism, it maps components to components; since $g = R_g(1_y) \in R_g(C_y)$ we have $C_x \cap R_g(C_y) \neq \emptyset$ and therefore $C_x = R_g(C_y)$. Hence $hg \in C_x$. So C is closed under multiplication. Taking $g \in C_x^y$ again, we have $1_y \in R_{g^{-1}}(C_x) \cap C_y$ so $R_{g^{-1}}(C_x) = C_y$ and hence $g^{-1} = 1_x g^{-1} \in C_y$, which proves that C is closed under inversion.

Let N be a tubular neighbourhood for 1_M, and regard N as a vector bundle over M with its zero section identified with $1 \colon M \to G$. Then $N \subseteq C$. Now the C_x, $x \in M$, are leaves of the foliation defined by $T^\alpha G = \ker T(\alpha)$ and so C is the union of the integral manifolds of $T^\alpha G$ which meet the open set N. Hence C is open and is therefore a Lie subgroupoid of G. □

It is implicit in this proof that the β–fibres C^y are the identity components of the β–fibres G^y of G, and that, for $g \in G_x^y$, the component of G_x containing g is $C_y g$ and the component of G^y containing g is $g C^x$. Clearly the various components of any one α–fibre need not be diffeomorphic.

If M is connected, then $C = \left(\bigcup_x C_x \right) \cup 1_M$ is connected, since each $C_x \cap 1_M$ is nonvoid. Conversely, if C is connected then $M = \beta(C)$ is connected.

Proposition 1.5.2 *Let* G *be a Lie groupoid on* M. *Then the identity–component subgroupoid* $C = C(G)$ *is open in* G.

Proof Let $\varphi \colon \mathbb{R}^p \times \mathbb{R}^q \to \mathscr{U} \subseteq G$ be a distinguished chart for the foliation induced by $T^\alpha G$, where $\mathscr{U} \cap 1_M \neq \emptyset$ and $\varphi(\{0\} \times \mathbb{R}^q) = \mathscr{U} \cap 1_M$. Then clearly $\mathscr{U} \subseteq C$. Taking the union of such $\mathscr{U}$ we obtain an open neighbourhood of 1_M in G which is contained in C. Now C is the union of those leaves of the foliation which intersect the open neighbourhood and so is itself open. □

Proposition 1.5.3 *Let* Ω *be a locally trivial Lie groupoid on a connected base* M. *Then* $C = C(\Omega)$ *is also locally trivial.*

Proof Let $\{\sigma_i\colon U_i \to \Omega_m\}$ be a section–atlas for Ω. We can assume the U_i are connected, so each $\sigma_i(U_i)$ lies in a single component C_i of Ω_m. Choose any $m_i \in U_i$ and define $\tau_i\colon U_i \to C_{\beta\sigma_i(m_i)}$ by $\tau_i(x) = \sigma_i(x)\sigma_i(m_i)^{-1}$.

Next consider any $\beta_x\colon \Omega_x \to M$. This is a surjective submersion and therefore open; hence $\beta_x(C_x)$ is open in M. Now the $\beta_x(C_x)$, as x ranges through M, are either disjoint or equal, and therefore are also closed. Since M is connected, we must have $\beta_x(C_x) = M$ for all $x \in M$.

For each i we can therefore choose a $\xi_i \in C_m$ with $\beta\xi_i = \beta\sigma_i(m_i)$. Then $\{\nu_i\colon U_i \to C_m\}$ defined by $\nu_i(x) = \tau_i(x)\xi_i$ gives a section-atlas for C. $\qquad\square$

Of course it is true in general that if a Lie groupoid has a subgroupoid which is locally trivial, then the ambient groupoid itself must be locally trivial. The construction of the identity component subgroupoid thus has a principal bundle analogue.

Consider a principal bundle $P(M, G, \pi)$ on a connected base. Choose any component $P_\circ$ of P. Define

$$H = \{g \in G \mid P_\circ g = P_\circ\}.$$

It is clear that H is an open subgroup of G. By an argument similar to that in 1.5.3, $\pi_\circ = \pi|_{P_\circ}\colon P_\circ \to M$ is surjective; it remains a surjective submersion since $P_\circ$ is open in P. Further, $P_\circ \times H \to P_\circ$ remains a free action, and its orbits are equal to the fibres of $\pi_\circ$, by an argument similar to that in 1.5.1. Thus $P_\circ(M, H, \pi_\circ)$ is a principal bundle, and a reduction of $P(M, G)$.

Definition 1.5.4 The principal bundle $P_\circ(M, H, \pi_\circ)$ is a *connected reduction of* $P(M, G)$.
$\qquad\square$

Example 1.5.5 Let P be the space $\mathbb{R} \times \mathbb{Z}$ and let G be the discrete space $\mathbb{Z} \times \mathbb{Z}$ with the semidirect product group structure

$$(m_1, n_1)(m_2, n_2) = (m_1 + m_2, (-1)^{m_2}n_1 + n_2).$$

Let G act on P to the right by

$$(x, p)(m, n) = (x + m, (-1)^m p + n)$$

and let $\pi\colon P \to M = \mathbb{S}^1$ be $\pi(x, p) = e^{2\pi i x}$. It is easy to verify that $P(M, G, \pi)$ is a principal bundle.

The component of P through $(0, 0)$ is $P_\circ = \mathbb{R} \times \{0\}$. Evidently,

$H = \mathbb{Z} \times \{0\}$. The bundle $P_o(M, H)$ can be identified with the covering $\mathbb{R}(\mathbb{S}^1, \mathbb{Z})$.

Note that $P(M, G)$ is the pullback of the universal cover $\mathbb{R}^2(K, G)$ of the Klein bottle K along the map $\mathbb{S}^1 = \mathbb{R}/\mathbb{Z} \to K = \mathbb{R}^2/G$ induced by $\mathbb{R} \to \mathbb{R}^2$, $x \mapsto (x, 0)$. $\boxtimes$

This example illustrates that H need not be normal in G. In general H is a union of cosets of G_o in G, and H is normal in G if and only if H/G_o is normal in $G/G_o = \pi_0(G)$. For most nondiscrete Lie groups encountered in familiar applications, $\pi_0(G)$ is abelian.

Proposition 1.5.6 *Let G be a Lie groupoid on M. Let $\mathcal{U}$ be a symmetric set (that is, $\mathcal{U} \supseteq 1_M$ and $\mathcal{U}^{-1} = \mathcal{U}$) such that each $\mathcal{U}_x$ is open in G_x. Then the subgroupoid H generated by $\mathcal{U}$ has H_x open in G_x for all $x \in M$.*

Proof Since $\mathcal{U}$ is symmetric, H is merely the set of all possible products of elements from $\mathcal{U}$. Choose $x \in M$. The set of all n–fold products $g_n \ldots g_1$ from $\mathcal{U}$ with $\alpha g_1 = x$ is the union of all $\mathcal{U}_{\beta j} j \subseteq G_x$ where j is an $(n-1)$–fold product from $\mathcal{U}$. Since $R_j \colon G_{\beta j} \to G_x$ is a diffeomorphism the set of all n–fold products from $\mathcal{U}$ which lie in G_x is open in G_x. Hence H_x is open in G_x. $\square$

A set $\mathcal{U}$ satisfying the conditions in 1.5.6 will be called a *symmetric α–neighbourhood* of 1_M (or, of the base) in G.

Proposition 1.5.7 *Let G be a Lie groupoid on M, and let H be a wide Lie subgroupoid of G. If each H_x is open in G_x, $x \in M$, then each H_x is also closed in G_x, $x \in M$.*

Proof The complement $G_x \backslash H_x$ is the union of all $H_{\beta g} g$ as g ranges over $G_x \backslash H_x$. Since $H_{\beta g}$ is open in $G_{\beta g}$ it follows that $H_{\beta g} g$ is open in G_x. $\square$

The following result is now immediate.

Proposition 1.5.8 *Let G be a Lie groupoid on M, and let $\mathcal{U}$ be a symmetric α–neighbourhood of 1_M in G. Then $\mathcal{U}$ generates the identity–component subgroupoid C of G.*

$*$ $*$ $*$ $*$ $*$

A groupoid $G \rightrightarrows M$ is transitive if and only if the anchor is surjective. At the other extreme, the following case is often encountered.

Definition 1.5.9 A groupoid $G \rightrightarrows M$ is *totally intransitive* if the image of (β, α) is the base subgroupoid Δ_M of $M \times M$.   □

For any $G \rightrightarrows M$, the image of the anchor $(\beta, \alpha) \colon G \to M \times M$ is an equivalence relation on M. The equivalence class of $x \in M$, denoted $\mathcal{O}_x = \mathcal{O}_x(G)$, is the *transitivity component of G through x*, or the *transitivity orbit of G through x*. Note that $\mathcal{O}_x = \beta_x(G_x) = \alpha^x(G^x)$. Except where confusion is possible, we will generally just write *orbit*.

Examples 1.5.10 The transitivity components of a fundamental groupoid $\Pi(M)$ are the connectedness components of M, and the transitivity components of an action groupoid $G \ltimes M$ are the orbits of the action. For a surjective submersion $q \colon M \to Q$, the transitivity components of $R(q) \subseteq M \times M$ are the fibres of q.

An LGB considered as a Lie groupoid is totally intransitive. Groups are both transitive and totally intransitive.   ⊠

For a set groupoid $G \rightrightarrows M$, each restriction $G_{\mathcal{O}}^{\mathcal{O}}$ to an orbit $\mathcal{O}$ is a transitive subgroupoid of G, and G can be regarded as the disjoint union of the transitive subgroupoids $G_{\mathcal{O}}^{\mathcal{O}}$, as $\mathcal{O}$ runs through the orbits. In the case of Lie groupoids, it is clear that the smooth structure cannot be obtained so simply. A good example to keep in mind throughout this section is the cotangent groupoid 1.1.17.

Theorem 1.5.11 *Let G be a Lie groupoid on M. Then for each $x \in M$, the orbit $\mathcal{O}_x = \beta_x(G_x)$ is a submanifold of M.*

Proof Denote G_x by P and G_x^x by W. Then, for the same reason as in Corollary 1.4.11, the restriction of the groupoid multiplication to $P \times W \to P$ is a smooth action of a Lie group on a manifold. It is easily seen to be proper: if $K, L \subseteq P$ are compact then $\{w \in W \mid Kw \cap L \neq \emptyset\}$ is the image under $P \times_{\beta_x} P \to W$, $(h, g) \mapsto h^{-1}g$, of the closed subset $K \times_{\beta_x} L = (K \times L) \cap (P \times_{\beta_x} P)$ of the compact set $K \times L$ and is therefore compact. Since the action is also free, it follows that $\{(gw, g) \mid g \in P,\ w \in W\}$ is a closed embedded submanifold of $P \times P$ and so there is a quotient manifold structure on P/W.

Define $i \colon P/W \to M$ by $i(gW) = \beta_x(g)$. Then i is smooth and injective. Since $P \to P/W$ is a submersion, $\mathrm{rk}_{gW}(i) = \mathrm{rk}_g(\beta_x)$, for all

$g \in P$, and so i is of constant rank with image $\mathscr{O}_x$. Now an injection of constant rank is an immersion. $\qquad\square$

In particular, $G_x(\mathscr{O}_x, G_x^x, \beta_x)$ is a principal bundle. It is clear that, as sets,

$$\frac{G_x \times G_x}{G_x^x} = G_{\mathscr{O}_x}^{\mathscr{O}_x}.$$

Since β_x is of constant rank it follows, as in 1.3.3, that the division map $G_x \times G_x \to G$ is also of constant rank. Hence $\frac{G_x \times G_x}{G_x^x} \to G$ is an injection of constant rank, and therefore an immersion. Thus $G_{\mathscr{O}_x}^{\mathscr{O}_x}$ is a submanifold of G, and we have the following result:

Theorem 1.5.12 *Let G be a Lie groupoid on M and let $\mathscr{O}$ be an orbit of M. Then there is a manifold structure on $G_{\mathscr{O}}^{\mathscr{O}}$ with respect to which it is a submanifold of G and a Lie groupoid on $\mathscr{O}$. Further, $G_{\mathscr{O}}^{\mathscr{O}}$ is locally trivial.*

Corollary 1.5.13 *A transitive Lie groupoid is locally trivial.*

For ease of reference we state explicitly the following results, proved above.

Corollary 1.5.14 *Let G be a Lie groupoid on M, and let $x \in M$. Then $\delta_x\colon G_x \times G_x \to G$ is of constant rank. Further, G is locally trivial if and only if δ_x is a surjective submersion.*

Example 1.5.15 Let $G \times M \to M$ be a smooth action of a Lie group on a manifold and let $G \lessdot M$ be the action groupoid. Applying 1.4.10 shows that each evaluation map is of constant rank, applying 1.5.11 that the orbits are submanifolds of M, and 1.5.13 that if G acts transitively, then M is equivariantly diffeomorphic to a homogeneous space; 1.5.13 also includes the existence of local cross–sections for closed subgroups of Lie groups. $\qquad\boxtimes$

To avoid misunderstanding, we will refer to 'the orbit foliation of the base' only when the foliation is known to be regular; in general we refer to 'the orbit partition of the base'.

We know that any Lie groupoid may be restricted to any open subset of the base, and that a Lie groupoid may be restricted to any of its orbits. The following general result concerning restrictions is easily proved.

Proposition 1.5.16 *Let $G \rightrightarrows M$ be a Lie groupoid and let $\iota \colon S \to M$ be a submanifold such that $\iota \times \iota \colon S \times S \to M \times M$ and $(\beta, \alpha) \colon G \to M \times M$ are transversal. Then G_S^S with its submanifold structure is a Lie subgroupoid of G.*

A submanifold satisfying the condition is said to be *transversal* to G.

1.6 Actions

Groupoids arise naturally as systems of automorphisms of fibre structures, and they accordingly act on spaces fibred over the base manifold.

Definition 1.6.1 Let G be a Lie groupoid on M and let $f \colon M' \to M$ be a smooth map. Let $G * M'$ denote the pullback manifold $\{(g, m') \in G \times M' \mid \alpha g = f(m')\}$. An *action of G on $f \colon M' \to M$* is a smooth map $G * M' \to M'$, $(g, m') \mapsto gm'$, such that

(i) $f(gm') = \beta(g)$, for all $g \in G$, $m' \in M'$ with $\alpha(g) = f(m')$;

(ii) $h(gm') = (hg)m'$, for all $(h, g) \in G * G$, and $m' \in M'$ such that $(h, m') \in G * M'$;

(iii) $1_{f(m')}m' = m'$, for all $m' \in M'$.

For $m' \in M'$, the subset $G[m'] = \{gm' \mid g \in G_{f(m')}\}$ is the *orbit* of m'.

□

If f is a surjective submersion, its fibres are closed embedded submanifolds of M' and each $g \in G_m^n$ induces a diffeomorphism $f^{-1}(m) \to f^{-1}(n)$. In general, not much can be said about the smoothness properties of the maps between individual fibres. It is nonetheless useful to have the full generality of 1.6.1.

We deal in §1.7 with the special features of linear actions on vector bundles, and of actions on Lie group bundles by Lie group isomorphisms. The case of groupoid actions on groupoids is found in §2.5.

Definition 1.6.2 Let G be a Lie groupoid on M, and let $G*M^1 \to M^1$ and $G * M^2 \to M^2$ be actions of G on smooth maps $f_i \colon M^i \to M$, $i = 1, 2$. Then a smooth map $\psi \colon M^1 \to M^2$ such that $f_2 \circ \psi = f_1$ is *G-equivariant* if $\psi(gm') = g\psi(m')$ for all $(g, m') \in G * M^1$.

Let H be a second Lie groupoid with base N, let $p \colon M' \to M$ and $q \colon N' \to N$ be smooth maps, let $G * M' \to M'$ and $H * N' \to N'$ be actions, let $\varphi \colon G \to H$ be a morphism of Lie groupoids over $f \colon M \to N$, and let $\psi\colon M' \to N'$ be a smooth map such that $q \circ \psi = f \circ p$. Then ψ is φ-*equivariant* if $\psi(gm') = \varphi(g)\psi(m')$, for all $(g, m') \in G * M'$. $\qquad\square$

The following examples are basic.

Example 1.6.3 Let G be a Lie groupoid on M and let $M \times F$ be a trivial fibre bundle. Then $g(\alpha g, a) = (\beta g, a)$, for $g \in G$, $a \in F$, is an action of G on $M \times F$, called the *trivial action*. $\qquad\boxtimes$

Example 1.6.4 Let $P(M, G, \pi)$ be a principal bundle and let $E = \frac{P \times F}{G}$ be an associated fibre bundle with respect to a smooth action $G \times F \to F$. Then

$$\frac{P \times P}{G} * \frac{P \times F}{G} \to \frac{P \times F}{G}, \qquad (\langle v, u \rangle, \lfloor u, a \rfloor) \mapsto \lfloor v, a \rfloor$$

is an action. $\qquad\boxtimes$

In fact all actions of locally trivial groupoids are of this type:

Theorem 1.6.5 *Let Ω be a locally trivial Lie groupoid on M, and let $\Omega * M' \to M'$ be an action of Ω on a surjective submersion $q\colon M' \to M$. Then (M', q, M) is a fibre bundle and, for any choice of $m \in M$ and writing $P = \Omega_m$, $G = \Omega_m^m$, $F = M'_m$, the map*

$$\frac{P \times F}{G} \to M', \quad \lfloor \xi, a \rfloor \mapsto \xi a,$$

is a diffeomorphism of fibrations over M and is equivariant with respect to the isomorphism $\frac{P \times P}{G} \to \Omega$ of 1.3.5. Here $\frac{P \times F}{G}$ is constructed with respect to the action of G on F which is the restriction $\Omega_m^m \times M'_m \to M'_m$.

Proof Take a section atlas $\{\sigma_i\colon U_i \to \Omega_m\}$ and with it define charts $\psi_i\colon U_i \times F \to M'_{U_i}$, $(x, a) \mapsto \sigma_i(x)a$. This proves that (M', q, M) is a fibre bundle. Define $P \times F \to M'$ by $(\xi, a) \mapsto \xi a$. In terms of the charts ψ_i for M' and $(x, g) \mapsto \sigma_i(x)g$ for Ω_m, this is $U_i \times G \times F \to U_i \times F$, $(x, g, a) \mapsto (x, ga)$, and is hence a surjective submersion. Hence $\frac{P \times F}{G} \to M$, $\lfloor \xi, a \rfloor \mapsto \xi a$ is a diffeomorphism. The other statements are easily proved. $\qquad\square$

Proposition 1.6.6 *Let $P(M, G)$ be a principal bundle and let B and B' be two associated fibre bundles corresponding to actions $G \times F \to F$ and $G \times F' \to F'$ of G on manifolds F and F'.*

(i) *If $f\colon F \to F'$ is a G–equivariant map then $\tilde{f}\colon B \to B'$ defined by $\lfloor u, a \rfloor \mapsto \lfloor u, f(a) \rfloor$ is a well–defined morphism of fibre bundles over M, and is $\frac{P \times P}{G}$–equivariant.*

(ii) *If $\varphi\colon B \to B'$ is a $\frac{P \times P}{G}$–equivariant morphism of fibre bundles over M, then $\varphi = \tilde{f}$ for some G–equivariant map f.*

Proof (i) is easy to verify. For (ii), observe that a map $f\colon P \times F \to F'$ can be defined by the condition that for any $u \in P$ and $a \in F$,

$$\varphi(\lfloor u, a \rfloor) = \lfloor u, f(u, a) \rfloor.$$

Now it is easy to see that equivariance with respect to $\frac{P \times P}{G}$ forces $f(u, a)$ to depend only on a. So we have $f\colon F \to F'$ and f must then be G–equivariant. $\square$

Example 1.6.7 Let Ω be a locally trivial Lie groupoid on M. Then the *inner automorphism action* is the map $\Omega * \mathscr{I}\Omega \to \mathscr{I}\Omega$, $(\xi, \lambda) \mapsto I_\xi(\lambda) = \xi\lambda\xi^{-1}$.

If Ω corresponds to a principal bundle $P(M, G)$, then Theorem 1.6.5 shows that $\mathscr{I}\Omega$ is equivariantly isomorphic as a LGB to the inner group bundle $\frac{P \times G}{G}$.   ⊠

Example 1.6.8 Let $f\colon M \to Q$ be a surjective submersion. Then there is a morphism of vector bundles $TM \to f^!(TQ)$ over M, which we denote $T(f)^!$, defined by $T(f)^!(X) = (f(m), T(f)(X))$ for $X \in T_m M$ (see §2.1). The kernel of $T(f)^!$ is $T^f M$, the vertical bundle of f. The corresponding normal bundle $TM/T^f M$ is accordingly isomorphic to the pullback bundle $f^!(TQ)$. Now the trivial action of $R(f)$ on $f^!(TQ)$ transfers to a canonical action of $R(f)$ on $TM/T^f M$.   ⊠

Example 1.6.9 Let $\sigma\colon G \times M \to M$ be a smooth action of a Lie group G on a manifold M. Let $\mathfrak{p}\colon T^*M \to \mathfrak{g}^*$ denote the dual of the infinitesimal action associated to σ. Thus

$$\langle \mathfrak{p}(\Phi), X \rangle = \langle \Phi, X^\dagger(x) \rangle,$$

for $\Phi \in T_x^* M$ and $X \in \mathfrak{g}$. Then the cotangent groupoid $T^*G \rightrightarrows \mathfrak{g}^*$ of 1.1.17 acts on $\mathfrak{p}$ by lifting the action of G on T^*M to T^*G. Explicitly, take $\varphi \in T_g^* G$ and $\Phi \in T_x^* M$ and define

$$\varphi \bullet \Phi = g\Phi = \Phi \circ T(\sigma_{g^{-1}})$$

where $\sigma_{g^{-1}}$ denotes $M \to M$, $y \mapsto g^{-1}y$. Since $\mathfrak{p}(g\Phi) = \mathrm{Ad}_g^*(\mathfrak{p}(\Phi))$, it follows that if $\widetilde{\alpha}(\varphi) = \mathfrak{p}(\Phi)$ then $\mathfrak{p}(\varphi \bullet \Phi) = \widetilde{\beta}(\varphi)$.

We call $\mathfrak{p}$ the *pith* of the action. ⊠

*　　*　　*　　*　　*

The construction of an action groupoid in 1.1.9 can be generalized: let G be a Lie groupoid on M and let $G * M' \to M'$ be an action of G on a smooth map $f\colon M' \to M$. Give $G * M'$ the structure of a groupoid on M' as follows: the projections are $\alpha'(g, m') = m'$, $\beta'(g, m') = gm'$, the object inclusion map is $m' \mapsto (1_{f(m')}, m')$, the multiplication is $(h, n')(g, m') = (hg, m')$, defined when $n' = gm'$, and the inversion is $(g, m')^{-1} = (g^{-1}, gm')$.

Now $\alpha'\colon G * M' \to M'$ is a surjective submersion, being the pullback of $\alpha\colon G \to M$. It is simple to check that the groupoid multiplication is smooth.

Definition 1.6.10 With the structure described above, $G * M'$ is denoted $G \lhd f$ or $G \lhd M'$ and is the *action groupoid* associated to the action of G on $f\colon M' \to M$. □

Note that the canonical map $G \lhd f \to G$, $(g, m') \mapsto g$, which we denote by $f_!$, is a morphism over $f\colon M' \to M$.

Example 1.6.11 Any Lie groupoid G acts on its own β–projection through the multiplication map $G * G \to G$. The action groupoid $G \lhd \beta$ is canonically isomorphic to the equivalence relation $G \times_\alpha G$ under the anchor of the action groupoid, $(h, g) \mapsto (hg, g)$.

In particular, for any Lie group G, the action groupoid $G \lhd G$ defined by right–translations is isomorphic to the pair groupoid $G \times G$. ⊠

Example 1.6.12 For any Lie groupoid $G \rightrightarrows M$, the Cartesian square $G \times G$ acts on the anchor (β, α) by $(\zeta_2, \zeta_1)(g) = \zeta_2 g \zeta_1^{-1}$. The action groupoid $(G \times G) \lhd (\beta, \alpha)$ is canonically isomorphic to the pullback groupoid $\alpha^{!!}G$ defined in §2.3 under the map $(\zeta_2, \zeta_1, g) \mapsto (\zeta_2 g \zeta_1^{-1}, \zeta_1, g)$. ⊠

Example 1.6.13 As noted in 1.3.4, the action groupoid $G \lhd (G/H)$, for a Lie group acting in the standard way on a homogeneous space, is canonically isomorphic to the gauge groupoid $\frac{G \times G}{H}$. ⊠

Action groupoids and the morphisms associated with them can be characterized intrinsically:

Definition 1.6.14 Let $\varphi\colon G' \to G$ be any morphism of Lie groupoids over a smooth map $f\colon M' \to M$. Then $f^!G = G * M'$ denotes the pullback manifold $\{(g, m') \in G \times M' \mid \alpha(g) = f(m')\}$ and $\varphi^!$ denotes the map $G' \to f^!G$, $g' \mapsto (\varphi(g'), \alpha'(g'))$. □

Of course in general $f^!G$ does not have a groupoid structure and $\varphi^!$ is not a morphism.

Definition 1.6.15 Let $\varphi\colon G' \to G$ be a morphism of Lie groupoids over $f\colon M' \to M$. Then φ is an *action morphism* if $\varphi^!\colon G' \to f^!G$ is a diffeomorphism. □

Given an action $G*M' \to M'$, the canonical morphism $f_!\colon G \lessdot f \to G$ is an action morphism.

Theorem 1.6.16 *Let $\varphi\colon G' \to G$ be an action morphism over a map $f\colon M' \to M$, and let $s\colon f^!G \to G'$ denote the inverse of $\varphi^!$. Then $\beta' \circ s\colon f^!G \to M'$ is an action of G on $f\colon M' \to M$.*

Proof We will show that $h(gx') = (hg)x'$ for $(h, g) \in G*G$ and $x' \in M'$ with $\alpha g = f(x')$; the other conditions are clear. First note that each $\varphi_{x'}\colon G'_{x'} \to G_{f(x')}$ is a bijection — it is easy to see that $\varphi^!$ maps $G'_{x'}$ onto $G_{f(x')} \times \{x'\}$. Thus $s(g, x')$ is the unique element of $G'_{x'}$ which is mapped by φ onto g. Write $y' = (\beta' \circ s)(g, x')$ and note that $f(y') = \beta g = \alpha h$. So $s(h, y')$ is defined and is the unique element of $G'_{y'}$ which is mapped by φ onto h. Since $\alpha'(s(h, y')) = y' = \beta'(s(g, x'))$, the product $s(h, y')s(g, x')$ is defined. Obviously it belongs to $G'_{x'}$ and is mapped by φ onto hg; it is therefore equal to $s(hg, x')$. That $h(gx') = (hg)x'$ now follows. □

The next result shows that these two constructions are indeed mutual inverses; the proof is straightforward.

Proposition 1.6.17 (i) *Let $\varphi\colon G' \to G$ be an action morphism, and let $G * M' \to M'$ be the associated action. Then, giving $G * M'$ the structure as the action groupoid $G \lessdot M'$, the map $\varphi^!\colon G' \to G \lessdot M'$ is an isomorphism of Lie groupoids over M' and $\mathrm{pr}_G \circ \varphi^! = \varphi$.*

(ii) *Let $G * M' \to M'$ be an action of a Lie groupoid on a smooth map $f: M' \to M$. Then the action of G on M' induced by the action morphism $\varphi^!: f^!G \to G$ is the original action.*

Let $\varphi: M^1 \to M^2$ be a G–equivariant map of two actions $G * M^1 \to M^1$ and $G * M^2 \to M^2$. Define $\widetilde{\varphi}: G * M^1 \to G * M^2$ by $(g, m_1) \mapsto (g, \varphi(m_1))$; this is a morphism of groupoids over φ, and a morphism of actions in the following sense.

Definition 1.6.18 Let $\varphi_1: G' \to G$ and $\varphi_2: G' \to G$ be action morphisms with the same target G. Then a *morphism of actions $\psi: \varphi_1 \to \varphi_2$ over G* is a morphism of Lie groupoids $\psi: G' \to G'$ such that $\varphi_2 \circ \psi = \varphi_1$. □

If $\psi: \varphi_1 \to \varphi_2$ is a morphism of actions over G, then $\psi_\circ: M' \to M'$ is G–equivariant with respect to the actions induced by the action morphisms.

One may express this by saying that the category of action morphisms over a fixed Lie groupoid G is equivalent to the category of actions of G and G–equivariant maps. The proof of the following result is straightforward.

Proposition 1.6.19 *Given an action morphism $\varphi_2: G' \to G'$, and an arbitrary morphism of Lie groupoids $\varphi_1: G \to G'$, the composite $\varphi_2 \circ \varphi_1$ is an action morphism if and only if φ_1 is.*

It is often useful to be able to deduce properties of an action from properties of its action groupoid. For a Lie group action $G \times M' \to M'$, the groupoid $G \ltimes M'$ is transitive if and only if the action is transitive, and the action is free if and only if the anchor $\chi: G \ltimes M' \to M' \times M'$ has trivial kernel in the sense that $\chi^{-1}(\Delta_{M'}) = 1_{M'}$.

For a general Lie groupoid $G \rightrightarrows M$ acting on a surjective submersion $f: M' \to M$, the action groupoid is transitive if and only if G is transitive and some (and hence every) vertex group action $G_m^m \times M'_m \to M'_m$ is transitive. Further, $G \ltimes M'$ is locally trivial if and only if G is locally trivial and some (and hence every) evaluation map $G_{f(m')} \to M'$, $g \mapsto gm'$, is a surjective submersion.

See also §2.5.

*   *   *   *   *

We now consider the orbit structure of action groupoids. The short-ness of the proof of the following result is a consequence of the action groupoid construction.

Theorem 1.6.20 *Let $G * M' \to M'$ be a smooth action of a Lie group-oid G on a smooth map $f\colon M' \to M$. Then:*

(i) *each orbit $G[m']$, $m' \in M'$, of G is a submanifold of M';*

(ii) *each evaluation map $G_{f(m')} \to M'$, $g \mapsto gm'$ is of constant rank;*

(iii) *if the action is transitive, then each evaluation map is a surjective submersion;*

(iv) *if G is locally trivial, then (M', f, M) is a fibre bundle and so too is each orbit $(G[m'], f, M)$;*

(v) *the set M'/G of orbits has the structure of a quotient manifold from M' if and only if the graph*

$$\Gamma = \{(m', gm') \in M' \times M' \mid m' \in M', \ g \in G_{f(m')}\}$$

of the action is a closed embedded submanifold of $M' \times M'$.

Proof (i) – (iii) follow from 1.5.11, 1.4.10, and 1.5.13.

(iv) follows as in 1.6.5.

(v) follows from the criterion of Godement and from the observation that $\bar{\alpha}\colon G * M \to M$ factors through the projection $\pi_1\colon \Gamma \to M$ via $(g, m') \mapsto (m', gm')$. $\square$

Definition 1.6.21 Let $\Omega * M' \to M'$ be an action of a locally trivial Lie groupoid $\Omega \rightrightarrows M$ on a surjective submersion (M', p, M). Then $\mu \in \Gamma M'$ is Ω–*deformable* if for all $x, y \in M$ there exists $\xi \in \Omega_x^y$ such that $\xi\mu(x) = \mu(y)$.

If $\mu \in \Gamma M'$ is Ω–deformable, then the *stabilizer subgroupoid* of Ω at μ is

$$\Omega\{\mu\} = \{\xi \in \Omega \mid \xi\mu(\alpha\xi) = \mu(\beta\xi)\}.$$

$\square$

Although the notion of deformable section can be given in greater generality, it is of little interest unless the acting groupoid is locally trivial.

A section μ is Ω–deformable if and only if its values lie in a single orbit; the condition ensures that the stabilizer subgroupoid is transitive. Note that $\Omega\{\mu\}$ is closed in Ω since M' is Hausdorff.

We now use the correspondence between deformable sections and their stabilizer subgroupoids to give a classification of those locally trivial subgroupoids of Ω which have a preassigned vertex group at a given point $m \in M$.

Definition 1.6.22 Let Ω be a locally trivial Lie groupoid on M. A *reduction* of Ω, or a *locally trivial Lie subgroupoid* of Ω, is a wide Lie subgroupoid (Ω', φ) such that Ω' is locally trivial. $\qquad\square$

Theorem 1.6.23 *Let Ω be a locally trivial Lie groupoid on M and let $\Omega * M' \to M'$ be a smooth action of Ω on the fibre bundle (M', q, M). Let $\mu \in \Gamma M'$ be an Ω–deformable section. Then the stabilizer groupoid $\Omega\{\mu\}$ of Ω at μ is a closed embedded reduction of Ω.*

Proof Since μ takes values in a single orbit of Ω, we can assume by Theorem 1.6.20 that Ω acts transitively on M'.

Define $f \colon \Omega \to M' \times_q M'$ by $\xi \mapsto (\mu(\beta\xi), \xi\mu(\alpha\xi))$. Then $\Omega(\mu) = f^{-1}(\Delta_{M'})$, which shows that $\Omega\{\mu\}$ is closed in Ω. We prove that f is transversal to $\Delta_{M'}$ in $M' \times_q M'$. For $\xi_0 \in \Omega\{\mu\}$, choose decomposing sections $U \to \Omega_m$ and $V \to \Omega_m$ in neighbourhoods U, V of $x_0 = \alpha\xi_0$, $y_0 = \beta\xi_0$; then f has the local expression

$$V \times G \times U \to (V \times F) * (V \times F), \qquad (y, g, x) \mapsto ((y, a'(y)), (y, ga(x)))$$

where $F = M'_m$, $G = \Omega^m_m$ and $a \colon U \to F$, $a' \colon V \to F$ are the local expressions of μ. Let ξ_0 correspond to (y_0, g_0, x_0), and note that $a'(y_0) = g_0 a(x_0)$ because $\xi_0 \in \Omega(\mu)$.

Given $X, Y \in T_{a'(y_0)}(F)$, there is a $W \in T_{g_0}(G)$ such that evaluation $G \to F$, $g \mapsto ga(x_0)$ maps W to $X - Y$. This is because the action $G \times F \to F$ is smooth and transitive and therefore each evaluation is a submersion. Hence, given also $Z \in T_{y_0}(V)$, we have $f_*(0 \oplus W \oplus 0) + (Z \oplus Y) \oplus (Z \oplus Y) = (Z \oplus Y) \oplus (Z \oplus X)$ and this proves that f is transversal to Δ_M. Hence $\Omega(\mu)$ is an embedded submanifold of Ω.

To show that $\Omega\{\mu\}$ is a Lie subgroupoid of Ω it is only necessary to show that the projections $\Omega\{\mu\} \to M$ are submersions. In fact $\beta \colon \Omega\{\mu\} \to M$ is the composite

$$\Omega(\mu) \xrightarrow{f} \Delta_{M'} \xrightarrow{\pi_1} M' \xrightarrow{q} M$$

in which each map is a submersion.

That $\Omega(\mu)$ is locally trivial follows from Corollary 1.5.13. $\qquad\square$

The following is a converse to Theorem 1.6.23.

Proposition 1.6.24 *Let* $\Omega \rightrightarrows M$ *be a locally trivial Lie groupoid and* Φ *a reduction of* Ω *for which the vertex groups are closed (and hence embedded) subgroups of the vertex groups of* Ω. *Choose* $m \in M$ *and write* $G = \Omega_m^m$, $H = \Phi_m^m$, $P = \Omega_m$. *Let* M' *be the fibre bundle* $\frac{P \times (G/H)}{G}$ *corresponding to the standard action of* G *on* G/H, *and let* $\Omega * M' \to M'$ *be the associated action* $\xi \lfloor \eta, gH \rfloor = \lfloor \xi\eta, gH \rfloor$ *(see Example 1.6.4, Theorem 1.6.5).*

For $x \in M$ *choose* $\zeta \in \Phi_m^x$ *and define* $\mu(x) = \lfloor \zeta, H \rfloor$. *Then* μ *is a well-defined, smooth,* Ω–*deformable section of* M', *and* $\Omega\{\mu\} = \Phi$. *In particular,* Φ *is a closed, embedded reduction of* Ω.

Proof That μ is smooth follows from the facts that $\beta_m \colon \Phi_m \to M$ is a surjective submersion, and Φ_m is a submanifold of Ω_m. That $\Omega\{\mu\} = \Phi$ is a trivial algebraic manipulation. $\square$

Proposition 1.6.24 shows that a closed reduction of a locally trivial Lie groupoid is an embedded submanifold.

Theorem 1.6.23 and Proposition 1.6.24 give a classification of those closed embedded reductions of a locally trivial Lie groupoid which have a specified vertex group at a chosen point of the base. A closed embedded reduction may of course fail to be trivializable over open subsets of the base over which the larger groupoid is trivialisable, as Example 1.2.12 shows.

*   *   *   *   *

There is of course a concept of right action, and all that has been said above concerning left actions applies equally well to right actions.

A *right action of* $G \rightrightarrows M$ *on* $f \colon M' \to M$ is defined as a map

$$M' \times_M G \to M', \quad (m', g) \mapsto m'g,$$

where now $M' \times_M G$ is the pullback of f and β, such that (i) $f(m'g) = \alpha(g)$; (ii) $(m'g_1)g_2 = m'(g_1g_2)$; (iii) $m'1_{f(m')} = m'$ for all compatible $g, g_2, g_1 \in G$, $m' \in M'$. If $(g, m') \mapsto gm'$ is a left action then the usual formula $m'g = g^{-1}m'$ defines a right action. Given a right action, we take the *right action groupoid* $f \gg G$ to be $M' \times_M G$ with $\alpha'(m', g) = m'g$, $\beta'(m', g) = m'$, and $(m', g_2)(m'g_2, g_1) = (m', g_2g_1)$. The natural map $(m', g) \mapsto g$ is again an action morphism. Given any action morphism $\varphi \colon G' \to G$, $f \colon M' \to M$ the formula $m'g = \beta'(\varphi^{!^{-1}}(m', g^{-1}))$

defines a right action with $G' \cong f \gg G$; indeed this is the right action corresponding to the left action defined by $\varphi \colon G' \to G$, $f \colon M' \to M$.

1.7 Linear actions and frame groupoids

Let (E, q, M) be a vector bundle and let $\Omega * E \to E$ be an action of a locally trivial Lie groupoid Ω on q. The action is *linear* if for each $\xi \in \Omega$, the diffeomorphism $E_{\alpha\xi} \to E_{\beta\xi}$ is a linear isomorphism.

Definition 1.7.1 Let Ω be a locally trivial Lie groupoid on M and let (E, q, M) be a vector bundle. Then a *representation of Ω on (E, q, M)* is a morphism $\rho \colon \Omega \to \Phi(E)$ of Lie groupoids over M. □

Proposition 1.7.2 *Let $\Omega \rightrightarrows M$ be a locally trivial Lie groupoid and let (E, q, M) be a vector bundle. If $\Omega * E \to E$ is a linear action of Ω on E then the associated map $\Omega \to \Phi(E)$, $\xi \mapsto (u \mapsto \xi u)$ is a representation; if $\rho \colon \Omega \to \Phi(E)$ is a representation then $(\xi, u) \mapsto \rho(\xi)(u)$ is a linear action.*

Proof One direction follows from noting that $\Phi(E) * E \to E$ is a linear action. For the converse, suppose we have a linear action $\Omega * E \to E$. It has to be proved that the set map $\Omega \to \Phi(E)$ is smooth. Take a chart $\psi \colon U \times V \to E_U$ for E and use it to write the map locally as $\Omega_U * (U \times V) \to U \times V$. It then follows easily that the associated map $\Omega_U \to U \times GL(V) \cong \Phi(E)_U$ is smooth. □

We now give some simple definitions and results on invariant sections.

Definition 1.7.3 Let $\Omega * M' \to M'$ be an action of a locally trivial Lie groupoid Ω on a smooth surjection (M', q, M). Then a section $\mu \in \Gamma M'$ is *Ω–invariant* if $\xi\mu(\alpha\xi) = \mu(\beta\xi)$, for all $\xi \in \Omega$. The set of Ω-invariant sections of M' is denoted $(\Gamma M')^{\Omega}$. □

If $\Omega * E \to E$ is a linear action on a vector bundle, then $(\Gamma E)^{\Omega}$ is an $\mathbb{R}$-vector space with respect to pointwise operations, but not usually a module over the ring of smooth functions on M. A general fibre bundle need not of course admit any (global) sections. In the case of a vector bundle and a linear action, $(\Gamma E)^{\Omega}$ may consist of the zero section alone (see Example 1.7.5 below).

Proposition 1.7.4 *Let* $\Omega \rightrightarrows M$ *be a locally trivial Lie groupoid and let* $\Omega * E \to E$ *be a linear action of* Ω *on a vector bundle* (E, q, M). *Choose* $m \in M$, *and write* V *for* E_m *and* G *for* Ω_m^m. *Then the evaluation map*

$$(\Gamma E)^\Omega \to V^G, \quad \mu \mapsto \mu(m)$$

is an isomorphism of $\mathbb{R}$-*vector spaces.*

Proof Obviously the map is injective. Given $v \in V^G$, define μ by $\mu(x) = \xi v$ where ξ is any element of Ω_m^x. Clearly $\mu(x)$ is well defined; μ is smooth because $\beta_m \colon \Omega_m \to M$ is a surjective submersion. $\square$

Example 1.7.5 Consider the principal bundle $SO(2)(S^1, \mathbb{Z}_2, p)$ where $\mathbb{Z}_2$ is embedded in $SO(2)$ as $\{1, -1\}$ and p is $z \mapsto z^2$. Let Ω be the gauge groupoid and let E be the vector bundle $SO(2) \times \mathbb{R}/\mathbb{Z}_2$, where $\mathbb{Z}_2$ acts on $\mathbb{R}$ by multiplication; this is the Möbius band. Then $(\Gamma E)^\Omega \cong \mathbb{R}^{\mathbb{Z}_2}$ is the zero space. $\boxtimes$

Proposition 1.7.6 *Let* $\Omega * E \to E$ *be a linear action of a locally trivial Lie groupoid* Ω *on a vector bundle* E. *For each* $x \in M$, *define* $E^{\mathscr{I}\Omega}\big|_x$ *to be*

$$E_x^{\Omega_x^x} = \{u \in E_x \mid \lambda u = u, \text{for all } \lambda \in \Omega_x^x\}.$$

Then $E^{\mathscr{I}\Omega}$ *is a subvector bundle of* E, *and there is a natural trivialization*

$$M \times V^G \to E^{\mathscr{I}\Omega}.$$

Proof Let $\{\sigma_i \colon U_i \to \Omega_m\}$ be a section–atlas for Ω, and write $V = E_m$, $G = \Omega_m^m$. Define $\psi_i \colon U_i \times V^G \to E^{\mathscr{I}\Omega}\big|_{U_i}$ by $(x, v) \mapsto \sigma_i(x)(v)$. Then $\psi_{i,x}$ maps V^G isomorphically onto $E^{\mathscr{I}\Omega}\big|_x$.

To define the trivialization, note that for $x \in M$ and any two $\xi, \xi' \in \Omega_m^x$, the maps $V^G \to E^{\mathscr{I}\Omega}\big|_x$, $v \mapsto \xi v$ and $v \mapsto \xi' v$ are identical. $\square$

The following series of examples is basic to connection theory.

Proposition 1.7.7 *Let* (E, q, M) *be a vector bundle, and let* $n \geqslant 1$.

(i) *The action* $\Phi(E) * \mathrm{Hom}^n(E; M \times \mathbb{R}) \to \mathrm{Hom}^n(E; M \times \mathbb{R})$ *defined by* $\xi\varphi = \varphi \circ (\xi^{-1})^n$ *is smooth.*

(ii) *The action* $\Phi(E) * \mathrm{Hom}^n(E; E) \to \mathrm{Hom}^n(E; E)$ *defined by* $\xi\varphi = \xi \circ \varphi \circ (\xi^{-1})^n$ *is smooth.*

(iii) *Let* (E', p', M) *be another vector bundle on the same base. Then the action*

$$(\Phi(E) \times_{M \times M} \Phi(E')) * \mathrm{Hom}(E; E') \to \mathrm{Hom}(E; E')$$

defined by $(\xi, \xi')\varphi = \xi' \circ \varphi \circ \xi^{-1}$ *is smooth.*

Proof We prove (ii) for $n = 1$; the other cases are similar. Given an atlas $\{\psi_i \colon U_i \times V \to E_{U_i}\}$ for E, we obtain charts for $\Phi(E)$ as in Example 1.1.12 and charts $\psi'_i \colon U_i \times \mathrm{Hom}^1(V, V) \to \mathrm{Hom}^1(E, E)_{U_i}$ for $\mathrm{Hom}^1(E, E)$ by $\psi'_i(x, f)(\varphi) = \psi_{i,x} \circ \varphi \circ \psi_{i,x}^{-1}$, where $\varphi \in \mathrm{Hom}^1(E, E)_x$. Now the smoothness of the action reduces to the smoothness of the standard action $GL(V) \times \mathrm{End}(V) \to \mathrm{End}(V)$. $\qquad\square$

In (iii) the product is the product of locally trivial Lie groupoids over a fixed base in the sense of the following result, which is easily proved.

Proposition 1.7.8 *Let* Ω *and* Ω' *be locally trivial Lie groupoids over the same base* M. *Let* $\Omega \times_{M \times M} \Omega'$ *be the pullback manifold of the two anchors* $(\beta, \alpha) \colon \Omega \to M \times M$ *and* $(\beta', \alpha') \colon \Omega' \to M \times M$. *Then, with the componentwise algebraic structure,* $\Omega \times_{M \times M} \Omega'$ *is a locally trivial Lie groupoid on* M *and the two projections* $\Omega \times_{M \times M} \Omega' \to \Omega$ *and* $\Omega \times_{M \times M} \Omega' \to \Omega'$ *are surjective submersions.*

The Lie groupoid structure on $\Omega \times_{M \times M} \Omega'$ can be defined for weaker local triviality conditions, but the full properties of a product in the categorical sense over the fixed base M will be lost.

$\mathrm{Hom}^n(E; E')$ of course denotes the vector bundle on M whose fibre over $x \in M$ is the space of n-multilinear maps $E_x \times \cdots \times E_x \to E'_x$ and whose bundle structure is induced from the bundle structure of E and E' in the usual way. The actions (i) and (ii) clearly restrict to the subbundles $\mathrm{Alt}^n(E; E')$ and $\mathrm{Sym}^n(E; E')$ of alternating and symmetric multilinear maps; further, $\mathrm{Hom}(E; E')$ in (iii) could be replaced by $\mathrm{Hom}^n(E; E')$ and the obvious action. Lastly, there are analogous actions of $\Phi(E)$ on the tensor bundles $\otimes_s^r E$. We take all these variations of Proposition 1.7.7 to be included in its statement.

Example 1.7.9 Let (E, q, M) be a vector bundle, and let $\langle \, , \, \rangle$ be a Riemannian structure in E, regarded as a section of $\mathrm{Hom}^2(E; M \times \mathbb{R})$. Then $\langle \, , \, \rangle$ is $\Phi(E)$–deformable with respect to the action of 1.7.7(i), since any two vector spaces of the same dimension with any positive-definite inner products, are isometric. Denote the stabilizer groupoid of

$\langle \, , \, \rangle$ by $\Phi_{\mathscr{O}}(E)$. By Theorem 1.6.23 it is a locally trivial Lie groupoid on M, the *Riemannian frame groupoid* or *orthonormal frame groupoid* of $(E, \langle \, , \, \rangle)$. A section $\sigma \colon U \to \Phi_{\mathscr{O}}(E)_m$ of $\Phi_{\mathscr{O}}(E)$ is a moving frame for E, and the local triviality of $\Phi_{\mathscr{O}}(E)$ is equivalent to the existence of moving frames in E.   ⊠

Example 1.7.10 A *complex structure* in a vector bundle (E, q, M) is an endomorphism $J \colon E \to E$ such that $J^2 = -\mathrm{id}$. Such a J has constant rank and so, regarded as a $(1,1)$ tensor field, J is $\Phi(E)$–deformable with respect to the action of 1.7.7(ii). Denote the stabilizer groupoid of J by $\Phi_{\mathbb{C}}(E)$. By Theorem 1.6.23 it is a locally trivial Lie groupoid on M, the *complex frame groupoid* of (E, J).

If E further has a Hermitian metric $\langle \, . \, \rangle$ for the given J, then the argument of 1.7.9 applies, and there is a locally trivial Lie groupoid $\Phi_U(E) \leqslant \Phi_{\mathbb{C}}(E)$ of Hermitian frames. $\Phi_U(E)$ is the *Hermitian frame groupoid* or *unitary frame groupoid* of $(E, J, \langle \, , \, \rangle)$.   ⊠

Example 1.7.11 Similarly, if Δ is a determinant function in a vector bundle (E, q, M), regarded as a never–zero section of $\mathrm{Alt}^r(E; M \times \mathbb{R})$ where $r = \mathrm{rank}\, E$, then Δ is $\Phi(E)$–deformable and the groupoid of orientation–preserving isomorphisms between the fibres of E is a closed embedded reduction of $\Phi(E)$, denoted $\Phi^+(E)$.   ⊠

Example 1.7.12 Let (L, q, M) be a vector bundle and let $[\, , \,]$ be a section of $\mathrm{Alt}^2(L; L)$ such that each $[\, , \,]_x \colon L_x \times L_x \to L_x$ is a Lie algebra bracket. We call such a section a *field of Lie algebra brackets* in L.

A field of Lie algebra brackets need not be $\Phi(L)$–deformable. For example, let $\mathfrak{g}$ be a non–abelian Lie algebra with bracket $[\, , \,]$ and in $L = \mathbb{R} \times \mathfrak{g}$ define $[\, , \,]_t = t[\, , \,]$. However Theorem 1.6.23 implies that if $[\, , \,]$ is a field of Lie algebra brackets in a vector bundle L and if the fibres of L are pairwise isomorphic as Lie algebras, then L admits an atlas of charts which fibrewise are Lie algebra isomorphisms. In this case, $(L, [\, , \,])$ is a Lie algebra bundle, as defined in 3.3.8, and we denote the stabilizer groupoid by $\Phi_{\mathrm{Aut}}(L)$.   ⊠

Example 1.7.13 Let μ be a section of a vector bundle E on a connected base M. Then, by a similar argument, E has an atlas of charts $U \times V \to E_U$ such that the local representatives $U \to V$ of μ are constant, if and only if μ is either never zero or always zero. This (trivial) result is well known in the case of tangent vector fields.   ⊠

A similar treatment may be applied to any tensor structure on a vector bundle.

It is worth noting that in 1.7.12 and 1.7.13 the condition of pairwise isomorphism, or of being never zero or always zero, need hold only on each component of the base separately. A similar comment applies to the next example.

Example 1.7.14 Let (E^ν, q_ν, M), $\nu = 1, 2$, be vector bundles on base M, and let $\varphi \colon E^1 \to E^2$ be a morphism, considered as a section of $\operatorname{Hom}(E^1; E^2)$. Then φ is $\Phi(E^1) \times_{M \times M} \Phi(E^2)$–deformable if and only if it is of constant rank. Now 1.6.23 shows that if this is the case, there are atlases $\{\psi_i^\nu \colon U_i \times V^\nu \to E_{U_i}^\nu\}$, $\nu = 1, 2$, and a linear map $f \colon V^1 \to V^2$ such that each $\varphi \colon E_{U_i}^1 \to E_{U_i}^2$ is represented by $(x, v) \to (x, f(v))$.   ⊠

This is a vector bundle version of the standard characterization of a subimmersion (though in order to apply to general subimmersions $M \to N$ it must be extended to varying base manifolds). This result will be useful in the abstract theory of transitive Lie algebroids.

*   *   *   *   *

Actions of Lie groupoids on bundles where the fibres have algebraic structures of other types can be handled in a similar way. We briefly consider the case of actions on Lie group bundles.

Consider an LGB (K, q, M) as in 1.1.19. An action of a locally trivial Lie groupoid $\Omega \rightrightarrows M$ on K is said to be an *action by Lie group isomorphisms* if each isomorphism $\xi \colon K_x \to K_y$, $\xi \in \Omega$, is a Lie group isomorphism.

Let $\Phi(K)$ denote the groupoid of all Lie group isomorphisms between the fibres of K. Using the standard Lie group structure on the automorphism group of the fibre type of K, and the method of 1.1.12, $\Phi(K)$ has a locally trivial Lie groupoid structure, and actions of a Lie groupoid Ω on K by Lie group isomorphisms can be identified with Lie groupoid morphisms $\Omega \to \Phi(K)$.

The following definition introduces the final example.

Definition 1.7.15 Let Ω be a locally trivial Lie groupoid on M and let (K, q, M) be an LGB. An *extension* of Ω by K is a sequence

$$K \overset{\iota}{\rightarrowtail} \Psi \overset{\pi}{\twoheadrightarrow} \Omega$$

in which Ψ is a locally trivial Lie groupoid on M, ι and π are groupoid morphisms over M, ι is an embedding, π is a surjective submersion, and $\mathrm{im}(\iota) = \ker(\pi)$. □

It is easy to see that the condition that Ψ be locally trivial is superfluous.

Example 1.7.16 Let $K \overset{\iota}{\rightarrowtail} \Psi \overset{\pi}{\twoheadrightarrow} \Omega$ be an extension as in 1.7.15 with K an abelian LGB. For $\xi \in \Omega_x^y$, $x, y \in M$, choose $\xi' \in \Psi_x^y$ with $\pi(\xi') = \xi$ and define $\rho(\xi) \colon K_x \to K_y$ as $\lambda \mapsto \xi' \lambda \xi'^{-1}$, the restriction of $I_{\xi'}$. It is clear that $\rho(\xi)$ is well defined. Now $I \colon \Psi \to \Phi(K)$ is smooth (as in 1.6.7) and π is an surjective submersion so $\rho \colon \Omega \to \Phi(K)$ is smooth. This ρ is the *representation associated to the extension* $K \rightarrowtail \Psi \twoheadrightarrow \Omega$. ⊠

1.8 Notes

An interesting algebraic theory of groupoids exists, having been begun by Brandt and by Baer in the 1920s, well before Ehresmann made the concept of groupoid central to his vision of differential geometry. However the algebraic theory depends substantially on various free constructions, which have no viable form in the Lie case, and we do not consider it in this book. See [Higgins, 1971] for a full account and further references, and [Brown, 1988] for an account which is more accessible to the non–algebraist, though less comprehensive than Higgins'. Much material on the algebraic theory of groupoids, from a different point of view to that of the work cited above, can be extracted from the book [Ehresmann, 1965].

The suffix *–oid* has no precise mathematical meaning, despite its long–established use in words such as *catenoid, ellipsoid* and *matroid*. It serves an important function here in making a semantic link between Lie groupoids and Lie algebroids, though these concepts differ from Lie groups and Lie algebras in different ways: a groupoid is often described as a 'group in which the multiplication is allowed to be only partial' whereas a Lie algebroid may be thought of as a generalized tangent bundle, or (in its algebraic form) as a Lie algebra over a module. What links these two usages of *–oid* is that they refer to extensions of familiar concepts to bundle contexts: a groupoid is the notion of group appropriate to actions on bundles and Lie algebroids are the notion of Lie algebra appropriate to bracket structures on vector bundles.

Topological groupoids. A theory of topological groupoids will differ greatly from the theory of Lie groupoids. Of course, topological group theory differs from Lie group theory, but the divergence in the case of groupoids will be much more marked. For a start, there is no longer the same need to require the source and target projections to be quotient maps. Secondly, there is no analogue of Theorem 1.4.10; a transitive topological groupoid may be very far from local triviality, in the same way that a transitive topological group action may fail to be a homogeneous space. The existence of bisections is no longer clear. Many other differences will occur to the reader as he or she progresses through the book.

For references on topological groupoids see the surveys [Brown and Hardy, 1976], [Brown et al., 1976]. A number of isolated but possibly interesting results about topological groupoids can be found in [Mackenzie, 1987a, Chapter II], which also contains results which are straightforward topological versions of Lie groupoid results.

My own view is that despite these various forays, a serious theory of topological groupoids still remains to be developed. Since both topological groups and pair groupoids admit natural uniformities, it may be more interesting to develop a theory of groupoids with uniform structures.

In [Mackenzie, 1987a] a Lie groupoid was always assumed to be locally trivial and the notion which in the present book is called Lie groupoid was there called a differentiable groupoid; *groupoïde différentiable* was the original terminology of Ehresmann [1959] and Pradines [1966].

A *smooth groupoid* in the sense of Connes [1994] is essentially the same as a Lie groupoid in the sense of the present book.

Throughout the book it is assumed that manifolds are Hausdorff, and second–countable. Some important aspects of groupoid theory, arising from foliations and related constructions, lead to structures where the groupoid space is non–Hausdorff but which otherwise satisfy the conditions of 1.1.3. References, and some account of the importance of this aspect, are given in the Appendix.

§1.1—§1.2. The definition of Lie groupoid given in Definition 1.1.3 is from [Pradines, 1966]. The original definition of Ehresmann [1959] required only a differentiable structure on G for which $g \mapsto 1_{\alpha(g)}$ and $g \mapsto 1_{\beta(g)}$ are subimmersions and for which the multiplication $G * G \to G$ is smooth; Kumpera and Spencer [1972] and ver Eecke [1981] merely require differentiable structures on G and M such that the projections and object inclusion map are smooth, $G * G$ is an embedded submanifold of $G \times G$ and multiplication is smooth. Ver Eecke [1981] proves that even in this more general case the smoothness of the inversion map follows from the other conditions, though not so easily.

In the light of Corollary 1.5.13 the condition that α and β be submersions seems strong. However, I know of no evidence that a more general concept is needed. None of [Ehresmann, 1959], [Kumpera and Spencer, 1972], [ver Eecke, 1981] develops in any substantial way the theory of groupoids which are differentiable in a more general sense, nor do they offer examples which do not satisfy 1.1.3.

A theory of Lie groupoids which did not assume that α and β are submersions would present significant difficulties: apart from the immediate problems with the domain of definition of the multiplication and with the nature of the α–fibres, it would no longer be clear that the Lie algebroid was a vector bundle.

I have followed Ehresmann in taking groupoid composition with the same conventions as functional composition (as indeed it is in a frame groupoid); thus hg is defined if $\alpha h = \beta g$ and then $\alpha(hg) = \alpha g$, et cetera. Many authors reverse this, so that hg is defined if and only if the source of g is equal to the target of h.

The convention of Weinstein, as in [da Silva and Weinstein, 1999], names the target map α and the source map β while otherwise following the Ehresmann conventions. As a result left–translations are maps of α–fibres, and so the Lie algebroid is constructed on what appears to be the same vector bundle as in Chapter 3 here, but using left–invariant vector fields.

Most writers call G_x^x the *isotropy group* at x. This usage can be distracting and I have retained the older term *vertex group*. However I have replaced the other use of the word *isotropy* with *stabilizer*.

Some authors denote G_x^y by $G(x,y)$, call G_x the *star* of G at x and denote it by $\mathrm{St}_G x$, and call G^y the *co-star* of G at y and denote it by $\mathrm{Cost}_G y$.

Pair groupoids have also been called *coarse groupoids*; other names include *fine* and *banal*.

The terminology 'α–connected' (and its cognates) is open to the criticism that it gives emphasis to the source projection; since this terminology goes back to Pradines and applies equally to both projections, perhaps we should say *pr–connected*, or even *pronnected*.

There are many books on algebraic topology which introduce the concept of fundamental group via that of the fundamental groupoid, but few make further use of the groupoid structure. For a detailed account of the modelling of elementary homotopy theory via the algebraic structure of $\Pi(M)$ see [Brown, 1988].

§1.3. In the older literature, transitive groupoids are sometimes called *connected groupoids* and totally intransitive groupoids called *totally disconnected groupoids.*

The concept of local triviality is due to Ehresmann [1959], as are the two constructions between locally trivial groupoids and principal bundles. Ehresmann [1959] also treated a weaker form of the notion in which the anchor is assumed only to be a submersion, not necessarily surjective. A Lie groupoid which satisfies this condition is easily seen to be the disjoint union of Lie groupoids which are locally trivial in the sense of Definition 1.3.2.

One sometimes sees principal bundles and vector bundles introduced as if they were slightly different forms of the one 'bundle' concept. This seems to me misleading; the relationship of principal bundles to vector bundles is like that of groups to vector spaces, and whatever formal algebraic similarities the definitions of these notions have, their roles in geometry are fundamentally different. The use of Lie groupoids eliminates this confusion.

It would be helpful to have an alternative terminology for local triviality in the context of groupoids (while retaining it for vector bundles); amongst the very broad expanse of groupoids, it seems unnatural to refer to the examples $M \times G \times M$ alone as 'trivial'. However this terminology is very firmly established.

The name *gauge groupoid* for the groupoid associated to a principal bundle comes from [Weinstein, 1987].

There is a straightforward topological version of the 'equivalence' 1.3.5 between locally trivial Lie groupoids and principal bundles. Considering the topological case, however, makes clear that this 'equivalence' does not depend on local triviality as such: Define a topological groupoid G on base M to be *principal* if it is transitive and if for any one, and hence every, $x \in M$, the map $\beta_x \colon G_x \to M$ and the map $\delta_x \colon G_x \times G_x \to G$, $(h, g) \mapsto hg^{-1}$, are identifications (and hence open maps). Then the correspondences 1.3.5 extend to principal topological groupoids and Cartan principal bundles. This observation is due to Dakin and Seda [1977], and may be found in [Mackenzie, 1987a].

That the dependence of the correspondences between principal bundles and locally trivial Lie groupoids upon reference points has consequences for their automorphism structures (see 1.3.7) was pointed out explicitly by Pradines around 1980.

The word *produced* as in 1.3.10 comes from [Mackenzie, 1987a]. It is more usual to call a produced principal bundle an 'extension' or 'prolongation', but both terms have other meanings within bundle and groupoid theory. The word *produced* is suitably antithetical to *reduced* and may remind the reader of the process in elementary geometry where 'to produce a line' means to continue the line in a manner that is already implicit.

Much of the basic work on locally trivial Lie groupoids and their Lie algebroids first appeared in work of Libermann [1964, 1971, 1972, 1973, 1974] and Ngô Van Quê [1967, 1968, 1969].

§1.4—§1.5. Some of the material of these sections comes from II§5, II§3 and parts of III§1 of [Mackenzie, 1987a]. The important Theorem 1.4.14 is due to Xu [1995].

The groupoid $TG \rightrightarrows TM$ is an example of an $\mathscr{L}\mathscr{A}$–groupoid; for the general theory of these see [Mackenzie, 1992] and [Mackenzie, 2000].

What have here been called bisections were used by Ehresmann [1967] in his construction of jet prolongation groupoids. They have usually been called *admissible sections*, β–*admissible* α–*sections*, or simply *sections*. They received a thorough treatment in [Kumpera and Spencer, 1972, Appendix], where they were used to give the adjoint formulas for Lie groupoids (see §3.7).

The terminology *bisection* was introduced in [Coste, Dazord, and Weinstein, 1987], for the image, rather than for the map. The use of the image has the advantage that the source and target projections remain symmetric.

Coste, Dazord, and Weinstein [1987] were concerned with Lagrangian bisections of symplectic groupoids in which the image of the bisection is a Lagrangian submanifold of the groupoid space. This combines the notion of bisection as a generalized element of an arbitrary groupoid with the notion [Weinstein, 1982] of the Lagrangian submanifolds of a symplectic manifold as 'generalized points' of the manifold.

In Example 1.4.4, the maps $\Gamma E \to \Gamma E$ induced by bisections of $\Phi(E)$ are *semilinear isomorphisms* of ΓE; see [Kosmann, 1972, 1976].

The results 1.4.10 and 1.4.11, here deduced from 1.4.9, were stated by Pradines [1966] and proved in [Mackenzie, 1987a]. The elegant proof of 1.4.9 given here is from [Moerdijk and Mrčun, 2003].

The results in §1.5 on the structure of a single orbit constitute the version of [Pradines, 1966, Théorème 4] appropriate to the category of pure, second countable, Hausdorff manifolds. Detailed proofs were given for the first time in [Mackenzie, 1987a], but the proofs here have been reworked. See also [Pradines, 1986a].

In fact the orbits of a Lie groupoid are not arbitrary immersed submanifolds, but are always quasiregular. This follows from showing that the orbits give a Stefan–Sussman foliation of the base manifold. For Stefan–Sussman foliations, see [Stefan, 1974, 1980], [Sussmann, 1973].

Given a simple foliation of a manifold M defined by a submersion $q\colon M \to Q$, the Lie groupoid $R(q)$ of 1.1.8 has orbits that are precisely the fibres of q. For general regular foliations the holonomy groupoid of the foliation is generally non-Hausdorff but is in other respects a Lie groupoid, the transitivity foliation of which is the given foliation; see the Appendix. It is not clear how to extend such results to general Stefan–Sussman foliations. It is nonetheless valid to regard any Lie groupoid as a desingularization of the Stefan–Sussman foliation defined by its transitivity orbits.

The reduction 1.5.4 of a principal bundle so that the total space is connected was given in [Mackenzie, 1988a].

Etale groupoids. Consider a pseudogroup $\mathscr{P}$ of local diffeomorphisms on a manifold M, and let $J^\lambda(\mathscr{P})$ denote the set of all germs $g_x\varphi$ of all elements φ of $\mathscr{P}$ at all points x of their domains. Then $J^\lambda(\mathscr{P})$ has a natural groupoid structure and, with the sheaf topology, satisfies the requirements for a Lie groupoid except that the space of arrows is usually non-Hausdorff.

Crainic and Moerdijk [2000] define an *étale groupoid* G on a (Hausdorff) manifold M to be a groupoid $G \rightrightarrows M$ together with a not necessarily Hausdorff smooth structure on G satisfying the conditions of 1.1.3, which is such that the α–fibres of G are Hausdorff, and target map are étale. Thus $J^\lambda(\mathscr{P})$ above is an étale groupoid. Furthermore, for any Lie groupoid $G \rightrightarrows M$ the set of germs of local bisections of G, with the evident groupoid structure and the sheaf topology, is an étale groupoid.

Etale groupoids arise naturally in foliation theory and provide elegant characterizations of quotient structures of various kinds. See [Moerdijk and Mrčun, 2003] and the references given there.

§1.6. The action groupoid $G \ltimes f$ of Example 1.1.9 and Definition 1.6.10 is often called a *transformation groupoid*. In the past the action groupoid has often been denoted $G \ltimes f$ but this causes confusion with the usual semi–direct product over a fixed base, and so we now write $\ltimes$.

The observation that groupoid actions could be characterized in terms of morphisms goes back at least to Ehresmann [1957]. The topological case of the equivalence was done by Brown et al. [1976]. Using fundamental groupoids, the correspondence between actions and action morphisms models the theory of coverings of topological spaces, and these authors therefore called $G \ltimes f$ a *covering groupoid* and the action morphism a *covering morphism*. The correspondence between groupoid coverings in this sense and the topological theory is dealt with in full in [Brown, 1988].

For cotangent actions and the notion of pith, see the Notes on §12.4.

The term Ω-*deformable* is adapted from [Greub et al., 1973, 8.2].

Theorem 1.6.23 is from [Quê, 1967, I.3.a]; the proof given there, however, seems to address only the local problem.

The principal bundle formulation of Theorem 1.6.5 is given in [Kobayashi and Nomizu, 1963, I.5.4 et seq.].

Proposition 1.6.24 goes back to Ehresmann [1959]. The principal bundle formulation is a standard result from [Kobayashi and Nomizu, 1963, I.5.6].

The study of wide Lie subgroupoids of general Lie groupoids has been begun very recently by [Moerdijk and Mrčun, arXiv:0406558].

For a notion of homogeneous space for Lie groupoid actions, see [Liu, Weinstein, and Xu, 1998].

Very recently, striking results on the linearization of proper groupoids and normal forms have been obtained; see [Weinstein, 2002] and [Zung, arXiv:0407208].

§1.7. For a different treatment of Examples 1.7.9 to 1.7.13 see [Greub et al., 1973, Chapter VIII].

[Greub et al., 1973, Chapter VIII] introduces a concept of Σ–bundle, a vector bundle E together with a finite set Σ of sections σ_i of various tensor bundles $\otimes_{s_i}^{r_i}(E)$, such that E admits an atlas with respect to which all the σ_i are constant. Their Theorem 1 follows from Theorem 1.6.23 by considering the natural action of $\Phi(E)$ on the direct sum of the relevant tensor bundles. It is interesting to compare the proof of their result with that of 1.6.23. The stabilizer groupoid corresponds to the G–structure defined by Σ.

Abelianity is not a condition that can be sensibly imposed on a groupoid. If $\Omega \rightrightarrows M$ is locally trivial and has vertex groups which are abelian then 1.7.16 applies and shows that $M \times M$ acts on $\mathscr{I}\Omega$; this action provides a canonical trivialization of $\mathscr{I}\Omega$. Groupoids in general are intrinsically more non–abelian than groups.

2

Lie groupoids:
Algebraic constructions

This chapter is concerned with quotients — in the most general sense, which includes descent constructions — with pullbacks, and with general semidirect products for Lie groupoids. Quotients and semidirect products in this sense are much more general, and of much wider importance, than the corresponding constructions for groups, in view of the possibility of changing the base manifold.

We begin in §2.1 by describing general quotients of vector bundles. This prefigures quotients of Lie algebroids as well as those of Lie groupoids. In §2.2 we briefly cover the case of base–preserving quotients of Lie groupoids; this very special case needs a separate treatment. In §2.3 we describe pullbacks of Lie groupoids; in principle the pullback construction allows most morphisms to be reduced to the base–preserving case.

The purpose of a notion of kernel is to characterize a class of morphisms (up to isomorphism) in terms of data entirely on their domain. The largest well behaved class of Lie groupoid morphisms for which this is possible is the fibrations; these are characterized by a lifting condition analogous to the classical notion of Hurewicz fibration. We prove in §2.4 that fibrations of Lie groupoids are characterized by what we call their kernel systems; for base–preserving fibrations, the kernel system is precisely the familiar kernel.

The notion of general semidirect product treated in §2.5 combines the action groupoid concept of Chapter 1 with the usual algebraic notion of semidirect product. This general notion of semidirect product corresponds to the notion of fibration: fibrations which are split in an appropriate sense correspond to general semidirect products.

§2.6 consists of a brief overview of the constructions of this chapter, and of the classes of morphism to which each corresponds.

2.1 Quotients of vector bundles

Given a vector bundle (E, q, M) and a smooth map $f: M' \to M$ we denote the inverse image bundle by $(f^! E, q^!, M')$ and the canonical projection $f^! E \to E$ by $f^!$. For $\mu \in \Gamma E$ we write $\mu^! \in \Gamma(f^! E)$ for the unique section such that $f^! \circ \mu^! = \mu \circ f$. The map

$$C^\infty(M') \otimes \Gamma E \to \Gamma(f^! E), \qquad u' \otimes \mu \mapsto u' \mu^!, \tag{1}$$

is an isomorphism of $C^\infty(M')$–modules, where the tensor product is over $C^\infty(M)$ and $C^\infty(M')$ is a $C^\infty(M)$–module under $uu' = (u \circ f)u'$.

The inverse image bundle $f^! E$ and the map $f^!$ have the universal property for vector bundle morphisms over f into E: namely, if $\varphi: E' \to E$ is any vector bundle morphism over f, then there is a unique vector bundle morphism $\varphi^!: E' \to f^! E$ over M' such that $\varphi = f^! \circ \varphi^!$. We thus often refer to $f^! E$ as the *pullback of E over f*. Any morphism $\psi: E' \to E$, $f: M' \to M$, which is a fibrewise bijection has this property and we usually identify such a E' and ψ with $f^! E$ and $f^!$ without comment. In particular, given any fibrewise bijection $\psi: E' \to E$ over $f: M' \to M$, and $\mu \in \Gamma E$, we may write $1 \otimes \mu$ or $\mu^!$ for the unique section of E' with $\psi \circ (1 \otimes \mu) = \mu \circ f$.

A fibrewise surjective morphism of vector bundles over a fixed base is determined (up to isomorphism) by its kernel. However this is no longer true for fibrewise surjections over general base maps.

Suppose $\varphi: E' \to E$ is a morphism of vector bundles over a surjective submersion $f: M' \to M$. If φ is fibrewise a surjection, then the union of the kernels $\cup_{m \in M} \ker(\varphi_m)$ will be precisely the kernel of the base–preserving morphism $\varphi^!: E' \to f^! E$. In particular, this union of the fibre kernels is a subbundle of E'.

This demonstrates that the kernel, defined as the union of the kernels of the maps of the fibres, cannot in general determine the image of a morphism. In order to describe a quotient appropriate to a change of base, extra data is needed.

In what follows we will restrict ourselves to morphisms the base maps of which are surjective submersions. This is the most general class of smooth maps which can usefully be regarded as quotient maps of manifolds. We recall the basic properties of quotient manifolds.

A surjective submersion $f: M' \to M$ determines an equivalence relation $R(f) = \{(n', m') \in M' \times M' \mid f(n') = f(m')\}$, sometimes called the *kernel pair* of f, which is a closed embedded regular submanifold of $M' \times M'$, and has the property that the projections $R(f) \to M'$ are surjective submersions. Conversely, given a manifold M', and an equivalence relation R whose graph is a closed embedded regular submanifold of $M' \times M'$ with projections $R \to M'$ which are submersions, there is a manifold $M = M'/R$ and a surjective submersion $f \colon M' \to M$ such that $R(f) = R$, and M and f are unique up to diffeomorphism with this property; this is the theorem of Godement.

Consider now a vector bundle morphism

$$
\begin{array}{ccc}
E' & \xrightarrow{\ \ \varphi\ \ } & E \\[2mm]
q' \downarrow & & \downarrow q \\[2mm]
M' & \xrightarrow{\ \ \ \ } & M \\
& f &
\end{array}
\qquad (2)
$$

and assume that f is a surjective submersion, and that φ is fibrewise surjective. Then $\varphi^! \colon E' \to f^! E$ is also fibrewise surjective, and so $E'/K \cong f^! E$, where $K = \ker(\varphi)$ is the kernel of $\varphi^!$. Denote the induced map $E'/K \to E$ by $\overline{\varphi}$, and the elements of E'/K by $\overline{e'}$, for $e' \in E'$.

The pullback bundle $f^! E$ is equipped with canonical identifications $f^! E|_{m'} \cong f^! E|_{n'}$ for all $(n', m') \in R(f)$; denote the map

$$
f^! E|_{m'} \cong f^! E|_{n'}, \qquad (m', e) \mapsto (n', e), \qquad (3)
$$

where $(n', m') \in R(f)$, by $\theta_0(n', m')$. It is clear that these maps constitute a linear action of the groupoid $R(f)$ on $f^! E$ in the sense of §1.7. For completeness, we repeat this case here.

Definition 2.1.1 Let (E', p', M') be a vector bundle and let $f \colon M' \to M$ be a surjective submersion. A *linear action of $R(f)$ on E'* is a collection of linear isomorphisms $\theta(n', m') \colon E'|_{m'} \to E'|_{n'}$, for $(n', m') \in R(f)$, such that:

(i) $\theta(m', m') = \mathrm{id}_{E'_{m'}}$ for all $m' \in M'$,

(ii) $\theta(m', n') = \theta(n', m')^{-1}$ for all $(n', m') \in R(f)$,

(iii) $\theta(p', n') \circ \theta(n', m') = \theta(p', m')$ for all $(p', n'), (n', m') \in R(f)$,

(iv) $R(f) \times_{M'} E' \to E'$, $((n', m'), a') \mapsto \theta(n', m')(a')$, is smooth, where the domain is the pullback manifold. $\qquad \square$

Theorem 2.1.2 *Consider a vector bundle (E', q', M') and a surjective submersion $f: M' \to M$. Let θ be a linear action of $R(f)$ on E'. Then there exists a vector bundle (E, q, M), unique up to isomorphism over M, such that $E' \cong f^! E$ as vector bundles over M' with respect to an isomorphism which carries θ to the canonical action (3) of $R(f)$ on $f^! E$.*

Proof Define an equivalence relation $\sim$ on E' by

$$a' \sim b' \quad \Longleftrightarrow \quad (q'b', q'a') \in R(f) \text{ and } b' = \theta(q'b', q'a')(a').$$

Let $W \subseteq E' \times E'$ be the graph of $\sim$. We need to show that W is a closed embedded submanifold of $E' \times E'$ and that $\mathrm{pr}_1: E' \times E' \to E'$, restricted to W, is a surjective submersion; from this it will follow that $E'/\sim$ is a quotient manifold. First consider $\mathrm{id} \times q': M' \times E' \to M' \times M'$ and define

$$W' = (\mathrm{id} \times q')^{-1}(R(f)).$$

Since $R(f)$ is a closed embedded submanifold of $M' \times M'$, the same is true of W' in $M' \times E'$. Now W is the image of $W' \to E' \times E'$, $(m', a') \mapsto (\theta(m', q'a')(a'), a')$, and this map is an immersion because the composite

$$W' \to E' \times E' \xrightarrow{q' \times \mathrm{id}} M' \times E'$$

is so. Indeed it also follows that $q' \times \mathrm{id}: W \to W'$ is a diffeomorphism and that W is an embedded submanifold of $E' \times E'$.

We can also regard W' as the pullback

$$
\begin{array}{ccc}
& \pi & \\
W' & \longrightarrow & E' \\
\mathrm{id} \times q' \downarrow & & \downarrow q' \\
R(f) & \longrightarrow & M' \\
& \mathrm{pr}_1 &
\end{array}
$$

where π is $(m', a') \mapsto a'$. Now $\mathrm{pr}_1: R(f) \to M'$ is a surjective submersion, so π is also, and therefore the required projection

$$W \overset{\cong}{\to} W' \overset{\pi}{\to} E'$$

is also a surjective submersion. Thus $E = E'/\sim$ has a manifold structure with respect to which the natural map $\natural: E' \to E$, $a' \mapsto \overline{a'}$ is a surjective submersion.

The remainder of the proof is straightforward: there is a map $q \colon E \to M$ with $q \circ \natural = f \circ q'$ and it is a surjective submersion. The restriction of $\natural$ to each $E'_{m'} \to E_{f(m')}$ is a bijection and so the fibres of q acquire vector space structures. Vector bundle charts for E may be constructed by taking a chart $\psi' \colon U' \times V' \to E'_{U'}$ for E' and a local section $\sigma \colon U = f(U') \to U'$ of f and defining $\psi \colon U \times V' \to E_U$ by

$$\psi(m, v') = \overline{\psi'(\sigma(m), v')}.$$

The isomorphism $f^! E \to E'$ is $(m', \overline{a'}) \mapsto \theta(m', q'a')(a')$ and clearly preserves the actions of $R(f)$. $\qquad\square$

The bundle (E, q, M) is the *quotient of* (E', q', M') *over* θ (or *over* $R(f)$), and is denoted E'/θ or $E'/R(f)$. This construction is also known as *descent*; we say that (E, q, M) is obtained from (E', q', M') by *descent over* $f \colon M' \to M$.

Since $\natural \colon E' \to E$ is a fibrewise bijection, each section $\mu \in \Gamma E$ induces a unique section $1 \otimes \mu$ of E' such that $\varphi \circ (1 \otimes \mu) = \mu \circ f$.

Lemma 2.1.3 *A section μ' of E' is of the form $1 \otimes \mu$ for some $\mu \in \Gamma E$ if and only if*

$$\theta(n', m')(\mu'(m')) = \mu'(n')$$

for all $(n', m') \in R(f)$.

A section satisfying this condition is said to be *θ–stable*. Thus $\mu' \in \Gamma E'$ is $\natural$–projectable if and only if it is *θ–stable*.

Given a morphism $\varphi \colon A \to B$ of vector bundles over M and any smooth map $f \colon M' \to M$, there is an induced morphism $f^!(\varphi) \colon f^! A \to f^! B$, $(m', a) \mapsto (m', \varphi(a))$ of the pullback bundles over M'. We now characterize morphisms which arise in this way (assuming that f is a surjective submersion).

Proposition 2.1.4 *Let A' and B' be vector bundles over M' with linear $R(f)$ actions θ and δ, where $f \colon M' \to M$ is a surjective submersion. Let A and B be the corresponding vector bundles on M. Then a morphism $\varphi' \colon A' \to B'$ of vector bundles over M' quotients (or descends) to a morphism $\varphi \colon A \to B$ over M if and only if φ' is $R(f)$–equivariant in the sense that*

$$\varphi'(\theta(n', m')(a')) = \delta(n', m')(\varphi'(a'))$$

for all $(n', m') \in R(f)$ and $a' \in A'$ with $q'a' = m'$.

Proof Define an $R(f)$ action η on $\mathrm{Hom}(A', B')$ by

$$\eta(n', m')(\psi) = \delta(n', m') \circ \psi \circ \theta(m', n').$$

Then the corresponding bundle on M is $\mathrm{Hom}(A, B)$ and the condition on φ asserts that it is η-stable as a section of $\mathrm{Hom}(A', B')$.   □

Now return to the vector bundle morphism (2): φ is fibrewise surjective and f is a surjective submersion. The kernel of $\varphi^!$ is denoted K. Define a linear action θ of $R(f)$ on E'/K by $\theta(n', m')(\overline{e'}) = \overline{g'}$, where g' is any element of $E'_{n'}$ with $\varphi(g') = \varphi(e')$.

We will regard not only K, but also $R(f)$ and θ, as constituting the kernel of (2). To distinguish this from the usual concept, we call $(K, R(f), \theta)$ the *kernel system* of (φ, f). In general we use the following terminology.

Definition 2.1.5 Let (E', q', M') be a vector bundle. A *subbundle system* $\mathscr{K} = (K, R(f), \theta)$ of (E', q', M') consists of a vector subbundle K of E', a surjective submersion $f\colon M' \to M$, and a linear action θ of $R(f)$ on E'/K.   □

Given a vector bundle (E', q', M') and a subbundle system $\mathscr{K} = (K, R(f), \theta)$ we can now form the quotient E'/K in the usual way and then apply 2.1.2 to obtain $E = (E'/K)/\theta$, a vector bundle over M. We denote the composition of the two quotient maps $E' \to E'/K$ and $E'/K \to (E'/K)/\theta$ by $\natural\colon E' \to E'/\mathscr{K}$. It is not hard to see that $\natural$ is fibrewise surjective and defines the kernel system $\mathscr{K}$.

The next result justifies calling $E = E'/\mathscr{K}$ the *quotient of E' over the subbundle system $\mathscr{K}$*.

Proposition 2.1.6 Let (E', q', M') be a vector bundle and let $\mathscr{K} = (K, R(f), \theta)$ be a subbundle system with quotient $E = E'/\mathscr{K}$ on base M. Let $\varphi\colon E' \to E''$ be any morphism of vector bundles over any smooth map $g\colon M' \to M''$ such that:

(i) $\varphi(K) = M'' \times \{0\}$,

(ii) $(g \times g)(R(f)) \subseteq \Delta_{M''}$,

(iii) if $\overline{\varphi}$ is the induced morphism $E'/K \to E''$, then

$$\overline{\varphi}(\theta(n', m')(\overline{e'})) = \overline{\varphi}(\overline{e'})$$

for all $(n', m') \in R(f)$ and $\overline{e'} \in E'/K$.

Then there is a unique vector bundle morphism $\psi\colon E \to E''$, $h\colon M \to M''$ such that $\psi \circ \natural = \varphi$ and $h \circ f = g$.

A morphism (φ, g) which satisfies the three conditions of 2.1.6 is said to *annul* $\mathcal{K}$.

Proof In (ii), $\Delta_{M''}$ is the diagonal of $M'' \times M''$. Condition (ii) implies that there is a well–defined set map $h\colon M \to M''$ with $h \circ f = g$; since f is a surjective submersion, h is smooth. Now the fibres of $\bar{\natural}\colon E'/K \to E$ are precisely the orbits of θ and so (iii) implies that there is a well–defined set map $\psi\colon E \to E''$ with $\psi \circ \bar{\natural} = \overline{\varphi}$. Again, ψ is smooth and it is routine to check that it is a vector bundle morphism over h. The uniqueness is clear. $\qquad\square$

Clearly these three annulment conditions are necessary for the existence of (ψ, h).

Example 2.1.7 Let $f\colon M' \to M$ be a surjective submersion. Then $T(f)$ is a fibrewise surjection $TM' \to TM$ and the kernel of $T(f)^!$ is the vertical bundle $T^f M'$ of f, defined by

$$T^f_{m'} M' = \{X' \in T_{m'}(M') \mid T(f)(X') = 0\}.$$

Now suppose that M' is the 3–sphere in $\mathbb{R}^4$, regarded as pairs (α, β) of complex numbers with $|\alpha|^2 + |\beta|^2 = 1$. We will show that there is a trivial rank–2 bundle on $\mathbb{S}^3$ which descends under the Hopf map to $T\mathbb{S}^2$.

Tangent vectors to $\mathbb{S}^3$ at a point (α, β) can be regarded as pairs (ξ, η) of complex numbers such that, in terms of the standard inner product on $\mathbb{R}^4$, $\langle (\alpha, \beta), (\xi, \eta) \rangle = 0$. In terms of $\mathbb{C}^2$ this inner product is given by

$$\langle (\alpha, \beta), (\alpha', \beta') \rangle = \mathscr{R}(\alpha \overline{\alpha'} + \beta \overline{\beta'}).$$

It follows that $T_{(\alpha, \beta)}(\mathbb{S}^3)$ is generated by $(\alpha i, -\beta i)$, $(-\beta, \alpha)$ and $(\beta i, \alpha i)$. These are linearly independent on all of $\mathbb{S}^3$, which shows that $T\mathbb{S}^3$ is trivializable.

Let $U(1) = \{z \in \mathbb{C} \mid |z| = 1\}$ act on $\mathbb{S}^3$ by

$$(\alpha, \beta)e^{it} = (\alpha e^{it}, \beta e^{-it})$$

and let $f\colon \mathbb{S}^3 \to M$ be the resulting quotient manifold and quotient map; M can be identified with $\mathbb{S}^2$.

At any point (α, β) of $\mathbb{S}^3$, the vectors tangent to the $U(1)$ orbit are generated by $(\alpha i, -\beta i)$. If we write E for the subbundle generated by $(-\beta, \alpha)$ and $(\beta i, \alpha i)$, it therefore follows that

$$T\mathbb{S}^3/T^f\mathbb{S}^3 \cong E.$$

Hence, by the discussion at the start of the example, $T(f)\colon E \to TM$ is a fibrewise isomorphism, and thus $T\mathbb{S}^2$ is obtained from E by descent. The action of $R(f)$ on E is

$$\theta((\alpha,\beta)e^{it},(\alpha,\beta))(\xi,\eta) = (\xi e^{it}, \eta e^{-it}),$$

for any $(\xi,\eta) \in E$.

The groupoid $R(f)$ also acts on E via the trivialization of $T\mathbb{S}^3$; this action of course causes E to descend to $\mathbb{S}^2 \times \mathbb{R}^2$. $\qquad\qquad\boxtimes$

Lastly, consider a general fibrewise surjection $\varphi\colon E' \to E$, $f\colon M' \to M$, with associated kernel system $\mathcal{K} = (K, R(f), \theta)$. A section μ' of E' is φ-*projectable* if there exists $\mu \in \Gamma E$ such that $\varphi \circ \mu' = \mu \circ f$. Define $\varphi' \in \Gamma E'$ to be θ-*stable* if $\overline{\mu'} \in \Gamma(E'/K)$ is θ-stable; that is, for all $(n', m') \in R(f)$, we have $\theta(n', m')(\overline{\mu'}(n')) = \overline{\mu'}(m')$. The next result is a simple development of 2.1.3.

Proposition 2.1.8 *A section of E' is φ-projectable if and only if it is θ-stable.*

2.2 Base–preserving quotients of groupoids

A morphism of groups may be factored into a surjective morphism, followed by an isomorphism, followed by an injective morphism. For base–preserving morphisms of groupoids, a similar decomposition holds. This is straightforward and often sufficient, so we present this case first, starting with the algebra.

Definition 2.2.1 Let G be a groupoid on M. A *normal subgroupoid* of G is a wide subgroupoid $N \subseteq \mathscr{I}G$ of the inner subgroupoid such that for any $\nu \in N$ and any $g \in G$ with $\alpha g = \alpha \nu = \beta \nu$, we have $g\nu g^{-1} \in N$. $\qquad\square$

Thus a normal subgroupoid is a collection of normal subgroups of the inner subgroupoid, which is invariant under the inner automorphisms of G.

Definition 2.2.2 Let $F\colon G \to G'$ be a base–preserving morphism of groupoids over M. Then the *kernel* of F is the set $\{g \in G \mid F(g) = 1_x \text{ for some } x \in M\}$. $\qquad\square$

Clearly the kernel of a base–preserving morphism is a normal subgroupoid. The following construction of quotient groupoids shows that every normal subgroupoid is the kernel of a base–preserving morphism.

Proposition 2.2.3 *Let N be a normal subgroupoid of a groupoid G on M. Define an equivalence relation, denoted $\sim$, on G by*

$$g \sim h \iff \exists \nu \in N \text{ such that } h\nu = g.$$

Denote the equivalence classes by $[g]$, $g \in G$, and the set of them by G/N.

Then the following defines the structure of a groupoid with base M on G/N : the source and target projections are $\overline{\alpha}([g]) = \alpha(g)$, $\overline{\beta}([g]) = \beta(g)$; the object inclusion map is $x \mapsto [1_x]$; and the product $[h][g]$, where $\alpha(h) = \beta(g)$, is defined as $[hg]$. The inverse of $[g]$ is $[g^{-1}]$.

The projection $\natural\colon g \mapsto [g]$, $G \to G/N$, is a groupoid morphism over M with kernel N.

The proof is a straightforward modification of the corresponding result for group quotients and is left to the reader.

Note the extreme cases: $G/1_M$ is isomorphic to G under $\natural$, and $G/\mathscr{I}G$ is isomorphic to the image of (β, α) in $M \times M$.

*   *   *   *   *

Now suppose that Ω is a locally trivial Lie groupoid on M and N is a closed embedded normal Lie subgroupoid of Ω. A section–atlas $\{\sigma_i\colon U_i \to \Omega_m\}$ for Ω induces charts ψ_i for $\mathscr{I}\Omega$ by $\psi_i(x, g) = \sigma_i(x) g \sigma_i(x)^{-1}$ where $g \in \Omega_m^m$, and the normality condition on N ensures that these restrict to $U_i \times N_m \to N_{U_1}$ and so N is a sub LGB of $\mathscr{I}\Omega$.

Let Γ denote the graph of the equivalence relation on Ω defined by N; thus

$$\Gamma = \{(\eta, \eta\nu) \mid \eta \in \Omega, \nu \in N, \alpha\eta = \beta\nu\}.$$

In place of the division map of 1.2.6, consider the other division map,

$$\delta'\colon \Omega \times_\beta \Omega \to \Omega, \quad (\eta, \xi) \mapsto \eta^{-1}\xi.$$

By a similar argument to 1.3.3, δ' is a surjective submersion, and so $\Gamma = \delta'^{-1}(N)$ is a closed embedded submanifold of $\Omega \times_\beta \Omega$, and hence of $\Omega \times \Omega$. Now the projection $\Gamma \xrightarrow{\text{pr}_2} \Omega$ is a surjective submersion, since composing it with $\Omega * N \to \Gamma$, $(\eta, \nu) \mapsto (\eta\nu, \eta)$, gives the projection $\Omega * N \to \Omega$ onto the first factor, and this latter map is a surjective submersion because in the pullback square defining $\Omega * N$, the projection $N \to M$ is a surjective submersion.

So, by Godement's criterion, the quotient manifold Ω/Γ, which is Ω/N, exists.

It remains to prove that Ω/N is a Lie groupoid. Since $\overline{\alpha} \circ \natural = \alpha$, and α is a surjective submersion, it follows that $\overline{\alpha}$ is also. To prove that composition in Ω/N is smooth, note first that

$$(\natural \times \natural)^{-1}(\Omega/N * \Omega/N) = \Omega * \Omega$$

and therefore the restriction of $\natural \times \natural$ to $\Omega * \Omega \to \Omega/N * \Omega/N$ is a surjective submersion. Since

$$
\begin{array}{ccc}
\Omega * \Omega & \longrightarrow & \Omega \\
\natural \times \natural \downarrow & & \downarrow \natural \\
\Omega/N * \Omega/N & \longrightarrow & \Omega/N
\end{array}
$$

commutes, it follows that the multiplication in Ω/N is smooth. Thus we have the following result.

Theorem 2.2.4 *Let Ω be a locally trivial Lie groupoid on M and let N be a closed embedded normal Lie subgroupoid. Then Ω/N, with the structure just defined, is a locally trivial Lie groupoid, and $\natural\colon \Omega \to \Omega/N$ is a surjective submersion and a morphism.*

Lastly we note the following results, which are often useful.

Proposition 2.2.5 *Let $F\colon \Omega \to \Omega'$ be a morphism of locally trivial Lie groupoids over a connected base M. Then:*

(i) *F is of constant rank;*

(ii) *if F_m^m is injective for some $m \in M$, then F is an injective immersion;*

(iii) *if F_m^m is surjective for some $m \in M$, then F is a surjective submersion.*

Proof Let $\{\sigma_i\colon U_i \to \Omega_m\}$ be a section–atlas for Ω and consider the section–atlas $\{\sigma_i' = \varphi \circ \sigma_i\colon U_i \to \Omega'_m\}$ for Ω'. With respect to the corresponding charts, F is locally $\mathrm{id} \times f \times \mathrm{id}\colon U_i \times G \times U_i \to U_i \times G' \times U_i$ where $f = F_m^m$. The results thus follow from the corresponding statements for Lie groups. $\square$

Example 2.2.6 Note that for general Lie groupoids a morphism may be a submersion into a connected codomain, but not be surjective; let M be the interval $(1, 2)$ (say) in $\mathbb{R}$, let G be the multiplicative group of positive reals, and let $F \colon M \times M \to G$ be $(y, x) \mapsto yx^{-1}$. ☒

The next result is proved using the same methods as for 2.2.4.

Proposition 2.2.7 *Let* $\varphi \colon \Omega \to \Omega'$ *be a morphism of locally trivial Lie groupoids over* M. *Then* $K = \ker \varphi$ *is a closed embedded submanifold of* $\mathscr{I}\Omega$, *and a Lie group subbundle of* $\mathscr{I}\Omega$. *Further,* $\operatorname{im}(\varphi)$ *is a submanifold of* Ω' *and a reduction of* Ω'.

For general Lie groupoids, not necessarily locally trivial, it would be useful to have a theory of quotients such that, for example, the quotient over the inner subgroupoid would be the image of the anchor with an appropriate structure. We do not attempt such a theory here.

2.3 Pullback groupoids

The general concept of morphism of groupoids was defined in §1.2. As a preliminary to the general quotients and semidirect products which follow, we express this in terms of pullbacks.

Consider a set groupoid $G \rightrightarrows M$ and any map $f \colon M' \to M$. Let $M' * G * M'$ denote the set of all $(y', g, x') \in M' \times G \times M'$ such that $f(y') = \beta g$, $\alpha g = f(x')$. Then there is an evident groupoid structure defined on $M' * G * M'$ by

$$\beta'(y', g, x') = y', \quad \alpha'(y', g, x') = x', \quad (z', h, y')(y', g, x') = (z', hg, x'),$$
$$1'_{x'} = (x', 1_{f(x')}, x'), \qquad (y', g, x')^{-1} = (x', g^{-1}, y').$$

We denote this by $f^{\downarrow\!\downarrow}G$ and call it the *pullback (set) groupoid of* G *over* f.

If G is a group, so that f is a constant map, then $f^{\downarrow\!\downarrow}G$ is the trivial groupoid $M' \times G \times M'$.

If G is now a Lie groupoid and f a smooth map, we clearly need to assume that $f \times f \colon M' \times M' \to M \times M$ and $(\beta, \alpha) \colon G \to M \times M$ are transversal, so that the pullback manifold in Figure 2.1 exists. If G is locally trivial, it is easily seen that the pullback groupoid structure on $f^{\downarrow\!\downarrow}G$ makes it a Lie groupoid, also locally trivial, with the appropriate universal property. However in general the transversality condition is

$$f^{\Downarrow}G \longrightarrow G$$

$$\downarrow \qquad\qquad \downarrow \quad (\beta, \alpha)$$

$$M' \times M' \xrightarrow[\; f \times f \;]{} M \times M$$

Fig. 2.1.

not sufficient, since the target projection $f^{\Downarrow}G \to M'$ may not be a submersion. The additional condition incorporated in the following result is often available.

Proposition 2.3.1 *Let* $G \rightrightarrows M$ *be a Lie groupoid and let* $f \colon M' \to M$ *be a smooth map. Suppose that the pullback manifold of* $(\beta, \alpha) \colon G \to M \times M$ *across the Cartesian square* $f \times f \colon M' \times M' \to M \times M$ *exists, and that the composition* $\overline{\beta} \colon f^! G \to M$, $(g, x') \mapsto \beta g$, *is a surjective submersion. Then with the pullback manifold structure and the groupoid structure above,* $f^{\Downarrow}G$ *is a Lie groupoid on* M' *and the natural projection* $f^{\Downarrow} \colon f^{\Downarrow}G \to G$ *is a morphism of Lie groupoids over* f.

Further, if $F \colon G' \to G$, $f \colon M' \to M$ *is a morphism of Lie groupoids, then* $F = f^{\Downarrow} \circ F^{\Downarrow}$ *where* $F^{\Downarrow} \colon G' \to f^{\Downarrow}(G)$ *is* $g' \mapsto (\beta'g', \varphi(g'), \alpha'g')$.

With this structure, $f^{\Downarrow}G$ is the *pullback Lie groupoid of* G *over* f.

Proof Construct the manifold $f^{\Downarrow}G$ as the pullback

$$f^{\Downarrow}G \longrightarrow f^! G$$

$$\downarrow \qquad\qquad \downarrow \;\; \overline{\beta}$$

$$M' \xrightarrow[\; f \;]{} M.$$

Since $\overline{\beta}$ is a surjective submersion by assumption, it follows that the projection $f^{\Downarrow}G \to M'$ is also, and this is the target projection of $f^{\Downarrow}G$. Now it is straightforward to see that inversion is a diffeomorphism and that the composition is smooth. The factorizability condition is immediate. $\qquad\square$

If f is a surjective submersion, then the projection $f^!G \to G$, and hence $\overline{\beta}$, is a surjective submersion, and so the additional condition above holds. This case, and the case in which G is locally trivial, are sufficient for most purposes.

In principle, 2.3.1 allows the study of arbitrary morphisms to be reduced to the base–preserving case. This will be useful when we come to consider morphisms of Lie algebroids.

Example 2.3.2 The frame groupoid $\Phi(f^!E)$ of a pullback vector bundle $f^!E$ is the pullback $f^{\Downarrow}\Phi(E)$ of the frame groupoid. $\boxtimes$

For pullbacks of general diagrams of Lie groupoids, see 2.4.14.

2.4 General quotients and fibrations

For set groupoids and Lie groupoids alike, the description of general groupoid morphisms involves phenomena similar to those which arose in the case of vector bundles in §2.1.

Certain problems arise already in the purely algebraic case. Firstly, the image of a groupoid morphism need not be a subgroupoid; it may happen that a product $F(h)F(g)$ is defined but the product hg is not and that another pair h_1, g_1 with $F(h_1) = F(h)$, $F(g_1) = F(g)$ and $h_1 g_1$ defined cannot be found. This can occur even for morphisms of trivial groupoids, such as that in Example 2.2.6.

Secondly, the usual concept of kernel, as applied to groupoid morphisms, does not adequately measure injectivity.

Definition 2.4.1 Let $F: G \to G'$, $f: M \to M'$, be a morphism of groupoids. Then the *kernel* of (F, f) is the set

$$\{g \in G \mid F(g) = 1_{x'}, \ \exists x' \in M'\}.$$

$\square$

If F is a surjective submersion, then the kernel is a closed embedded submanifold of the domain.

Example 2.4.2 Let $P(M, G, \pi)$ be a principal bundle and consider the gauge groupoid $\Omega = \frac{P \times P}{G}$ constructed in §1.3. It is easy to see that the map $F: P \times P \to \Omega$, $(u_2, u_1) \mapsto \langle u_2, u_1 \rangle$, is a morphism of Lie groupoids over $\pi: P \to M$, where $P \times P$ has the pair groupoid structure of 1.1.7.

The kernel is the diagonal Δ_P of P, which is the base subgroupoid of $P \times P$.   ⊠

This example shows that surjective groupoid morphisms are not determined by their kernels: both the morphism in 2.4.2 and $\mathrm{id}_{P \times P}$ are surjective morphisms with kernel Δ_P.

This section is devoted to extending the notion of kernel so as to restore a form of the familiar 'first isomophism theorem'. A notion of kernel, however generalized, must be expressed in terms of data on the domain. We are accordingly seeking a general class of groupoid morphisms which are determined, up to isomorphism, by data on their domains. There are several candidates, the most inclusive of which is the class of fibrations.

Definition 2.4.3 Let $F\colon G \to G'$, $f\colon M \to M'$ be a morphism of Lie groupoids. Then (F, f) is a *fibration* if both f and $F^!\colon G \to f^! G'$ are surjective submersions.   □

Fibrations are closed under composition.

The set fibration condition is precisely what is needed to guarantee that given any elements h, g of G such that $F(h)F(g)$ is defined, there exists $h_1 \in G$ such that $h_1 g$ is defined and $F(h_1) = F(h)$; then

$$F(h)F(g) = F(h_1)F(g) = F(h_1 g)$$

shows that the product $F(h)F(g)$ is determined by the domain groupoid and the map.

The surjective submersion condition on $F^!$ ensures that the kernel K of F is a Lie subgroupoid of G. Regard K as the preimage of

$$\{(m, 1'_{f(m)}) \mid m \in M\} \subseteq f^! G'$$

under $F^!$. Then $\alpha\colon K \to M$ corresponds to the restriction of $F^!$ to the complete inverse image K. It follows that $\alpha\colon K \to M$ is a surjective submersion and K is a Lie subgroupoid of G.

Before proceeding we need a lemma. Call the commutative square of smooth maps in Figure 2.2(a) *versal* if the pullback $M \times_{M'} B'$ exists (as a submanifold of the product) and the induced map $B \to M \times_{M'} B'$ is a surjective submersion. In this section we use pr_f to denote the projection $R(f) \to M$, $(y, x) \mapsto x$.

Lemma 2.4.4 *Given a commutative diagram as in Figure 2.2(a) in*

$$\begin{array}{ccc} B & \xrightarrow{F} & B' \\ p \downarrow & & \downarrow p' \\ M & \xrightarrow[f]{} & M' \end{array} \qquad\qquad \begin{array}{ccc} R(F) & \xrightarrow{\mathrm{pr}_F} & B \\ p \times p \downarrow & & \downarrow p \\ R(f) & \xrightarrow[\mathrm{pr}_f]{} & M \end{array}$$

(a) (b)

Fig. 2.2.

which each of the maps F, f, p and p' is a surjective submersion, the diagram is versal if and only if the diagram of Figure 2.2(b) *is versal. When this is so the restriction of $p \times p$ to $R(F) \to R(f)$ is a surjective submersion.*

Proof The conditions on p, F and f ensure that $R(F)$ and $R(f)$ are closed embedded submanifolds of $B \times B$ and $M \times M$, with pr_F and pr_f surjective submersions, and that the pullbacks $M \times_{M'} B'$ and $F(f) \times_M B$ exist. On the set level, the result is easy: if $R(F) \to R(f) \times_M B$ is surjective then, given $(m, b') \in M \times_{M'} B'$ there is $b \in B$ with $F(b) = b'$ and the element $((m, pb), b)$ of $R(f) \times_M B$ lifts to some $(a, b) \in R(F)$; now $a \in B$ is a lift of (m, b') as required.

Conversely, if $B \to M \times_{M'} B'$ is surjective then, given $((n, m), b) \in R(f) \times_M B$ we have $pb = m$ and so $(n, F(b))$ lies in $M \times_{M'} B'$ and therefore has a lift $a \in B$; now (a, b) lies in $R(F)$ and is a lift of $((n, m), b)$ as required. Lastly, if $R(F) \to R(f) \times_M B$ and $p\colon B \to M$ are surjective, it is clear that $p \times p\colon R(F) \to R(f)$ is surjective also.

$$\begin{array}{ccc} T_{(a,b)}(R(F)) & \xrightarrow{T_{(a,b)}(\mathrm{pr}_F)} & T_b(B) \\ \downarrow & & \downarrow \\ T_{(m,n)}(R(f)) & \xrightarrow[T_{(m,n)}(\mathrm{pr}_f)]{} & T_n(M) \end{array}$$

Fig. 2.3.

Now for each $(a, b) \in R(F)$ there are corresponding diagrams of linear

$$
\begin{array}{ccc}
T_b(B) & \xrightarrow{\;T_b(F)\;} & T_{F(b)}(B') \\
\downarrow & & \downarrow \\
T_n(M) & \xrightarrow[T_n(f)]{} & T_{f(n)}(M')
\end{array}
$$

Fig. 2.4.

maps as shown in Figures 2.3 and 2.4, where $m = pa$, $n = pb$. Here

$$
T_{(a,b)}(R(F)) = \{(Y, X) \in T_a(B) \times T_b(B) \mid T(F)(Y) = T(F)(X)\},
$$

with a similar equation for $T_{(m,n)}(R(f))$. The same argument applied on the tangent level then completes the proof.   □

We proceed to the notion of kernel system via an intermediate notion of congruence.

Definition 2.4.5 Let G be a Lie groupoid on M. A *(smooth) congruence* on G is a pair (S, R) where $S \subseteq G \times G$ and $R \subseteq M \times M$ satisfy the following conditions:

(C1) S is a closed, embedded wide Lie subgroupoid of the pair groupoid $G \times G$ on G and R is a closed, embedded wide Lie subgroupoid of the pair groupoid $M \times M$ on M;

(C2) S is a Lie subgroupoid with base R of the Cartesian product groupoid $G \times G$ on $M \times M$;

(C3) the square in (4) below is versal.

$$
\begin{array}{ccc}
S & \xrightarrow{\;\mathrm{pr}_S\;} & G \\
{\scriptstyle \alpha \times \alpha}\downarrow & & \downarrow{\scriptstyle \alpha} \\
R & \xrightarrow[\mathrm{pr}_R]{} & M
\end{array}
\qquad (4)
$$

□

Theorem 2.4.6 *If* $F\colon G \to G'$, $f\colon M \to M'$, *is a fibration of Lie groupoids, then* $(R(F), R(f))$ *is a congruence on* G. *Conversely, given a Lie groupoid* G *on* M *and a congruence* (S, R) *on* G, *there is a unique*

Lie groupoid structure on the quotient sets $G' = G/S$, $M' = M/R$ such that the natural projections $F\colon G \to G'$, $f\colon M \to M'$ form a morphism of Lie groupoids. Further, this (F, f) is a fibration and is universal for morphisms $\Phi\colon G \to H$, $\varphi\colon M \to N$ such that $\Phi \times \Phi$ maps S to the diagonal of H and $\varphi \times \varphi$ maps R to the diagonal of N.

Proof Assume $F\colon G \to G'$, $f\colon M \to M'$, is a fibration of Lie groupoids. Then $R(F)$ and $R(f)$ satisfy (C1) of 2.4.5, and 2.4.4 applied to F, f, pr_F and pr_f yields (C3) and the fact that $\alpha \times \alpha\colon R(F) \to R(f)$ is a surjective submersion. The remaining conditions in (C2) then follow easily.

Conversely, consider a congruence (S, R) on a Lie groupoid $G \rightrightarrows M$. By (C1) and the Godement criterion, there are manifold structures on the quotient sets $G' = G/S$, $M' = M/R$ such that the natural projections $F\colon G \to G'$, $f\colon M \to M'$ are surjective submersions. Define $\alpha'\colon G' \to M'$ by $\alpha' \circ F = f \circ \alpha$; since α, F and f are surjective submersions, it follows that α' is smooth and itself a surjective submersion. Likewise define β' by $\beta' \circ F = f \circ \beta$. Given $g', h' \in G'$ with $\alpha'(g') = \beta'(h')$, choose any $h \in G$ with $F(h) = h'$ and then choose $g \in G$ with $F(g) = g'$ and $\alpha(g) = \beta(h)$; that this is possible follows from (C3) and 2.4.4. We can now define $g'h' = F(gh)$, for (C2) implies that the product is unambiguous and defines a groupoid structure on G' with base M'. Certainly $F\colon G \to G'$, $f\colon M \to M'$ is a morphism (of set groupoids) and, by (C3) and 2.4.4, the diagram

$$
\begin{array}{ccc}
& F & \\
G & \longrightarrow & G' \\
\alpha \downarrow & & \downarrow \alpha' \\
M & \longrightarrow & M' \\
& f &
\end{array}
$$

is versal. It remains to prove that the groupoid structure in G' is Lie.

Let $\delta\colon G \times_\alpha G \to G$ denote the division $\delta(g, h) = gh^{-1}$ as in 1.2.6. To prove that the corresponding division map $\delta'\colon G' \times_{\alpha'} G' \to G'$ is smooth, it suffices to show that the restriction $F \times F\colon G \times_\alpha G \to G' \times_{\alpha'} G'$ is a surjective submersion, and this now follows by applying 2.4.4 to the diagram in Figure 2.5.

The universal property and the uniqueness are easily verified. $\qquad\square$

Theorem 2.4.6 provides a complete characterization of fibrations of

$$
\begin{array}{ccccc}
G \times_\alpha G & \xrightarrow{\ \mathrm{pr}_\alpha\ } & G & \xrightarrow{\ \alpha\ } & M \\
\Big\downarrow{\scriptstyle F\times F} & & \Big\downarrow{\scriptstyle F} & & \Big\downarrow{\scriptstyle f} \\
G' \times_{\alpha'} G' & \xrightarrow{\ \mathrm{pr}_{\alpha'}\ } & G' & \xrightarrow{\ \alpha'\ } & M'.
\end{array}
$$

Fig. 2.5.

Lie groupoids in terms of data from the domain groupoid. We now reformulate this notion of congruence so that its relation to the usual notion of kernel can be seen clearly.

Consider a Lie groupoid $G \rightrightarrows M$ and any closed, embedded wide Lie subgroupoid N of G (not necessarily normal). Define

$$E = \{(g,h) \mid \alpha g = \alpha h, \ gh^{-1} \in N\} \subseteq G \times G.$$

Then $E = \delta^{-1}(N)$ where δ is the division map. Since $\alpha\colon G \to M$ and δ are surjective submersions, it follows that E is a closed, embedded submanifold of $G \times G$. Further,

$$
\begin{array}{ccc}
E & \xrightarrow{\ \mathrm{pr}_E\ } & G \\
\Big\downarrow{\scriptstyle \delta} & & \Big\downarrow{\scriptstyle \beta} \\
N & \xrightarrow{\ \ \alpha\ \ } & M
\end{array}
$$

is a pullback of manifolds, and since $\alpha\colon N \to M$ is a surjective submersion by assumption, it follows that pr_E is also. The Godement criterion now implies that the set of one–sided cosets

$$Ng = \{\nu g \mid \alpha\nu = \beta g, \ \nu \in N\}$$

as g ranges through G, which we denote $G \,\dot{.}\, N$, has a smooth manifold structure such that $q\colon G \to G \,\dot{.}\, N$, $g \mapsto Ng$, is a surjective submersion. Note that $G \,\dot{.}\, N$ is usually not a groupoid, even if N is normal, but that $\alpha\colon G \to M$ quotients to a well–defined surjective submersion $G \,\dot{.}\, N \to M$, which we also denote by α.

Definition 2.4.7 A *normal subgroupoid system* in $G \rightrightarrows M$ is a triple

$\mathscr{N} = (N, R, \theta)$ where N is a closed, embedded, wide Lie subgroupoid of G, where R is a closed, embedded, wide, Lie subgroupoid of the pair groupoid $M \times M$, and θ is an action of R on the map $\alpha \colon G \cdot^{\cdot} N \to M$ just described, such that the following conditions hold.

(N1) Consider $(n, m) \in R$ and $Ng \in G \cdot^{\cdot} N$ with $\alpha(Ng) = m$. Then, writing $\theta(n, m)(Ng) = Nh$, we have $(\beta h, \beta g) \in R$.

(N2) For $(n, m) \in R$ we have $\theta(n, m)(N1_m) = N1_n$.

(N3) Consider $(n, m) \in R$ and $Ng \in G \cdot^{\cdot} N$ with $\alpha(Ng) = m$, and $h \in G$ with $\alpha h = \beta g$. Then if

$$\theta(n, m)(Ng) = Ng_1 \quad \text{and} \quad \theta(\beta g_1, \beta g)(Nh) = Nh_1,$$

we have $\theta(n, m)(Nhg) = Nh_1 g_1$. $\qquad\square$

For $\nu \in N$ it follows from (N1) that $(\beta\nu, \alpha\nu) \in R$ and it then follows from (N2) and (N3) that if $Ng \in G \cdot^{\cdot} N$ has $\alpha(Ng) = \alpha\nu$, then $\theta(\beta\nu, \alpha\nu)(Ng) = Ng\nu^{-1}$. In particular, if $\nu \in IN$, $g \in G$ and $g\nu g^{-1}$ is defined, then $g\nu g^{-1} \in N$.

Theorem 2.4.8 (i) *Let $\mathscr{N} = (N, R, \theta)$ be a normal subgroupoid system on G, and define*

$$S = \{(h, g) \in G \times G \mid (\alpha h, \alpha g) \in R \text{ and } \theta(\alpha h, \alpha g)(Ng) = Nh\}.$$

Then (S, R) is a congruence on G.

(ii) *Let (S, R) be a congruence on G. Then $N = \{g \in G \mid (g, 1_{\alpha g}) \in S\}$ is a closed, embedded, Lie subgroupoid of G. Define an action θ of R on $G \cdot^{\cdot} N$ by*

$$\theta(n, m)(Ng) = Nh,$$

where $m = \alpha g$ and h is any element of G with $\alpha h = n$ and $(h, g) \in S$. Then (N, R, θ) is a normal subgroupoid system on G.

These two constructions are mutually inverse.

Proof (i) The algebraic conditions are easily verified. The smoothness conditions reduce to proving that three maps are surjective submersions, namely $\mathrm{pr}_S \colon S \to G$, $\alpha \times \alpha \colon S \to R$ and the map

$$S \to R \times_\alpha G, \qquad (h, g) \mapsto ((\alpha h, \alpha g), g)$$

of (C3). The first two of these follow easily from the last. To prove the last, consider the diagram

$$
\begin{array}{ccc}
S & \xrightarrow{\ \subseteq\ } & G \times G \\[2mm]
\downarrow & & \downarrow \quad q \times \mathrm{id} \\[2mm]
R \times_\alpha G & \xrightarrow[\ \widehat{\theta}\ ]{} & (G \mathbin{.\!\!\cdot} N) \times G
\end{array}
$$

where $\widehat{\theta}((n,m),g) = (\theta(n,m)(Ng),g)$. One proves that this is a pullback of manifolds by identifying $(((n,m),g),(h',g'))$, where $(n,m) \in R$, $n = \alpha g$, $g = g'$, and $\theta(n,m)(Ng) = Nh'$, with $(h',g') \in S$. It then follows that because $q \times \mathrm{id}$ is a surjective submersion, the desired map $S \to R \times_\alpha G$ is also.

(ii) First observe that $\{((m,m),1_m) \mid m \in M\}$ is a closed embedded submanifold of $R \times_\alpha G$ and so its preimage under $S \to R \times_\alpha G$, namely $L = \{(h,g) \in S \mid g = 1_{\alpha h}\}$, is a closed embedded submanifold of S. By considering

$$
G \to G \times_\alpha G, \qquad g \mapsto (g, 1_{\alpha g}),
$$

it follows that N is a closed embedded submanifold of G and that $\alpha \colon N \to M$, which corresponds to the restriction of $S \to R \times_\alpha G$ to $L \to M$, is a surjective submersion. Thus N is also a Lie subgroupoid of G.

That θ is well–defined and an action follow from the algebraic conditions in (C1) – (C3). To prove that it is smooth consider the diagram

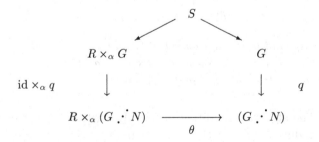

in which $S \to G$ is $(h,g) \mapsto h$ and $S \to R\times_\alpha G$ is $(h,g) \mapsto ((\alpha h, \alpha g), g)$. It is easily proved by a direct calculation with tangent vectors that $\mathrm{id}\times_\alpha q$ is a surjective submersion, and since $S \to R\times_\alpha G$ is also, the smoothness of θ follows from that of $S \to G \mathbin{.\!\!\cdot} N$.

That the two constructions are mutually inverse is easily verified. $\quad\square$

Putting 2.4.6 and 2.4.8 together, we obtain a bijective correspondence between normal subgroupoid systems of G and fibrations of Lie groupoids with domain G.

Given a fibration $F\colon G \to G'$, $f\colon M \to M'$, let K be the kernel of F as in 2.4.1. Then $F^!$ induces a diffeomorphism $\overline{F}\colon G \mathbin{.^\ast} K \to f^! G'$ and we use $\overline{F}$ to transport the canonical action $\theta_0(n,m)(m,g') = (n,g')$ of $R(f)$ on $f^! G' \to M$ to an action θ of $R(f)$ on $\alpha\colon G \mathbin{.^\ast} K \to M$. It is easy to check that $(K, R(f), \theta)$ is the normal subgroupoid system corresponding to $(R(F), R(f))$ under 2.4.8; we call it the *kernel system of* (F, f). Note that this notion of kernel system is not defined for arbitrary morphisms.

Conversely, let $\mathcal{N} = (N, R, \theta)$ be a normal subgroupoid system on G. Then $R = R(f)$ for some surjective submersion $f\colon M \to M'$. Let G' denote the set of orbits of $G \mathbin{.^\ast} N$ under θ, with $\langle Ng \rangle$ denoting the orbit of Ng. Write $F\colon G \to G'$ for $F(g) = \langle Ng \rangle$. One may prove that G' has a quotient manifold structure, by the methods of 2.4.6 and 2.4.8; see also 2.1.2. Define groupoid projections on G', with base M', by $\alpha' \circ F = f \circ \alpha$, $\beta' \circ F = f \circ \beta$. Define composition for $\langle Nh \rangle, \langle Ng \rangle \in G'$ with $\alpha'(\langle Nh \rangle) = \beta'(\langle Ng \rangle)$ by

$$\langle Nh \rangle \langle Ng \rangle = \langle Nh_1 g \rangle$$

where h_1 is any element of G with $Nh_1 = \theta(\beta g, \alpha h)(Nh)$. One can prove directly now that G' is a Lie groupoid on M' with (F, f) a fibration, and that G' is the Lie groupoid which corresponds under 2.4.6 to the congruence on G defined by $\mathcal{N}$. We write $G' = G/\mathcal{N}$ and call it the *quotient Lie groupoid of G over the normal subgroupoid system $\mathcal{N}$*. We call (F, f) the *quotient fibration corresponding to $\mathcal{N}$*. The universal property for F is as follows: the proof is immediate.

Theorem 2.4.9 *Let $F\colon G \to G'$, $f\colon M \to M'$, be a fibration of Lie groupoids, with kernel system $\mathcal{K} = (K, R(f), \theta)$.*

Suppose that $\Phi\colon G \to H$ is any morphism of Lie groupoids over a smooth map $\varphi\colon M \to P$ which annuls $\mathcal{K}$ in the sense that:

(i) *for all $k \in K$, $\Phi(k)$ is an identity of H;*

(ii) *$\varphi \times \varphi$ maps $R(f)$ into the diagonal of $P \times P$;*

(iii) *if $\overline{\Phi}$ is the induced map $G \mathbin{.^\ast} K \to H$, then*

$$\overline{\Phi}(\theta(n,m)(Kg)) = \overline{\Phi}(Kg)$$

for all $(n, m) \in R(f)$ *and* $Kg \in G .^{\cdot} K$ *with* $\alpha g = m$.

Then there is a unique morphism of Lie groupoids $\Psi \colon G' \to H$ over $\psi \colon M' \to P$ such that $\Psi \circ F = \Phi$ and $\psi \circ f = \varphi$. In particular, if (Φ, φ) is a fibration and $\mathscr{K}$ is its kernel system, then (Ψ, ψ) is an isomorphism of Lie groupoids.

Example 2.4.10 Let $P(M, G, \pi)$ be a principal bundle. Continuing 2.4.2, the map $F \colon P \times P \to \Omega$ is a fibration over π. The kernel pair $R(\pi) = P \times_\pi P$ acts on $P \times P$ by

$$(ug, u) {\cdot} (v, u) = (vg, ug).$$

The orbits of this action coincide with the orbits of the diagonal G action on $P \times P$ and the quotient is indeed the gauge groupoid. ⊠

Example 2.4.11 If N is a normal subgroupoid of $\mathscr{I}G$ in the sense of §2.2 then taking $R = \Delta$ and θ to be the trivial representation gives a normal subgroupoid system. ⊠

Example 2.4.12 The projection $p_G \colon TG \to G$, for any Lie groupoid $G \rightrightarrows M$, is a fibration over $p_M \colon TM \to M$. Elements of the conventional kernel $K = T_1 G$ are of the form $X + T(1)(x)$ where $X \in AG$, $x \in TM$ lie over the same point of M. (We are using the basic properties of the Lie algebroid, for which see §3.5.) Applying 1.4.14 we have

$$(Y + T(1)(y)) \bullet (X + T(1)(x)) = Y + X + T(1)(x)$$

where $y = aX + x$. So $T_1 G$ is the action Lie groupoid $A_{vb}G \prec TM$ where $A_{vb}G$ denotes the additive groupoid defined by the vector bundle structure of AG and the action is $(X, x) \mapsto aX + x$.

Given $\xi \in T_g G$, applying 1.4.14 again, the coset defined by ξ consists of all elements of the form

$$(Y + T(1)(y)) \bullet \xi = \xi + \overrightarrow{Y}(g)$$

where $y = T(\beta)(\xi)$ and $Y \in A_{\beta g}G$. It is now clear directly that the map induced by p_G, which maps the coset $\langle \xi \rangle$ to $(T(\alpha)(\xi), g) \in p_M^! G$, is a diffeomorphism. There is no canonical way to describe the action of $R(p_M) = TM \oplus TM$ on $TG .^{\cdot} T_1 G$ in terms of representatives. ⊠

The notion of fibration arises naturally in the context of direct products and general pullbacks in the category of Lie groupoids.

Example 2.4.13 For Lie groupoids $G_1 \rightrightarrows M_1$ and $G_2 \rightrightarrows M_2$ there is a natural Cartesian product Lie groupoid $G_1 \times G_2 \rightrightarrows M_1 \times M_2$, and the projections $G_1 \times G_2 \to G_1$ and $G_1 \times G_2 \to G_2$ are fibrations. ☒

For pullbacks of general diagrams of Lie groupoid morphisms, it is natural to assume that the pullback of the base manifolds exists.

Proposition 2.4.14 *Let $F_1 \colon G_1 \to G$ and $F_2 \colon G_2 \to G$ be fibrations of Lie groupoids over $f_1 \colon M_1 \to M$ and $f_2 \colon M_2 \to M$. Suppose that the pullback manifold $\overline{M}$ of f_1 and f_2 exists. Then the pullback manifold $\overline{G}$ of F_1 and F_2 exists and is an embedded Lie subgroupoid of the Cartesian product groupoid $G_1 \times G_2$. Furthermore, $\overline{G}$ is the pullback of F_1 and F_2 in the category of Lie groupoids.*

Proof The algebraic requirements are easily verified. The smoothness considerations will follow easily once it is established that the source map $\overline{\alpha} \colon \overline{G} \to \overline{M}$ is a surjective submersion.

Let f denote the diagonal map $\overline{M} \to M$, $(m_1, m_2) \mapsto f_1(m_1) = f_2(m_2)$, and observe that

$$f^! G \to f_1^! G \times f_2^! G, \quad ((m_1, m_2), g) \mapsto ((m_1, g), (m_2, g))$$

represents $f^! G$ as an embedded submanifold of $f_1^! G \times f_2^! G$. Let F denote the map $F_1^! \times F_2^! \colon G_1 \times G_2 \to f_1^! G \times f_2^! G$. Then $\overline{G}$ is precisely $F^{-1}(f^! G)$ and since F is a surjective submersion, it follows that $\overline{G}$ is an embedded submanifold of $G_1 \times G_2$, and that the restriction

$$\overline{G} \to f^! G, \quad (g_1, g_2) \mapsto ((\alpha_1 g_1, \alpha_2 g_2), F_1(g_1))$$

is still a surjective submersion. So it suffices to prove that the projection $f^! G \to \overline{M}$ is a surjective submersion. Since it is the left–hand side of the pullback diagram

$$
\begin{array}{ccc}
f^! G & \longrightarrow & G \\
\downarrow & & \downarrow{\scriptstyle \alpha} \\
\overline{M} & \longrightarrow & M
\end{array}
$$

and α is a surjective submersion, it follows that $f^! G \to \overline{M}$ is also. □

* \quad * \quad * \quad * \quad *

If $F\colon G \to G'$, $f\colon M \to M'$, is a fibration of Lie groupoids, corresponding to a normal subgroupoid system $\mathscr{N} = (N, R(f), \theta)$ for which N consists entirely of identity elements, then $G \cdot^{\cdot} N = G$ and the diffeomorphism between $G \cdot^{\cdot} N$ and $f^!G'$ can be identified with $F^!$. Thus F is an action morphism corresponding to an action of G' on f.

We now consider the other extreme, in which both $R(f)$ and θ are determined by N. This case includes quotients which preserve the base, and in general behaves in a more familiar way.

Definition 2.4.15 Let $G \rightrightarrows M$ be a Lie groupoid.

(i) A normal subgroupoid system $\mathscr{N} = (N, R, \theta)$ on G is *uniform* if the anchor of G, restricted to $N \to R$, is a surjective submersion.

(ii) A congruence (S, R) on G is *uniform* if the map from S to

$$(R \times R) \times_{M \times M} G = \{(n_2, n_1, m_2, m_1, g) \in R \times R \times G \mid n_1 = \beta g,\ m_1 = \alpha g\}$$

given by $(h, g) \mapsto (\beta h, \beta g, \alpha h, \alpha g, g)$, is a surjective submersion.

(iii) A morphism of Lie groupoids $F\colon G \to G', f\colon M \to M'$, is a *uniform fibration* if both f and $F^{\Updownarrow}\colon G \to f^{\Updownarrow}G'$ are surjective submersions. (Here $f^{\Updownarrow}G$ is the pullback groupoid of §2.3.) □

If $\mathscr{N}$ is uniform, then θ is entirely determined by the fact that

$$\theta(\beta\nu, \alpha\nu)(Ng) = Ng\nu^{-1}$$

for $\nu \in N$ with $\alpha g = \alpha\nu$. Thus N determines θ. Since N also determines R, as the image of its anchor, both $\mathscr{N}$ and $G/\mathscr{N}$ are entirely determined by N.

$$
\begin{array}{ccc}
N & \longrightarrow & S \\
\chi \downarrow & & \downarrow \chi \times \chi \\
R & \longrightarrow & R \times R
\end{array}
$$

Fig. 2.6.

Theorem 2.4.16 *Let $\mathscr{N} = (N, R, \theta)$ be a normal subgroupoid system on a Lie groupoid G, with $F\colon G \to G' = G/\mathscr{N}$, $f\colon M \to M'$, the quotient fibration, and corresponding congruence (S, R).*

Then $\mathscr{N}$ *is uniform if and only if* (S, R) *is uniform, and this is so if and only if* (F, f) *is a uniform fibration.*

Proof Suppose that (F, f) is a uniform fibration. Then we apply 2.4.4 to

$$
\begin{array}{ccccc}
S = R(F) & \xrightarrow{\hspace{2cm}} & G & \xrightarrow{\hspace{1cm}F\hspace{1cm}} & G' \\
\downarrow{\chi \times \chi} & & \downarrow{\chi} & & \downarrow{\chi'} \\
R \times R = R(f \times f) & \xrightarrow{\hspace{2cm}} & M \times M & \xrightarrow[\hspace{1cm}f \times f\hspace{1cm}]{} & M' \times M'
\end{array}
$$

with a minor modification since the anchors χ and χ' of G and G' need not be surjective submersions. It follows that (S, R) is uniform.

Next, if (S, R) is uniform then the diagram in Figure 2.6 is a pullback, where $N \to S$ is $\nu \mapsto (\nu, 1_{\alpha\nu})$ and $R \to R \times R$ is $(n, m) \mapsto (n, m, m, m)$. It follows that $\chi : N \to R$ is a surjective submersion, so $\mathscr{N}$ is uniform.

The proof that uniformity of $\mathscr{N}$ implies uniformity of (F, f) can be carried out by using 2.4.4. $\qquad\square$

2.5 General semidirect products

We come now to consider actions of one groupoid on another in such a way that structure is preserved. This requires a more elaborate setting than that of §1.6.

Definition 2.5.1 Let $G \rightrightarrows M$ and $W \rightrightarrows M'$ be Lie groupoids and let $f : M' \to M$ be a surjective submersion such that $f \circ \alpha_W = f \circ \beta_W : W \to M$. Write $p = f \circ \alpha_W$. Then G *acts smoothly on the Lie groupoid* W *via* f if there is given a smooth map $(g, w) \mapsto gw$ from $G * W = \{(g, w) \in G \times W \mid \alpha g = pw\}$ to W such that:

- $p(gw) = \beta g$ for all $(g, w) \in G * W$;
- for each $g \in G$, the map $w \mapsto gw$, $p^{-1}(\alpha g) \to p^{-1}(\beta g)$, is an isomorphism of Lie groupoids;
- $h(gw) = (hg)w$ whenever both hg and gw are defined;
- $1_{pw}w = w$ for all $w \in W$.

$\qquad\square$

We will use α, β for the source and target on both G and W unless there is a real likelihood of confusion. Note that the condition $f \circ \alpha_W = f \circ \beta_W = p$ on f is the condition that (p, f) is a groupoid morphism from W to the base groupoid 1_M. In particular, f is constant on each transitivity component of W and if W is transitive, G must be a Lie group.

The requirement that f (or equivalently p) be a surjective submersion ensures that each $p^{-1}(m)$, $m \in M$, is a closed embedded Lie subgroupoid of W on base $f^{-1}(m)$.

Consider now an action $G * W \to W$ as in 2.5.1. There is an action of G on the map $f \colon M' \to M$ in the sense of §1.6 determined by

$$1'_{gm'} = g1'_{m'}.$$

The diffeomorphism $f^{-1}(\alpha g) \to f^{-1}(\beta g)$ induced by $g \in G$ is the base map corresponding to the isomorphism $p^{-1}(\alpha g) \to p^{-1}(\beta g)$ induced by g. In particular, the source and target projections $W \to M'$ are G–equivariant.

We now define a groupoid structure on $G * W$ with base M'. The source and target maps are

$$\alpha'(g, w) = \alpha(w), \qquad \beta'(g, w) = g\beta(w),$$

and the composition is

$$(g_2, w_2)(g_1, w_1) = (g_2 g_1, (g_1^{-1} w_2)w_1),$$

defined if $\alpha(w_2) = g_1\beta(w_1)$. The identity element corresponding to m' is $(1_{f(m')}, 1'_{m'})$ and the inverse of (g, w) is $(g^{-1}, g(w^{-1}))$. It is routine to verify that this is a Lie groupoid structure on $G * W$. We denote it by $G \sqsubset W$ and call it the *general semidirect product*.

The construction clearly includes the usual semidirect product arising from a Lie group action on another Lie group by Lie group automorphisms and also, when W is a base groupoid, the action groupoid of §1.6.

There are natural morphisms $\iota \colon W \to G \sqsubset W$, $w \mapsto (1_{pw}, w)$ and $\pi \colon G \sqsubset W \to G$, $(g, w) \mapsto g$. Here ι is a morphism over M' and π is a fibration over f. One expects π to split in some sense, but it is clear that π cannot have a true right–inverse in general. What does exist is a morphism

$$\sigma_0 \colon G \lessdot M' \to G \sqsubset W, \quad (g, m') \mapsto (g, 1'_{m'})$$

over M', which is right–inverse to both $G \mathbin{\square\!\!\!\!\square} W \to G \lessdot M'$, $(g,w) \mapsto$ $(g, \alpha(w))$ and $G \mathbin{\square\!\!\!\!\square} W \to G \lessdot M'$, $(g,w) \mapsto (g, \beta(w))$. The first of these is $\pi^!$ in the notation of 2.4.3; we temporarily denote the second by π'. It should be noted that, although both $G \mathbin{\square\!\!\!\!\square} W$ and $G \lessdot M'$ have groupoid structures, neither $\pi^!$ nor π' is usually a morphism. Nonetheless the image of σ_0 is a Lie subgroupoid of $G \mathbin{\square\!\!\!\!\square} W$ naturally isomorphic to $G \lessdot M'$ and is a complement to the normal subgroupoid $\iota(W)$ in the sense that each element of $G \mathbin{\square\!\!\!\!\square} W$ is uniquely a product $\mu\nu$ where $\mu \in \sigma_0(G \lessdot M')$ and $\nu \in \iota(W)$.

Definition 2.5.2 A *split fibration* of Lie groupoids consists of a fibration $F \colon G' \to G$, $f \colon M' \to M$, together with an action of G on f and a morphism $\sigma \colon G \lessdot M' \to G'$ which is right–inverse to $F^!$. □

Theorem 2.5.3 (i) *Let a Lie groupoid* $G \rightrightarrows M$ *act on a Lie groupoid* $W \rightrightarrows M'$ *via a map* $f \colon M' \to M$. *Then* $\pi \colon G \mathbin{\square\!\!\!\!\square} W \to G$, $(g,w) \mapsto g$, *together with the given action of* G *on* M' *and* $\sigma_0 \colon G \lessdot M' \to G \mathbin{\square\!\!\!\!\square} W$, $(g, m') \mapsto (g, 1'_{m'})$, *is a split fibration.*

(ii) *Let* $F \colon G' \to G$, $f \colon M' \to M$, *be a split fibration with kernel* W. *Then there is an action of* G *on* W, *namely*

$$gw = \sigma(g, \beta'(w))w\, \sigma(g, \alpha'(w))^{-1},$$

and a unique isomorphism $\Psi \colon G \mathbin{\square\!\!\!\!\square} W \to G'$ *such that* $\Psi \circ \pi^! = F^!$ *and* $\Psi \circ \sigma_0 = \sigma$, *namely* $(g,w) \mapsto \sigma(g, \beta'(w))w$.

These constructions are mutually inverse, up to natural isomorphism.

We omit the proof, which uses techniques familiar from the preceding sections.

PBG–groupoids

The following description of extensions of Lie groupoids illustrates several of the constructions of this chapter. This material will be used in Chapters 6 and 8.

Consider an extension of locally trivial Lie groupoids

$$\Lambda \succ\!\!\longrightarrow \Phi \overset{\pi}{\longrightarrow\!\!\!\!\gg} \Omega. \tag{5}$$

Choose $m \in M$ and let $Q(M, H, q)$ and $P(M, G, p)$ denote the vertex bundles of Φ and Ω. Write $N = \Lambda_m$. The corresponding extension of principal bundles is therefore

$$N \rightarrowtail Q(M,H) \longrightarrow\!\!\!\!\!\rightarrow P(M,G). \qquad (6)$$

In what follows we will identify Φ and Ω with the gauge groupoids of Q and P.

Now $Q(P,N,\pi)$ is itself a principal bundle. Form the gauge groupoid $\Upsilon = \frac{Q \times Q}{N}$. For clarity we denote elements of Υ by $\langle v_2, v_1 \rangle_N$, elements of Φ by $\langle v_2, v_1 \rangle_H$ and elements of Ω by $\langle u_2, u_1 \rangle_G$.

The group G does not generally act on H, much less on Q. It does, however, act on Υ by

$$\langle v_2, v_1 \rangle_N g = \langle v_2 h, v_1 h \rangle_N,$$

where h is any element of H with $\pi(h) = g$. This is an action of G on $\Upsilon \rightrightarrows P$ by Lie groupoid automorphisms over the given principal action $P \times G \to P$, and is an action in the sense of 2.5.1 via the constant map $M \to \{\cdot\}$. This action is free and it is easy to see that the quotient manifold Υ/G, with its quotient groupoid structure in the sense of §2.4, is isomorphic to Φ under the map induced by $\Upsilon \to \Phi$, $\langle v_2, v_1 \rangle_N \mapsto \langle v_2, v_1 \rangle_H$.

Furthermore, this map $\Upsilon \to \Phi$ is an action morphism over $p \colon P \to M$. The induced action of Φ on P is

$$\langle v_2, v_1 \rangle_H \, (u) = \pi(v_2)g^{-1}$$

where g is determined by $\pi(v_1) = ug$. Thus Υ can be constructed directly from (5) and $P = \Omega_m$. The action of G in terms of $\Upsilon = \Phi \ltimes P$ is the canonical action $(\zeta, u)g = (\zeta, ug)$.

This process can be reversed.

Definition 2.5.4 Let $P(M,G,p)$ be a principal bundle. A *PBG–groupoid* on $P(M,G)$ is a locally trivial Lie groupoid $\Upsilon \rightrightarrows P$ together with a right action of G on Υ by Lie groupoid automorphisms over the principal action $P \times G \to P$. $\qquad \square$

Notice that the action of G on Υ must necessarily be free.

Proposition 2.5.5 *Given a PBG–groupoid $\Upsilon \rightrightarrows P(M,G)$, the quotient manifold Υ/G exists.*

Proof Since the quotient is defined by a group action, it suffices to show that the graph

$$\Gamma' = \{(v, vg) \mid v \in \Upsilon,\ g \in G\}$$

is a closed submanifold of $\Upsilon \times \Upsilon$. Denote the anchor $\Upsilon \to P \times P$ by χ and write Γ for the graph of the diagonal action of G on $P \times P$. Then $\Gamma' \subseteq (\chi \times \chi)^{-1}(\Gamma)$. Since Γ is a closed submanifold of P^4 and $\chi \times \chi$ is a surjective submersion, it suffices to show that Γ' is a closed submanifold of $(\chi \times \chi)^{-1}(\Gamma)$.

Define $f \colon (\chi \times \chi)^{-1}(\Gamma) \to \mathscr{I}\Upsilon$, $(v, v') \mapsto v'(vg)^{-1}$, where g is determined by v and v'. Clearly, $f^{-1}(1_P) = \Gamma'$, so it suffices to show that f is a surjective submersion. Since Υ is locally trivial, the division map $\delta \colon \Upsilon \times_\alpha \Upsilon \to \Upsilon$ is a surjective submersion, and so its restriction to $\delta^{-1}(\mathscr{I}\Upsilon) \to \mathscr{I}\Upsilon$ is also. Now it is only necessary to incorporate the effect of translation by the group action. $\qquad \square$

Consider a PBG–groupoid $\Upsilon \rightrightarrows P(M, G, p)$. Denote the quotient manifold Υ/G by Φ, and elements of Φ by $\|v\|$ for $v \in \Upsilon$. Then φ has a groupoid structure on base M with source and target maps

$$\beta(\|v\|) = p(\beta(v)), \qquad \alpha(\|v\|) = p(\alpha(v))$$

and multiplication of $\|v_2\|$ and $\|v_1\|$ with $\alpha(\|v_2\|) = \beta(\|v_1\|)$ defined by

$$\|v_2\|\,\|v_1\| = \|v_2(v_1 g)\|$$

where $g \in G$ is determined by $\alpha(v_2) = \beta(v_1)g$. A modification of the argument in §1.3 shows that Φ is a locally trivial Lie groupoid and the canonical map $\natural \colon \Upsilon \to \Phi$, $v \mapsto \|v\|$, is a morphism. Furthermore, $\natural$ is an action morphism.

The short exact sequence $\mathscr{I}\Upsilon \rightarrowtail \Upsilon \twoheadrightarrow P \times P$ now quotients to an extension of Lie groupoids

$$\frac{\mathscr{I}\Upsilon}{G} \rightarrowtail \Phi \twoheadrightarrow \frac{P \times P}{G}.$$

Thus extensions (5) of locally trivial Lie groupoids can be studied in terms of the single Lie groupoid $\Upsilon \rightrightarrows P$ and the action of G upon it. This correspondence emphasizes the significance of the distinction between automorphisms of a principal bundle and automorphisms of its gauge groupoid.

2.6 Classes of morphisms

Many of the general algebraic constructions possible for groupoids may be characterized in terms of the properties of associated morphisms. This is a peculiarity of groupoids (and of Lie algebroids, and other related concepts), with no precursor in the theory of groups.

The definitions which follow have been given in earlier sections, but are collected here for the purpose of comparison. For simplicity, we consider only morphisms over base–maps which are surjective submersions.

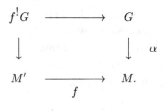

Fig. 2.7.

Consider a morphism of Lie groupoids $F\colon G' \to G$, $f\colon M' \to M$, and form the pullback manifold as in Figure 2.7. Denote by $F^!\colon G' \to f^!G$ the induced map $g' \mapsto (\alpha'g', F(g'))$.

Then:

- (F, f) is a *fibration* if $F^!$ is a surjective submersion;
- (F, f) is an *action morphism* if $F^!$ is a diffeomorphism.

Now form the pullback groupoid, as in §2.3,

$$
\begin{array}{ccc}
f^{\Downarrow}G & \longrightarrow & G \\
\downarrow & & \downarrow \chi \\
M' \times M' & \longrightarrow & M \times M.
\end{array}
$$
$$
f \times f
$$

There is an induced map $F^{\Downarrow}\colon G' \to f^{\Downarrow}G$, $g' \mapsto (\beta'g', F(g'), \alpha'g')$, and

- (F, f) is a *uniform fibration* if $F^{\Downarrow}$ is a surjective submersion,
- (F, f) is an *inductor* if $F^{\Downarrow}$ is a diffeomorphism.

These four classes of morphism may be regarded as giving four notions of surjectivity (or epimorphism) for groupoid morphisms. The fibrations are the largest class for which the kernel of 2.4.1 is a Lie subgroupoid of G'.

We saw in §1.6 that action morphisms correspond to, and characterize, actions of the target groupoid on the base map. In §2.4 of the present chapter we showed that fibrations correspond to, and characterize, quotients over general normal subgroupoid systems, and that uniform fibrations likewise correspond to quotients over a specific class of normal subgroupoid systems.

It is easy to see, in a similar way, that inductors correspond to, and characterize, pullback groupoids. They can be characterized amongst the fibrations as those uniform fibrations whose kernel may be identified with an equivalence relation on the base manifold; this equivalence relation is then the kernel pair $R(f)$, and the inductor is equivalent to the morphism $F^{\Downarrow}$.

2.7 Notes

§2.1. For the basic quotienting processes for manifolds, I follow [Serre, 1992, III§12]. For the basic facts about vector bundles see [Dieudonné, 1972] or [Greub et al., 1972].

    The descent process given in this section is as used in [Higgins and Mackenzie, 1990a, §4]. I do not know of a book account of descent for C^{∞} vector bundles but the material of this section is surely folklore.

§2.2. The results of this section on the base–preserving case are taken from [Mackenzie, 1987a, I§2]. For groupoids which are locally trivial, the smooth case is a simple analogue of base–preserving quotients for set groupoids, which were dealt with by Higgins [1971]. In [Mackenzie, 1987a], N was allowed to contain elements outside $\mathscr{I}G$; however, only the elements of $N \cap \mathscr{I}G$ were used in the definition or the quotienting process.

§2.3. Pullbacks of Lie groupoids seem to have first been considered by Pradines [1986b] and in [Mackenzie, 1987c]. For set groupoids, pullbacks always exist, of course. In that category they provide a concept dual to the universal morphisms of Higgins [1971].

    The need for the additional condition in 2.3.1 was pointed out by Pradines [1989].

§2.4. Most of this section has been taken directly from [Higgins and Mackenzie, 1990b]. The terminology *weak pullback* from [Higgins and Mackenzie, 1990b] has been replaced by *versal* as used by Pradines [1986a]. Further, I have replaced the uses of *regular* in [Higgins and Mackenzie, 1990b] by *uniform* (see below).

    For the notion of fibration of set groupoids see [Brown, 1970], where it is treated explicitly as an algebraic model of the path–lifting property for fibrations of topological spaces.

    Pradines [1986b] introduced a notion of *extenseurs réguliers* and gave a first isomorphism theorem for them. These results were then recovered in [Higgins and Mackenzie, 1990b, §4] as a special case of the treatment given here.

A *congruence* (2.4.5) on a Lie groupoid $G \rightrightarrows M$ is equivalent to a Lie double subgroupoid of the double Lie groupoid $(G \times G; G, M \times M; M)$; see [Higgins and Mackenzie, 1990b, p.105].

Interesting examples of fibrations arise in the theory of double Lie groupoids. In [Brown and Mackenzie, 1992], for example, a double version of local triviality is defined using a fibration condition, and it is shown that double groupoids for which this fibration is split can be described by an explicit construction in terms of crossed modules. In [Mackenzie, 1992, §3] it is noted that (under a weak extra condition) the maps from a double groupoid to its component groupoids are fibrations; this is useful in the description of affinoids (see also [Mackenzie, 2000a]).

Proposition 2.4.14 is from [Brown and Mackenzie, 1992].

§2.5. The general notion of groupoid action and the *general semidirect products* of this section were introduced for set groupoids by Brown [1970, 1972], following work by Frölich; the treatment given here follows [Higgins and Mackenzie, 1990a, §3]. General semidirect products have usually been denoted by the standard semidirect product symbol $\ltimes$; I have introduced $\sqsubset$ here to keep the distinctions clear.

A still more general notion of action of one groupoid on another is used by Xu [1992].

The notion of PBG–groupoid comes from [Mackenzie, 1988b] and [Mackenzie, 1987b]. This concept is essentially of double nature: a PBG–groupoid is a principal G–bundle object in the category of locally trivial Lie groupoids. The terminology is meant to suggest that Υ is not merely a G–groupoid (a groupoid with a group action from G) but a principal bundle with respect to that action.

§2.6. The general scheme followed in this section is from [Pradines, 1986b], but the account here (which broadly follows [Higgins and Mackenzie, 1990b]) varies many details. The *extenseurs réguliers* of [Pradines, 1986b] are here called *uniform fibrations*, since the meaning given to 'regular' in [Higgins and Mackenzie, 1990b] does not correspond to Pradines' 'régulier'. The *inducteurs surmersifs* of [Pradines, 1986b] are here called inductors. Note that in [Pradines, 1986b], the notation $f_0^* G$ refers to what is here denoted $f^{\sharp\!\sharp} G$.

There is an interesting analysis of general morphisms of set groupoids in [Higgins, 1971]; the results summarized in §2.6 may be regarded as the smooth version of (a part of) this.

Occasions arise on which it would be useful to have versions of the results of §2.4 and §2.5 without the assumption that the base–maps are surjective submersions. In that generality the framework which is used here is not available (because the pullback groupoids generally do not exist) and it is generally necessary to proceed on an *ad hoc* basis.

Notions of equivalence. An inductor $F \colon G' \to G$, $f \colon M' \to M$, with the further property that the map $\bar{\beta} \colon f^! G \to M$, $(g, m') \mapsto \beta g$, is a surjective submersion, is generally known as a *weak equivalence*. Inverting the weak equivalences gives the notion of Morita equivalence for Lie groupoids. See [Moerdijk and Mrčun, 2003, §5.4] for a clear recent account. Morita equivalence is fundamental in several areas where groupoids are applied: in C^*–algebra theory [Muhly et al., 1987], in aspects of symplectic and Poisson geometry [Xu, 1991a,b], [Landsman, 2002], and in foliation theory [Moerdijk and Mrčun, 2003].

3

Lie algebroids:
Fundamental theory

The definition of a Lie algebroid is given in §3.3. We spend the first
two sections of this chapter in constructing the Atiyah sequence of a
principal bundle. Atiyah sequences were the first and remain one of the
most important of the major classes of examples of Lie algebroids. The
account we give in §3.1 and §3.2 makes no use of groupoid methods or
of the general quotients of §2.1. A reader who is interested in principal bundles and their infinitesimal theory can therefore read most of
Chapters 3 and 4 independently of the rest of the book.

On the other hand, a reader who is mainly interested in the general
theory of Lie algebroids can omit §3.1 and §3.2. After the basic definitions in §3.3 we give in §3.4 a thorough account of the canonical Lie
algebroid $\mathfrak{D}(E)$ which plays for E the role played for a vector space by
its endomorphism Lie algebra. Sections of $\mathfrak{D}(E)$ are called *derivations
of E* or *derivative endomorphisms of ΓE*; in the past we have called
them covariant differential operators and denoted $\mathfrak{D}(E)$ by CDO(E).
Derivations on E may be identified with linear vector fields on E, and
§3.4 therefore includes a detailed account of the double structure associated with the tangent of a vector bundle; the material of §3.4 is needed
from Chapter 9 onwards, but not elsewhere in Chapters 3 or 4.

In §3.5 we construct the Lie algebroid of a Lie groupoid and calculate
the basic examples.

In §3.6 and §3.7 we give the Lie groupoid versions of the exponential map and adjoint formulas. The exponential map for a Lie groupoid $G \rightrightarrows M$ associates to a section X of the Lie algebroid AG, a
1–parameter family Exp tX of local bisections of G. This reflects the
fact that local left–translations, which constitute the flows of right invariant vector fields, are defined not by elements of G, as is the case
with a group, but by local bisections.

85

The exponential map for a Lie groupoid is used in much the same way, and for much the same purposes, as the exponential map of a Lie group. The first use, in 3.6.6, is to calculate the Lie algebroid $A\Phi(E)$ for a vector bundle E, in terms of differential operators on E. From this we calculate in 3.6.9 the Lie algebroid of a stabilizer subgroupoid. In 3.6.10 we use the exponential to differentiate the standard representations of the frame groupoid $\Phi(E)$ of a vector bundle E on the associated bundles such as $\mathrm{Hom}^n(E; M \times \mathbb{R})$ and $\mathrm{Hom}^n(E; E)$. From 3.6.9 and 3.6.10 it follows that, for a vector bundle E and a geometric structure on E defined by tensor fields, the Lie algebroid of the frame groupoid corresponds to differential operators with respect to which the tensor fields are constant (or parallel). This result encapsulates a number of calculations commonly regarded as part of connection theory.

3.1 Quotients of vector bundles by group actions

In §2.1 we gave a general descent process for vector bundles over groupoid actions. For the construction of the Atiyah sequence of a principal bundle, and elsewhere in Lie algebroid theory, we need the particular case of a quotient over a well–behaved group action and the alternative treatment which we give here is more suitable for this case.

Throughout this section $P(M, G, \pi)$ is a given principal bundle.

Proposition 3.1.1 *Let (E, p^E, P) be a vector bundle over P, on which G acts to the right*

$$E \times G \to E, \quad (\xi, g) \mapsto \xi g,$$

with the following two properties:

(i) *G acts on E by vector bundle isomorphisms; that is, each map $\xi \to \xi g, E \to E$ is a vector bundle isomorphism over the right translation $R_g \colon P \to P$;*

(ii) *E is covered by the ranges of equivariant charts, that is, around each $u_0 \in P$ there is a π–saturated open set $\mathscr{U} = \pi^{-1}(U)$, where $U \subseteq M$ is open, and a vector bundle chart*

$$\psi \colon \mathscr{U} \times V \to E_{\mathscr{U}}$$

for E, which is equivariant in the sense that for all $u \in \mathscr{U}, v \in V$ and $g \in G$,

$$\psi(ug, v) = \psi(u, v)g.$$

Then the orbit set E/G has a unique vector bundle structure over M such that the natural projection $\natural\colon E \to E/G$ is a surjective submersion, and a vector bundle morphism over $\pi\colon P \to M$. Further,

$$
\begin{array}{ccc}
E & \xrightarrow{\;\natural\;} & E/G \\
\scriptstyle p^E \downarrow & & \downarrow \scriptstyle p^{E/G} \\
P & \xrightarrow{\;\pi\;} & M
\end{array}
\tag{1}
$$

is a pullback.

We call $(E/G, p^{E/G}, M)$ the *quotient vector bundle* of (E, p^E, P) by the action of G.

Proof Denote the orbit of $\xi \in E$ by $\langle\xi\rangle$. Define $\overline{p} = p^{E/G}\colon E/G \to M$ by $\overline{p}(\langle\xi\rangle) = \pi(p^E(\xi))$; it is clear from (i) that $\overline{p}$ is well defined. We will give $(E/G, \overline{p}, M)$ the structure of a vector bundle by constructing local charts for it, and this simultaneously gives the manifold structure for E/G.

Firstly, make each $E/G|_x = \overline{p}^{-1}(x), x \in M$, into a vector space: if $\langle\xi\rangle, \langle\eta\rangle \in E/G|_x$ then $p^E(\xi), p^E(\eta)$ lie in the same fibre of P so there exists a unique $g \in G$ such that $p^E(\eta) = p^E(\xi)g$. Define

$$\langle\xi\rangle + \langle\eta\rangle = \langle\xi g + \eta\rangle$$

and, for $t \in \mathbb{R}$,

$$t\langle\xi\rangle = \langle t\xi\rangle.$$

It is easily verified that these operations are well defined and make $E/G|_x$ a vector space. The restriction $\natural_u$ of $\natural$ to $E_u \to E/G|_x$, where $x = \pi(u)$, is clearly linear; it is in fact an isomorphism. For if $\xi, \eta \in E_u$ and $\langle\xi\rangle = \langle\eta\rangle$ then there exists $g \in G$ such that $\eta = \xi g$ and, by (i), it follows that $u = ug$. So $g = 1$ and $\eta = \xi$.

Given $x_0 \in M$, choose $u_0 \in \pi^{-1}(x_0)$ and let $\psi_i\colon \mathscr{U}_i \times V \to E_{\mathscr{U}_i}$ be an equivariant chart for E defined around u_0. Assume, by shrinking $\mathscr{U}_i$ if necessary, that $\mathscr{U}_i$ is the range of a chart $U_i \times G \to \mathscr{U}_i$ for $P(M, G)$, so that there is a section $\sigma_i\colon U_i \to \mathscr{U}_i \subseteq P$.

Define

$$\psi_i^{/G}\colon U_i \times V \to E/G|_{U_i}, \qquad (x, v) \mapsto \langle\psi_i(\sigma_i(x), v)\rangle.$$

Then $\psi_{i,x}^{/G}\colon V \to E/G|_x$ is the composite

$$V \xrightarrow{\;\psi_{i,\sigma_i(x)}\;} E_{\sigma_i(x)} \xrightarrow{\;\natural_{\sigma_i(x)}\;} E/G|_x$$

and is thus an isomorphism. If $\psi_j^{/G} \colon U_j \times V \to E/G|_{U_j}$ is another chart constructed in the same way from an equivariant chart ψ_j for E and a section σ_j of P, then it is easily seen that

$$(\psi_{i,x}^{/G})^{-1} \circ (\psi_{j,x}^{/G}) = \psi_{i,u}^{-1} \circ \psi_{j,u} \text{ where } u = \sigma_j(x).$$

Thus the charts $\{\psi_i^{/G}\}$ define smooth transition functions

$$x \mapsto (\psi_{i,x}^{/G})^{-1} \circ (\psi_{j,x}^{/G})$$

and so there is a unique manifold structure on E/G with respect to which $(E/G, \bar{p}, M)$ is a vector bundle with the $\psi_i^{/G}$ as charts.

We next prove that $\natural \colon E \to E/G$ is a surjective submersion. Let $\psi \colon \mathscr{U} \times V \to E_{\mathscr{U}}$ be an equivariant chart for E, let $\bar{\sigma} \colon U \to \mathscr{U}$ be a section of P, and let $\psi^{/G} \colon U \times V \to E/G|_U$ be the chart for E/G constructed as above. Then

$$
\begin{array}{ccc}
E_{\mathscr{U}} & \xleftarrow{\quad \psi \quad} & \mathscr{U} \times V \\
{\scriptstyle \natural_{\mathscr{U}}}\big\downarrow & & \big\downarrow{\scriptstyle \pi \times \mathrm{id}_V} \\
E/G|_U & \xleftarrow{\quad \psi^{/G} \quad} & U \times V
\end{array}
$$

commutes: for if $(u, v) \in \mathscr{U} \times V$, then

$$\psi^{/G}((\pi \times \mathrm{id}_V)(u, v)) = \langle \psi(\sigma(\pi(u)), v) \rangle = \langle \psi(ug, v) \rangle,$$

where $g \in G$ is such that $ug = \sigma(\pi(u))$, and this is then equal to $\langle \psi(u, v)g \rangle = \langle \psi(u, v) \rangle = \natural(\psi(u, v))$. Since $\pi \times \mathrm{id}_V$ is smooth, and a submersion, it follows that $\natural$ is smooth, and a submersion. It is clear that $\bar{p} \circ \natural = \pi \circ p^E$, so $\natural$ is a vector bundle morphism. Since $\natural_u$ is an isomorphism for each $u \in P$ it follows that $(\natural, \pi)$ is a pullback (see the remarks at the start of §2.1).

The uniqueness assertion follows from the facts that there is at most one manifold structure on the range of a surjection which makes it a submersion and at most one vector space structure on the range of a surjection which makes it linear. $\square$

Proposition 3.1.2 (i) *If (E, p^E, P) is a vector bundle over $P(M, G)$ together with an action of G on E which satisfies the conditions of 3.1.1, and $(E', p^{E'}, M')$ is any vector bundle, then given a vector bundle morphism $\varphi \colon E \to E'$ over a map $f \colon P \to M'$ such that $\varphi(\xi g) = \varphi(\xi)$ for all $\xi \in E$, $g \in G$, and $f(ug) = f(u)$, for all $u \in P, g \in G$, there is* *a unique vector bundle morphism*

$$
\begin{array}{ccc}
E/G & \xrightarrow{\ \varphi^{/G}\ } & E' \\
{\scriptstyle p^{E/G}}\big\downarrow & & \big\downarrow{\scriptstyle p^{E'}} \\
M & \xrightarrow{\ f^{/G}\ } & M'
\end{array}
$$

such that $\varphi = \varphi^{/G} \circ \natural$ *and* $f = f^{/G} \circ \pi$.

(ii) *Consider vector bundles* (E, p^E, P) *and* $(E', p^{E'}, P')$ *over principal bundles* $P(M, G)$ *and* $P'(M', G')$, *respectively, together with actions of* G *and* G' *on* E *and* E', *respectively, which satisfy the conditions of* 3.1.1. *If* $\Phi \colon E \to E'$ *is a vector bundle morphism over a principal bundle morphism* $F(f, \varphi)$ *from* $P(M, G)$ *to* $P'(M', G')$ *which is equivariant in the sense that* $\Phi(\xi g) = \Phi(\xi)\varphi(g)$ *for all* $\xi \in E, g \in G$, *then there is a unique morphism of vector bundles*

$$
\begin{array}{ccc}
E/G & \xrightarrow{\ \varphi^{/G}\ } & E'/G' \\
\big\downarrow{\scriptstyle p^{E/G}} & & \big\downarrow{\scriptstyle p^{E'/G'}} \\
M & \xrightarrow{\ \ f\ \ } & M'
\end{array}
$$

such that $\Phi^{/G} \circ \natural = \natural' \circ \Phi$.

Proof We prove (ii) only; (i) is a special case of (ii).

Define $\Phi^{/G} \colon E/G \to E'/G'$ by $\langle \xi \rangle \mapsto \langle \Phi(\xi) \rangle$. That $\Phi^{/G}$ is well defined and fibrewise linear is clear. Clearly $\Phi^{/G} \circ \natural = \natural' \circ \Phi$ so, since $\natural' \circ \Phi$ is smooth, and $\natural$ is a surjective submersion, $\Phi^{/G}$ is smooth. That $p^{E'/G'} \circ \Phi^{/G} = f \circ p^{E/G}$ is immediate, and since $\natural$ is onto, the condition $\Phi^{/G} \circ \natural = \natural' \circ \Phi$ determines $\Phi^{/G}$ uniquely. $\qquad\square$

Remark 3.1.3 (i) This quotienting process includes the construction of an associated vector bundle $\frac{P \times V}{G}$ for a representation $\rho \colon G \to GL(V)$ of G on a vector space V. Namely, let G act on the product bundle $E = P \times V$ over P by $(u, v)g = (ug, \rho(g^{-1})v)$; this action clearly satisfies (i) of 3.1.1. If $\sigma \colon U \to P$ is a section of P and $\psi \colon U \times G \to \mathscr{U} = \pi^{-1}(U)$ is the associated chart $\psi(x, g) = \sigma(x)g$, then it is easy to verify that the chart $\mathscr{U} \times V \to E_{\mathscr{U}}$ defined by

$$
\begin{array}{ccc}
\mathscr{U} \times V & \xrightarrow{\hspace{3cm}} & E_{\mathscr{U}} = \mathscr{U} \times V \\
{\scriptstyle \psi \times \mathrm{id}_V}\big\uparrow & & \big\uparrow{\scriptstyle \psi \times \mathrm{id}_V} \\
U \times G \times V & \xrightarrow{\ (x,g,v) \mapsto (x,g,\rho(g^{-1})v)\ } & U \times G \times V
\end{array}
$$

is equivariant. (Although $E \to P$ is a product bundle, it does not admit global equivariant charts in general.)

Clearly the quotient vector bundle $E/G \to M$ coincides with the associated vector bundle $\frac{P \times V}{G} \to M$.

(ii) The construction 3.1.1 and the universality property 3.1.2 can easily be extended to equivariant actions of G on general fibre bundles over P. If this is done, then the construction includes all associated fibre bundles $\frac{P \times F}{G} \to M$.

Proposition 3.1.4 *Let (E, p, P) be a vector bundle over P together with an action of G on E which satisfies the conditions of 3.1.1. Denote by $\Gamma^G E$ the set of (global) sections X of E which are invariant in the sense that for all $u \in P, g \in G$,*

$$X(ug) = X(u)g.$$

Then $\Gamma^G E$ is a $C^\infty(M)$–module where $fX = (f \circ \pi)X$, for all $f \in C^\infty(M), X \in \Gamma^G E$, and the map

$$X \mapsto \overline{X}, \qquad \Gamma(E/G) \to \Gamma^G E, \qquad \overline{X}(u) = (\natural_u)^{-1}(X(\pi u))$$

is an isomorphism of $C^\infty(M)$–modules with inverse

$$X \mapsto \underline{X}, \qquad \Gamma^G E \to \Gamma(E/G)$$

where $\underline{X}(x) = \natural_u(X(u)) = \langle X(u) \rangle$ for any $u \in \pi^{-1}(x)$.

Proof For $X \in \Gamma(E/G)$ the pullback section across (1) is precisely $\overline{X}$; it follows that $X \mapsto \overline{X}$ is a $C^\infty(M)$–morphism $\Gamma(E/G) \to \Gamma E$. It is easily checked that, in fact, $\overline{X} \in \Gamma^G E$. Given $X \in \Gamma^G E$ it is clear that $\underline{X}$ is well defined, and since $\underline{X} \circ \pi = \natural \circ X$ and π is a surjective submersion, $\underline{X}$ is smooth. It is straightforward to check that $X \mapsto \overline{X}$ and $X \mapsto \underline{X}$ are mutual inverses. $\qquad \square$

This result can of course be localized: if $\mathcal{U} = \pi^{-1}(U)$ is a saturated open subset of P, where $U \subseteq M$ is open, then the same formulas for $\overline{X}$ and $\underline{X}$ define mutually inverse $C^\infty(U)$–isomorphisms between $\Gamma_U(E/G)$ and $\Gamma^G_{\mathcal{U}}(E)$.

3.2 The Atiyah sequence of a principal bundle

Throughout this section $P(M, G, \pi)$ is a principal bundle.

Proposition 3.2.1 (i) *The action of G on the tangent bundle $TP \to P$ induced by the action of G on P, namely*

$$X_g = T_u(R_g)(X), \quad X \in T_u(P),$$

satisfies the conditions of 3.1.1.

(ii) *The action of G on $TP \to P$ restricts to an action of G on the vertical subbundle $T^\pi P \to P$, and this action also satisfies the conditions of* 3.1.1.

Proof (i) It is clear that G acts on TP by vector bundle isomorphisms. To construct equivariant charts for TP, let $\varphi \colon U \times G \to \mathscr{U} = \pi^{-1}(U)$ be a chart for $P(M, G)$ in which U is the range of a chart $\theta \colon \mathbb{R}^n \to U$ for the manifold M. Now $T(U) \cong U \times \mathbb{R}^n$ and $T(G) \cong G \times \mathfrak{g}$, so $T(\varphi) \colon T(U) \times T(G) \to T_{\mathscr{U}}(P)$ can be regarded as a map $(U \times G) \times (\mathbb{R}^n \times \mathfrak{g}) \to T_{\mathscr{U}}(P)$ and, identifying $U \times G$ with $\mathscr{U}$ by φ, this gives the required equivariant chart.

IN precise terms, define

$$(U \times G) \times (\mathbb{R}^n \times \mathfrak{g}) \to T_{\mathscr{U}}(P)$$

$$(x, g, \mathbf{t}, X) \mapsto T_{(x,g)}(\varphi)(T_{\theta^{-1}(x)}(\theta)(\mathbf{t}), T_1(R_g)(X))$$

and define $\psi \colon \mathscr{U} \times V \to T_{\mathscr{U}}(P)$, where $V = \mathbb{R}^n \times \mathfrak{g}$, as the composition of $(\varphi^{-1} \times \mathrm{id}_V)$ and this map. To show that $\psi(uh, v) = \psi(u, v)h$, it suffices to show that

$$T_{(x,gh)}(\varphi)(Y, T_1(R_{gh})(X)) = T_{\varphi(x,g)}(R_h)(T_{(x,g)}(\varphi)(Y, T_1(R_g)(X)))$$

where $Y \in T_x(U)$, and this is the derivative of the identity $\varphi(x, R_h(g)) = R_h(\varphi(x, g))$.

(ii) That $T(R_g) \colon TP \to TP$ sends $T^\pi P$ to $T^\pi P$ follows from $\pi \circ R_g = \pi$. In the notation above, an equivariant chart for $T^\pi P$ over $\mathscr{U} \cong U \times G$ is the composite of $\varphi^{-1} \times \mathrm{id}_{\mathfrak{g}}$ with

$$(U \times G) \times \mathfrak{g} \to T^\pi_{\mathscr{U}}(P)$$
$$(x, g, X) \mapsto T_{(x,g)}(\varphi)(0_x, T_1(R_g)(X)).$$

$\square$

Note that in both cases the identification of $T(G)$ with $G \times \mathfrak{g}$ must be made using right translations.

The inclusion map $T^\pi P \overset{\subseteq}{\to} TP$ is manifestly equivariant so it induces, by 3.1.2(ii), a morphism $\frac{T^\pi P}{G} \to \frac{TP}{G}$ of vector bundles over M, which is clearly injective, and which we also regard as an inclusion.

On the other hand, from $\pi \circ R_g = \pi$ it follows that $T_{ug}(\pi) \circ T_u(R_g) = T_u(\pi)$, where $u \in P, g \in G$, so by 3.1.2(i) it follows that the vector bundle morphism

$$
\begin{array}{ccc}
TP & \xrightarrow{\ T(\pi)\ } & TM \\
\downarrow & & \downarrow \\
P & \xrightarrow{\ \pi\ } & M
\end{array}
$$

quotients to a map $\pi_* = T(\pi)^{/G} \colon \frac{TP}{G} \to TM$ which is a vector bundle morphism over M. It is clear that π_*, like $T(\pi)$, is fibrewise surjective and therefore a surjective submersion. Alternatively, it is easy to see that π_* is given locally by

$$
\begin{array}{ccc}
\frac{TP}{G}\big|_U & \xrightarrow{\qquad \pi_* \qquad} & T_U(M) \\
{\scriptstyle \psi^{/G}}\uparrow & & \uparrow{\scriptstyle T(\theta)} \\
U \times (\mathbb{R}^n \times \mathfrak{g}) & \xrightarrow{(x,\mathbf{t},X) \mapsto (x,\mathbf{t})} & U \times \mathbb{R}^n
\end{array}
$$

where the notation is that of 3.2.1.

The kernel of $\pi_* \colon \frac{TP}{G} \to TM$ is clearly $\frac{T^\pi P}{G}$, since $T^\pi P$ is the kernel of $T(\pi)$, and so we have proved that $\frac{T^\pi P}{G} \overset{\subseteq}{\rightarrowtail} \frac{TP}{G} \overset{\pi_*}{\twoheadrightarrow} TM$ is an exact sequence of vector bundles over M.

This may be regarded as the Atiyah sequence of $P(M,G)$ but in practice it is generally easier to work with a slight reformulation in which $\frac{T^\pi P}{G}$ is replaced by the bundle $\frac{P \times \mathfrak{g}}{G} \to M$ associated to $P(M,G)$ by the adjoint action of G on $\mathfrak{g}$.

Proposition 3.2.2 *The map* $j \colon \frac{P \times \mathfrak{g}}{G} \longrightarrow \frac{TP}{G}$ *induced by*

$$
P \times \mathfrak{g} \mapsto TP, \quad (u, X) \mapsto T_1(m_u)(X),
$$

where $m_u \colon G \to P$, $g \mapsto ug$, *is a vector bundle isomorphism over* M *of* $\frac{P \times \mathfrak{g}}{G}$ *onto* $\frac{T^\pi P}{G} \subseteq \frac{TP}{G}$.

Proof We regard $\frac{P \times \mathfrak{g}}{G} \to M$ as the quotient of the product bundle $P \times \mathfrak{g} \to P$ over the action $(u, X)g = (ug, \mathrm{Ad}_{g^{-1}} X)$ as in 3.1.3(i). That the map $P \times \mathfrak{g} \to TP$ is smooth can be seen by reformulating $T_1(m_u)(X)$ as $T_{(u,1)}(m)(0_u, X_1)$, where $m \colon P \times G \to P$ is the action.

Thus $P \times \mathfrak{g} \to TP$ is the composite

$$P \times \mathfrak{g} \to TP \times TG \xrightarrow{\ T(m)\ } TP$$

where $P \to TP$ is the zero section and $\mathfrak{g} = T_1(G) \to T(G)$ the inclusion. It is clearly a vector bundle morphism over P.

Now $T_1(m_{ug})(\mathrm{Ad}_{g^{-1}}X) = T_1(m_{ug} \circ I_g^{-1})(X)$ and $m_{ug} \circ I_g^{-1} = R_g \circ m_u$. Thus $(ug, \mathrm{Ad}_{g^{-1}}X)$ is mapped to $T_u(R_g)(T_1(m_u)(X))$, which proves that $P \times \mathfrak{g} \to TP$ is G–equivariant and so quotients, by 3.1.2(ii), to a vector bundle morphism over M

$$\frac{P \times \mathfrak{g}}{G} \to \frac{TP}{G}, \qquad \langle u, X \rangle \mapsto \langle T_1(m_u)(X) \rangle$$

which we denote by j.

That $T_1(m_u)(X) \in T^\pi P$ for all $u \in P$, $X \in \mathfrak{g}$ follows from the fact that $\pi \circ m_u$ is constant. On the other hand, $P \times \mathfrak{g} \to T^\pi P \subseteq TP$ is clearly injective (because each m_u is so) and since $T^\pi P$ and $P \times \mathfrak{g}$ have the same rank, namely $\dim \mathfrak{g}$, it follows that $P \times \mathfrak{g} \to T^\pi P$ is a fibrewise isomorphism. Clearly j inherits this property. $\qquad\square$

Summarizing, we have proved

Proposition 3.2.3 $\frac{P \times \mathfrak{g}}{G} \overset{j}{\rightarrowtail} \frac{TP}{G} \overset{\pi_*}{\twoheadrightarrow} TM$ *is an exact sequence of vector bundles over* M.

The bundle $\frac{P \times \mathfrak{g}}{G} \to M$ is the *adjoint bundle* of $P(M, G)$.

Remark 3.2.4 Alternatively, one can check that

$$T^\pi P \overset{\subseteq}{\rightarrowtail} TP \overset{\pi^!}{\twoheadrightarrow} \pi^! TM$$

is an exact sequence of vector bundles over P, where $\pi^! TM$ is the pullback and $\pi^!$ is the map $X_u \to (u, T_u(\pi)(X_u))$. Then $\pi^! TM$ admits a natural G–action, satisfying the conditions of 3.1.1, and $\pi^!$ is equivariant and $\frac{\pi^! TM}{G} \cong TM$. Lastly, the map $\frac{TP}{G} \to TM$ induced by $\pi^!$ is π_*.

We proceed now to define a bracket of Lie algebra type on $\Gamma(\frac{TP}{G})$; with this additional structure the exact sequence of 3.2.3 will be the Atiyah sequence of $P(M, G)$.

From 3.1.4 we know that $\Gamma(\frac{TP}{G})$ is isomorphic as a $C^\infty(M)$–module to $\Gamma^G TP$. Now $X \in \mathcal{X}(P)$ is in $\Gamma^G TP$ precisely if X is R_g–related to itself for all $g \in G$. It therefore follows that $\Gamma^G TP$ is closed under the bracket of vector fields and so we can define a bracket on $\Gamma(\frac{TP}{G})$ by

$$\overline{[X,Y]} = [\overline{X}, \overline{Y}].$$

This bracket on $\Gamma(\frac{TP}{G})$ inherits the Jacobi identity from the bracket on $\mathscr{X}(P)$, and also the property of being alternating. For $f \in C^\infty(M)$,

$$\overline{[X, fY]} = [\overline{X}, (f \circ \pi)\overline{Y}] = (f \circ \pi)[\overline{X}, \overline{Y}] + \overline{X}(f \circ \pi)\overline{Y}$$

Recall that a vector field $\mathscr{X}$ on P is called π–projectable if there is a vector field $\mathscr{Y}$ on M such that $\mathscr{X}$ is π–related to $\mathscr{Y}$; that is, such that $T_u(\pi)(\mathscr{X}(u)) = \mathscr{Y}(\pi(u))$ for all $u \in P$, or, equivalently, such that $\mathscr{X}(f \circ \pi) = \mathscr{Y}(f) \circ \pi$ for all $f \in C^\infty(M)$. It is clear from the definition of π_* that $\overline{X} \in \Gamma^G TP$ is π–related to $\pi_*(X) \in \mathscr{X}(M)$ and so

$$\overline{X}(f \circ \pi) = \pi_*(X)(f) \circ \pi.$$

We therefore have,

$$\overline{[X, fY]} = (f \circ \pi)\overline{[X,Y]} + (\pi_*(X)(f) \circ \pi)\overline{Y}$$

and so, for $X, Y \in \Gamma(\frac{TP}{G})$, $f \in C^\infty(M)$,

$$[X, fY] = f[X,Y] + \pi_*(X)(f)Y. \tag{2}$$

A bracket on the module of global sections of any vector bundle A over M, which has the property (2) with respect to a morphism $\pi_* : A \to TM$, can be localized to sections over any open subset of the base (see 3.3.2). In the present case the resulting bracket $\Gamma_U(\frac{TP}{G}) \times \Gamma_U(\frac{TP}{G}) \to \Gamma_U(\frac{TP}{G})$ where $U \subseteq M$ is open, is easily seen to be equal to that obtained by transporting the bracket on $\Gamma^G_{\pi^{-1}(U)} TP$ to $\Gamma_U(\frac{TP}{G})$ via the $C^\infty(U)$–isomorphism $\Gamma_U(\frac{TP}{G}) \to \Gamma^G_{\pi^{-1}(U)} TP$.

From the fact that $\overline{X}$ is π–related to $\pi_*(X)$ it also follows that, for $X, Y \in \Gamma(\frac{TP}{G})$, the bracket $[\overline{X}, \overline{Y}]$ is π–related to $[\pi_*(X), \pi_*(Y)]$ so since $[\overline{X}, \overline{Y}] = \overline{[X,Y]}$ is also π–related to $\pi_*([X,Y])$, and π is onto, it follows that for all $X, Y \in \Gamma(\frac{TP}{G})$.

$$[\pi_*(X), \pi_*(Y)] = \pi_*([X,Y]). \tag{3}$$

From (3) it follows that the bracket on $\Gamma(\frac{TP}{G})$ restricts to a bracket on $\Gamma(\frac{P \times \mathfrak{g}}{G})$. For, given $V, W \in \Gamma(\frac{P \times \mathfrak{g}}{G})$, we have

$$\pi_*([jV, jW]) = [\pi_* jV, \pi_* jW] = 0$$

and so there exists a unique section $[V, W]$ of $\frac{P \times \mathfrak{g}}{G}$ such that for $V, W \in \Gamma(\frac{P \times \mathfrak{g}}{G})$,

$$[j(V), j(W)] = j[V, W].$$

This restricted bracket is of course also alternating and satisfies the Jacobi identity. And, for $f \in C^\infty(M)$,

$$j[V, fW] = f[j(V), j(W)] + \pi_*(j(V))(f)j(W)$$

so, since $\pi_* \circ j = 0$, it follows that the bracket on $\Gamma(\frac{P \times \mathfrak{g}}{G})$, unlike that on $\Gamma(\frac{TP}{G})$, is actually bilinear over $C^\infty(M)$ and so defines a tensor field, and restricts to each fibre. Since each fibre $\left.\frac{P \times \mathfrak{g}}{G}\right|_x$ is isomorphic to $\mathfrak{g}$, the question arises as to whether the bracket in $\left.\frac{P \times \mathfrak{g}}{G}\right|_x$ is induced by that in $\mathfrak{g}$.

Proposition 3.2.5 *For $V \in \Gamma(\frac{P \times \mathfrak{g}}{G})$, denote by $\widetilde{V} \in C_G^\infty(P, \mathfrak{g})$ the corresponding equivariant function $P \to \mathfrak{g}$ for which $V(\pi(u)) = \langle u, \widetilde{V}(u) \rangle$ for all $u \in P$. Then for all $V, W \in \Gamma(\frac{P \times \mathfrak{g}}{G})$ and $u \in P$,*

$$\widetilde{[V, W]}(u) = [\widetilde{V}(u), \widetilde{W}(u)]_R,$$

where the bracket on the RHS is the right–hand bracket in $\mathfrak{g}$.

Remark: This result may be expressed as follows: For $V, W \in \Gamma(\frac{P \times \mathfrak{g}}{G})$ and $x \in M$, the value at x of the Lie algebroid bracket $[V, W]$ is $[V(x), W(x)]_x$ where $[\ ,\ ]_x$ is the restriction of $[\ ,\ ]$ to $\left.\frac{P \times \mathfrak{g}}{G}\right|_x$. Proposition 3.2.5 now states that for $u \in \pi^{-1}(x)$ and $X, Y \in \mathfrak{g}$,

$$[\langle u, X \rangle, \langle u, Y \rangle]_x = \langle u, [X, Y]_R \rangle.$$

That this bracket is well defined follows from the fact that Ad_g is a Lie algebra automorphism for all $g \in G$.

Proof First note that $\overline{j(V)} \in \Gamma^G TP$ is $u \mapsto T_1(m_u)(\widetilde{V}(u))$. We prove that this vector field has a global flow, namely $\varphi_t(u) = u \exp t\widetilde{V}(u)$ for $t \in \mathbb{R}, u \in P$. First, we see at once that $\frac{d}{dt}\varphi_t(u)\big|_0 = T_1(m_u)(\widetilde{V}(u))$. For the group law $\varphi_t \circ \varphi_s = \varphi_{t+s}$ we have, using $\tilde{V}(ug) = \mathrm{Ad}_{g^{-1}}\tilde{V}(u)$,

$$\begin{aligned}
\varphi_t(\varphi_s(u)) &= u \exp s\tilde{V}(u) . \exp t\tilde{V}(u \exp s\tilde{V}(u)) \\
&= u \exp s\tilde{V}(u) . \exp t\mathrm{Ad}_{\exp s\tilde{V}(u)^{-1}}\tilde{V}(u) \\
&= u \exp t\tilde{V}(u) \exp s\tilde{V}(u) \\
&= \varphi_{t+s}(u).
\end{aligned}$$

So

$$\begin{aligned}
[\overline{jV}, \overline{jW}](u) &= -\frac{d}{dt}(T(\varphi_t)(\overline{jW}(\varphi_{-t}(u))))\Big|_0 \\
&= -\frac{d}{dt}(T_1(\varphi_t \circ m_{\varphi_{-t}(u)})(\widetilde{W}(\varphi_{-t}(u))))\Big|_0 .
\end{aligned}$$

Now by using the equivariance of $\tilde{V}$ in a similar manner to the proof of the group law, one sees that $\varphi_t \circ m_{\varphi_{-t}(u)} = m_u$ and so this last expression is actually

$$\begin{aligned}
-\frac{d}{dt}(T_1(m_u)(\widetilde{W}(\varphi_{-t}(u))))\Big|_0 &= T_1(m_u)\left(-\frac{d}{dt}\widetilde{W}(u \exp -t\tilde{V}(u))\Big|_0\right) \\
&= T_1(m_u)\left(-\frac{d}{dt}\mathrm{Ad}_{\exp t\tilde{V}(u)}\widetilde{W}(u)\Big|_0\right) \\
&= T_1(m_u)([\tilde{V}(u), \widetilde{W}(u)]_R).
\end{aligned}$$

We therefore have

$$\begin{aligned}
T_1(m_u)([\tilde{V}(u), \widetilde{W}(u)]_R) &= [\overline{jV}, \overline{jW}](u) \\
&= \overline{j([V, W])}(u) \\
&= T_1(m_u)(\widetilde{[V, W]}(u))
\end{aligned}$$

and the result follows. □

That the bracket on $\Gamma(\frac{P \times \mathfrak{g}}{G})$ corresponds to the right–hand bracket in $\mathfrak{g}$ is a consequence of the fact that $\Gamma(\frac{P \times \mathfrak{g}}{G})$ is embedded in $\mathfrak{X}(P)$ as the set of *right*–invariant vertical vector fields. Throughout the book we use only this right–hand bracket on the Lie algebra of a Lie group and so drop the subscript 'R'.

This completes the construction of the Atiyah sequence.

Definition 3.2.6 The exact sequence of vector bundles

$$\frac{P \times \mathfrak{g}}{G} \xrightarrow{\ j\ } \frac{TP}{G} \xrightarrow{\ \pi_*\ } TM,$$

together with the bracket structures on $\Gamma(\frac{TP}{G})$ and $\Gamma(\frac{P\times\mathfrak{g}}{G})$ just defined, is the *Atiyah sequence of* $P(M,G,\pi)$.   □

In the remainder of this section we use the various properties of the brackets, proved above, without comment. They are precisely the properties that demonstrate that the Atiyah sequence is a (transitive) Lie algebroid on M as defined in §3.3. We may refer to $\frac{TP}{G}$ itself, with its bracket structure and the map π_*, as the *Atiyah–Lie algebroid of* $P(M,G,\pi)$.

The following description of the flows of right–invariant vector fields is often needed.

Proposition 3.2.7 (i) *Given* $\overline{X} \in \Gamma^G TP$ *and* $u_0 \in P$ *there is a local flow* $\{\varphi_t\}$ *for* $\overline{X}$ *around* u_0 *defined on a* π–*saturated open set* $\mathcal{U} = \pi^{-1}(U)$, $U \subseteq M$ *open, for which* $\varphi_t(ug) = \varphi_t(u)g$ *for all* $u \in \mathcal{U}$, $g \in G$, *nd all* t.

(ii) *Given* $\overline{X} \in \Gamma^G TP$ *with local flow* $\{\varphi_t\}$ *as in* (i), *the vector field* $\pi_*(X)$ *on* M *has local flow* $\{\psi_t\}$ *on* U *determined by* $\varphi_t \circ \pi = \pi \circ \varphi_t$.

(iii) *For* $V \in \Gamma(\frac{P\times\mathfrak{g}}{G})$, *the vector field* $\overline{j(V)} \in \Gamma^G TP$ *is complete and has the global flow* $\varphi_t(u) = u \exp t\widetilde{V}(u)$.

Proof (i) Let $\{\varphi_t\}$ be a local flow for $\overline{X}$ defined on open $\mathcal{O} \subseteq P$ around u_0; write $U = \pi(\mathcal{O})$ and $\mathcal{U} = \pi^{-1}(U)$. It is easy to verify that, for any given $g \in G$, $\{R_g \circ \varphi_t \circ R_g^{-1}\}$ is a local flow for $\overline{X}$ on $\mathcal{O}g$. By the uniqueness of local flows it follows that $\{\varphi_t\}$ and $\{R_g \circ \varphi_t \circ R_g^{-1}\}$ must coincide on $\mathcal{O} \cap \mathcal{O}g$. We can thus extend φ_t smoothly to the whole of $\mathcal{U} = \cup_{g\in G}\mathcal{O}g$ and a repetition of the argument now shows that $R_g \circ \varphi_t \circ R_g^{-1} = \varphi_t$, for all $g \in G$ and t.

(ii) is straightforward and (iii) was proved in the course of 3.2.5.   □

The proof of (i) of course shows that any local flow for $\overline{X}$ which is defined on a π–saturated open set commutes with right–translations. We call such a flow a *saturated local flow*

Lastly, we describe the morphism of Atiyah sequences induced by a morphism of principal bundles $F(\mathrm{id},\varphi)\colon P(M,G) \to P'(M,G')$ over a fixed base M. It is easily checked that $TF\colon TP \to TP'$ satisfies the conditions of 3.1.2(ii) and so induces a morphism $F_* = (TF)^{/G}\colon \frac{TP}{G} \to \frac{TP'}{G'}$ of vector bundles over M. It is also straightforward to check that F_* commutes with π_* and therefore restricts to the adjoint bundles; we denote the restriction by F_*^+. One checks that $F_*^+\colon \frac{P\times\mathfrak{g}}{G} \to \frac{P'\times\mathfrak{g}}{G'}$, is

$$\frac{P \times \mathfrak{g}}{G} \; \overset{j}{>\!\!-\!\!\!-\!\!\!>} \; \frac{TP}{G} \; \overset{\pi_*}{-\!\!\!-\!\!\!\twoheadrightarrow} \; TM$$

$$\Big\downarrow F_*^+ \qquad\qquad \Big\downarrow F_* \qquad\qquad \Big\|$$

$$\frac{P' \times \mathfrak{g}'}{G'} \; \overset{j'}{>\!\!-\!\!\!-\!\!\!>} \; \frac{TP'}{G'} \; \overset{\pi'_*}{-\!\!\!-\!\!\!\twoheadrightarrow} \; TM.$$

Fig. 3.1.

$\langle u, X \rangle \mapsto \langle F(u), \varphi_*(X) \rangle$ and we have the commutative diagram shown in Figure 3.1.

Now $TF \colon TP \to TP'$ preserves the bracket of vector fields in the sense that if $X \in \mathcal{X}(P)$ and $X' \in \mathcal{X}(P')$ are F-related, and Y and Y' similarly, then $[X, Y]$ and $[X', Y']$ are F-related. For $X \in \Gamma(\frac{TP}{G})$ and $X' \in \Gamma(\frac{TP'}{G'})$ it is easily verified that $\overline{X}$ and $\overline{X}'$ are F-related if and only if $X' = F_*(X)$, and so it follows that for $X, Y \in \Gamma(\frac{TP}{G})$,

$$F_*([X, Y]) = [F_*(X), F_*(Y)].$$

Since j and j' are injective, the map $\Gamma(\frac{P \times \mathfrak{g}}{G}) \to \Gamma(\frac{P' \times \mathfrak{g}'}{G'})$ also preserves the brackets.

Definition 3.2.8 The map $F_* \colon \frac{TP}{G} \to \frac{TP'}{G'}$ is the *morphism of Atiyah sequences induced by* $F(\mathrm{id}, \varphi)$. $\square$

Note that $\pi(\mathrm{id}, k) \colon P(M, G) \to M(M, \{1\})$ is a morphism of principal bundles, where $k \colon G \to \{1\}$ is the constant morphism onto the trivial group. The induced morphism of Atiyah sequences is $\pi_* \colon \frac{TP}{G} \to TM$.

Example 3.2.9 Let H be a closed subgroup of a Lie group G and consider the homogeneous bundle $G(G/H, H)$. Its Atiyah sequence is

$$\frac{G \times \mathfrak{h}}{H} \; \overset{j}{>\!\!-\!\!\!-\!\!\!>} \; \frac{TG}{H} \; \overset{\pi_*}{-\!\!\!-\!\!\!\twoheadrightarrow} \; T(G/H) \qquad\qquad (4)$$

where $j(\langle g, X \rangle) = \langle T_1(L_g)X \rangle$ and $\pi_*(\langle X \rangle) = T(\pi)(X)$. There are two alternative formulations of this sequence.

Firstly, the vector bundle isomorphism $G \times \mathfrak{g} \to TG$, $(g, X) \mapsto T_1(L_g)(X)$, respects the right actions of H and so quotients to a vector bundle isomorphism $\mathscr{L} \colon \frac{G \times \mathfrak{g}}{H} \to \frac{TG}{H}$, where $\frac{G \times \mathfrak{g}}{H}$ is the bundle associated to $G(G/H, H)$ through the adjoint action of H on $\mathfrak{g}$. Likewise there is a vector bundle isomorphism $\mathscr{M} \colon \frac{G \times (\mathfrak{g}/\mathfrak{h})}{H} \to T(G/H)$ defined by $\langle g, X + \mathfrak{h} \rangle \mapsto T_1(\pi \circ L_g)(X)$, where H acts on the vector space $\mathfrak{g}/\mathfrak{h}$ by $h(X + \mathfrak{h}) = \mathrm{Ad}_h X + \mathfrak{h}$. We will show that $\mathcal{M}$ is injective; that it is well defined and a smooth and surjective vector bundle morphism are easily verified. Suppose $\langle g, X + \mathfrak{h} \rangle$ and $\langle g', X' + \mathfrak{h} \rangle$ have $T_1(\pi \circ L_g)(X) = T_1(\pi \circ L_{g'})(X')$. Then $\pi(g) = \pi(g')$ so there exists $h \in H$ such that $g' = gh$. Now $\pi = \pi \circ R_{h^{-1}}$ so we have

$$T_g(\pi)\{T_{gh}(R_{h^{-1}})\, T_1(L_{gh})(X') - T_1(L_g)(X)\} = 0.$$

Thus $T_1(L_g)\{\mathrm{Ad}_h X' - X\}$ is vertical. But L_g is precisely the map m_g (in the notation of 3.2.2) so $\mathrm{Ad}_h X' - X \in \mathfrak{h}$. This shows that

$$\langle g', X' + \mathfrak{h} \rangle = \langle g, \mathrm{Ad}_h X' + \mathfrak{h} \rangle = \langle g, X + \mathfrak{h} \rangle$$

as required.

Thus the sequence of vector bundles (4) can be written as

$$\frac{G \times \mathfrak{h}}{H} \overset{j_1}{>\!\!-\!\!\!\longrightarrow} \frac{G \times \mathfrak{g}}{H} \overset{a_1}{-\!\!\!\longrightarrow\!\!\!\gg} \frac{G \times (\mathfrak{g}/\mathfrak{h})}{H} \tag{5}$$

where $j_1(\langle g, X \rangle) = \langle g, X \rangle$ and $a_1(\langle g, X \rangle) = \langle g, X + \mathfrak{h} \rangle$.

Secondly, the map $G \times \mathfrak{g} \to (G/H) \times \mathfrak{g}$, $(g, X) \mapsto (gH, \mathrm{Ad}_g X)$, which is a vector bundle morphism over $\pi\colon G \to G/H$, respects the action of H on $G \times \mathfrak{g}$ and so induces a vector bundle morphism $\frac{G \times \mathfrak{g}}{H} \to (G/H) \times \mathfrak{g}$, which is easily seen to be an isomorphism. So (4) can also be written as

$$\frac{G \times \mathfrak{h}}{H} \overset{j_2}{>\!\!-\!\!\!\longrightarrow} (G/H) \times \mathfrak{g} \overset{a_2}{-\!\!\!\longrightarrow\!\!\!\gg} T(G/H) \tag{6}$$

where $j_2(\langle g, X \rangle) = (gH, \mathrm{Ad}_g X)$ and $a_2(gH, X) = T_1(\pi \circ R_{g^{-1}})(X)$.

Taking $H = \{1\}$ the composite isomorphism from $G \times \mathfrak{g}$ in (6) to TG in (4) is $\mathscr{R}\colon (g, X) \mapsto T_1(R_g)(X)$. If $k_X\colon G \to \mathfrak{g}$ temporarily denotes the map constant at X then $\mathscr{R} \circ k_X$ is $\overrightarrow{X}$, the right–invariant vector field corresponding to X. Since we are using the right bracket on $\mathfrak{g}$, the transported bracket on $\Gamma(G \times \mathfrak{g})$ has $[k_X, k_Y] = k_{[X,Y]}$. Now using standard properties of vector fields on G we have

$$[uk_X, vk_Y] = uvk_{[X,Y]} + u\overrightarrow{X}(v)k_Y - v\overrightarrow{Y}(u)k_X \tag{7}$$

for $u, v \in C^\infty(G)$. Extending by additivity gives the bracket on (6) for $H = \{1\}$ and the argument easily extends to general H. $\boxtimes$

3.3 Lie algebroids

This section introduces the concept of Lie algebroid, preparatory to the construction in §3.5 of the Lie algebroid of a Lie groupoid.

Definition 3.3.1 Let M be a manifold. A *Lie algebroid* on M is a vector bundle (A, q, M) together with a vector bundle map $a\colon A \to TM$ over M, called the *anchor* of A, and a bracket $[\,,\,]\colon \Gamma A \times \Gamma A \to \Gamma A$ which is $\mathbb{R}$–bilinear and alternating, satisfies the Jacobi identity, and is such that

$$[X, uY] = u[X, Y] + a(X)(u)Y, \tag{8}$$
$$a([X, Y]) = [a(X), a(Y)],, \tag{9}$$

for all $X, Y \in \Gamma A$, $u \in C^\infty(M)$. The Lie algebroid A is *transitive* if a is fibrewise surjective, *regular* if a is of locally constant rank, and *totally intransitive* if $a = 0$. The manifold M is the *base* of A.

Let A' be a second Lie algebroid, on the same base M. Then a *morphism of Lie algebroids* $\varphi\colon A \to A'$ over M, or a *base–preserving morphism of Lie algebroids*, is a vector bundle morphism such that $a' \circ \varphi = a$ and $\varphi[X, Y] = [\varphi(X), \varphi(Y)]$, for all $X, Y \in \Gamma A$. $\square$

The simplest examples of Lie algebroids are Lie algebras, Lie algebra bundles, and the tangent bundle to a manifold. We showed in §3.2 that the Atiyah sequence of a principal bundle is a transitive Lie algebroid.

The anchor of a Lie algebroid A encodes its geometric properties. If A is transitive then right inverses to the anchor are connections in A (see Chapter 5). If A is regular then the image of the anchor defines a foliation of the base manifold, the *characteristic foliation of A* and over each leaf of this foliation, the Lie algebroid is transitive.

We deal in Chapter 4 with the general notion of morphism for Lie algebroids.

Proposition 3.3.2 *Let A be a Lie algebroid on M, and let $U \subseteq M$ be an open subset. Then the bracket $\Gamma A \times \Gamma A \to \Gamma A$ 'restricts' to $\Gamma_U A \times \Gamma_U A \to \Gamma_U A$ and makes A_U a Lie algebroid on U, called the restriction of A to U.*

Proof It suffices to show that if $X, Y \in \Gamma A$ and Y vanishes on the open set $U \subseteq M$, then $[X, Y]$ vanishes on U. For $x_0 \in U$ take $u\colon M \to \mathbb{R}$ with $u(x_0) = 0$ and $u(M \backslash U) = \{1\}$; then $[X, Y](x_0) = [X, uY](x_0) = u(x_0)[X, Y](x_0) + a(X)(u)(x_0)Y(x_0) = 0$. $\square$

For the restriction of Lie algebroids to more general submanifolds of the base, see §4.3.

The following examples are basic.

Example 3.3.3 Let M be a manifold and let $\mathfrak{g}$ be a Lie algebra. On $TM \oplus (M \times \mathfrak{g})$ define an anchor $a = \pi_1 \colon TM \oplus (M \times \mathfrak{g}) \to TM$ and a bracket

$$[X \oplus V, \ Y \oplus W] = [X, Y] \oplus \{X(W) - Y(V) + [V, W]\}.$$

Then $TM \oplus (M \times \mathfrak{g})$ is a transitive Lie algebroid on M, called the *trivial Lie algebroid* on M with structure algebra $\mathfrak{g}$.

Let $\varphi \colon TM \oplus (M \times \mathfrak{g}) \to TM \oplus (M \times \mathfrak{g}')$ be a morphism of trivial Lie algebroids on M. Then the condition $a' \circ \varphi = a$ implies that φ has the form

$$\varphi(X \oplus V) = X \oplus (\omega(X) + \varphi^+(V)) \tag{10}$$

where $\omega \colon TM \to M \times \mathfrak{g}'$ is a $\mathfrak{g}'$–valued 1–form on M and $\varphi^+ \colon M \times \mathfrak{g} \to M \times \mathfrak{g}'$ is a vector bundle morphism. Writing out the equation

$$[\varphi(X \oplus V), \ \varphi(Y \oplus W)] = \varphi[X \oplus V, \ Y \oplus W]$$

and setting firstly $X = Y = 0$, then $V = W = 0$, and lastly $V = 0$, we obtain, successively

$$[\varphi^+(V), \ \varphi^+(W)] = \varphi^+([V, W]), \tag{11}$$

$$\delta\omega(X, Y) + [\omega(X), \omega(Y)] = 0, \tag{12}$$

$$X(\varphi^+(W)) - \varphi^+(X(W)) + [\omega(X), \varphi^+(W)] = 0. \tag{13}$$

In the terminology of 3.3.8 below, (11) is the condition that φ^+ be an LAB morphism, and (12) is the condition that ω be a Maurer–Cartan form. We call (13) the *compatibility condition*.

It is easy to see that, conversely, a Maurer–Cartan form $\omega \in \Omega^1(M, \mathfrak{g}')$ and an LAB morphism $\varphi^+ \colon M \times \mathfrak{g} \to M \times \mathfrak{g}'$ which satisfy (13), define by (10) a morphism of Lie algebroids $TM \oplus (M \times \mathfrak{g}) \to TM \oplus (M \times \mathfrak{g}')$.

This decomposition does not correspond exactly to the decomposition in Example 1.2.5 of morphisms of trivial groupoids. Given a morphism $F \colon M \times G \times M \to M \times G' \times M$ over M, if (unlike 1.2.5) we define $\widetilde{F} \colon M \times G \to M \times G'$ by $(x, g) \mapsto (x, \pi_2 \circ \varphi(x, g, x))$ and $\theta \colon M \to G'$ by $\theta(x) = \pi_2 \circ \varphi(x, 1, m)$, where m is fixed, then θ and $\widetilde{F}$ must satisfy a compatibility condition similar to (13). See Example 3.5.13 and §9.7.

Contrariwise, it is possible to describe $\varphi \colon TM \oplus (M \times \mathfrak{g}) \to TM \oplus (M \times \mathfrak{g}')$ in terms of ω and a single morphism of Lie algebras $f = \varphi^+|_m \colon \mathfrak{g} \to \mathfrak{g}'$. Here, however, f and ω must still obey a compatibility condition; see §6.5. ⊠

Example 3.3.4 Let E be a vector bundle on M. A *zeroth–order differential operator on E* is a $C^\infty(M)$–linear endomorphism $\Gamma E \to \Gamma E$, and thus corresponds to a vector bundle endomorphism $E \to E$. A *first–order differential operator on E* is an $\mathbb{R}$–linear map $D\colon \Gamma E \to \Gamma E$ such that for each $u \in C^\infty(M)$ the map

$$\Gamma E \to \Gamma E, \qquad \mu \mapsto D(u\mu) - uD(\mu)$$

is a zeroth–order differential operator. (Strictly speaking we should say that D is a first– or zeroth–order differential operator.) The first–order differential operators are the sections of a vector bundle $\mathrm{Diff}^1(E)$ on M. Define a vector bundle morphism $\sigma\colon \mathrm{Diff}^1(E) \to \mathrm{Hom}(T^*M, \mathrm{End}(E))$ by

$$\sigma(D)(v\delta u)(\mu) = v(D(u\mu) - uD(\mu)),$$

and call $\sigma(D)$ the *symbol of D*. The symbol map σ is surjective and has as kernel the zeroth–order differential operators. There is thus a short exact sequence of vector bundles over M

$$\mathrm{End}(E) \overset{\subseteq}{\rightarrowtail} \mathrm{Diff}^1(E) \overset{\sigma}{\twoheadrightarrow} \mathrm{Hom}(T^*(M), \mathrm{End}(E)).$$

Map TM injectively into $\mathrm{Hom}(T^*(M), \mathrm{End}(E))$ by

$$X \mapsto (\omega \mapsto (\mu \mapsto \langle \omega, X \rangle \mu)) \qquad \omega \in \Gamma T^*(M), \mu \in \Gamma E$$

and construct the pullback vector bundle over M,

$$
\begin{array}{ccc}
\mathfrak{D}(E) & \longrightarrow & TM \\
\downarrow & & \downarrow \\
\mathrm{Diff}^1(E) & \underset{\sigma}{\longrightarrow} & \mathrm{Hom}(T^*(M), \mathrm{End}(E)).
\end{array}
$$

The inverse image exists because σ is a surjective submersion. Since the right–hand vertical arrow is an injective immersion, it follows as usual that the left–hand arrow is also and we can therefore regard $\mathfrak{D}(E)$ as a vector subbundle of $\mathrm{Diff}^1(E)$. Similarly, because σ is a surjective submersion, it follows that the top arrow is too; we denote it by a. Clearly the kernel of a is still $\mathrm{End}(E)$, and we have an exact sequence of vector bundles over M,

$$\mathrm{End}(E) \rightarrowtail \mathfrak{D}(E) \overset{a}{\twoheadrightarrow} TM$$

where the sections of $\mathcal{D}(E)$ are those first–order differential operators $D\colon \Gamma E \to \Gamma E$ for which there exists a vector field $X = a(D)$ on M such that for all $u \in C^\infty(M), \mu \in \Gamma E$,

$$D(u\mu) = uD(\mu) + a(D)(u)\mu. \tag{14}$$

A first–order differential operator D with this property is a *derivation* on E or a *derivative endomorphism* of ΓE. For any connection ∇ in E and vector field X on M, the covariant derivative ∇_X obeys (14).

For derivations D, D' on E, the bracket

$$[D, D'] = D \circ D' - D' \circ D$$

is also a derivation on E, with $a([D, D']) = [a(D), a(D')]$. One can also easily check from (14) that, for all $D, D' \in \Gamma \mathcal{D}(E)$ and $u \in C^\infty(M)$,

$$[D, uD'] = u[D, D'] + a(D)(u)D'$$

and so $\mathcal{D}(E)$ is a transitive Lie algebroid on M, the *Lie algebroid of derivations on* E.

Now consider a trivial vector bundle $E = M \times V$ and define a morphism from the trivial Lie algebroid $TM \oplus (M \times \mathfrak{gl}(V))$ into $\mathcal{D}(M \times V)$ by

$$(X \oplus \varphi)(\mu) = X(\mu) + \varphi(\mu)\colon M \to V,$$

where $X(\mu)$ is the Lie derivative. It is easy to check that this is an isomorphism of Lie algebroids over M.

In general, $\mathcal{D}(E)$ plays the role for E that is played by the general linear Lie algebra $\mathfrak{gl}(V)$ of a vector space V. This will become apparent in the course of the following sections. $\boxtimes$

Example 3.3.5 An involutive distribution (without singularities) Δ on a manifold M is a regular Lie algebroid on M with respect to the inclusion $\Delta \subseteq TM$ as anchor and the standard bracket of vector fields.

$\boxtimes$

Example 3.3.6 Let $M = \mathbb{R}$ and define a bracket $[\,,\,]'$ on TM by

$$\left[\xi\frac{d}{dt}, \eta\frac{d}{dt}\right]' = t\left(\frac{d\eta}{dt}\xi - \frac{d\xi}{dt}\eta\right)\frac{d}{dt}, \qquad \xi, \eta\colon M \to \mathbb{R}$$

and an anchor $a\colon TM \to TM$ by $a(\xi\frac{d}{dt}) = t\xi\frac{d}{dt}$. It is straightforward to check that, with this structure, TM is a Lie algebroid on M, and that it is not regular. $\boxtimes$

Example 3.3.7 Let $X \mapsto X^\dagger$ be an action of a Lie algebra $\mathfrak{g}$ on a manifold M. Thus $X^\dagger$ is a vector field on M, for all $X \in \mathfrak{g}$, and $X \mapsto X^\dagger$ is $\mathbb{R}$-linear, and preserves brackets, $[X, Y]^\dagger = [X^\dagger, Y^\dagger]$ for all $X, Y \in \mathfrak{g}$. Extend the dagger notation to maps $V \colon M \to \mathfrak{g}$ so that $V^\dagger$ is the vector field $V^\dagger(m) = V(m)^\dagger(m)$ on M. Then the trivial vector bundle $M \times \mathfrak{g}$ on M acquires a Lie algebroid structure with anchor $a \colon M \times \mathfrak{g} \to TM$ defined by $a(m, X) = X^\dagger(m)$, and bracket

$$[V, W] = V^\dagger(W) - W^\dagger(V) + [V, W]^\bullet, \tag{15}$$

where $V^\dagger(W)$ denotes the Lie derivative with respect to $V^\dagger$ of the vector valued function W, and $[V, W]^\bullet$ is the pointwise bracket of maps into $\mathfrak{g}$. This is the *action Lie algebroid* corresponding to $X \mapsto X^\dagger$, $\mathfrak{g} \to \mathfrak{X}(M)$, and we denote it by $\mathfrak{g} \ltimes M$.

Notice that if $V, W \colon M \to \mathfrak{g}$ are constant maps, then the Lie algebroid bracket $[V, W]$ is constant also, and is constant at the bracket in $\mathfrak{g}$ of the values of V and W. In fact this property, together with the anchor $a \colon \mathfrak{g} \ltimes M \to TM$ and (8), determines (15).

We prove in 3.5.14 that the Lie algebroid of an action Lie groupoid is the action Lie algebroid for the corresponding infinitesimal action. ☒

More extensive families of examples follow in §3.5 and in subsequent chapters. We deal now with two extreme cases that arise repeatedly.

Lie algebra bundles

Definition 3.3.8 A *Lie algebra bundle*, or *LAB*, is a vector bundle (L, q, M) together with a field of Lie algebra brackets $[\ ,\ ] \colon \Gamma L \times \Gamma L \to \Gamma L$ (see 1.7.12) such that L admits an atlas $\{\psi_i \colon U_i \times \mathfrak{g} \to L_{U_i}\}$ in which each $\psi_{i,x}$ is a Lie algebra isomorphism.

A *morphism of LABs* $\varphi \colon L \to L'$, $\varphi_0 \colon M \to M'$, is a morphism of vector bundles such that each $\varphi_x \colon L_x \to L'_{\varphi_0(x)}$ is a Lie algebra morphism. □

An LAB is clearly a totally intransitive Lie algebroid. However, a totally intransitive Lie algebroid may be merely a vector bundle with a field of Lie algebra brackets, as in 1.7.12, and need not be an LAB.

We collect here some necessary observations about LABs.

Many constructions in the category of Lie algebras carry over to LABs. Examples we will need include the centre LAB ZL of an LAB L, the derived LAB $[L, L]$, the LAB of derivations $\mathrm{Der}(L)$, and the adjoint

LAB $\mathrm{ad}(L)$. In the first two cases, the following construction principle applies. The proof is immediate.

Proposition 3.3.9 *Let L be an LAB on M with fibre type $\mathfrak{g}$. Let $\mathfrak{h}$ be a characteristic subalgebra of $\mathfrak{g}$; that is, $\varphi(\mathfrak{h}) = \mathfrak{h}$ for all $\varphi \in \mathrm{Aut}(\mathfrak{g})$. Then there is a well–defined sub–LAB K of L such that any LAB chart $\psi \colon U \times \mathfrak{g} \to L_U$ for L restricts to an LAB chart $U \times \mathfrak{h} \to K_U$ for K.*

Taking $\mathfrak{h} = \mathfrak{z}$, the centre of $\mathfrak{g}$, the resulting LAB, denoted ZL, clearly has fibres $ZL|_x$ which are the centres of the fibres L_x of L. Further, for any open $U \subseteq M$, the (infinite–dimensional) $\mathbb{R}$–Lie algebra $\Gamma_U(ZL)$ is the centre of $\Gamma_U L$. The LAB ZL is called the *centre* of L. The *derived sub LAB* $[L, L]$ is obtained in the same way.

For $\mathrm{Der}(L)$, consider first a vector bundle E on M. The vector bundle $\mathrm{End}(E)$ is the unique vector bundle with fibres $\mathrm{End}(E)_x = \mathrm{End}(E_x)$, for $x \in M$, and charts $\overline{\psi} \colon U \times \mathfrak{gl}(V) \to \mathrm{End}(E)_U$ induced from charts $\psi \colon U \times V \to E_U$ for E by $\overline{\psi}_x(A) = \psi_x \circ A \circ \psi_x^{-1}$. Here V is the fibre type of E and ψ_x is the isomorphism $V \to E_x$ obtained by restricting ψ. It follows that $\mathrm{End}(E)$ is an LAB with respect to these charts.

Now given an LAB L, with fibre type $\mathfrak{g}$, observe that the Lie subalgebra $\mathrm{Der}(\mathfrak{g})$ of $\mathfrak{gl}(\mathfrak{g})$ is invariant under automorphisms of $\mathfrak{gl}(\mathfrak{g})$ of the form $A \mapsto s \circ A \circ s^{-1}$, where $s \colon \mathfrak{g} \to \mathfrak{g}$ is a Lie algebra automorphism. Applying the same method of proof as for 3.3.9, it follows that $\mathrm{Der}(L)$ is a sub LAB of $\mathrm{End}(L)$. It is called the *LAB of derivations* of L.

Proposition 3.3.10 *Let L be an LAB on M. Then the LAB morphism* $\mathrm{ad} \colon L \to \mathrm{Der}(L)$, *defined as being fibrewise the adjoint map* $\mathrm{ad}_x \colon L_x \to \mathrm{Der}(L_x)$ *of the fibres of L, is of locally constant rank.*

Proof To prove that ad is smooth, note that $[V, W] = \mathrm{ad}(V)(W)$ is smooth whenever V and W are smooth sections of L. To prove that it is of locally constant rank, note that ad is locally $\mathrm{id}_U \times \mathrm{ad}_{\mathfrak{g}} \colon U \times \mathfrak{g} \to U \times \mathrm{Der}(\mathfrak{g})$, with respect to an LAB chart of L and the corresponding chart for $\mathrm{Der}(L)$. $\qquad\square$

In fact it will follow from §6.4 that ad is locally constant as a morphism of LABs.

It follows that $\mathrm{ad}(L)$, the image of ad, is a sub–LAB of $\mathrm{Der}(L)$, called the *adjoint LAB* of L. It is also an ideal of $\mathrm{Der}(L)$, in the sense of the following definition, and is also called the *ideal of inner derivations* of L.

Definition 3.3.11 Let L be an LAB on M and let K be a sub LAB of L. Then K is an *ideal* of L, denoted $K \trianglelefteq L$, if K_x is an ideal of L_x, for all $x \in B$. □

Given K an ideal of L, a *quotient* LAB L/K can be constructed in an obvious fashion. Its elements will be written $V + K$ or $\overline{V}$.

The terminology of Lie algebra theory will be taken over without comment: an LAB is *reductive, semisimple, nilpotent, abelian*, et cetera, if each of its fibres has the corresponding property. Many results of the structure theory of Lie algebras generalize effortlessly: for example, a reductive LAB L is the direct sum $ZL \oplus [L, L]$.

Transitive Lie algebroids

Let A be a transitive Lie algebroid on M. The kernel L of $a\colon A \to TM$ inherits the bracket structure of A by (9) and is itself a totally intransitive Lie algebroid on M. We usually write a transitive Lie algebroid in the form

$$L \rightarrowtail^{\ j\ } A \overset{a}{\longrightarrow\!\!\!\!\gg} TM \tag{16}$$

The notation j allows L to be any totally intransitive Lie algebroid isomorphic to the kernel of a; for example, in the case of the Atiyah sequence of a principal bundle $P(B, G, \pi)$, the kernel $\frac{T^{\pi}P}{G}$ of π_* is usually replaced by $\frac{P \times \mathfrak{g}}{G}$ (see 3.2.2).

We call L the *adjoint bundle* of A. In 6.5.1 we will prove that L is actually an LAB.

A morphism $\varphi\colon A \to A'$ of transitive Lie algebroids over M obeys $a' \circ \varphi = a$ and therefore induces a morphism of the adjoint bundles $\varphi^+\colon L \to L'$. This is a morphism of totally intransitive Lie algebroids. In §6.5 we show that φ^+, and hence φ, is of locally constant rank.

The following version of the 5–lemma of category theory is extremely useful.

Proposition 3.3.12 *Let $\varphi\colon A \to A'$ be a morphism of transitive Lie algebroids over M. Then φ is a surjection, injection or bijection if and only if $\varphi^+\colon L \to L'$ is, respectively, a surjection, injection or bijection.*

Proof This is a diagram chase in

$$
\begin{array}{ccccc}
L & \xrightarrow{\;j\;} & A & \xrightarrow{\;a\;} & TM \\[2pt]
\downarrow{\varphi^+} & & \downarrow{\varphi} & & \| \\[2pt]
L' & \xrightarrow{\;j'\;} & A' & \xrightarrow{\;a'\;} & TM.
\end{array}
$$

$\square$

If A is a regular Lie algebroid on M then $I = \operatorname{im} a$ is an involutive distribution on M and $L = \ker a$ is still a vector bundle. Such a Lie algebroid could be written in the form

$$
L \xrightarrow{\;j\;} A \xrightarrow{\;a\;} I
$$

and a version of Proposition 3.3.12 would continue to hold. However in this case L need not be an LAB and morphisms need not be of locally constant rank (as the totally intransitive case shows).

$$
\qquad\qquad*\qquad*\qquad*
$$

We conclude this section with some basic algebraic definitions which are needed later in the chapter.

Definition 3.3.13 Let A be a Lie algebroid on M and let E be a vector bundle, also on M. A *representation* of A on E is a morphism of Lie algebroids over M,

$$
\rho\colon A \to \mathfrak{D}(E).
$$

Let $\rho'\colon A \to \mathfrak{D}(E')$ be a second representation of A. Then a vector bundle morphism $\varphi\colon E \to E'$ is *A–equivariant* if $\varphi(\rho(X)(\mu)) = \rho(X)(\varphi(\mu))$ for $X \in \Gamma A$, $\mu \in \Gamma E$.

Let $\rho^i\colon A^i \to \mathfrak{D}(E^i)$, $i = 1, 2$, be representations of Lie algebroids over M on vector bundles over M, let $\varphi\colon A^1 \to A^2$ be a morphism of Lie algebroids over M, and let $\psi\colon E^1 \to E^2$ be a morphism of vector bundles. Then ψ is *φ–equivariant* if $\psi(\rho^1(X)(\mu)) = \rho^2(\varphi(X))(\psi(\mu))$. for all $X \in \Gamma A^1$, $\mu \in \Gamma E^1$. $\square$

In §6.5 it will be proved that if A is transitive, then equivariant morphisms are of locally constant rank.

Definition 3.3.14 Let A be a Lie algebroid on M and let V be a vector space. The *trivial representation* of A on $M \times V$ is

$$\rho^0(X)(f) = a(X)(f), \qquad f \colon M \to V, \ X \in \Gamma A.$$

□

Example 3.3.15 Let A be a transitive Lie algebroid on M. The *adjoint representation* of A is the representation

$$\mathrm{ad} \colon A \to \mathfrak{D}(L)$$

of A on its adjoint bundle L defined by

$$\mathrm{ad}(X)(V) = [X, V], \qquad X \in \Gamma A, V \in \Gamma L.$$

We return to the adjoint representation in §3.6 and §3.7.   ⊠

Example 3.3.16 Let $P(M, G)$ be a principal bundle and let $E = \frac{P \times V}{G}$ be an associated vector bundle. Given $X \in \Gamma(\frac{TP}{G})$ denote by $\overline{X}$ the corresponding G–invariant vector field on P; and given $\mu \in \Gamma E$ denote by $\widetilde{\mu} \colon P \to V$ the corresponding G–equivariant map.

Lemma 3.3.17 If $\mu \in \Gamma E$ and $X \in \Gamma(\frac{TP}{G})$ then $\overline{X}(\widetilde{\mu}) \in C^\infty(P, V)$ is G–equivariant.

Proof Let $\{\varphi_t\}$ be a saturated local flow for $\overline{X}$ defined in a neighbourhood of $u \in P$, as in 3.2.7. Then, for all $g \in G$,

$$\overline{X}(\widetilde{\mu})(ug) = \left.\frac{d}{dt}\widetilde{\mu}(\varphi_t(ug))\right|_0 = \left.\frac{d}{dt}\widetilde{\mu}(\varphi_t(u)g)\right|_0 =$$
$$\left.\frac{d}{dt}g^{-1}\widetilde{\mu}(\varphi_t(u))\right|_0 = g^{-1}\overline{X}(\widetilde{\mu})(u),$$

as required.   □

Denote the section of E corresponding to $\overline{X}(\widetilde{\mu})$ by $\rho_X(\mu)$. Then the map $(X, \mu) \mapsto \rho_X(\mu)$ has the following properties:

$$\rho_{X_1 + X_2}(\mu) = \rho_{X_1}(\mu) + \rho_{X_2}(\mu), \qquad \rho_X(\mu_1 + \mu_2) = \rho_X(\mu_1) + \rho_X(\mu_2),$$
$$\rho_X(f\mu) = f\rho_X(\mu) + \pi_*(X)(f)\mu, \qquad \rho_{fX}(\mu) = f\rho_X(\mu),$$
$$\rho_{[X_1, X_2]}(\mu) = \rho_{X_1}(\rho_{X_2}(\mu)) - \rho_{X_2}(\rho_{X_1}(\mu)).$$

Here $X, X_1, X_2 \in \Gamma(\frac{TP}{G})$, $\mu, \mu_1, \mu_2 \in \Gamma E$, and $f \in C^\infty(M)$.

The proofs are straightforward. We prove the fourth as example.

Recalling that $\widetilde{f\mu} = (f \circ \pi)\widetilde{\mu}$, that $\overline{fX} = (f \circ \pi)\overline{X}$, and that $\overline{X}(f \circ \pi) = \pi_*(X)(f) \circ \pi$, we have

$$\widetilde{\rho_X(f\mu)} = \overline{X}((f \circ \pi)\widetilde{\mu}) = (f \circ \pi)\overline{X}(\widetilde{\mu}) + \overline{X}(f \circ \pi)\widetilde{\mu} = \widetilde{f\rho_X(\mu)} + \widetilde{\pi_*(X)(f)\mu},$$

whence the result.

It now follows that the value of $\rho_X(\mu)$ at a point $x \in M$ depends only on the value of X at x and the values of μ in a neighbourhood of x.

This result shows that ρ is a representation of the Lie algebroid $\frac{TP}{G}$ on E. ◻

Example 3.3.18 Let E be a vector bundle on M, and let ∇ be a flat connection in E (see §5.2). Then $X \mapsto \nabla_X$ is a representation of TM on E. ◻

Definition 3.3.19 An *extension of Lie algebroids* over M is a sequence of morphisms of Lie algebroids over M

$$A' \stackrel{\iota}{\rightarrowtail} A \stackrel{\pi}{\twoheadrightarrow} A''$$

that is exact as a sequence of vector bundles over M. ◻

The Lie algebroid A' must be totally intransitive, for $a' = a \circ \iota = (a'' \circ \pi) \circ \iota = 0$. The basic example is of course $L \stackrel{j}{\rightarrowtail} A \stackrel{a}{\twoheadrightarrow} TM$ for A a transitive Lie algebroid.

The proof of the following result is simple.

Proposition 3.3.20 *Let* $E \stackrel{\iota}{\rightarrowtail} A' \stackrel{\pi}{\twoheadrightarrow} A$ *be an extension of Lie algebroids over* M *with* E *abelian. Then*

$$\iota(\rho(X)(\mu)) = [X', \iota(\mu)], \qquad X \in \Gamma A, \; \mu \in \Gamma E,$$

where $X' \in \Gamma A'$ *has* $\pi(X') = X$, *defines a representation of* A *on* E.

In particular, if A is transitive with abelian adjoint bundle L, there is a natural representation of TM on — or a flat connection in — L.

Definition 3.3.21 Let A be a Lie algebroid on M. A *wide Lie subalgebroid* of A is a Lie algebroid A' on M together with an injective morphism $A' \to A$ of Lie algebroids over M.

If A is transitive, a *reduction* of A is a Lie subalgebroid of A which is itself transitive. ◻

Proposition 3.3.22 *Let A and A' be Lie algebroids on M and let A be transitive. Let $A \oplus_{TM} A'$ denote the inverse image vector bundle over M of the diagram*

$$
\begin{array}{ccc}
A \oplus_{TM} A' & \longrightarrow & A' \\
\downarrow & & \downarrow\ a' \\
A & \longrightarrow & TM.
\end{array}
$$
$$a$$

Let $\bar{a}: A \oplus_{TM} A' \to TM$ be the diagonal composition and define a bracket on $\Gamma(A \oplus_{TM} A')$ by

$$[X \oplus X', Y \oplus Y'] = [X, Y] \oplus [X', Y'].$$

Then $A \oplus_{TM} A'$ is a Lie algebroid on M, and the diagram above is now a pullback in the category of Lie algebroids over M. Lastly, $A \oplus_{TM} A'$ is transitive if and only if A' (as well as A) is transitive and this is so if and only if $A \oplus_{TM} A' \to A$ (as well as $A \oplus_{TM} A' \to A'$) is surjective.

The proof is straightforward. We call $A \oplus_{TM} A'$ the *direct sum Lie algebroid of A and A' over TM*. Note that the trivial Lie algebroid $TM \oplus (M \times \mathfrak{g})$ is not a direct sum of TM and $M \times \mathfrak{g}$.

3.4 Linear vector fields

In 3.3.4 we defined a derivation D on a vector bundle (E, q, M) to be an $\mathbb{R}$-linear map $\Gamma E \to \Gamma E$ such that

$$D(f\mu) = fD(\mu) + a(D)(f)\mu, \qquad f \in C^\infty(M),\ \mu \in \Gamma E, \qquad (17)$$

where $a(D)$ is a vector field on M. The derivations form the module of sections of a transitive Lie algebroid $\mathfrak{D}(E)$ on M.

There is a further interpretation of derivations on a vector bundle which is extremely useful. This identifies derivations with vector fields on the total space of the bundle which are linear in a natural sense.

Applying the tangent functor to the vector bundle operations in E, as in 1.1.16, yields a vector bundle $(TE, T(q), TM)$ which we call the *tangent prolongation of E*. Together with the standard structure $TE \to$

E this forms the *tangent double vector bundle*

$$
\begin{array}{ccc}
TE & \xrightarrow{\;T(q)\;} & TM \\
\Big\downarrow{\scriptstyle p_E} & & \Big\downarrow{\scriptstyle p} \\
E & \xrightarrow{\;\;q\;\;} & M
\end{array}
\tag{18}
$$

of E. (The general concept of double vector bundle is defined in Chapter 9.) In E and TM, we use standard notation. The zero of E over $m \in M$ is 0_m^E, and the zero of TM over m is 0_m^T.

We denote elements of TE by $\xi, \eta, \zeta \ldots$, and we write $(\xi; e, x; m)$ to indicate that $e = p_E(\xi)$, $x = T(q)(\xi)$, and $m = p(T(q)(\xi)) = q(p_E(\xi))$. With respect to the tangent bundle structure (TE, p_E, E), we use standard notation: $+$ for addition, $-$ for subtraction and juxtaposition for scalar multiplication. The notation $T_e(E)$ will always denote the fibre $p_E^{-1}(e)$, for $e \in E$, with respect to this bundle. The zero element in $T_e(E)$ is denoted $\widetilde{0}_e$. We refer to this bundle structure as the *standard tangent bundle structure*.

In the *prolonged tangent bundle structure*, $(TE, T(q), TM)$, we use $\mathbin{+\!\!\!+}$ for addition, $\mathbin{-\!\!\!-}$ for subtraction, and $\cdot$ for scalar multiplication. The fibre over $x \in TM$ will always be denoted $T(q)^{-1}(x)$, and the zero element of this fibre is $T(0)(x)$. If we consider elements ξ of TE as derivatives of paths in E and write

$$
\xi = \left.\frac{d}{dt} e_t\right|_0,
$$

where e_t denotes a path in E defined in a neighbourhood of $0 \in \mathbb{R}$, then $p_E(\xi) = e_0$ and $T(q)(\xi) = \left.\frac{d}{dt} q(e_t)\right|_0$. If $\xi, \eta \in TE$ have $T(q)(\xi) = T(q)(\eta)$, then we can arrange that $\xi = \left.\frac{d}{dt} e_t\right|_0$ and $\eta = \left.\frac{d}{dt} h_t\right|_0$, where $q(e_t) = q(h_t)$ for all t in a neighbourhood of $0 \in \mathbb{R}$ and then

$$
\xi \mathbin{+\!\!\!+} \eta = \left.\frac{d}{dt}(e_t + h_t)\right|_0, \qquad \lambda \cdot \xi = \left.\frac{d}{dt} \lambda e_t\right|_0.
$$

For each $m \in M$, the tangent space $T_{0_m}(E_m)$ identifies canonically with E_m; we denote the element of $T_{0_m}(E_m)$ corresponding to $e \in E_m$ by $\overline{e}$ and call it the *core element* corresponding to e. Note that, for $e, h \in E_m$ and $t \in \mathbb{R}$,

$$
\overline{e} + \overline{h} = \overline{e + h} = \overline{e} \mathbin{+\!\!\!+} \overline{h}, \qquad t\overline{e} = \overline{te} = t \cdot \overline{e}.
$$

Regarding $T(q)$ in (18) as a morphism over q, the induced map $T(q)^{!} \colon TE \to q^{!}TM$ is a surjective submersion. The kernel over $e \in E_m$

consists of the vertical tangent vectors in $T_e(E)$. We identify $T_e(E_m)$ with $\{e\} \times E_m$. Thus the kernel of $T(q)^!$ is identified with $q^! E$, with the injection $q^! E \to TE$ mapping $(e, h) \in E_m \times E_m$ to the vector in E_m, denoted $h \,@\, e$, which has tail at e and is parallel to h. Denote this map by τ. In terms of the prolongation structure, $\tau(e, h) = \tilde{0}_e \mathbin{+\!\!\!+} \bar{h}$. Thus we have a short exact sequence of vector bundles over E,

$$ q^! E \,\rightarrowtail^{\tau}\, TE \,\xrightarrow{T(q)^!}\!\!\!\!\!\!\twoheadrightarrow\, q^! TM. \tag{19} $$

Similarly there is a short exact sequence of vector bundles over TM,

$$ p^! E \,\rightarrowtail^{v}\, TE \,\xrightarrow{p_E^!}\!\!\!\!\!\!\twoheadrightarrow\, p^! E, \tag{20} $$

where the inclusion $v \colon p^! E \to TE$ is $(x, h) \mapsto T(0)(x) + \bar{h}$.

We refer to (19) as the *core sequence for p_E*, and to (20) as the *core sequence for $T(q)$*.

Given $\mu \in \Gamma E$ define a vector field $\mu^{\uparrow}$ on E, the *vertical lift of μ*, by $\mu^{\uparrow}(e) = \tilde{0}_e \mathbin{+\!\!\!+} \overline{\mu(qe)}$, $e \in E$. Then

$$ \mu^{\uparrow}(F)(e) = \frac{d}{dt} F(e + t\mu(qe)) \Big|_0 $$

for $F \in C^{\infty}(E), e \in E$, and so

$$ \mu^{\uparrow}(f \circ q) = 0, \qquad \mu^{\uparrow}(\ell_{\varphi}) = \langle \varphi, \mu \rangle \circ q, \qquad [\mu^{\uparrow}, \nu^{\uparrow}] = 0, \tag{21} $$

for $f \in C^{\infty}(M), \mu, \nu \in \Gamma E, \varphi \in \Gamma(E^*)$. Here $\ell_{\varphi} \in C^{\infty}(E)$ is the function $e \mapsto \langle \varphi(qe), e \rangle$. Note also that $(f\mu)^{\uparrow} = (f \circ q)\mu^{\uparrow}$ and $(\mu + \nu)^{\uparrow} = \mu^{\uparrow} + \nu^{\uparrow}$.

A section $\mu \in \Gamma E$ also induces a section $\hat{\mu}$ of $T(q)$ by $\hat{\mu}(x) = T(0)(x) + \overline{\mu(px)}$. Note that

$$ (\mu + \nu)^{\widehat{}} = \hat{\mu} \mathbin{+\!\!\!+} \hat{\nu}, \qquad (f\mu)^{\widehat{}} = (f \circ p) \cdot \hat{\mu}, $$

for $\mu, \nu \in \Gamma E, f \in C^{\infty}(M)$. (There are no bracket relations.)

If E is a trivial vector bundle $M \times V \to M$, where V is a vector space, we denote elements of $TE = TM \times V \times V$ by $\xi = (x, e, e')$, where if m_t is a path in M with $x = \dfrac{d}{dt} m_t \Big|_0$, we identify (x, e, e') with $\dfrac{d}{dt}(m_t, e + te') \Big|_0$. The operations in the prolongation bundle are then given by $T(q)(x, e, e') = x$, and

$$
\begin{aligned}
(x, e, e') +\!\!+\ (x, h, h') &= (x, e + h, e' + h'), \\
t \cdot (x, e, e') &= (x, te, te'), \\
T(0)(x) &= (x, 0, 0),
\end{aligned}
\tag{22}
$$

where $x \in TM$, $e, e', h, h' \in V$, $t \in \mathbb{R}$. The operations in the standard tangent bundle are of course given by $p_{M \times V}(x, e, e') \mapsto (px, e)$ and

$$
\begin{aligned}
(x, e, e') + (y, e, h') &= (x + y, e, e' + h'), \\
t(x, e, e') &= (tx, e, te'), \\
\widetilde{0}_{(m, e)} &= (0^T_m, e, 0),
\end{aligned}
\tag{23}
$$

where $e, e', h, h' \in V$, $t \in \mathbb{R}$ and $x, y \in TM$ satisfy $p(x) = p(y)$. Given $(m, e) \in M \times V$, the corresponding core element of $TM \times V \times V$ is

$$
\overline{(m, e)} = (0^T_m, 0, e).
$$

Definition 3.4.1 A *linear vector field on* E is a pair (ξ, x), where ξ is a vector field on E, and x is a vector field on M, such that

$$
\begin{array}{ccc}
E & \xrightarrow{\ \xi\ } & TE \\
{\scriptstyle q}\downarrow & & \downarrow{\scriptstyle T(q)} \\
M & \xrightarrow{\ x\ } & TM
\end{array}
\tag{24}
$$

is a morphism of vector bundles. $\qquad\qquad\square$

In particular, ξ projects under q to x, and

$$
\xi(e + e') = \xi(e) +\!\!+\ \xi(e'), \qquad \xi(te) = t \cdot \xi(e),
$$

for all $e, e' \in E$, $t \in \mathbb{R}$, where $+\!\!+$ and $\cdot$ are the prolonged operations in TE. The sum of two linear vector fields is a linear vector field, and a scalar multiple of a linear vector field is a linear vector field. Given a connection in E, the horizontal lift of any vector field on the base is a linear vector field.

In the following proposition, $C^\infty_{\mathrm{lin}}(E)$ denotes the subring of $C^\infty(E)$ consisting of the fibrewise linear functions $E \to \mathbb{R}$. Given $\varphi \in \Gamma E^*$, the function $\ell_\varphi \colon E \to \mathbb{R}$ is $e \mapsto \langle \varphi(qe), e \rangle$.

Proposition 3.4.2 Let ξ be a vector field on E. The following are equivalent:

(i) (ξ, x) is a linear vector field on E for some vector field x on M;

(ii) $\xi\colon C^\infty(E) \to C^\infty(E)$ *sends* $C^\infty_{\mathrm{lin}}(E)$ *into* $C^\infty_{\mathrm{lin}}(E)$ *and sends* $q^*C^\infty(M)$ *into* $q^*C^\infty(M)$;

(iii) ξ *has flows* F_t *which are (local) vector bundle morphisms* $E \to E$ *over a flow* f_t *on* M.

Proof (i) $\Longrightarrow$ (ii). Suppose first that $\xi, \eta \in T(E)$ have $T(q)(\xi) = T(q)(\eta)$, and that $F \in C^\infty_{\mathrm{lin}}(E)$. Write $\xi = \frac{d}{dt}e_t\big|_0$, $\eta = \frac{d}{dt}h_t\big|_0$, where $q(e_t) = q(h_t)$ for t close to $0 \in \mathbb{R}$. Then

$$(\xi \mathbin{+\!\!\!+} \eta)(F)(e_0 + h_0) = \frac{d}{dt}F(e_t + h_t)\bigg|_0$$

$$= \frac{d}{dt}F(e_t)\bigg|_0 + \frac{d}{dt}F(h_t)\bigg|_0 = \xi(F)(e_0) + \eta(F)(h_0).$$

From this, and the similar result for scalar multiplication, it follows that a linear vector field ξ maps $C^\infty_{\mathrm{lin}}(E)$ into $C^\infty_{\mathrm{lin}}(E)$. Since ξ is projectable under q, it is clear that ξ maps $q^*C^\infty(M)$ into $q^*C^\infty(M)$.

(ii) $\Longrightarrow$ (iii). From the assumption that ξ sends $q^*C^\infty(M)$ into itself it follows that ξ projects to a vector field x on M, and hence $q \circ F_t = f_t \circ q$ for all t, where F_t and f_t are flows for ξ and x respectively. It remains to prove that the F_t are linear and since this is a local question, we may assume that $E = M \times V$ and that ξ and x have global flows. The details are straightforward.

(iii) $\Longrightarrow$ (i). Suppose, for simplicity, that ξ has a global flow F_t by vector bundle morphisms over a global flow f_t on M. Then ξ certainly projects under q to the vector field x generated by f_t. For $e, h \in E$ with $q(e) = q(h)$,

$$\begin{aligned}
\xi(e + h) &= T_0(t \mapsto F_t(e + h))(1) \\
&= T_0(t \mapsto F_t(e) + F_t(h))(1) \\
&= T(+)(T_0(t \mapsto F_t(e))(1), T_0(t \mapsto F_t(h))(1)) \\
&= T(+)(\xi(e), \xi(h)) \\
&= \xi(e) \mathbin{+\!\!\!+} \xi(h),
\end{aligned}$$

and similarly one proves that $\xi(te) = t \cdot \xi(e)$ for $t \in \mathbb{R}$. $\qquad\square$

Corollary 3.4.3 *If* (ξ, x) *and* (η, y) *are linear vector fields, then so too is* $([\xi, \eta], [x, y])$.

Consider a linear vector field (ξ, x). Since $C^\infty_{\mathrm{lin}}(E)$ is canonically isomorphic (as a $C^\infty(M)$ module) to ΓE^*, condition (ii) of 3.4.2 shows that ξ induces a map

$$D_\xi^{(*)}\colon \Gamma E^* \to \Gamma E^*, \quad \text{such that} \quad \ell_{D_\xi^{(*)}(\varphi)} = \xi(\ell_\varphi). \qquad (25)$$

Clearly, $D_\xi^{(*)}$ is additive, and it is easily checked that $D_\xi^{(*)}(f\varphi) = fD_\xi^{(*)}(\varphi) + x(f)\varphi$ for $f \in C^\infty(M)$. Thus $D_\xi^{(*)}$ is a derivation on E^*, and we have a map, which is linear over $C^\infty(M)$ and bracket–preserving,

$$(\xi, x) \mapsto D_\xi^{(*)}, \quad \Gamma^{\mathrm{LIN}} TE \to \Gamma \mathfrak{D}(E^*), \qquad (26)$$

where $\Gamma^{\mathrm{LIN}} TE$ is the $C^\infty(M)$–module of linear vector fields on E.

By Corollary 3.4.3, $\Gamma^{\mathrm{LIN}} TE$ is closed under the bracket on $\mathfrak{X}(E)$. Let a denote the map $(\xi, x) \mapsto x$. Then the kernel of a consists of those vertical vector fields on E which are linear. Any vertical vector field ξ can be written as $\xi(e) = \tau(e, \mathsf{X}(e))$, where $\mathsf{X}\colon E \to E$ has $q \circ \mathsf{X} = q$, and ξ is linear if and only if X is a vector bundle morphism. Thus the kernel of a can be identified with $\Gamma \operatorname{End}(E)$. Since this is the module of sections of a vector bundle over M, it follows that $\Gamma^{\mathrm{LIN}} TE$ is also the module of sections of a vector bundle on M; we denote this by $T^{\mathrm{LIN}} E$. We can now regard (26) as a morphism of Lie algebroids from $T^{\mathrm{LIN}} E$ to $\mathfrak{D}(E^*)$.

Proposition 3.4.4 *The morphism $T^{\mathrm{LIN}} E \to \mathfrak{D}(E^*)$ just defined is an isomorphism of Lie algebroids.*

Proof By 3.3.12 it suffices to prove that the restriction of this map to the kernels of the anchor maps (the adjoint bundles) is an isomorphism, and this restriction is the canonical identification of $\operatorname{End}(E)$ with $\operatorname{End}(E^*)$. □

Now each derivation $D^{(*)}$ on E^* corresponds to a derivation D on E by

$$\langle D^{(*)}(\varphi), \mu \rangle = a(D)(\langle \varphi, \mu \rangle) - \langle \varphi, D(\mu) \rangle,$$

where $\varphi \in \Gamma E^*$, $\mu \in \Gamma E$. This defines an isomorphism of Lie algebroids $\mathfrak{D}(E) \cong \mathfrak{D}(E^*)$. Letting D_ξ denote the element of $\Gamma \mathfrak{D}(E)$ corresponding to $D_\xi^{(*)}$, we have

$$\langle \varphi, D_\xi(\mu) \rangle = x(\langle \varphi, \mu \rangle) - \xi(\ell_\varphi) \circ \mu. \qquad (27)$$

We have proved the first part of the following theorem.

Theorem 3.4.5 *The maps described above are bracket–preserving isomorphisms of $C^\infty(M)$–modules between linear vector fields on E, and derivations on either E or E^*. Let (ξ, x) be a linear vector field on E and let $D = D_\xi$ be the corresponding element of $\Gamma\mathfrak{D}(E)$. Then, under this isomorphism, for all $\mu \in \Gamma E$ and $m \in M$,*

$$D(\mu)^\uparrow(\mu(m)) = T(\mu)(x(m)) - \xi(\mu(m)). \tag{28}$$

If φ_t is a flow for ξ near $\mu(m)$ and f_t the corresponding flow for x near m, then

$$D(\mu)^\frown(x(m)) = \frac{d}{dt}\Big(\mu(f_t(m)) - \varphi_t(\mu(m))\Big)\Big|_0. \tag{29}$$

Lastly,

$$[\xi, \mu^\uparrow] = D(\mu)^\uparrow. \tag{30}$$

Proof Let $e = D(\mu)(m)$ and let Z be the RHS of the first equation. It suffices to prove that $\tau(\mu(m), e) = \tau(\mu(m), Z)$, both of which are vertical tangent vectors to E at $\mu(m)$. The functions on E are generated by those of the form ℓ_φ for $\varphi \in \Gamma E^*$ and those of the form $f \circ q$ for $f \in C^\infty(M)$. Since both vectors are vertical, they coincide on all $f \circ q$. Now, for any $\varphi \in \Gamma E^*$,

$$\tau(\mu(m), e)(\ell_\varphi) = \langle \varphi, e \rangle = x(m)(\langle \varphi, \mu \rangle) - \xi(\ell_\varphi)(\mu(m))$$

and

$$\tau(\mu(m), Z)(\ell_\varphi) = T(\mu)(x(m))(\ell_\varphi) - \xi(\mu(m))(\ell_\varphi),$$

whence the result.

For the second equation, note first that $T(\mu)(x(m))$ and $\xi(\mu(m))$ have the same two projections, and therefore

$$v(x(m), D(\mu)(m)) = T(\mu)(x(m)) - \xi(\mu(m)).$$

From this the second equation follows.

For the third equation, it suffices to verify equality on functions of the form ℓ_φ and $f \circ q$, using (21) and (27). $\square$

From the first equation it also follows that

$$T(\mu)(x(m)) = \xi(\mu(m)) + D(\mu)^\uparrow(x(m))$$

since $T(\mu)(x(m))$ and $\xi(\mu(m))$ have both projections the same.

If, in Proposition 3.4.5, $\mu(m_0) = 0$ for a specific $m_0 \in M$, then $D(\mu)(m_0)$ depends only on x, μ and m_0 and can be identified with

$T(\mu)(x(m_0))$; the map $T_{m_0}(M) \to E_{m_0}$, $x \mapsto T(\mu)(x)$, is then the intrinsic derivative of μ. If, in addition, $E = TM$, this map is the linearization of the vector field $\mu = X$ at the singularity m_0.

$$* \quad * \quad * \quad * \quad *$$

Notice that there is now a bijective correspondence between linear vector fields on E and linear vector fields on E^*, mediated by the derivative endomorphisms. This will be clarified by introducing a canonical pairing between TE and $T(E^*)$ as vector bundles over TM.

Suppose given $\mathfrak{X} \in T(E^*)$ and $\xi \in TE$ with $T(q)(\mathfrak{X}) = T(q_*)(\xi)$. Then we can write $\mathfrak{X} = \dfrac{d}{dt}\varphi_t\Big|_0 \in T(E^*)$ and $\xi = \dfrac{d}{dt}e_t\Big|_0 \in TE$ where $e_t \in E$ and $\varphi_t \in E^*$ have $q_*(\varphi_t) = q(e_t)$ for t near zero. Now define the *tangent pairing* $\langle\!\langle\,,\,\rangle\!\rangle$ by

$$\langle\!\langle \mathfrak{X}, \xi \rangle\!\rangle = \frac{d}{dt}\langle \varphi_t, e_t \rangle\Big|_0. \tag{31}$$

To show that this is non–degenerate it is sufficient to work locally. Suppose, therefore, that $E = M \times V$. Regard ξ as $(x_0, v_0, w_0) \in T_{m_0}M \times V \times V$ and $\mathfrak{X}$ as $(x_0, \varphi_0, \psi_0) \in T_{m_0}M \times V^* \times V^*$. Then

$$\mathfrak{X} = \frac{d}{dt}(m_t, \varphi_0 + t\psi_0)\Big|_0, \qquad \xi = \frac{d}{dt}(m_t, v_0 + tw_0)\Big|_0,$$

where $\dfrac{d}{dt}m_t\Big|_0 = x = T(q)(\mathfrak{X}) = T(q_*)(\xi)$. So

$$\langle\!\langle \mathfrak{X}, \xi \rangle\!\rangle = \frac{d}{dt}\langle \varphi_0 + t\psi_0, v_0 + tw_0 \rangle\Big|_0$$

Expanding out the RHS, the constant term and the quadratic term vanish in the derivative, and we are left with

$$\langle\!\langle \mathfrak{X}, \xi \rangle\!\rangle = \langle \psi_0, v_0 \rangle + \langle \varphi_0, w_0 \rangle$$

from which it is clear that $\langle\!\langle\,,\,\rangle\!\rangle$ is non–degenerate.

Proposition 3.4.6 *Given* $(\xi; \mu(m), x; m) \in TE$ *and* $(\mathfrak{X}; \varphi_m, x; m) \in T(E^*)$, *let* $\mu \in \Gamma(E)$ *and* $\varphi \in \Gamma(E^*)$ *be any sections taking the values* $\mu(m)$ *and* φ_m *at* m. *Then*

$$\langle\!\langle \mathfrak{X}, \xi \rangle\!\rangle = \mathfrak{X}(\ell_\mu) + \xi(\ell_\varphi) - x(\langle \varphi, \mu \rangle). \tag{32}$$

Proof Again we work locally. Regard $\varphi \in \Gamma(M \times V^*)$ as a function $M \to V^*$ with $\varphi(m_0) = \varphi_0$. Then $\ell_\varphi \colon M \times V \to \mathbb{R}$ is $(m, v) \mapsto \langle \varphi(m), v \rangle$. So

$$\xi(\ell_\varphi) = \frac{d}{dt}\ell_\varphi(m_t, v_0 + tw_0)\bigg|_0 = \frac{d}{dt}\langle \varphi(m_t), v_0 + tw_0 \rangle\bigg|_0$$

$$= \frac{d}{dt}\langle \varphi(m_t), v_0 \rangle\bigg|_0 + \langle \varphi_0, w_0 \rangle.$$

Similarly regarding $\mu \in \Gamma(M \times V)$ as a function $M \to V$ with $\mu(m_0) = v_0$, we get

$$\mathfrak{X}(\ell_\mu) = \frac{d}{dt}\langle \varphi_0, \mu(m_t) \rangle\bigg|_0 + \langle \psi_0, v_0 \rangle.$$

Lastly,

$$x(\langle \varphi, \mu \rangle) = \frac{d}{dt}\langle \varphi(m_t), \mu(m_t) \rangle\bigg|_0 = \frac{d}{dt}\langle \varphi(m_t), v_0 \rangle\bigg|_0 + \frac{d}{dt}\langle \varphi_0, \mu(m_t) \rangle\bigg|_0.$$

$\square$

Suppose (ξ, x) is a linear vector field on E and that (ξ_*, x) is the linear vector field on E^* with the same derivation. Then, from 3.4.6 and (27), it follows that

$$\langle\!\langle \xi_*(\varphi), \xi(e) \rangle\!\rangle = 0$$

for all $(\varphi, e) \in E^* \times_M E$.

Proposition 3.4.7 *Let (ξ, x) be a linear vector field on E. For $\varphi \in E^*_m$, $m \in M$, the value $H = \xi_*(\varphi)$ of the corresponding linear vector field on E^* is the unique element of the form $(H; \varphi, x(m); m) \in T(E^*)$ for which*

$$\langle\!\langle H, \xi(e) \rangle\!\rangle = 0$$

for all $e \in E_m$.

Proof Given the form of H, it can be written as

$$H = \xi_*(\varphi) \mathbin{+\!\!\!+} (T(0_*)(x(m)) + \overline{\omega})$$

for some $\omega \in E^*_m$. Here 0_* is the zero section in E^*. Now, remembering that $\mathbin{+\!\!\!+}$ is the addition with respect to which the tangent pairing is bilinear,

$$\langle\!\langle H, \xi(e) \rangle\!\rangle = \langle\!\langle \xi_*(\varphi), \xi(e) \rangle\!\rangle + \langle\!\langle T(0_*)(x(m)) + \overline{\omega}, \xi(e) \rangle\!\rangle.$$

Now we use (32) with the zero section 0_* in E^* and any section μ of E for which $\mu(m) = e$. Two terms of (32) vanish and we have

$$\langle\!\langle H, \xi(e) \rangle\!\rangle = (T(0_*)(x(m)) + \overline{\omega})(\ell_\mu) = T(0_*)(x(m))(\ell_\mu) + \overline{\omega}(\ell_\mu) = \langle \omega, e \rangle.$$

This can be zero for all $e \in E_m$ if and only if $\omega = 0$. $\qquad\square$

From now on we denote the $C^\infty(M)$–module $\Gamma^{\mathrm{LIN}}TE$ by $\mathfrak{X}^{\mathrm{LIN}}(E)$. It follows from 3.4.7 that if $(\xi, x) \in \mathfrak{X}^{\mathrm{LIN}}(E)$ and $(\xi_*, x) \in \mathfrak{X}^{\mathrm{LIN}}(E^*)$ correspond to the same derivations, and if φ_t is a (local) flow for ξ, then φ^*_{-t} is a (local) flow for ξ_*.

Example 3.4.8 For a Lie algebroid A, each $X \in \Gamma A$ induces a derivation $\mathfrak{L}_X \colon \Gamma A \to \Gamma A$, $\mathfrak{L}_X(Y) = [X, Y]$. (The map $X \mapsto \mathfrak{L}_X$ is not $C^\infty(M)$–linear unless A is totally intransitive, so $\mathfrak{L}$ is not otherwise a representation of A.) Denote the corresponding vector fields on A and A^* by $\widetilde{X}$ and H_X; thus

$$\widetilde{X}(\ell_\varphi) = \ell_{\mathfrak{L}_X(\varphi)}, \qquad\qquad \widetilde{X}(f \circ q) = a(X)(f) \circ q,$$
$$H_X(\ell_Y) = \ell_{[X,Y]}, \qquad\qquad H_X(f \circ q_*) = a(X)(f) \circ q_*,$$

for $\varphi \in \Gamma A^*$, $Y \in \Gamma A$, $f \in C^\infty(M)$. Here $\mathfrak{L}_X(\varphi)$ is defined by a natural extension of the usual Lie derivative formula; see 7.1.2. From these it easily follows that

$$[\widetilde{X}, \widetilde{Y}] = \widetilde{[X, Y]}, \qquad [\widetilde{X}, Y^\uparrow] = [X, Y]^\uparrow, \qquad [X^\uparrow, Y^\uparrow] = 0,$$
$$[H_X, H_Y] = H_{[X,Y]}, \qquad [H_X, \varphi^\uparrow] = \mathfrak{L}_X(\varphi)^\uparrow, \qquad [\varphi^\uparrow, \psi^\uparrow] = 0,$$

where $X, Y \in \Gamma A$, and $\varphi, \psi \in \Gamma A^*$.

In the case $A = TM$, these $\widetilde{X}$ and H_X are the classical *complete lifts* to TM and T^*M of the vector field X. The complete lifts, together with the vertical lifts $X^\uparrow$ or $\varphi^\uparrow$, generate all vector fields on TM or T^*M. For a general Lie algebroid A this is not the case: for example, if A is totally intransitive, then all the $\widetilde{X}$ and H_X are vertical. $\qquad\boxtimes$

3.5 The Lie algebroid of a Lie groupoid

This section gives the construction of the Lie algebroid of a Lie groupoid and works the basic examples. The construction follows very closely the construction of the Lie algebra of a Lie group. However since the right translations on a groupoid, the $R_\xi \colon G_{\beta\xi} \to G_{\alpha\xi}$, are diffeomorphisms of the α–fibres only and not of the whole groupoid, a vector field X on a Lie groupoid G is defined to be right–invariant only if it is tangent to the α–fibres. Having noted this, a right–invariant vector field on G is determined by its values on the unities 1_x, $x \in M$, and we define the Lie algebroid of G to be $AG = \cup_{x \in M} T_{1_x}(G_x)$ with the natural vector bundle structure over M which it inherits from TG. The Lie bracket is placed on the module of sections of AG (not on AG itself) via the correspondence between sections of AG and right–invariant vector fields on G. This bracket is not bilinear with respect to the module structure on ΓAG but obeys

$$[X, uY] = u[X, Y] + a(X)(u)Y, \qquad u \in C^\infty(M), \ X, Y \in \Gamma AG,$$

where $a\colon AG \to TM$ is a vector bundle morphism over M which maps each $X \in \Gamma AG$ to the β–projection of the corresponding right–invariant vector field. The map a, the anchor of AG, ties the bracket structure on ΓAG to its module structure. It is the only feature of the Lie algebroid concept which does not appear in the case of Lie algebras.

Let G be a Lie groupoid on a manifold M.

Definition 3.5.1 The vector bundle $AG \to M$ is the pullback across the embedding $1\colon M \to G$ of $T^\alpha G \to G$. Thus

$$
\begin{array}{ccc}
AG & \longrightarrow & T^\alpha G \\
\downarrow & & \downarrow \\
M & \overset{1}{\longrightarrow} & G
\end{array}
\tag{33}
$$

is a pullback diagram.   □

Since 1 is an embedding we will usually regard AG as the restriction of $T^\alpha G$ to M and we identify the fibres $AG|_x$ with the tangent spaces $T_{1_x}(G_x)$, $x \in M$, and regard $AG \to T^\alpha G$ as an inclusion.

Definition 3.5.2 A *vertical vector field* on G is a vector field X which is vertical with respect to α; that is, $X(g) \in T_g(G_{\alpha g})$ for all $g \in G$.

A vector field X on G is *right–invariant* if it is vertical and $X(hg) = T_h(R_g)(X(h))$ for all $(h, g) \in G * G$.   □

A trivial manipulation shows that a vertical vector field X is right–invariant if and only if $X(g) = T(R_g)(X(1_{\beta g}))$ for all $g \in G$. Thus a right–invariant vector field is determined by its values on the submanifold of identity elements.

Proposition 3.5.3 *The vector bundle morphism*

$$
\begin{array}{ccc}
T^{\alpha}G & \xrightarrow{\mathscr{R}} & AG \\
\downarrow & & \downarrow \\
G & \xrightarrow{\beta} & M
\end{array}
\tag{34}
$$

defined on fibres by $\mathscr{R}_g = T_g(R_{g^{-1}})\colon T_g(G_{\alpha g}) \to T_{1_{\beta g}}(G_{\beta g})$, *is a fibre-wise isomorphism.*

Proof Recall the division map $\delta\colon G \times_\alpha G \to G$ of 1.2.6. The tangent of δ applied to some $(Y_h, 0_g)$ with $T(\alpha)(Y) = 0$ yields $T(R_{g^{-1}})(Y_h)$. Composing this restriction with $T^{\alpha}G \to TG \times_{T(\alpha)} TG$, $Y_g \mapsto (Y_g, 0_g)$ gives a smooth map $T^{\alpha}G \to TG$ taking values in $T^{\alpha}G$ and with base map $g \mapsto 1_{\beta g}$. It can therefore be factored over (33), and gives $\mathscr{R}$, which is therefore smooth.

Each $\mathscr{R}_g$ is clearly an isomorphism of vector spaces. □

Corollary 3.5.4 *Given* $X \in \Gamma AG$, *the formula*

$$
\overrightarrow{X}(g) = T(R_g)(X(\beta g)), \qquad g \in G,
$$

defines a right–invariant vector field on G. *The map*

$$
C^\infty(G) \otimes_{C^\infty(M)} \Gamma AG \to \Gamma T^{\alpha}G, \qquad f \otimes X \mapsto f\overrightarrow{X}
$$

is an isomorphism of $C^\infty(G)$*–modules.*

Proof It need only be proved that $\overrightarrow{X}$ is smooth and, in the notation of §2.1, $\overrightarrow{X} = X^! = 1 \otimes X$ is the unique section of $T^{\alpha}G$ projecting under $\mathscr{R}$ to X. The final statement is an instance of (1) in §2.1. □

For a Lie group G this isomorphism reduces to $\mathscr{X}(G) \cong C^\infty(G) \otimes_{\mathbb{R}} \mathfrak{g}$ and since Lie algebras are free as vector spaces it follows that any vector field X on G can be written as

$$
X = f_1\overrightarrow{X_1} + \cdots + f_n\overrightarrow{X_n}
\tag{35}
$$

where $\{X_1, \ldots, X_n\}$ is any chosen basis for $\mathfrak{g}$. For a vertical vector field $X \in \Gamma T^{\alpha}G$, Corollary 3.5.4 states that there are $X_i \in \Gamma AG$ for which (35) holds, but since ΓAG is in general only a projective $C^\infty(M)$–module, and not free, the X_i and n may vary with X.

The two pullback diagrams (33) and (34) show that $T^{\alpha}G \to G$ is trivializable if and only if $AG \to M$ is so; this generalizes the fact that the tangent bundle of a Lie group is trivializable.

Denote the set of right–invariant vector fields on G by $\Gamma^{RI}T^\alpha G$. It is a $C^\infty(M)$–module under the multiplication $fX = (f \circ \beta)X$ and the maps

$$\Gamma^{RI}T^\alpha G \to \Gamma AG, \ X \mapsto X \circ 1; \qquad \Gamma AG \to \Gamma^{RI}T^\alpha G, \ X \mapsto \overrightarrow{X}, \quad (36)$$

are mutually inverse $C^\infty(M)$–module isomorphisms.

Lemma 3.5.5 *The module $\Gamma^{RI}T^\alpha G$ is closed under the bracket of vector fields.*

Proof A vector field $X \in \Gamma TG$ is vertical if and only if $X \overset{\alpha}{\sim} 0 \in \Gamma TM$ and $X \in \Gamma T^\alpha G$ is right–invariant if and only if $X|_{G_y} \overset{R_g}{\sim} X|_{G_x}$ for each $g \in G_x^y$, $x, y \in M$. The result now follows from the fact that for any smooth map $\varphi \colon M \to M'$, and any vector fields X, Y on M and X', Y' on M', if $X \overset{\varphi}{\sim} X'$ and $Y \overset{\varphi}{\sim} Y'$, then $[X, Y] \overset{\varphi}{\sim} [X', Y']$. $\qquad\square$

The bracket of vector fields on $\Gamma^{RI}T^\alpha G$ may therefore be transferred to ΓAG. We define

$$[X, Y] = [\overrightarrow{X}, \overrightarrow{Y}] \circ 1 \qquad X, Y \in \Gamma AG. \quad (37)$$

This bracket is alternating, $\mathbb{R}$–bilinear and satisfies the Jacobi identity; these follow immediately from the corresponding properties of the bracket of vector fields. For a function $f \in C^\infty(M)$ and $X, Y \in \Gamma AG$ we have

$$\begin{aligned}
\overrightarrow{[X, fY]} &= [\overrightarrow{X}, (f \circ \beta)\overrightarrow{Y}] \\
&= (f \circ \beta)[\overrightarrow{X}, \overrightarrow{Y}] + \overrightarrow{X}(f \circ \beta)\overrightarrow{Y} \\
&= \overrightarrow{f[X, Y]} + \overrightarrow{X}(f \circ \beta)\overrightarrow{Y}.
\end{aligned}$$

Now because $\beta \colon G \to M$ is a surjective submersion and $\beta \circ R_g = \beta$ for all $g \in G$, every right–invariant vector field $\overrightarrow{X}$ is β–projectable; that is, there is a vector field X' on M such that $X'(f) \circ \beta = \overrightarrow{X}(f \circ \beta)$ for all $f \in C^\infty(M)$, and in terms of X' we obtain

$$[X, fY] = f[X, Y] + X'(f)Y.$$

This X' is the β–projection of the right–invariant vector field associated to X, and is described more simply as follows.

Definition 3.5.6 The *anchor* $a = a_G \colon AG \to TM$ of AG is the composite of the vector bundle morphisms

$$
\begin{array}{ccccccc}
AG & \longrightarrow & T^\alpha G & \overset{\subseteq}{\longrightarrow} & TG & \overset{T(\beta)}{\longrightarrow} & TM \\
\downarrow & & \downarrow & & \downarrow & & \downarrow \\
M & \overset{1}{\longrightarrow} & G & =\!=\!=\!= & G & \overset{\beta}{\longrightarrow} & M.
\end{array}
\tag{38}
$$

$\square$

Since $\beta \circ 1 = \mathrm{id}_M$, it follows that a is a morphism over M.

Lemma 3.5.7 *For* $X \in \Gamma AG$, $\overrightarrow{X}$ *is* β–*related to* $a(X)$.

Proof Clearly $T(\beta)(\overrightarrow{X}(g)) = T(\beta \circ R_g)(X(\beta g)) = T(\beta)(X(\beta g))$, and so $a_x = T_{1_x}(\beta_x) \colon T_{1_x}(G_x) \to T_x(M)$. $\square$

It now follows that for any $X, Y \in \Gamma AG$ and any $f \in C^\infty(M)$,

$$
[X, fY] = f[X, Y] + a(X)(f)Y.
\tag{39}
$$

Further, since $\overrightarrow{X}$ is β–related to $a(X)$ and $\overrightarrow{Y}$ is β–related to $a(Y)$, it follows that $[\overrightarrow{X}, \overrightarrow{Y}]$ is β–related to $[a(X), a(Y)]$. But $[\overrightarrow{X}, \overrightarrow{Y}] = \overrightarrow{[X, Y]}$ is also β–related to $a([X, Y])$ and since β is surjective it follows that

$$
a[X, Y] = [a(X), a(Y)].
$$

This completes the construction of the Lie algebroid AG.

Definition 3.5.8 The *Lie algebroid of* G is the vector bundle $AG \to M$ defined in 3.5.1 together with the bracket $[\,,\,]$ defined in (37) and the anchor a defined in 3.5.6. $\square$

Note that the vector bundle $AG \to M$ is always locally trivial and that this is not related to any local triviality property of G. The local triviality of AG goes back ultimately to the assumption that β is a surjective submersion.

$$
* \qquad * \qquad * \qquad * \qquad *
$$

We need to make a few comments on the local version of the correspondence between sections of AG and right–invariant vector fields on G. For $\mathscr{U} \subseteq G$ open, a *local right–invariant vector field* on $\mathscr{U}$ is a vertical vector field on $\mathscr{U}$ such that $X(hg) = T(R_g)(X(h))$ whenever $\alpha h = \beta g$ and both hg and h are in $\mathscr{U}$. If $X \in \Gamma_U AG$, where $U \subseteq M$ is open, then 3.5.3 shows that $\overrightarrow{X}(g) = T(R_g)(X(\beta g))$ defines a smooth local right–invariant vector field on G^U. On the other hand, we have:

Lemma 3.5.9 *Let* $X \in \Gamma_{\mathscr{U}}^{RI} T^{\alpha}G$ *be a local right–invariant vector field on an open set* $\mathscr{U} \subseteq G$. *For* $x \in \beta(\mathscr{U})$ *choose any* $g \in \mathscr{U}^x$ *and define*

$$\mathbf{X}(x) = T(R_{g^{-1}})(X(g)).$$

Then $\mathbf{X}$ *is a well–defined and smooth local section of* AG *on* $\beta(\mathscr{U})$.

Proof That $\mathbf{X}$ is well defined follows from the right–invariance of X. Let $U = \beta(\mathscr{U})$ and note that $\mathscr{R}$ restricts to $T^{\alpha}G|_{\mathscr{U}} \to AG|_U$. Now $\mathbf{X} \circ \beta = \mathscr{R} \circ X$ is smooth, and since β is a submersion it follows that $\mathbf{X}$ is smooth. □

If we now take $X \in \Gamma_{\mathscr{U}}^{RI} T^{\alpha}G$, where $\mathscr{U} \subseteq G$ is open, and apply first 3.5.9 to get $\mathbf{X} \in \Gamma_{\beta(\mathscr{U})} AG$ and then Corollary 3.5.4 to get $\overrightarrow{\mathbf{X}} \in \Gamma_{\beta^{-1}(\beta(\mathscr{U}))}^{RI} T^{\alpha}G$, then we obtain a right–invariant vector field defined on the β–saturation of $\mathscr{U}$ which is equal to X on $\mathscr{U}$. We call $\overrightarrow{\mathbf{X}}$ the β–*saturation* of X and any right–invariant vector field defined on a β–saturated open subset of G will be called a β–*saturated right–invariant vector field*. The above shows that we need not consider right–invariant vector fields defined on more general open sets. Clearly $\overrightarrow{\mathbf{X}}$ is the only right–invariant vector field on $\beta^{-1}(\beta(\mathscr{U}))$ which coincides with X on $\mathscr{U}$.

If $\mathscr{U}$ is itself β–saturated, say $\mathscr{U} = \beta^{-1}(U)$ where U is open in M, and $X \in \Gamma_{\mathscr{U}}^{RI} T^{\alpha}G$, then $\mathbf{X}$ is just the restriction of X to $\mathscr{U} \cap M = U$. In this case we write $X \circ 1_U$, rather than $\mathbf{X}$. There are mutually inverse $C^{\infty}(U)$–module isomorphisms

$$X \mapsto \overrightarrow{X}, \qquad \Gamma_U AG \to \Gamma_{\beta^{-1}(U)}^{RI} T^{\alpha}G$$

and

$$X \mapsto X \circ 1_U, \qquad \Gamma_{\beta^{-1}(U)}^{RI} T^{\alpha}G \to \Gamma_U AG.$$

It is straightforward to show that the bracket on $\Gamma_U AG$ transported from $\Gamma_{\beta^{-1}(U)}^{RI} T^{\alpha}G$ via these isomorphisms coincides with the bracket induced from the bracket on ΓAG by the method of Proposition 3.3.2.

Morphisms

Consider a morphism $F \colon G' \to G$ of Lie groupoids over $f \colon M' \to M$. We construct what we call the *induced morphism* $A(F) = F_* \colon AG' \to AG$ of Lie algebroids.

Since $\alpha \circ F = f \circ \alpha'$, the vector bundle morphism $T(F)$ restricts to $T^\alpha(F)\colon T^\alpha G' \to T^\alpha G$. Since $F \circ 1' = 1 \circ f$, the composition

$$
\begin{array}{ccccc}
AG' & \longrightarrow & T^\alpha G' & \overset{T^\alpha(F)}{\longrightarrow} & T^\alpha G \\
\downarrow & & \downarrow & & \downarrow \\
M' & \overset{1'}{\longrightarrow} & G' & \overset{F}{\longrightarrow} & G
\end{array}
\tag{40}
$$

is a vector bundle morphism over $1 \circ f\colon M' \to G$, and so there is a unique vector bundle morphism $F_*\colon AG' \to AG$ over f such that the composition

$$
\begin{array}{ccccc}
AG' & \overset{F_*}{\longrightarrow} & AG & \longrightarrow & T^\alpha G \\
\downarrow & & \downarrow & & \downarrow \\
M' & \overset{f}{\longrightarrow} & M & \overset{1}{\longrightarrow} & G
\end{array}
\tag{41}
$$

is equal to (40).

From $\beta \circ F = f \circ \beta'$ and the definition of the anchors on AG' and AG it follows immediately that $a \circ F_* = T(f) \circ a'$. We next prove a bracket preservation property for F_*. Note first that an $X' \in \Gamma AG'$ and an $X \in \Gamma AG$ have $F_* \circ X' = X \circ f$ if and only if $\overrightarrow{X'} \overset{F}{\sim} \overrightarrow{X}$.

Proposition 3.5.10 *Take $X', Y' \in \Gamma AG'$ and $X, Y \in \Gamma A\Omega$ such that $F_* \circ X' = X \circ f$, $F_* \circ Y' = Y \circ f$. Then $F_* \circ [X', Y'] = [X, Y] \circ f$.*

Proof We have $\overrightarrow{X'} \overset{F}{\sim} \overrightarrow{X}$ and $\overrightarrow{Y'} \overset{F}{\sim} \overrightarrow{Y}$, so $\overrightarrow{[X', Y']} = [\overrightarrow{X'}, \overrightarrow{Y'}] \overset{F}{\sim} [\overrightarrow{X}, \overrightarrow{Y}] = \overrightarrow{[X, Y]}$ from which it follows that $F_* \circ [X', Y'] = [X, Y] \circ f$. $\qquad\square$

If $M = M'$ and $f = \mathrm{id}_M$ this can be stated more simply as

$$
F_*[X, Y] = [F_* X, F_* Y], \qquad X, Y \in \Gamma AG
$$

and shows that F_* is a morphism of Lie algebroids over M as defined in 3.3.1. In this case the constructions $G \longmapsto AG$ and $F \longmapsto F_*$ constitute a functor from the category of Lie groupoids on the given base M and smooth morphisms of Lie groupoids over M to the category of Lie algebroids on M and Lie algebroid morphisms over M. This is the *Lie functor* for the fixed base M.

For general M and M', we have not yet defined a morphism of Lie algebroids; in particular the bracket preservation property 3.5.10 is not sufficient. Although the construction of F_* in the general case is as we have given it here, it is not a simple matter to give a general definition of an abstract Lie algebroid morphism. See §4.3.

Examples

We now give the simplest examples of the construction of the Lie algebroid of a Lie groupoid. Many further examples are developed in the following sections.

Example 3.5.11 The Lie algebroid of a pair groupoid $M \times M$ is naturally isomorphic to TM. For any Lie groupoid G, the anchor map $(\beta, \alpha) \colon G \to M \times M$, considered as a morphism over M, induces the anchor $a \colon AG \to TM$ of AG. ⊠

Example 3.5.12 Let (K, q, M) be a Lie group bundle. The *Lie algebra bundle associated to* K is denoted K_* and defined as follows.

Let K_* be the inverse image of $T^q K \to K$ across the identity section $M \to K$, $x \mapsto 1_x$, of K. The fibre of K_* over x is then the Lie algebra of K_x. Let $\{\psi_i \colon U_i \times G \to K_{U_i}\}$ be an LGB atlas for K; then the induced chart for $T^q K$ is the composite of $\psi_i^{-1} \times \mathrm{id} \colon K_{U_i} \times \mathfrak{g} \to U_i \times G \times \mathfrak{g}$ followed by $U_i \times G \times \mathfrak{g} \to T(U_i \times G)$, $(x, g, X) \mapsto 0_x \oplus T(R_g)(X)$, followed by $T(\psi_i) \colon T(U_i \times G) \to TK$. The restriction of this chart to the identity section $1_{U_i} \subseteq K_{U_i}$ is therefore

$$(x, X) \mapsto T(\psi_i)(0_x \oplus X) = T(\psi_{i,x})(X)$$

and is denoted $\psi_{i*} \colon U_i \times \mathfrak{g} \to K_*|_{U_i}$; fibrewise ψ_{i*} is the Lie algebra isomorphism $\mathfrak{g} \to K_*|_x$ induced by $\psi_{i,x} \colon G \to K_x$. Thus K_* with the Lie algebra bracket on each fibre induced from the Lie group structures on the fibres of K, is a Lie algebra bundle.

It is clear that the construction of K_* as an LAB coincides with the construction of the Lie algebroid AK of K considered as a totally intransitive Lie groupoid. ⊠

Example 3.5.13 Let M be a manifold, G a Lie group, and Ω the trivial Lie groupoid $M \times G \times M$.

The source map α is the projection $\pi_3 \colon M \times G \times M \to M$, so $T^\alpha \Omega$ is the Whitney sum $\mathrm{pr}_1^! TM \oplus \mathrm{pr}_2^! TG$ of the inverse image bundles $\mathrm{pr}_1^! TM$ and $\mathrm{pr}_2^! TG$ over $M \times G \times M$. Since the compositions $\mathrm{pr}_1 \circ 1$ and $\mathrm{pr}_2 \circ 1$ are $x \mapsto x$ and $x \mapsto 1$ respectively, the inverse image of $T^\alpha \Omega$ over 1 is $(\mathrm{pr}_1 \circ 1)^! TM \oplus (\mathrm{pr}_2 \circ 1)^! TG = TM \oplus (M \times \mathfrak{g})$, where the Whitney sum is now over M. This is the vector bundle $A\Omega$.

We will write a general vector field on Ω in the form $X \oplus V \oplus Y$ where X, V, Y are sections of $\mathrm{pr}_1^! TM$, $\mathrm{pr}_2^! TG$, $\mathrm{pr}_3^! TM$, respectively. Such a field is vertical if and only if $Y = 0$. The right–translation $R_{(y,h,z)} \colon M \times G \times \{y\} \to M \times G \times \{z\}$ has tangent

$$T_{(x,g,y)}(R_{(y,h,z)}) \colon X_x \oplus V_g \oplus 0_y \mapsto X_x \oplus T(R_h)(V_g) \oplus 0_z$$

and so $X \oplus V \oplus 0$ is right–invariant if and only if

$$V(x, gh, z) = T_g(R_h)(V(x, g, y)) \quad \text{and} \quad X(x, gh, z) = X(x, g, y)$$

identically in x, y, z, g, h. This is so if and only if $V(x, g, y)$ is independent of y and right–invariant in g, and $X(x, g, y)$ depends only on x. When this is so, V can be identified with the function $x \longmapsto V(x, 1, x)$, $M \to \mathfrak{g}$, also denoted by V, and X can be identified with the vector field $x \longmapsto X(x, 1, x)$ on M, also denoted by X; with this notation $X \oplus V$ is the section $(X \oplus V \oplus 0) \circ 1$ of $TM \oplus (M \times \mathfrak{g})$. Conversely, given $X \in \Gamma TM$ and $V \colon M \to \mathfrak{g}$, the right–invariant vector field $\overrightarrow{X \oplus V}$ is $(x, g, y) \longmapsto X_x \oplus T(R_g)(V(x)) \oplus 0_y$.

To simplify the calculation of the bracket on $A\Omega$, temporarily denote by $\overrightarrow{V}$ and $\overrightarrow{X}$ the right–invariant vector fields $\overrightarrow{0 \oplus V}$ and $\overrightarrow{X \oplus 0}$. With this notation, for any $V, W \colon M \to \mathfrak{g}$ and $X, Y \in \Gamma TM$,

$$[\overrightarrow{X \oplus V}, \overrightarrow{Y \oplus W}] = [\overrightarrow{X} + \overrightarrow{V}, \overrightarrow{Y} + \overrightarrow{W}] = [\overrightarrow{X}, \overrightarrow{Y}] + [\overrightarrow{X}, \overrightarrow{W}] - [\overrightarrow{Y}, \overrightarrow{V}] + [\overrightarrow{V}, \overrightarrow{W}].$$

That $\overrightarrow{V}$ has the global flow $\psi_t(y, g, x) = (y, \exp tV(y)g, x)$ is easily verified. Clearly if $\{\varphi_t\}$ is a local flow for X on $U \subseteq M$ then $\{\overrightarrow{\varphi_t}\}$ defined by $\overrightarrow{\varphi_t}(y, g, x) = (\varphi_t(y), g, x)$ is a local flow for $\overrightarrow{X}$ on $U \times G \times M$. Using these flows it is straightforward to show that $[\overrightarrow{V}, \overrightarrow{W}] = \overrightarrow{[V, W]}$, where $[V, W]$ is the map $x \mapsto [V(x), W(x)]$, $M \to \mathfrak{g}$; that $[\overrightarrow{X}, \overrightarrow{W}] = \overrightarrow{X(W)}$, where $X(W)$ is the Lie derivative of the vector–valued function W; and that $[\overrightarrow{X}, \overrightarrow{Y}] = \overrightarrow{[X, Y]}$. Consequently the bracket on $TM \oplus (M \times \mathfrak{g})$ is given by

$$[X \oplus V, Y \oplus W] = [X, Y] \oplus \{X(W) - Y(V) + [V, W]\}. \qquad (42)$$

The anchor $a \colon TM \oplus (M \times \mathfrak{g}) \to TM$ is the projection $X \oplus V \longmapsto X$ and the adjoint bundle is, by the formula for the bracket in $TM \oplus (M \times \mathfrak{g})$, the trivial Lie algebra bundle $M \times \mathfrak{g}$. Thus $A(M \times G \times M)$ is the trivial Lie algebroid on M with structure algebra $\mathfrak{g}$ constructed in 3.3.3.

Now consider a morphism $F \colon M \times G \times M \to M \times H \times M$ of trivial Lie groupoids over M. Applying 1.2.5, F can be written in the form

$$F(y, g, x) = (y, \theta(y)s(g)\theta(x)^{-1}, x)$$

for some morphism of Lie groups $s\colon G \to H$ and smooth map $\theta\colon M \to H$. In terms of this description $T(F)(X \oplus V \oplus Y)$ is given by

$$X \oplus \{T(R_{s(g)\theta(y)^{-1}})T(\theta)(X) + T(L_{\theta(x)})T(R_{\theta(y)^{-1}})T(s)(V)$$
$$- T(L_{\theta(x)s(g)\theta(y)^{-1}})T(R_{\theta(y)^{-1}})T(\theta)(Y)\} \oplus Y,$$

where $X \in T_x(M)$, $V \in T_g(G)$, $Y \in T_y(M)$. Setting $Y = 0$ and $y = x$, $g = 1$, this reduces to $X \oplus \{\Delta(\theta)(X) + \mathrm{Ad}_{\theta(x)}s_*(V)\} \oplus 0$, where Δ is the Darboux derivative (see §5.1). Hence $F_*\colon TM \oplus (M \times \mathfrak{g}) \to TM \oplus (M \times \mathfrak{h})$ is

$$F_*(X \oplus V) = X \oplus \{\Delta(\theta)(X) + \mathrm{Ad}(\theta)s_*(V)\}.$$

In terms of the description in 3.3.3, F_* is formed from the Maurer–Cartan form $\Delta(\theta)$ and the LAB morphism $\mathrm{Ad}(\theta) \circ s_*$. The compatibility condition for $\Delta(\theta)$ and $\mathrm{Ad}(\theta) \circ s_*$ can be proved directly: see §5.1. ⊠

Example 3.5.14 Let $m\colon G \times M \to M$ be a smooth action of a Lie group G on a manifold M, and let Υ be the corresponding action groupoid $G \ltimes M$. Then, as in the preceding example, $T^\alpha \Upsilon = \mathrm{pr}_1^! TG \oplus 0$ and $A\Upsilon = M \times \mathfrak{g}$ as a vector bundle on M. Regarding vertical vector fields as $\mathscr{V}\colon G \times M \to TG$, such a $\mathscr{V}$ is right–invariant if and only if $\mathscr{V}(gh, y) = T(R_h)(\mathscr{V}(g, hy))$ identically in g, h, y. For $V\colon M \to \mathfrak{g}$ the corresponding right–invariant vector field $\overrightarrow{V}$ is $\overrightarrow{V}(g, x) = T(R_g)(V(gx))$. The anchor $a\colon M \times \mathfrak{g} \to TM$ is $(x, X) \to T_1(m(-, x))(X) = X^\dagger(x)$, where $X^\dagger$ is the fundamental vector field generated by X. We extend this notation to any $V\colon M \to \mathfrak{g}$, so that $V^\dagger(x) = V(x)^\dagger(x)$.

For the bracket, notice that $F\colon G \ltimes M \to M \times G \times M$, $(g, x) \mapsto (gx, g, x)$, is a morphism over M into the trivial groupoid. The induced morphism $F_*\colon M \times \mathfrak{g} \to TM \oplus (M \times \mathfrak{g})$ is $(x, X) \mapsto X^\dagger(x) \oplus (x, X)$ or, in terms of sections, $F_*(V) = V^\dagger \oplus V$. Now using (42) it follows that, for $V, W\colon M \to \mathfrak{g}$,

$$[V, W] = V^\dagger(W) - W^\dagger(V) + [V, W]^\bullet,$$

where we write for clarity $[,]^\bullet$ for the pointwise bracket.

Thus $A(G \ltimes M)$ is the action Lie algebroid $\mathfrak{g} \ltimes M$ of 3.3.7 corresponding to the infinitesimal action of $\mathfrak{g}$ on M. ⊠

We now want to verify that locally trivial Lie groupoids correspond to transitive Lie algebroids. The following more general result comes at no extra cost.

Proposition 3.5.15 *Let* $F\colon G \to G'$ *be a morphism of Lie groupoids over* M. *Then:*

(i) *F is a submersion if and only if each $F_x\colon G_x \to G'_x$, $x \in M$, is a submersion, and if and only if $F_*\colon AG \to AG'$ is a fibrewise surjection;*

(ii) *F is an immersion if and only if each F_x is an immersion, and if and only if F_* is a fibrewise injection;*

(iii) *F is of locally constant rank if and only if F_* is.*

Proof We prove (iii); the other results can be proved by the same method.

($\Longrightarrow$) Let the connectedness components of G be C_i and let the rank of F on C_i be k_i. Take $g \in G$ and write $x = \alpha g$, $y = \beta g$. From the diagram

$$
\begin{array}{ccccc}
T_g(G_x) & \rightarrowtail & T_g(G) & \overset{T(\alpha)}{\relbar\joinrel\twoheadrightarrow} & T_x(M) \\
{\scriptstyle T_g(F_x)}\big\downarrow & & {\scriptstyle T_g(F)}\big\downarrow & & \| \\
T_{F(g)}(G'_x) & \rightarrowtail & T_{F(g)}(G') & \overset{T(\alpha')}{\relbar\joinrel\twoheadrightarrow} & T_x(M)
\end{array}
$$

it follows that $\mathrm{rk}_g(F_{\alpha g}) = k_i - \dim M$, for $g \in C_i$. Now for any $g \in G$ we have $F_{\beta g} \circ R_g = R_{F(g)} \circ F_{\alpha g}$ so $\mathrm{rk}_{1_{\beta g}}(F_{\beta g}) = \mathrm{rk}_g(F_{\alpha g})$. Therefore, $\mathrm{rk}_{1_x}(F_x) = k_i - \dim M$ is constant for $x \in \beta(C_i)$ and since β is an open map, this shows that $\mathrm{rk}_{1_x}(F_x)$ is a locally constant function of x.

($\Longleftarrow$) Let the components of M be M_i and let the rank of F_* on M_i be k_i. By the same argument it follows that $\mathrm{rk}_g(F_{\alpha g}) = \mathrm{rk}_{1_{\beta g}}(F_{\beta g}) = k_i$ for $g \in G^{M_i}$ and so $\mathrm{rk}_g(F) = k_i + \dim M$ for $g \in G^{M_i}$. Since G^{M_i} is open in G, this completes the proof. $\qquad\square$

Recall from 2.2.5 that every base–preserving morphism of locally trivial Lie groupoids is of locally constant rank, and that such a morphism is a submersion if it is surjective and is an immersion if it is injective. It is not true that every base–preserving morphism of general Lie groupoids is of locally constant rank, even if the base is connected: let G be an action groupoid $H \ltimes M$ where $H \times M \to M$ is a smooth action of the Lie group H with a fixed point $x_\circ$ and all other orbits of positive dimension. Then (β, α) is $H \times M \to M \times M$, $(h, x) \mapsto (hx, x)$ and the rank is not constant in any neighbourhood of $(1, x_\circ)$. On the other hand, $(\beta, \alpha)_x = \beta_x$ is the evaluation $H \to M$ at x and hence is of locally constant rank, so this example also shows that 3.5.15(iii) cannot be strengthened to make it similar in form to (i) and (ii).

The following result is used in §6.2.

Proposition 3.5.16 *Let* $F: G \to G'$ *be a morphism of Lie groupoids over* M. *If* G' *is* α*–connected and* $F_*: AG \to AG'$ *is fibrewise surjective, then* F *is surjective.*

Proof By 3.5.15, F is a submersion and so $F(G)$ is a symmetric α–neighbourhood of the base in G'. By 1.5.8, $F(G)$ generates G'; since $F(G)$ is a subgroupoid of G' it follows that $F(G) = G'$. □

We come now to the exactness result for the Lie functor. Extensions of Lie groupoids were defined in 1.7.15, and extensions of Lie algebroids in 3.3.19.

Proposition 3.5.17 *Let* $K \rightarrowtail^{\iota} G' \xrightarrow{\pi} G$ *be an extension of Lie groupoids over* M. *Then* $AK \rightarrowtail^{\iota_*} AG' \xrightarrow{\pi_*} AG$ *is an extension of Lie algebroids over* M.

Proof All that need be proved is that the sequence of vector spaces

$$T_{1_x}(K_x) \rightarrowtail^{T_{1_x}(\iota_x)} T_{1_x}(G'_x) \xrightarrow{T_{1_x}(\pi_x)} T_{1_x}(G_x)$$

is exact, and this follows immediately from the fact that $\pi_x: G'_x \to G_x$ is a surjective submersion with $(\pi_x)^{-1}(1_x) = K_x$. □

Corollary 3.5.18 *Let* G *be a Lie groupoid on* M. *If* G *is locally trivial then* AG *is transitive. If* AG *is transitive and* M *is connected, then* G *is locally trivial.*

Proof Since $(\beta, \alpha)_* = a: AG \to TM$, the first result follows from 3.5.15(i) and the second from 3.5.16. □

If Ω is a locally trivial Lie groupoid on M then $\mathscr{I}\Omega$ is a LGB on M and

$$\mathscr{I}\Omega \rightarrowtail \Omega \xrightarrow{(\beta,\alpha)} M \times M$$

is an extension of Lie groupoids over M. Since $(\beta, \alpha)_* = a: A\Omega \to TM$, the kernel of a is the Lie algebroid of $\mathscr{I}\Omega$ as a Lie groupoid; by 3.5.12 this coincides with the LAB associated to $\mathscr{I}\Omega$.

Definition 3.5.19 *Let* Ω *be a locally trivial Lie groupoid. The LAB associated to* $\mathscr{I}\Omega$ *is denoted* $L\Omega$ *and called the* adjoint Lie algebra bundle *or simply* adjoint bundle *of* Ω. □

Proposition 3.5.20 *Let Ω be a locally trivial Lie groupoid. For $\xi \in \Omega$ define*

$$\mathrm{Ad}_\xi = T_{1_{\alpha\xi}}(I_\xi)\colon L\Omega|_{\alpha\xi} \to L\Omega|_{\beta\xi}.$$

Then $\mathrm{Ad}\colon \Omega \to \Phi_{\mathrm{Aut}}(L\Omega)$ *is a representation of Ω on the LAB $L\Omega$, the* adjoint representation.

Proof The proof is straightforward, using the smoothness of the action of Ω on $\mathscr{I}\Omega$. □

If $\mathscr{I}\Omega$ is abelian then Ad_ξ depends only on $\alpha\xi$ and $\beta\xi$ and so there is a global chart $M \times \mathfrak{g} \to L\Omega$, where $\mathfrak{g}$ is the Lie algebra of a vertex group of Ω.

Left–invariant vector fields

Given $X \in \Gamma AG$, define $\overleftarrow{X} \in \mathscr{X}(G)$ by

$$\overleftarrow{X}(g) = T(L_g)T(\iota)(X(\alpha g)), \qquad g \in G.$$

We call $\overleftarrow{X}$ the *left–invariant vector field on G corresponding to $X \in \Gamma AG$*. Note that, when G is a group, $\overleftarrow{X}(g) = -T(L_g)(X)$; this is *not* what is usually meant by 'the left–invariant vector field on G corresponding to $X \in \mathfrak{g}$.'

Since $L_g \circ \iota = \iota \circ R_{g^{-1}}$, it follows that $\overleftarrow{X} = \iota_*(\overrightarrow{X})$ and so

$$[\overleftarrow{X}, \overleftarrow{Y}] = \overleftarrow{[X, Y]}$$

for $X, Y \in \Gamma AG$. Notice also that $\overleftarrow{X} \overset{\beta}{\sim} 0$ and $\overleftarrow{X} \overset{\alpha}{\sim} a(X)$. The flow of $\overleftarrow{X}$ is $R_{\mathrm{Exp}\,-tX}$ (see §3.6).

The symmetric construction

Equally, one may define the *left Lie algebroid* $A_L G$ of a Lie groupoid $G \rightrightarrows M$ to be the pullback of $T^\beta G$ over $1\colon M \to G$, with the bracket and anchor now defined by left–invariant vector fields. Namely, given $\xi \in \Gamma A_L G$, there is a left–invariant vector field $\widetilde{\xi}$ defined by $\widetilde{\xi}(g) = T(L_g)(\xi(\alpha g))$. (Thus $\widetilde{\xi} = \overleftarrow{X}$ where $X \in \Gamma AG$ is defined by $X(m) = T(\iota)(\xi(m))$.) The bracket and anchor for $A_L G$ are now defined by

$$\widetilde{[\xi, \eta]_L} = [\widetilde{\xi}, \widetilde{\eta}], \qquad \widetilde{\xi} \overset{\alpha}{\sim} a_L(\xi).$$

Finally, the Lie algebroid of a Lie groupoid $G \rightrightarrows M$ can be regarded as the normal bundle in G of the base manifold. Let $\mathscr{N} = T_{1_M}(G)/T(1)(TM)$ denote this normal bundle and write elements of $\mathscr{N}$ as $\overline{X}$ where $X \in T_{1_m}G$ for some $m \in M$, and $\overline{X} = \overline{X + T(1)(x)}$ for all $x \in T_m(M)$.

Given $X \in T_{1_m}G$, there is a unique $X^R \in T_{1_m}^{\alpha}G$ with $\overline{X^R} = \overline{X}$, namely $X^R = X - T(1)T(\alpha)(X)$, and X^R depends only on $\overline{X} \in \mathscr{N}$. The map $\overline{X} \mapsto X^R$ is a vector bundle isomorphism over M and the Lie algebroid structure of AG can therefore be transported to $\mathscr{N}$. We denote the anchor by

$$a_R \colon \mathscr{N} \to TM, \qquad a_R(\overline{X}) = T(\beta)(X) - T(\alpha)(X)$$

and the bracket by $[\,,\,]_R$. This is the *symmetric Lie algebroid of the Lie groupoid* G.

Similarly there is a canonical vector bundle isomorphism from $\mathscr{N}$ to $A_L G$.

3.6 The exponential map

In this section we define the exponential map for a Lie groupoid, and use it to calculate the Lie algebroids of the various frame groupoids introduced in §1.7.

Let G be a Lie groupoid on M and take $X \in \Gamma AG$. Consider a local flow $\{\varphi_t \colon \mathscr{U} \to \mathscr{U}_t\}$ for $\overrightarrow{X} \in \Gamma TG$. Since $\overrightarrow{X}$ is α–vertical, we have $\alpha \circ \varphi_t = \alpha$ so for $x \in M$ with $\mathscr{U} \cap G_x \neq \emptyset$, each φ_t restricts to $\mathscr{U} \cap G_x \to \mathscr{U}_t \cap G_x$. For $\xi \in G_x^y$ where $\mathscr{U} \cap G_x \neq \emptyset$, $\mathscr{U} \cap G_y \neq \emptyset$, we know $(R_\xi)_*(\overrightarrow{X}|_{G_y}) = \overrightarrow{X}|_{G_x}$ so $\varphi_t \circ R_\xi = R_\xi \circ \varphi_t$ for all t.

Now define $U = \beta(\mathscr{U}), U_t = \beta(\mathscr{U}_t)$ and $\psi_t \colon U \to U_t$ so that

$$
\begin{array}{ccc}
\mathscr{U} & \xrightarrow{\varphi_t} & \mathscr{U}_t \\
\downarrow{\scriptstyle\beta} & & \downarrow{\scriptstyle\beta} \\
U & \xrightarrow{\psi_t} & U_t
\end{array}
$$

commutes; that is, $\psi_t(x) = \beta(\varphi_t(h))$ for any $h \in \mathscr{U}$ with $\beta h = x$. Since $\varphi_t(h\xi) = \varphi_t(h)\xi$, this is well defined, and since β is a submersion ψ_t is smooth. Since $(\varphi_t)^{-1} = \varphi_{-t}$ has the same properties as φ_t, it follows that $(\psi_{-t} \circ \psi_t) \circ \beta = \beta$ and ψ_t is therefore a local diffeomorphism. Now, for $x \in U$,

$$\frac{d}{dt}\psi_t(x)\Big|_0 = \frac{d}{dt}\beta(\varphi_t(h))\Big|_0 = \beta_*(\overrightarrow{X}(h))$$

for any $h \in \mathcal{U} \cap G^x$, and so $\{\psi_t \colon U \to U_t\}$ is a local flow for $a(X) = \beta_*(\overrightarrow{X})$.

Lastly, $\varphi_t \colon \mathcal{U} \to \mathcal{U}_t$ and $\psi_t \colon U \to U_t$ satisfy the conditions of 1.4.12 and so each φ_t is the restriction of a unique local left–translation $L_{\sigma_t} \colon G^U \to G^{U_t}$, where $\sigma_t \in \Gamma_U G$ is defined by $\sigma_t(x) = \varphi_t(h)h^{-1}$, where h is any element of $\mathcal{U} \cap G^x$. This proves the following result.

Proposition 3.6.1 *Let G be a Lie groupoid on M, let $W \subseteq M$ be an open subset, and take $X \in \Gamma_W AG$. Then for each $x_0 \in W$ there is an open neighbourhood U of x_0 in W, called a* flow neighbourhood *for X, an $\varepsilon > 0$, and a unique smooth family of local bisections $\mathrm{Exp}\, tX \in \Gamma_U G$, $|t| < \varepsilon$, such that;*

(i) $\dfrac{d}{dt}\mathrm{Exp}\, tX\Big|_0 = X$,

(ii) $\mathrm{Exp}\, 0X = \mathrm{id} \in \Gamma_U G$,

(iii) $\mathrm{Exp}(t+s)X = (\mathrm{Exp}\, tX) \star (\mathrm{Exp}\, sX)$, *whenever* $|t|, |s|, |t+s| < \varepsilon$,

(iv) $\mathrm{Exp}{-t}X = (\mathrm{Exp}\, tX)^{-1}$,

(v) $\{\beta \circ \mathrm{Exp}\, tX \colon U \to U_t\}$ *is a local 1–parameter group of transformations for $a(X) \in \Gamma_W TM$ in U.*

Property (iii) expands out to:

$$(\mathrm{Exp}\,(t+s)X)(x) = (\mathrm{Exp}\, tX)(\beta(\mathrm{Exp}\, sX(x)))\mathrm{Exp}\, sX(x)$$

where $x \in U$ and s is such that $\beta(\mathrm{Exp}\, sX(x)) \in U$ (in addition to $|s|, |t|, |s+t| < \varepsilon$).

The map $t \mapsto \mathrm{Exp}\, tX$ is smooth in the sense that $\mathbb{R} \times U \rightsquigarrow G$, $(t,x) \mapsto \mathrm{Exp}\, tX(x)$ is smooth; this follows from the smoothness of the local flow $\mathbb{R} \times G \to G$, $(t,\xi) \mapsto \varphi_t(\xi) = \mathrm{Exp}\, tX(\beta\xi)\xi$ for $\overrightarrow{X}$.

A well–defined *exponential map* $X \mapsto \mathrm{Exp}X$ may be constructed on the sheaf of germs of local sections of AG with values in the sheaf of germs of local bisections of G. We will use the term 'exponential map' in the obvious loose sense.

Examples 3.6.2 Let $\Omega = M \times M$ and let $X \in \Gamma A\Omega = \mathcal{X}(M)$ have a local flow $\{\varphi_t \colon U \to U_t\}$. Then $\{\varphi_t \times \mathrm{id}_M \colon U \times M \to U_t \times M\}$ is a local flow for $\overrightarrow{X} = X \oplus 0$ and $\mathrm{Exp}\, tX$ can be identified with $\varphi_t \in \Gamma_U(M \times M)$ (see 1.4.3).

Let K be an LGB on M, and let $X \in \Gamma(K_*)$. Then $\mathrm{Exp}\, tX$ is the global section $x \mapsto \exp tX(x)$ of K whose value at $x \in M$ is the Lie algebra exponential of $X(x) \in K_*|_x$.

Let $P(M, G, \pi)$ be a principal bundle and take $X \in \Gamma(\frac{TP}{G})$. It is shown in 3.2.7 that the local flows of $\overline{X} \in \Gamma^G TP$ are of the form $\varphi_t(\psi_t, \mathrm{id}_G)$, where $\{\psi_t \colon U \to U_t\}$ is a local flow for $\pi_*(X)$ and the φ_t are defined on the π–saturations $\pi^{-1}(U) \to \pi^{-1}(U_t)$. It is easy to see that if X is regarded as in $\Gamma A(\frac{P \times P}{G})$ then $\varphi_t(\psi_t, \mathrm{id}_G)$ corresponds to $\mathrm{Exp}\, tX$ under 1.4.7.   ⊠

Proposition 3.6.3 *Let G be a Lie groupoid on a connected manifold M. Then the set of values $\mathrm{Exp}\, tX(m)$, for all $X \in \Gamma AG$ and possible $t \in \mathbb{R}$ and $m \in M$, is the identity–component subgroupoid $C(G)$.*

Proof Given $\mathrm{Exp}\, tX(m)$, letting $t \to 0$ defines a curve within G_m to 1_m. On the other hand, the image of each $\mathrm{Exp}\, tX$ is an open set in G. So, for fixed X, the union of these images, as t varies through possible values, is an α–connected open set which intersects 1_M. Now use §1.5. □

Theorem 3.6.4 *Let G be a Lie groupoid on M, let $X \in \Gamma AG$, and let $g_0 \in G$ with $\beta g_0 = y_0$. Then the integral curve for $\overrightarrow{X}$ through g_0 is infinitely extendable in both directions if and only if the integral curve for $a(X)$ through y_0 is infinitely extendable in both directions. In particular, $\overrightarrow{X}$ is complete if and only if $a(X)$ is complete.*

Proof Since $\overrightarrow{X}$ is β–projectable to $a(X)$, each integral curve of $\overrightarrow{X}$ projects under β to an integral curve of $a(X)$, and the 'only if' statements are immediate.

For the converse, recall first that for any vector field X on any manifold, if $X(x_0) \neq 0$ for a specific point x_0, then X is nowhere zero on the integral curve through x_0.

Let $\gamma \colon \mathbb{R} \to G_{x_0}$ be the integral curve for $a(X)$ through x_0. Assume first that $a(X)(x_0) \neq 0$. It follows that $a(X)(\gamma(t)) \neq 0$ for all $t \in \mathbb{R}$, so γ is an immersion. Write $\overline{\gamma} \colon (a, b) \to G_{x_0}$ for the integral curve of $\overrightarrow{X}$ through 1_{x_0}. We must prove that neither a nor b can be finite. Note that $\beta(\overline{\gamma}(t)) = \gamma(t)$ for $t \in (a, b)$.

Suppose that b is finite. If $G_{x_0}^{\gamma(b)} \neq \emptyset$ take $g \in G_{x_0}^{\gamma(b)}$ and let the integral curve for $\overrightarrow{X}$ be $\delta^g \colon (b - \varepsilon, b + \varepsilon) \to G_{x_0}$ with $\delta^g(b) = g$. Again $\beta(\delta^g(t)) = \gamma(t)$ for $t \in (b - \varepsilon, b + \varepsilon)$. Since γ is an immersion we can choose ε so that γ is injective on $(b - \varepsilon, b + \varepsilon)$. Since $\overrightarrow{X}$ is right–invariant it follows that for any $\ell \in G_{x_0}^{x_0}$ the curve $\delta^{g\ell} = R_\ell \circ \delta^g$ is the integral curve of $\overrightarrow{X}$ with $\delta^{g\ell}(b) = g\ell$. Each $\delta^{g\ell}$ is injective since γ is injective on $(b - \varepsilon, b + \varepsilon)$.

Now write

$$S = \{\gamma(t) \mid t \in (b - \varepsilon, b + \varepsilon)\}, \qquad \overline{S} = \{h \in G_{x_0} \mid \beta h \in S\}.$$

We claim that each $h \in \overline{S}$ is $\delta^{g\ell}(t)$ for a unique $t \in (b - \varepsilon, b + \varepsilon)$ and $\ell \in G_{x_0}^{x_0}$. First suppose that $\delta^{g\ell_1}(t_1) = \delta^{g\ell_2}(t_2)$. Applying β gives $\gamma(t_1) = \gamma(t_2)$, so $t_1 = t_2$. Since both $\delta^{g\ell_1}$ and $\delta^{g\ell_2}$ are integral curves we must have $\delta^{g\ell_1} = \delta^{g\ell_2}$. Evaluating at b gives $g\ell_1 = g\ell_2$ so $\ell_1 = \ell_2$. To prove existence, take $h \in \overline{S}$, say $h \in G_{x_0}^{\gamma(t)}$ where $t \in (b - \varepsilon, b + \varepsilon)$. Since $\delta^g(t) \in G_{x_0}^{\gamma(t)}$ also, there exists ℓ such that

$$h = \delta^g(t)\ell = \delta^{g\ell}(t).$$

This proves the claim. Now take any $t \in (b - \varepsilon, b)$. Then $\overline{\gamma}(t) \in \overline{S}$ and so $\overline{\gamma}(t) = \delta^{g\ell}(t')$ for some $t' \in (b - \varepsilon, b + \varepsilon)$ and $\ell \in G_{x_0}^{x_0}$. As before, $t = t'$. So $\overline{\gamma}$ and $\delta^{g\ell}$ have the same value at t and so must coincide on $(b - \varepsilon, b)$. So $\delta^{g\ell}$ extends $\overline{\gamma}$ to $(a, b + \varepsilon)$, and b cannot be maximal. Likewise a finite a cannot be minimal.

Lastly, consider the case $a(X)(x_0) = 0$. Then $\gamma(t) = x_0$ for all $t \in \mathbb{R}$. Furthermore, $X(x_0)$ is in the Lie algebra of $G_{x_0}^{x_0}$, and on $G_{x_0}^{x_0}$ the vector field $\overrightarrow{X}$ restricts to the right–invariant vector field corresponding to $X(x_0)$. Thus we are in the standard Lie group case.

The case of the integral curve passing through a general g_0 follows by right–translation. $\qquad\qquad\qquad\qquad\qquad\qquad\qquad\qquad\qquad\qquad\qquad\qquad\square$

Proposition 3.6.5 *Let $F\colon G \to G'$ be a morphism of Lie groupoids over M. Then if U is a flow neighbourhood for $X \in \Gamma AG$, it is also a flow neighbourhood for $F_*(X)$ and*

$$\widetilde{F}(\operatorname{Exp} tX) = \operatorname{Exp} tF_*(X)$$

for all t for which $\operatorname{Exp} tX$ is defined.

Proof It is easy to verify that $t \mapsto \widetilde{F}(\operatorname{Exp} tX)$ has the properties which characterize $t \mapsto \operatorname{Exp} tF_*(X)$. $\qquad\qquad\qquad\qquad\qquad\qquad\qquad\qquad\square$

Theorem 3.6.6 *Let (E, q, M) be a vector bundle. For $X \in \Gamma A\Phi(E)$ define $\mathscr{D}(X)\colon \Gamma E \to \Gamma E$ by*

$$\mathscr{D}(X)(\mu)(x) = -\frac{d}{dt}\overline{\operatorname{Exp} tX}(\mu)(x)\bigg|_0$$

where $\mu \in \Gamma E, x \in M$, *and the exponential is taken in a flow neighbourhood of* x. *(The bar notation is defined in 1.4.4.)*

Then $\mathscr{D}(X) \in \Gamma\mathfrak{D}(E)$, *and* $\mathscr{D}\colon \Gamma A\Phi(E) \to \Gamma\mathfrak{D}(E)$ *defines an isomorphism* $A\Phi(E) \to \mathfrak{D}(E)$ *of Lie algebroids over* M.

Proof In more detail, $\mathscr{D}(X)(\mu)(x)$ is defined as

$$\frac{d}{dt}\operatorname{Exp} tX(x)^{-1}(\mu)((\beta \circ \operatorname{Exp} tX)(x))\Big|_0$$

$$= -\frac{d}{dt}\operatorname{Exp} tX((\beta \circ \operatorname{Exp} tX)^{-1}(x))(\mu)((\beta \circ \operatorname{Exp} tX)^{-1}(x))\Big|_0 .$$

Choose $m \in M$, write $P = \Phi(E)_m$ and $V = E_m$ and for $\mu \in \Gamma E$ define $\widetilde{\mu}\colon P \to V$ by $\widetilde{\mu}(\xi) = \xi^{-1}\mu(\beta\xi)$. For $X \in \Gamma A\Phi(E)$, let $\overline{X}$ denote the restriction of $\overrightarrow{X} \in \Gamma^{RI}T^\alpha\Phi(E)$ to P. Then, comparing the flows, it is straightforward to verify that

$$\mathscr{D}(X)(\mu)^{\widetilde{}} = \overline{X}(\widetilde{\mu}).$$

Now for $f\colon M \to \mathbb{R}$ it follows that

$$\mathscr{D}(X)(f\mu)^{\widetilde{}} = \overline{X}(f \circ \beta_m)(\widetilde{\mu}) = (f \circ \beta_m)\overline{X}(\widetilde{\mu}) + \overline{X}(f \circ \beta_m)\widetilde{\mu}$$

and hence that $\mathscr{D}(x)(f\mu) = f\mathscr{D}(X)(\mu) + a(X)(f)\mu$. This shows that $\mathscr{D}(X)$ is a first– or zeroth–order differential operator and, moreover, is an element of $\Gamma\mathfrak{D}(E)$.

Similarly it is easy to verify that for $X, Y \in \Gamma A\Phi(E)$ and $f\colon M \to \mathbb{R}$,

$$\mathscr{D}(X + Y) = \mathscr{D}(X) + \mathscr{D}(Y), \quad \text{and} \quad \mathscr{D}(fX) = f\mathscr{D}(X),$$

so $\mathscr{D}$ induces a morphism $\mathscr{D}\colon A\Phi(E) \to \mathfrak{D}(E)$ of vector bundles over M. It follows from what we have already done that $\mathscr{D}$ respects the anchors on $A\Phi(E)$ and $\mathfrak{D}(E)$. As for the bracket condition, for $X, Y \in \Gamma A\Phi(E)$ and $\mu \in \Gamma E$,

$$\mathscr{D}([X,Y])(\mu)^{\widetilde{}} = [\overline{X,Y}](\widetilde{\mu}) = [\overline{X}, \overline{Y}](\widetilde{\mu}) = \overline{X}(\overline{Y}(\widetilde{\mu})) - \overline{Y}(\overline{X}(\widetilde{\mu}))$$

$$= (\mathscr{D}(X)(\mathscr{D}(Y)(\mu)) - \mathscr{D}(Y)(\mathscr{D}(X)(\mu)))^{\widetilde{}}$$

and so $\mathscr{D}$ is a morphism of Lie algebroids over M.

Lastly, the map $\mathscr{D}^+\colon L\Phi(E) \to \operatorname{End}(E)$ is fibrewise the realization of $T_{\mathrm{id}}(GL(E_x))$ as $\mathfrak{gl}(E_x)$, and is therefore a vector bundle isomorphism. By Proposition 3.3.12 it follows that $\mathscr{D}$ is an isomorphism. $\qquad\square$

In what follows we will identify $A\Phi(E)$ with $\mathfrak{D}(E)$ via this isomorphism, usually without comment.

Consider a section D of $\mathfrak{D}(E)$. As a section of the Lie algebroid of $\Phi(E)$, it has an exponential flow $\operatorname{Exp} tD$ and these bisections act, as in 1.4.4, on E. Write $F_t(e) = \operatorname{Exp} tX(qe)e$. Then F_t is a flow by vector bundle isomorphisms on E and hence, by 3.4.2, induces a linear vector field ξ_D on E. It is easily checked directly that

$$\xi_D(\ell_\varphi) = \ell_{D^{(*)}(\varphi)}, \qquad \xi_D(f \circ q) = a(X)(f) \circ q,$$

for $\varphi \in \Gamma E^*$, $f \in C^\infty(M)$, and so by (25) we have the following useful observation.

Proposition 3.6.7 *Let D be a derivation on the vector bundle E. Then, in the notation of (27), $D = D_\xi$ where ξ is the vector field with flow $F_t(e) = \operatorname{Exp} tX(qe)e$.*

The various formulations of derivations and linear vector fields may of course be carried over to any representation.

Definition 3.6.8 *Let $\rho\colon \Omega \to \Phi(E)$ be a representation of a locally trivial Lie groupoid Ω on a vector bundle E. Then the induced representation ρ_* of $A\Omega$ on E is*

$$\rho_*(X)(\mu) = \rho_X(\mu) = -\frac{d}{dt}\overline{\rho}(\operatorname{Exp} tX)(\mu)\bigg|_0,$$

for $X \in \Gamma A\Omega$, $\mu \in \Gamma E$.

If $E * \Omega \to E$, $(e, \xi) \mapsto e\xi$, is a right linear action of Ω on E, then the induced representation of $A\Omega$ on E is

$$\rho_*(X)(\mu)(x) = \frac{d}{dt}\mu((\beta \circ \operatorname{Exp} tX)(x)) \operatorname{Exp} tX(x)\bigg|_0,$$

for $X \in \Gamma A\Omega$, $\mu \in \Gamma E$. □

For the notation $\overline{\rho}$, see §1.4.

$$* \qquad * \qquad * \qquad * \qquad *$$

Finally in this section we calculate the Lie algebroids of stabilizer subgroupoids. The next result is a generalization of a simple formula for Lie groups and Lie algebras. In the present generality it is needed in connection theory.

Theorem 3.6.9 *Let* $\rho\colon \Omega \to \Phi(E)$ *be a representation of a locally trivial Lie groupoid* Ω *on a vector bundle* E. *Let* $\mu \in \Gamma E$ *be* Ω*–deformable and let* $\Upsilon = \Omega(\mu)$ *be the stabilizer groupoid at* μ. *Then*

$$\Gamma A\Upsilon = \{X \in \Gamma A\Omega \mid \rho_*(X)(\mu) = 0\}.$$

Proof If $X \in \Gamma A\Upsilon$ then each $\mathrm{Exp}\, tX$ takes values in Υ. It follows that $\overline{\rho}(\mathrm{Exp}\, tX)(\mu) = \mu$ for all t and so $\rho_*(X)(\mu) = 0$.

Conversely, take $X \in \Gamma A\Omega$ and suppose $\rho_*(X)(\mu) = 0$. Since $X = \frac{d}{dt}\mathrm{Exp}\, tX|_0$, and Υ is an embedded submanifold of Ω, it suffices to show that each $\mathrm{Exp}\, tX$ takes values in Υ. Take $x \in M$ and $\mathrm{Exp}\, tX$ defined in a neighbourhood of x. Consider the curve

$$c(t) = \rho((\mathrm{Exp}\, tX(x))^{-1})\mu((\beta \circ \mathrm{Exp}\, tX)(x))$$

in E_x; that is,

$$c(t) = \overline{\rho}(\mathrm{Exp}\, -tX)(\mu)(x).$$

Now, fixing t_0, we have

$$
\begin{aligned}
\frac{d}{dt}c(t_0 + t)\Big|_{t=0} &= \frac{d}{dt}\overline{\rho}(\mathrm{Exp}\, -(t_0 + t)X)(\mu)(x)\Big|_0 \\
&= \overline{\rho}(\mathrm{Exp}\, -t_0 X)\left(\frac{d}{dt}\overline{\rho}(\mathrm{Exp}\, -tX)(\mu)\Big|_0\right)(x) \\
&= \overline{\rho}(\mathrm{Exp}\, -t_0 X)(\rho_*(X)(\mu))(x) \\
&= 0,
\end{aligned}
$$

so $\dfrac{dc}{dt}(t_0) = 0$ for all t_0 and hence c is constant at $c(0)$. Therefore

$$\rho(\mathrm{Exp}\, tX(x)^{-1})\mu((\beta \circ \mathrm{Exp}\, tX)(x)) = \mu(x),$$

which shows that $\mathrm{Exp}\, tX(x) \in \Upsilon$, as required. $\square$

The proof of 3.6.9 of course relies on the fact that Υ is already known to be a Lie (in fact, a locally trivial) subgroupoid of Ω. In the applications of 3.6.9 we need the following formulas for induced representations.

Theorem 3.6.10 *Let* E *be a vector bundle on* M.
(i) *Let* $\Phi(E) * \mathrm{Hom}^n(E; M \times \mathbb{R}) \to \mathrm{Hom}^n(E; M \times \mathbb{R})$ *be the action of* 1.7.7(i). *The induced representation of* $\mathfrak{D}(E)$ *on* $\mathrm{Hom}^n(E; M \times \mathbb{R})$ *is given by*

$$X(\varphi)(\mu_1, \ldots, \mu_n) = a(X)(\varphi(\mu_1, \ldots, \mu_n)) - \sum_{i=1}^{n}\varphi(\mu_1, \ldots, X(\mu_i), \ldots, \mu_n).$$

(ii) *Let* $\Phi(E) * \mathrm{Hom}^n(E; E) \to \mathrm{Hom}^n(E; E)$ *be the action of* 1.7.7(ii). *The induced representation of* $\mathfrak{D}(E)$ *on* $\mathrm{Hom}^n(E; E)$ *is given by*

$$X(\varphi)(\mu_1, \ldots, \mu_n) = X(\varphi(\mu_1, \ldots, \mu_n)) - \sum_{i=1}^{n} \varphi(\mu_1, \ldots, X(\mu_i), \ldots, \mu_n).$$

(iii) *Let* E' *be a second vector bundle on the same base, and let*

$$(\Phi(E) \times_{M \times M} \Phi(E')) * \mathrm{Hom}(E; E') \to \mathrm{Hom}(E; E')$$

be the action of 1.7.7(iii). *The Lie algebroid of* $(\Phi(E) \times_{M \times M} \Phi(E'))$ *may be identified with the direct sum Lie algebroid* $\mathfrak{D}(E) \oplus_{TM} \mathfrak{D}(E')$, *and the induced representation on* $\mathrm{Hom}(E; E')$ *is then given by*

$$(X \oplus X')(\varphi)(\mu) = X'(\varphi(\mu)) - \varphi(X(\mu)).$$

Remark. These formulas generalize results which are well known in the case of general linear groups of vector spaces. As with 1.7.7, we take 3.6.10 to include the restrictions of (i) and (ii) to Alt^n and Sym^n and the corresponding formulas for general tensor bundles, exterior algebra bundles, and symmetric bundles.

Proof To illustrate the use of the groupoid exponential, we prove (ii) with $n = 1$. The adaptation of the proof to the other cases is straightforward.

Take $X \in \Gamma\mathfrak{D}(E)$, $\varphi \in \Gamma\mathrm{Hom}(E; E)$, $\mu \in \Gamma E$ and $x \in M$. Write $x_t = (\beta \circ \mathrm{Exp}\, tX)^{-1}(x)$ where $\mathrm{Exp}\, tX$ is defined in a neighbourhood of x. Then (all limits are taken as $t \to 0$)

$$\begin{aligned}
X(\varphi)(x) &= \left. -\frac{d}{dt} \overline{\mathrm{Exp}\, tX}(\varphi) \right|_0 \\
&= -\lim \frac{1}{t} \{ \mathrm{Exp}\, tX(x_t)(\varphi(x_t)) - \varphi(x) \} \\
&= -\lim \frac{1}{t} \{ \mathrm{Exp}\, tX(x_t) \circ \varphi(x_t) \circ (\mathrm{Exp}\, tX(x_t))^{-1} - \varphi(x) \}
\end{aligned}$$

so

$$X(\varphi)(\mu)(x) =$$
$$- \lim \frac{1}{t} \{ \mathrm{Exp}\, tX(x_t) \circ \varphi(x_t) \circ (\mathrm{Exp}\, tX(x_t))^{-1})(\mu(x)) - \varphi(x)(\mu(x)) \}.$$

On the other side we have

$$X(\varphi)(\mu)(x) = -\lim \frac{1}{t} \{ (\mathrm{Exp}\, tX(x_t) \circ \varphi(x_t))(\mu(x_t)) - \varphi(x)(\mu(x)) \}$$

and

$$\varphi(X(\mu))(x) = -\lim\frac{1}{t}\{(\varphi(x) \circ \operatorname{Exp} tX(x_t))(\mu(x_t)) - \varphi(x)(\mu(x))\}.$$

Define a curve $c(t)$ for t near $0 \in \mathbb{R}$ with values in $\operatorname{Hom}(E_x; E_x)$ by

$$c(t) = \operatorname{Exp}tX(x_t) \circ \varphi(x_t) \circ (\operatorname{Exp}tX(x_t))^{-1}$$

and a curve $f(t)$ for t near $0 \in \mathbb{R}$ with values in E_x by

$$f(t) = \operatorname{Exp}tX(x_t)(\mu(x_t)).$$

Note $c(0) = \varphi(x)$ and $f(0) = \mu(x)$. Then the LHS of the equation is

$$-\lim\frac{1}{t}\{c(t)f(0) - c(0)f(0)\}$$

and the RHS is

$$-\lim\frac{1}{t}\{c(t)f(t) - c(0)f(t)\};$$

both limits being taken in the one vector space E_x. It is elementary that these limits are equal. $\qquad\square$

Corollary 3.6.11 (i) *Let E be a vector bundle on M, and let $\langle \, , \, \rangle$ be a Riemannian structure in E. Then the Lie algebroid $\Gamma A \Phi_{\mathcal{O}}(E)$ of the Riemannian frame groupoid $\Phi_{\mathcal{O}}(E)$ (see 1.7.9) is given by*

$$\{X \in \Gamma A\Phi(E) \mid a(X)(\langle \mu, \nu \rangle) = \langle X(\mu), \nu \rangle + \langle \mu, X(\nu) \rangle \text{ for all } \mu, \nu \in \Gamma E\}.$$

In particular, the fibres of the adjoint bundle $L\Phi(E)$ of $A\Phi(E)$ are the Lie algebras $\mathfrak{so}(E_x)$ for $x \in M$.

(ii) *Let L be a Lie algebra bundle on M. Then the Lie algebroid $\Gamma A \Phi_{\operatorname{Aut}}(L)$ of the LAB frame groupoid $\Phi_{\operatorname{Aut}}(L)$ (see 1.7.12) is given by*

$$\{X \in \Gamma A\Phi(L) \mid X([\mu, \nu]) = [X(\mu), \nu] + [\mu, X(\nu)] \text{ for all } \mu, \nu \in \Gamma L\}.$$

In particular the fibres of the adjoint bundle $L\Phi_{\operatorname{Aut}}(L)$ of $A\Phi_{\operatorname{Aut}}(L)$ are the Lie algebras $\operatorname{Der}(L_x)$ for $x \in M$.

Strictly speaking this is a corollary of 1.6.23 as well as of 3.6.10.

Given a Riemannian vector bundle $(E, \langle \, , \, \rangle)$ we denote the transitive Lie algebroid of 3.6.11(i) by

$$\mathfrak{so}(E) \rightarrowtail \mathfrak{D}_{\mathfrak{so}}(E) \twoheadrightarrow TM.$$

In what follows we will regard this equally as a Lie subalgebroid of $\mathfrak{D}(E)$ and as the Lie algebroid of $\Phi_{\mathcal{O}}(E)$. Sections of $\mathfrak{D}_{\mathfrak{so}}(E)$ may be called *metric derivations* on E, or *skew–adjoint derivative endomorphisms* of ΓE.

Likewise, given an LAB L we denote the transitive Lie algebroid of 3.6.11(ii) by

$$\text{Der}(L) \!>\!\!-\!\!\!-\!\!\!> \mathfrak{D}_{\text{Der}}(L) \longrightarrow\!\!\!\gg TM$$

and we will regard this equally as a Lie subalgebroid of $\mathfrak{D}(E)$ and as the Lie algebroid of $\Phi_{\text{Aut}}(L)$. Sections of $\mathfrak{D}_{\text{Der}}(L)$ are called *bracket derivations* of L.

The constructions just given may of course be modified to apply to any tensor structure on a vector bundle.

3.7 Adjoint formulas

We turn now to the adjoint representations of Lie groupoids and Lie algebroids. We establish that the representation $\text{Ad}\colon \Omega \to \Phi_{\text{Aut}}(L\Omega)$, for Ω locally trivial, induces $\text{ad}\colon A\Omega \to A\,\Phi_{\text{Aut}}(L\Omega)$. In fact we prove a stronger version of this result.

Let G be a Lie groupoid on M. Let $\sigma \in \Gamma_U G$ be a bisection with $(\beta \circ \sigma)(U) = V$. Then $I_\sigma \colon G_U^U \to G_V^V$ is a morphism of Lie groupoids over $\beta \circ \sigma \colon U \to V$ and we define

$$\text{Ad}(\sigma) = (I_\sigma)_* \colon AG|_U \to AG|_V.$$

Proposition 3.7.1 *With the notation just introduced:*

(i) *for $X, Y \in \Gamma_U AG$, $\text{Ad}(\sigma)[X, Y] = [\text{Ad}(\sigma)X, \text{Ad}(\sigma)Y]$;*

(ii) *if $X \in \Gamma_U AG$ and U is a flow neighbourhood for X, then $V = (\beta \circ \sigma)(U)$ is a flow neighbourhood for $\text{Ad}(\sigma)X$ and, for $|t|$ sufficiently small,*

$$\text{Exp}\, t\text{Ad}(\sigma)X = \widetilde{I}_\sigma(\text{Exp}\, tX),$$

where $\widetilde{I}_\sigma(\text{Exp}\, tX) \in \Gamma_V G$ is

$$y \mapsto I_\sigma(\text{Exp}\, tX((\beta \circ \sigma)^{-1}(y)));$$

(iii) *if $X, Y \in \Gamma_U AG$ and U is a flow neighbourhood for X, then*

$$[X, Y] = -\frac{d}{dt}\text{Ad}(\text{Exp}\, tX)(Y)\Big|_0.$$

Proof (i) is the result 3.5.10 for $F = I_\sigma$.

(ii) is a version of 3.6.5 for the case in which f is a general diffeomorphism, and is easily established.

(iii) can be deduced easily from the formula for $[\overrightarrow{X}, \overrightarrow{Y}]$ as a Lie derivative of $\overrightarrow{Y}$ with respect to the flow $L_{\operatorname{Exp} tX}$ of $\overrightarrow{X}$.   □

The results 3.7.1 generalize well–known identities for the relationship
between vector fields and their flows. For example, (ii) generalizes the
following: if φ_t is a local flow for a vector field X on a manifold M and
$\varphi\colon M \to M$ is a diffeomorphism, then $\{\varphi \circ \varphi_t \circ \varphi^{-1}\}$ is a local flow for
$\varphi_*(X)$. In turn, (ii) can be deduced from this result by applying it to
the right–invariant vector field $\overrightarrow{X}$ corresponding to $X \in \Gamma_U A\Omega$.

Similarly we obtain the following formula for 'canonical co–ordinates
of the second kind' on a Lie groupoid G.

Proposition 3.7.2 *Let $X_1, \ldots, X_r$ be a local basis for AG on an open
set $U \subseteq M$ and suppose that U is a flow neighbourhood for each X_i.
Then the map*

$$(t_1, \ldots, t_r) \mapsto (\operatorname{Exp} t_1 X_1 \star \operatorname{Exp} t_2 X_2 \star \ldots \star \operatorname{Exp} t_r X_r)(x),$$

*where $x \in U$ is fixed, is a diffeomorphism of an open neighbourhood of
$\mathbf{0} \in \mathbb{R}^r$ onto an open neighbourhood of 1_x in G_x.*

Proof This follows immediately from the result that if $X_1, \ldots, X_r$ are
linearly independent vector fields on an open subset U of a manifold M,
and if for each i, $\{\varphi_t^i\}$ is a local flow for X_i on U, then the map

$$(t_1, \ldots, t_r) \mapsto (\varphi_{t_1}^1 \circ \varphi_{t_2}^2 \circ \ldots \circ \varphi_{t_r}^r)(x),$$

where $x \in U$ is fixed, is a diffeomorphism of an open neighbourhood of
$\mathbf{0} \in \mathbb{R}^r$ onto an open neighbourhood of x in M.   □

If U itself is the domain of a chart for M, we obtain coordinates
$\mathbb{R}^n \times \mathbb{R}^r \cong U \times \mathbb{R}^r \rightsquigarrow \Omega_U$.

Compared with the corresponding result for Lie groups, the proof
of this result is immediate. However the purpose of the more delicate
analysis in the case of Lie groups is to prove the existence of analytic
co–ordinates; such coordinates do not usually exist for Lie groupoids.

Proposition 3.7.3 *Let $X, Y \in \Gamma_U AG$ where U is a flow neighbourhood
for both X and Y. Then, on U :*
  (i) *if $[X, Y] = 0$, then $\operatorname{Exp} t(X + Y) = \operatorname{Exp} tX \star \operatorname{Exp} tY$;*
  (ii) $\dfrac{d}{dt}(\operatorname{Exp} - \sqrt{t}Y \star \operatorname{Exp} - \sqrt{t}X \star \operatorname{Exp} \sqrt{t}Y \star \operatorname{Exp} \sqrt{t}X)\Big|_0 = [X, Y].$

Proof (i) From $[X, Y] = 0$ it follows that $[\vec{X}, \vec{Y}] = 0$. Hence the local flows $\varphi_t(\xi) = \mathrm{Exp}\, tX(\beta\xi)\xi$ and $\psi_t(\xi) = \mathrm{Exp}\, tY(\beta\xi)\xi$ commute. It is now easy to check that $\theta_t = \varphi_t \circ \psi_t$ is a local 1–parameter group of local transformations, and that $\frac{d}{dt}\theta_t(\xi)\big|_0 = \vec{X}(\xi) + \vec{Y}(\xi)$. Lastly, θ_t, being a composition of left–translations, is itself a left–translation and corresponds to $\mathrm{Exp}\, tX \star \mathrm{Exp}\, tY$.

(ii) follows, in the same way as does (i), from the corresponding result for general vector fields. □

From 3.7.1(iii) the following result is immediate.

Proposition 3.7.4 *Let Ω be a locally trivial Lie groupoid on M. Then*

$$\mathrm{Ad}_* = \mathrm{ad}\colon A\Omega \to A\,\Phi_{\mathrm{Aut}}(L\Omega).$$

For Lie groups, the formula in 3.7.1(iii) follows as a special case from 3.6.5. However in the groupoid setting, we cannot write $[X, Y] = \mathrm{ad}(X)(Y)$ for $X, Y \in \Gamma A\Omega$, and get a Lie algebroid representation and so this method is not available.

Lastly, we note the following for future reference.

Proposition 3.7.5 *Let E be a vector bundle on M. Then the adjoint map $\mathrm{Ad}\colon \Phi(E) \to \Phi(\mathrm{End}(E))$ is given by*

$$\mathrm{Ad}(\xi)(\varphi) = \xi \circ \varphi \circ \xi^{-1}$$

for $\xi \in \Phi(E)_x^y$ and $\varphi\colon E_x \to E_y$, where $x, y \in M$.

Proof The ma[$I_\xi\colon GL(E_x) \to GL(E_y)$ is the restriction to open sets of the linear map $\mathfrak{gl}(E_x) \to \mathfrak{gl}(E_y)$, $\varphi \longmapsto \xi \circ \varphi \circ \xi^{-1}$ and is therefore its own derivative. (The change of sign in the identification of $T_{\mathrm{id}}(GL(V))$ with $\mathfrak{gl}(V)$ of course cancels out.) □

3.8 Notes

§3.1 — §3.2. The material of these two sections has been taken from [Mackenzie, 1987a, App.A§2–§3] with minor revision.

The Atiyah sequence of a principal bundle was constructed by Atiyah [1957], specifically for use in connection theory. Nonetheless virtually all work involving general connection theory since the mid 1960s has been based on the account in the massively influential book of Kobayashi and Nomizu [1963]. The account in [Mackenzie, 1987a], from which these sections have been taken, was the first detailed exposition of the relationships between the Atiyah sequence approach and the standard treatment of connection theory.

One point that should be noted is that Kobayashi and Nomizu [1963] and most other accounts take the bracket on the Lie algebra of the structure group to be defined by left–invariant vector fields. Since the group action in a principal bundle is from the right, this leads to various avoidable sign problems, especially in the case of homogeneous bundles $G(G/H, H)$ — see Example 3.2.9. Here, as in [Mackenzie, 1987a], I use the bracket defined by right–invariant vector fields. Thus, for example, the bracket induced on $\Gamma(\frac{P \times \mathfrak{g}}{G})$ from the Atiyah sequence coincides with the pointwise bracket.

Note also that the map $P \times \mathfrak{g} \to TP$ of 3.2.2, which is the infinitesimal action corresponding to the right action of G on P, needs to be distinguished from the 'fundamental vector field map' $(u, X) \mapsto X^*(u)$ of [Kobayashi and Nomizu, 1963, p.51], defined in terms of the corresponding left action.

The two reformulations given in 3.2.9 of the Atiyah sequence of a homogeneous bundle include the descriptions given in [Greub et al., 1973, 5.11].

§3.3. The concept of Lie algebroid, as a generalization of the concept of Lie algebra, and obtained from a Lie groupoid by a process clearly generalizing that by which a Lie algebra is obtained from a Lie group, is due to Pradines [1967b]. It was not used by Ehresmann [1951, 1956] in his definitions of a connection. As already noted, Atiyah [1957] had earlier constructed from a principal bundle the exact sequence (3.2.6) of vector bundles which is the Lie algebroid of the corresponding groupoid.

Pradines [1967b] named the map $a\colon A \to TM$ the 'flèche' of the Lie algebroid. The usual translation, 'arrow', could hardly be given a second meaning in the context of groupoids. 'Anchor' suggests the tie which may or may not exist between the bracket structure on the Lie algebroid and that of the smooth structure on the base.

I have not been concerned to state minimal axioms for Lie algebroids. The condition (9) actually follows from (8) by expanding out $[X, [Y, fZ]]$ in two ways. This was noted by Herz [1953a] and by Kosmann–Schwarzbach and Magri [1990]. See Grabowski [2003] for a recent account of redundancies in the definition.

Lie pseudoalgebras   Related to the concept of Lie algebroid is the purely algebraic concept of Lie pseudoalgebra over a (commutative and unitary) ring, which stands in the same relationship to the concept of Lie algebroid as does that of module over a ring to the concept of vector bundle. Given a (commutative and unitary) algebra $\mathscr{C}$ over a ring R, a *Lie pseudoalgebra* is a module $\mathscr{A}$ together with an R–Lie algebra structure $[\ ,\ ]$ and an anchor $a\colon \mathscr{A} \to \mathrm{Der}(\mathscr{C})$ which satisfy the two equations (8) and (9) for $X, Y \in \mathscr{A}$ and $u \in \mathscr{C}$.

The concept of Lie pseudoalgebra has been introduced — under a great many different names — by a long series of authors. In chronological order:

- Herz [1953a,b]: *Pseudo–algèbre de Lie;*
- Hochschild [1955, §3]: *regular restricted Lie algebra extension;*
- Palais [1961]: *Lie d–ring;*
- Rinehart [1963]: *(R, C)–Lie algebra;*
- de Barros [1964]: *(R, C)–espace d'Elie Cartan régulier et sans courbure;*
- Bkouche [1966]: *(R, C)–algèbre de Lie;*
- Hermann [1967]: *Lie algebra with an associated module structure;*
- Nelson [1967]: *Lie module;*
- Kostant and Sternberg (see [Ne'eman, unpublished] and [Kostant and Sternberg, 1990]): *($\mathscr{A}, \mathscr{C}$) system;*
- Kamber and Tondeur [1971]: *sheaf of twisted Lie algebras;*
- Illusie [1972]: *algèbre de Lie sur C/R;*
- Teleman [1972]: *Lie algebra extension;*
- Kastler and Stora [1985] (see also [Jadczyk and Kastler, 1987a,b]): *Lie–Cartan pair;*
- Beilinson and Schechtmann [1988] (see also Manin [1988]): *Atiyah algebra;*
- Huebschmann [1990]: *Lie–Rinehart algebra;*
- Kosmann–Schwarzbach and Magri [1990]: *differential Lie algebra.*

Most of the early writers in this list were concerned primarily with unifying de Rham cohomology and Lie algebra cohomology. The only deep work prior to the consideration in [Pradines, 1968] of the integrability of Lie algebroids was Rinehart's Poincaré–Birkhoff–Witt theorem for projective Lie pseudoalgebras [Rinehart, 1963].

In an extensive body of work, Huebschmann has developed many aspects of the algebraic theory of Lie pseudoalgebras, using the terminology *Lie–Rinehart algebras*. See in particular [Huebschmann, 1998, 1999a,b].

For the concept of Schouten algebra or Gerstenhaber algebra, which may be regarded as a graded generalization of Lie algebroid or Lie pseudoalgebra, see the notes to Chapter 7. Sweedler and Takeuchi [1986], in their study of algebraic analogues of the Frobenius theorem and Poincaré lemma in positive characteristic, use notions close to that of Lie pseudoalgebra.

There are variations in the concepts defined by the authors listed above: Kamber and Tondeur [1971] define a form of Lie pseudoalgebra appropriate to the language of sheaves, and Beilinson and Schechtmann [1988] impose an additional linearity condition. Several writers require that the ring R be a field. Hochschild [1955] defines a form of the concept appropriate to working over a field of positive characteristic. Such variations are minor.

The future importance of Lie pseudoalgebras surely lies in the study of singular and infinite dimensional systems. Where the concept of Lie algebroid can be used — which is to say, within C^∞ differential geometry — it is important to establish that one is working with Lie algebroids and not merely with the associated Lie pseudoalgebras. This usually requires checking of locally constant rank conditions. See §6.5 for examples of what is required.

There is however a considerable body of purely algebraic work which uses the concept of Lie pseudoalgebra. If K is a suitable field extension of k, then Lie subpseudoalgebras of the Lie pseudoalgebra of derivations of K over k are implicit in [Jacobson, 1944], and the general concept (with the modification appropriate to positive characteristic) is explicit in [Hochschild, 1955, §3]. See also [Fel'dman, 1982] and [Malliavin, 1988].

An extended discussion of the relationship between Lie algebroids and Lie pseudoalgebras is given in [Mackenzie, 1995].

In an entirely different direction, Lie algebroids have recently been used in supergeometry. This aspect lies outside the present book, but see in particular [Voronov, 2002] and references given there.

Actions It is a curious fact that the notion of an action Lie algebroid in 3.3.7, and the calculation 3.5.14 were not established until the mid to late 1980s. In late 1986 Weinstein pointed out to the author the formula (15), from which the relevant parts of [Mackenzie, 1988a] and [Mackenzie, 1987c] quickly followed. See the Notes to Chapter 4 for more on the history of action Lie algebroids.

Derivations and derivative endomorphisms For background on differential operators as used in 3.3.4 see, for example, [Palais, 1965].

Derivations on a vector bundle E — or derivative endomorphisms in ΓE — have a considerable history. They appear, in purely algebraic settings, in papers of Jacobson [1935, 1937] and Herz [1953a,b], but they seem to have not been much developed at that time.

In the paper [Atiyah, 1957, §4] in which he introduced what is now known as the Atiyah sequence, Atiyah defined a bundle equivalent to $\mathfrak{D}(E)$. This $\mathfrak{D}(E)$ (or rather its module of sections) has since sometimes been called the *Atiyah algebra* of E.

The *quasi–scalar differential operators* of Palais [1968] are in first order precisely the derivative endomorphisms. A little later, Kosmann [1976] used the Lie algebra structure on such first–order operators, at first calling them *first–order operators with scalar symbol*, and then introducing the terminology *derivative endomorphisms* which is used here.

In [Mackenzie, 1987a] I used the terminology *covariant differential operator*, on the ground that, for any connection ∇ in E and any vector field X on M, ∇_X is such an operator. The terminology 'covariant differential operator' led to the awkward notation $CDO(E)$, which was hard to adapt to vector bundles with additional structure.

For a more detailed account of the various approaches to these matters, see [Kosmann–Schwarzbach and Mackenzie, 2002].

Regular Lie algebroids　The study of regular Lie algebroids may to some extent be reduced to the transitive case, since the restriction of a regular Lie algebroid to a leaf of the foliation defined by the image of its anchor is transitive. However there remains the question of how the restrictions are bound together, and we will not address this problem. See [Nistor, 2000]. and, for a general technique for working with regular Lie groupoids, [Moerdijk, 2003].

§3.4.　The material in §3.4 is taken from the treatments in [Mackenzie and Xu, 1994] and [Mackenzie and Xu, 1998] with light revision. The definition (31) was given in [Mackenzie and Xu, 1994, 5.3] and the intrinsic formula (3.4.6) was Lemma 6.3 of [Mackenzie and Xu, 1994]. Example 3.4.8 comes from [Mackenzie and Xu, 1998, §6].

The correspondence between linear vector fields and derivative endomorphisms was given, in a more general form, by Kosmann–Schwarzbach [1978, 1980]. Let (F, q, M) be a general fibre bundle. Then [Kosmann–Schwarzbach, 1980] establishes a bijective correspondence between projectable vector fields on F and certain first–order differential operators from F to $T^q F$. The correspondence (27) is an immediate consequence.

The tangent prolongation of a vector bundle has been dealt with in [Besse, 1978], [Pradines, 1974a] and [Mackenzie, 1992, §1]. An account of linear vector fields is also given in [Kolář et al., 1993, §47]. The vertical lift and complete lifts of a vector field form a small part of the considerable apparatus developed in [Yano and Ishihara, 1973]. The tangent pairing is also dealt with systematically in the work of Tulczyjew; see [Tulczyjew, 1989] for an overview.

Regarding Proposition 3.4.5, see [Golubitsky and Guillemin, 1973] for the intrinsic derivative and [Abraham and Marsden, 1985, p.72] for the linearization of a vector field at a singularity.

§3.5　The construction of the Lie algebroid of a Lie groupoid is due to Pradines [1967b]. The account given in this section is a light revision of [Mackenzie, 1987a, III§3].

For the Darboux derivative, see §5.1.

The 'symmetric' approach given at the end of §3.5 is due to Coste, Dazord and Weinstein [1987].

The notion of Lie group bundle and its associated Lie algebra bundle is developed in depth in [Douady and Lazard, 1966].

§3.6　The definition and basic properties of the exponential map, the fundamental theorems 3.6.6 and 3.6.4, and the adjoint formulas in 3.7.1 are based on the accounts of Kumpera [1971]. (Except for 3.6.6, this material also appears in the more readily available [Kumpera and Spencer, 1972, Appendix].) A proof of 3.6.6 was provided at about the same time by Libermann [1973]. The content of 3.6.6, that the derivations on a vector bundle are the infinitesimal generators of 1–parameter families of vector bundle automorphisms, was noted by Kosmann [1976]; see [Kosmann–Schwarzbach and Mackenzie, 2002] for more detail on the the relationship between the various approaches. The proof of 3.6.6 given here is the simplification from [Mackenzie, 1987a] of the proof in [Kumpera, 1971].

A similar construction to that in 3.6.6 occurs in the context of principal bundles in [Kobayashi and Nomizu, 1963, p.115].

Apart from these, the material in §3.6 and §3.7 first appeared in [Mackenzie, 1987a, III§4].

There is another concept of exponential map, which maps $A\Omega$ into Ω itself and is étale on a neighbourhood of the zero section of $A\Omega$; it however depends on a choice of connection in the vector bundle $A\Omega$. For this see [Pradines, 1967a].

Theorem 3.6.9 gives quick proofs of various standard results of connection theory: see §6.4.

Given a Riemannian bundle E or a LAB L, it is possible to construct $A\,\Phi_{\mathcal{O}}(E)$ and $A\,\Phi_{\mathrm{Aut}}(L)$ without use of the underlying groupoids if one proves the existence of Riemannian and Lie connections in E and L (respectively) directly by partitions of unity.

An abstract setting for tensor structures is given in [Greub et al., 1973, Chapter VIII]. A Σ–bundle in their sense is a vector bundle E together with a finite set Σ of sections of various associated tensor bundles, such that, for any two points of M, the structures at the two points are isomorphic. Theorems 3.6.9 and 3.6.10 can be used in the same way as in §3.6 to calculate the Lie algebroid of the Lie groupoid of Σ–preserving isomorphisms. The LAB described in [Greub et al., 1973, 8.4] is the adjoint bundle of this Lie groupoid.

§3.7 It is possible to overcome the difficulties which the general adjoint involves by lifting to the 1–jet prolongation groupoid $\mathscr{J}^1\Omega$ (see [Kumpera and Spencer, 1972] or [Kumpera, 1975] for the definition): The adjoint map defined in 3.7.1 is well defined as a map

$$\mathscr{J}^1\Omega \to \Phi(A\Omega)$$

$$j_x^1(\sigma) \longmapsto (\mathrm{Ad}(\sigma)|_x : A\Omega|_x \to A\Omega|_{(\beta \circ \sigma)(x)}),$$

and gives a smooth representation of $\mathscr{J}^1\Omega$ on the vector bundle $A\Omega$, which we denote by Ad^1. Now, by [Kumpera, 1975, §18], $A(\mathscr{J}^1\Omega)$ is naturally isomorphic to the natural Lie algebroid structure on $J^1(A\Omega)$ and 3.7.1(iii) then states that the induced representation $(\mathrm{Ad}^1)_* : J^1(A\Omega) \to \mathfrak{D}(A\Omega)$ is

$$j^1 X \longmapsto (Y \longmapsto [X, Y]), \qquad X, Y \in \Gamma A\Omega$$

Thus 3.7.1(iii) can be rewritten as $[X, Y] = (\mathrm{Ad}^1)_*(j^1 X)(Y)$.

4

Lie algebroids:
Algebraic constructions

This chapter is concerned with the general concept of morphism of Lie algebroids. It is unusual for the definition of a concept of morphism to require much attention. However the bracket of a Lie algebroid is defined not on the vector bundle but on its module of sections, and a morphism of vector bundles does not generally induce a map of sections. We handle this problem by considering first the case in which the target Lie algebroid may be pulled back across the base map; when this is possible, the definition of morphism presents itself naturally, and may then be reformulated so as to be viable when the pullback does not exist.

This procedure requires that pullbacks of Lie algebroids be defined before a concept of morphism can be formulated. The pullback notion for Lie groupoids of §2.3 has a clear analogue for Lie algebroids, and we proceed from this, justifying it retrospectively in 4.3.6 once the general notion of morphism has been defined.

The key to this chapter, as with Chapter 2, is the fact that the principal algebraic constructions for Lie algebroids may be characterized in terms of morphisms of specific types. This makes it possible to bypass a considerable number of explicit calculations. Rather than explicitly differentiating the algebraic constructions of Chapter 2, we proceed by analogy — based on diagrams and some very basic categorical principles — to define corresponding notions for abstract Lie algebroids. The characterization of these notions in terms of classes of morphisms, together with the functoriality of the process of taking the Lie algebroid, then gives immediately that differentiation — the Lie functor — maps the Lie groupoid constructions to their Lie algebroid analogues.

The chapter begins with infinitesimal actions of Lie algebroids. This case can be treated without a general notion of morphism and indeed provides the best basis on which to introduce the general notion. §4.2 is a rapid treatment of direct products and pullbacks preparatory to the definition of general morphisms in §4.3. The remaining sections treat general quotients, generalized actions and the most general notion of semidirect product.

4.1 Actions of Lie algebroids

The first step is to extend the definition of an infinitesimal action of a Lie algebra on a manifold (as in 3.3.7) to Lie algebroids.

Definition 4.1.1 Let A be a Lie algebroid on M, and let $f\colon M' \to M$ be a smooth map. Then an *infinitesimal action of A on f*, or *on M'*, is a map $X \mapsto X^\dagger$, $\Gamma A \to \mathfrak{X}(M')$ satisfying the four conditions:

$$(X + Y)^\dagger = X^\dagger + Y^\dagger \tag{1}$$

$$(uX)^\dagger = (u \circ f)X^\dagger \tag{2}$$

$$[X, Y]^\dagger = [X^\dagger, Y^\dagger] \tag{3}$$

$$X^\dagger \overset{f}{\sim} a(X) \tag{4}$$

for $X, Y \in \Gamma A$, $u \in C^\infty(M)$. □ □

The last condition is that $X^\dagger$, a vector field on M', projects under $f\colon M' \to M$ to $a(X)$, a vector field on M. Equivalently, $X^\dagger(u \circ f) = a(X) \circ f$ for all $u \in C^\infty(M)$.

We drop the word 'infinitesimal' except when emphasis is needed.

This definition is justified by two observations: firstly, an action of a Lie groupoid G, as defined in 1.6.1, induces an action of AG in this sense (see 4.1.6); and, secondly, there is an abstract version of the correspondence of §1.6 between groupoid actions and action morphisms. We establish the second point first.

Suppose given a Lie algebroid A on M, a smooth map $f\colon M' \to M$, and an infinitesimal action $X \mapsto X^\dagger$ as in 4.1.1. We will define a Lie algebroid structure on the pullback vector bundle $A' = f^!A$, regarding the sections of $f^!A$ as sums $\sum u_i' \otimes X_i$ where $u_i' \in C^\infty(M')$, $X_i \in \Gamma A$ (see §2.1). Here the tensor product is over $C^\infty(M)$ and the isomorphism of $C^\infty(M')$–modules $C^\infty(M') \otimes \Gamma A \cong \Gamma(f^!A)$ is $u' \otimes X \mapsto (u' \circ f)X^!$, where $X^!$ is the pullback of the section X.

Proposition 4.1.2 *Let A be a Lie algebroid on M, and $f\colon M' \to M$ a* *smooth map. Let* $X \mapsto X^{\dagger}$ *be an infinitesimal action as in 4.1.1. Then an anchor and a bracket structure are defined on* $f^{!}A$ *by*

$$a' \left(\sum u'_i \otimes X_i \right) = \sum u'_i X_i^{\dagger} \tag{5}$$

$$\left[\sum u'_i \otimes X_i, \sum v'_j \otimes Y_j \right] =$$
$$\sum u'_i v'_j \otimes [X_i, Y_j] + \sum u'_i X_i^{\dagger}(v'_j) \otimes Y_j - \sum v'_j Y_j^{\dagger}(u'_i) \otimes X_i, \tag{6}$$

where $u'_i, v'_j \in C^{\infty}(M')$, $X_i, Y_j \in \Gamma A$. *With this structure,* $f^{!}A$ *is a Lie algebroid on* M'.

Proof Note first that if $w \in C^{\infty}(M)$ then $a'(u' \otimes (wX)) = u'(w \circ f)X^{\dagger} = a'(u'(w \circ f) \otimes X)$, using (2), so a' is well defined. It is $C^{\infty}(M')$–linear and so defines a vector bundle morphism $f^{!}A \to TM'$.

Now fix $u'_i \in C^{\infty}(M')$ and $X_i \in \Gamma A$. Define a map

$$E \colon C^{\infty}(M') \times \Gamma A \to C^{\infty}(M') \otimes \Gamma A$$

by

$$E(v', Y) = \sum u'_i v' \otimes [X_i, Y] + \sum u'_i X_i^{\dagger}(v') \otimes Y - \sum v' Y^{\dagger}(u'_i) \otimes X_i.$$

It is easy to see that $E(v', wY) = E(v'(w \circ f), Y)$ for $w \in C^{\infty}(M)$, and so E defines a $C^{\infty}(M')$–linear map $\widetilde{E} \colon C^{\infty}(M') \otimes \Gamma A \to C^{\infty}(M') \otimes \Gamma A$ by

$$\widetilde{E} \left(\sum v'_j \otimes Y_j \right) = \sum E(v'_j, Y_j).$$

Thus the bracket in (6) is well defined in its second variable. A similar argument applies to the first.

To verify the Leibniz condition it suffices, since both sides are $\mathbb{R}$–bilinear, to consider elements $u' \otimes X$. Then

$$[u' \otimes X, w'(v' \otimes Y)] = [u' \otimes X, (v'w') \otimes Y]$$
$$= u'v'w' \otimes [X, Y] + u'X^{\dagger}(v'w') \otimes Y - v'w'Y^{\dagger}(u') \otimes X$$
$$= w'[u' \otimes X, v' \otimes Y] + u'v'X^{\dagger}(w') \otimes Y$$
$$= w'[u' \otimes X, v' \otimes Y] + a'(u' \otimes X)(w').(v' \otimes Y), \tag{7}$$

as required. The Jacobi identity and bracket–preservation for a' are proved in the same way. $\quad\square$

We denote $f^!A$ with this structure by $A \triangleleft f$ or $A \triangleleft M'$; it is the *action Lie algebroid* or *transformation Lie algebroid* corresponding to $X \mapsto X^\dagger$.

Example 4.1.3 Let A be any Lie algebroid on a connected base M and let $f \colon \widetilde{M} \to M$ be any covering. Then there is a canonical action of A on f in which $X^\dagger$ is the π–invariant lift of $a(X)$ to $\widetilde{M}$, where π is the group of the covering.

For $A = TM$ the canonical map $T(\widetilde{M}) \to TM \triangleleft f$ is an isomorphism of Lie algebroids. $\boxtimes$

Examples 4.1.4 Any Lie groupoid $G \rightrightarrows M$ acts on its target projection by the groupoid multiplication. The action Lie algebroid $AG \triangleleft \beta$ is canonically isomorphic to the distribution $T^\alpha G$ under the map $\mathscr{R}$ of 3.5.3 which sends $1 \otimes X \in \Gamma(AG \triangleleft \beta)$ to $\overrightarrow{X}$. Compare 1.6.11.

In particular, for any Lie group G, let $\mathfrak{g}$ act on the manifold G by $X^\dagger = \overrightarrow{X}$. Then the anchor $\mathfrak{g} \triangleleft G \to TG$ of the action Lie algebroid is an isomorphism onto the standard Lie algebroid TG. This is equivalent to the fact that if, for any $X \in \mathfrak{X}(G)$, the function $X^R \colon G \to \mathfrak{g}$ is defined by $X^R(g) = T(R_{g^{-1}})(X(g))$, then

$$[X, Y]^R = \mathfrak{L}_X(Y^R) - \mathfrak{L}_Y(X^R) + [X^R, Y^R]^\bullet$$

where $[\ ,\ ]^\bullet$ is the pointwise bracket.

Similarly, for any principal bundle $P(M, G)$, the standard Lie algebroid TP is isomorphic to the action Lie algebroid $\frac{TP}{G} \triangleleft P$. $\boxtimes$

We now consider a Lie algebroid version of 1.6.16. Let A and A' be Lie algebroids on bases M and M', and suppose that

$$
\begin{array}{ccc}
A' & \xrightarrow{\ \varphi\ } & A \\[2mm]
{\scriptstyle q'}\Big\downarrow & & \Big\downarrow{\scriptstyle q} \\[2mm]
M' & \xrightarrow[\ f\ ]{} & M
\end{array}
\qquad (8)
$$

is a pullback of vector bundles. Thus the vector bundle A' is isomorphic to the pullback vector bundle $f^!A \to M'$ under the map $\varphi^! \colon A' \to f^!A$, $X' \mapsto (q'X', \varphi X')$. For $X \in \Gamma A$, denote by $X^!$ the unique section of A' with $\varphi \circ X^! = X \circ f$. If M is singleton, then the sections $X^!$ are the constant sections of the trivial(izable) bundle A'. Generalizing that case, let us impose on (8) the two conditions that

$$T(f) \circ a' = a \circ \varphi, \tag{9}$$

$$[X^!, Y^!] = [X, Y]^! \quad \text{for all } X, Y \in \Gamma A. \tag{10}$$

We will see shortly that, in this case, these are precisely the conditions that (φ, f) is a Lie algebroid morphism.

In the case where M was singleton, the original action could be recovered from the anchor. In the present situation we therefore define a map $\Gamma A \to \mathfrak{X}(M')$ by $X \mapsto X^\dagger = a'(X^!)$. It is straightforward to verify that conditions (1) – (4) hold and we have the following result.

Proposition 4.1.5 *Let A and A' be Lie algebroids on M and M' and let $\varphi \colon A' \to A$, $f \colon M' \to M$ be a pullback of vector bundles satisfying (9) and (10). Then, with the above notation, $X^\dagger = a'(X^!)$ defines an infinitesimal action of A on f, and $\varphi^! \colon A' \to A \blacktriangleleft f$ is an isomorphism of Lie algebroids over M'.*

There is thus established a bijective correspondence between actions of A on the map $f \colon M' \to M$ and maps $\varphi \colon A' \to A$, $f \colon M' \to M$, which are pullback morphisms of vector bundles, which commute with the anchors on A' and A, and which satisfy (10). In §4.3 such maps (φ, f) will re-emerge as *action morphisms* of Lie algebroids.

Consider now a Lie groupoid action $G * M' \to M'$ of a Lie groupoid $G \rightrightarrows M$ on a smooth map $f \colon M' \to M$. Then, as in §1.6, the canonical map $F \colon G \blacktriangleleft f \to G$, $(g, m') \mapsto g$, is a morphism of Lie groupoids and so induces, by 3.5.10, a vector bundle morphism AF from $A(G \blacktriangleleft f)$ to AG which satisfies the two conditions (9) and (10). Applying 4.1.5, there is an infinitesimal action of AG on f given for $X \in \Gamma AG, m' \in M'$ by

$$X^\dagger(m') = T_{1_{f(m')}}(g \mapsto gm')(X(f(m'))) \tag{11}$$

and we have the following theorem.

Theorem 4.1.6 *Let $G \rightrightarrows M$ be a Lie groupoid acting on a smooth map $f \colon M' \to M$. Let $F \colon G \blacktriangleleft f \to G$ be the corresponding action morphism. Then $(AF)^! \colon A(G \blacktriangleleft f) \to (AG) \blacktriangleleft f$ is an isomorphism of Lie algebroids, where the action of AG on f is given by (11).*

When a Lie groupoid acts linearly on a vector bundle there are now two induced Lie algebroid phenomena: the induced representation and the infinitesimal action. These are related in the natural way:

Proposition 4.1.7 *Let ρ be a linear action of a Lie groupoid $G \rightrightarrows M$ on a vector bundle E. Then the induced representation $\rho_* \colon AG \to \mathfrak{D}(E)$ and the infinitesimal action $X \mapsto X^\dagger$, $\Gamma AG \to \mathfrak{X}(E)$ are related by*

$$\langle \varphi, \rho_X(\mu) \rangle = a(X)\langle \varphi, \mu \rangle - X^\dagger(\ell_\varphi) \circ \mu$$

for $\varphi \in \Gamma E^*$, $\mu \in \Gamma E$.

Proof This is simply the statement that, for all $X \in \Gamma AG$, the derivation corresponding to the linear vector field $(X^\dagger, a(X))$ is ρ_X (see 3.4.5). ☐

The three equations of 3.4.5 now also apply to ρ_X and $X^\dagger$. In particular, for $X \in \Gamma AG$, $\mu \in \Gamma E$,

$$X^\dagger(\mu(m)) = T(\mu)(a(X)(m)) - \rho_X(\mu)^\dagger(\mu(m)), \tag{12}$$

$$\rho_X(\mu)^\dagger = [X^\dagger, \mu^\dagger]. \tag{13}$$

Of course, these formulas also give a correspondence between representations $\rho \colon A \to \mathfrak{D}(E)$ of abstract Lie algebroids and infinitesimal actions $\Gamma A \to \mathfrak{X}(E)$ by linear vector fields.

Example 4.1.8 Let $\mathfrak{g}$ be a Lie algebra and $E \to M$ a vector bundle. A *derivative representation* of $\mathfrak{g}$ on E is a Lie algebra morphism $\kappa \colon \mathfrak{g} \to \Gamma \mathfrak{D}(E)$. Composing with the anchor $\mathfrak{D}(E) \to TM$, a derivative representation induces an infinitesimal action of $\mathfrak{g}$ on M. If κ is now lifted to $\mathfrak{g} \ltimes M$, we have a representation (in the sense of 3.3.13) of the action Lie algebroid on E. Conversely, any infinitesimal action of $\mathfrak{g}$ on M and a representation $\mathfrak{g} \ltimes M \to \mathfrak{D}(E)$ define a derivative representation by restricting to constant sections. ☒

Use of connections

There is an alternative form for the bracket formula (6), obtained by using an auxiliary connection in the vector bundle underlying A.

Start with an arbitrary Lie algebroid A on M and a Koszul connection ∇ in A (see §5.2). The usual notion of torsion for connections in tangent bundles extends to this case, and we define T_∇, the *torsion of* ∇ for $X, Y \in \Gamma A$ by

$$T_\nabla(X, Y) = \nabla_{aX}(Y) - \nabla_{aY}(X) - [X, Y]. \tag{14}$$

As a map $\Gamma A \times \Gamma A \to \Gamma A$ this is bilinear over $C^\infty(M)$ and hence defines $T_\nabla \colon A \oplus A \to A$ which is alternating.

Let $f \colon M' \to M$ be an arbitrary smooth map and denote by $\overline{\nabla}$ the pullback connection in $f^!A$. Thus

$$\overline{\nabla}_{V'}\left(\sum u_i' \otimes X_i\right) = \sum u_i' \otimes \nabla_{T(f)(V')}(X_i) + \sum V'(u_i') \otimes X_i$$

where $V' \in TM'$, $u_i' \in C^\infty(M')$, $X_i \in \Gamma A$. A priori, we do not have a Lie algebroid structure in $f^!A$, so there is no concept of torsion for $\overline{\nabla}$. However we pull back T_∇ to $f^!A$, obtaining $\overline{T}_\nabla \colon f^!A \oplus f^!A \to f^!A$. The following is now a simple check.

Proposition 4.1.9 *Let A act on $f \colon M' \to M$ by $X \mapsto X^\dagger$. Then the bracket* (6) *in* $\Gamma(A \triangleleft f)$ *is*

$$[C, D] = \overline{\nabla}_{\overline{a}(C)}(D) - \overline{\nabla}_{\overline{a}(D)}(C) - \overline{T}_\nabla(C, D) \tag{15}$$

where $C, D \in \Gamma(A \triangleleft f)$, and $\overline{a}$ is the anchor $A \triangleleft f \to TM'$. The Lie algebroid torsion of $\overline{\nabla}$ coincides with $\overline{T}_\nabla$.

The advantage of this formulation is that, once ∇ is chosen, each term on the RHS is well defined and there is no further need to deal with the representation of elements of $\Gamma(f^!A)$ as tensor products.

Example 4.1.10 Let G be a Lie group and H a closed subgroup. Let G act on $M = G/H$ in the standard way, and let $\mathfrak{g} \triangleleft M$ be the action Lie algebroid. The vector bundle $\mathfrak{g} \to \{*\}$ has a unique (trivial) connection, the torsion of which, according to (14), is $T(X, Y) = -[X, Y]$ for $X, Y \in \mathfrak{g}$. The pullback connection is the standard flat connection ∇^0 in $M \times \mathfrak{g}$ and we have the bracket formula for $\mathfrak{g} \triangleleft M$

$$[V, W] = V^\dagger(W) - W^\dagger(V) + [V, W]^\bullet,$$

as in (15) of Chapter 3, with the Lie algebroid torsion being the negative of the pointwise bracket.

When H is the trivial subgroup, the Lie algebroid torsion reduces to the usual concept and (15) is the usual formula for the torsion of the right–invariant flat connection in G. $\boxtimes$

4.2 Direct products and pullbacks of Lie algebroids

Consider Lie algebroids $A^1 \to M^1$ and $A^2 \to M^2$. We want to define a Lie algebroid structure on $A^1 \times A^2 \to M^1 \times M^2$ in such a way that the bracket of two sections which both come from A^1 is determined by their bracket in $\Gamma(A^1)$, likewise for A^2, and such that the bracket of two sections, one from A^1 and one from A^2, is zero. More care needs to be taken with the formulation of these requirements than is the case with Lie algebras — by 'a section which comes from A^1' we must mean not merely a section which takes values in A^1, but one which also does not depend on M^2. However we will see that these requirements, together with the Leibniz condition, determine a Lie algebroid structure on $A^1 \times A^2$.

Denote the projections from $M^1 \times M^2$ to M^1, M^2 by $\mathrm{pr}_1, \mathrm{pr}_2$. The product vector bundle $A^1 \times A^2 \to M^1 \times M^2$ can be regarded as the Whitney sum over $M^1 \times M^2$ of the pullback vector bundles $\mathrm{pr}_1^! A^1$ and $\mathrm{pr}_2^! A^2$. Sections of $\mathrm{pr}_1^! A^1$ are of the form $\sum u_i \otimes X_i^1$ where $u_i \in C^\infty(M^1 \times M^2)$ and $X_i^1 \in \Gamma(A^1)$. Likewise, we write sections of $\mathrm{pr}_2^! A^2$ in the form $\sum u_j' \otimes X_j^2$ where $u_j' \in C^\infty(M^1 \times M^2)$ and $X_j^2 \in \Gamma(A^2)$.

The tangent bundle $T(M^1 \times M^2)$ may in the same way be regarded as the Whitney sum $\mathrm{pr}_1^!(TM^1) \oplus \mathrm{pr}_2^!(TM^2)$. We define the anchor $a \colon A^1 \times A^2 \to T(M^1 \times M^2)$ by using these descriptions and extending a_1 and a_2 linearly; thus

$$a\left(\sum(u_i \otimes X_i^1) \oplus \sum(u_j' \otimes X_j^2)\right) = \sum(u_i \otimes a_1(X_i^1)) \oplus \sum(u_j' \otimes a_2(X_j^2)).$$

Now we impose the conditions

$$[1 \otimes X^1, 1 \otimes Y^1] = 1 \otimes [X^1, Y^1], \qquad [1 \otimes X^1, 1 \otimes Y^2] = 0, \qquad (16)$$
$$[1 \otimes X^2, 1 \otimes Y^2] = 1 \otimes [X^2, Y^2], \qquad [1 \otimes X^2, 1 \otimes Y^1] = 0,$$

for $X^1, Y^1 \in \Gamma(A^1)$, $X^2, Y^2 \in \Gamma(A^2)$. It follows that for

$$X = \sum(u_i \otimes X_i^1) \oplus \sum(u_j' \otimes X_j^2) \quad \text{and} \quad Y = \sum(v_k \otimes Y_k^1) \oplus \sum(v_\ell' \otimes Y_\ell^2),$$

we have, using the Leibniz condition,

$$[X, Y] = \left\{\sum u_i v_k \otimes [X_i^1, Y_k^1] + \sum u_i a_1(X_i^1)(v_k) \otimes Y_k^1 \right.$$
$$\left. - \sum v_k a_1(Y_k^1)(u_i) \otimes X_i^1\right\}$$
$$\oplus \left\{\sum u_j' v_\ell' \otimes [X_j^2, Y_\ell^2] + \sum u_j' a_2(X_j^2)(v_\ell') \otimes Y_\ell^2 \right.$$
$$\left. - \sum v_\ell' a_2(Y_\ell^2)(u_j') \otimes X_j^2\right\} \quad (17)$$

where we write $a_1(X^1)(v)$ as shorthand for the action of $a_1(X^1)^!$ on $v \in C^\infty(M^1 \times M^2)$.

It is now a straightforward exercise to verify that this bracket and anchor are well defined and make $A^1 \times A^2$ a Lie algebroid over $M^1 \times M^2$, which we call the *direct product Lie algebroid*. We postpone a more formal statement to §4.3 since we do not yet have a concept of morphism. Note however, that if $A^1 = TM^1$, $A^2 = TM^2$, then the product Lie algebroid is canonically isomorphic (over $M^1 \times M^2$) to $T(M^1 \times M^2)$.

$$* \qquad * \qquad * \qquad * \qquad *$$

We now turn to pullbacks of Lie algebroids over smooth maps. Given a Lie algebroid A on M and a smooth map $f \colon M' \to M$, the two maps in the diagram

$$
\begin{array}{c}
A \\
\\
\downarrow a \\
\\
TM' \xrightarrow{\quad\quad} TM \\
T(f)
\end{array}
\qquad (18)
$$

should be included in any definition of Lie algebroid morphism. Suppose that the vector bundle pullback $TM' \oplus_{TM} A$ of (18) exists. Thus its elements are $x' \oplus X$ where $x' \in TM'$ and $X \in A$ are subject to $T(f)(x') = a(X)$. When dealing with sections, it is convenient to regard $TM' \oplus_{TM} A$ as the pullback of $T(f)^! \colon TM' \to f^! TM$ and $f^!(a) \colon f^!(A) \to f^!(TM)$ in the category of vector bundles over M'. Thus sections of $TM' \oplus_{TM} A$ are expressions

$$x' \oplus \left(\sum u_i' \otimes X_i \right), \qquad x' \in \mathfrak{X}(M'),\ u_i' \in C^\infty(M'),\ X_i \in \Gamma A,$$

such that $T(f)(x')(m') = \sum u_i'(m') a(X_i(f(m')))$ for all $m' \in M'$.

On the basis that a pullback should be a subobject of the direct product, define an anchor a' by

$$a' \left(x' \oplus \left(\sum u_i' \otimes X_i \right) \right) = x', \qquad (19)$$

and a bracket by

$$\left[x' \oplus \left(\sum u'_i \otimes X_i\right), y' \oplus \left(\sum v'_j \otimes Y_j\right)\right] =$$
$$[x', y'] \oplus \left(\sum u'_i v'_j \otimes [X_i, Y_j] + \right.$$
$$\left. \sum a'(x')(v'_j) \otimes Y_j - \sum a'(y')(u'_i) \otimes X_i\right). \quad (20)$$

Again, it is a straightforward task to verify that, with this structure, $TM' \oplus_{TM} A$ is a Lie algebroid on M'. We postpone a formal statement to 4.3.6. With this structure, we denote $TM' \oplus_{TM} A$ by $f^{!!}A$ and call it the *pullback Lie algebroid of A over f.*

The pullback Lie algebroid always exists if A is transitive, or if f is a surjective submersion.

Example 4.2.1 Taking $A = \mathfrak{g}$ to be a Lie algebra, the pullback of $\mathfrak{g}$ to a manifold M' is the trivial Lie algebroid of 3.3.3. ⊠

4.3 Morphisms of Lie algebroids

A general vector bundle map

$$
\begin{array}{ccc}
A' & \xrightarrow{\ \varphi\ } & A \\
q' \downarrow & & \downarrow q \\
M' & \xrightarrow{\ f\ } & M
\end{array}
\qquad (21)
$$

does not induce a map from the sections of A' to the sections of A, and the simple definition 3.3.1 of a morphism of Lie algebroids available in the base preserving case does not apply here.

Suppose first of all that the pullback $f^{!!}A$ of the target exists. Then, by analogy with the decomposition of Lie groupoid morphisms in §2.3, we could define φ to be a morphism if

$$a \circ \varphi = T(f) \circ a', \quad (22)$$

and if the map

$$\varphi^{!!} \colon A' \to f^{!!}A, \qquad X' \mapsto a'(X') \oplus \varphi(X'), \quad (23)$$

is a morphism (over M') into $f^{!!}A$. By (19), (20) this last condition is equivalent to the condition that, if $X', Y' \in \Gamma A'$ are such that $\varphi^!(X') = \sum_i u'_i \otimes X_i$ and $\varphi^!(Y') = \sum_j v'_j \otimes Y_j$, then

$$\varphi^!([X', Y']) = \sum_{i,j} u'_i v'_j \otimes [X_i, Y_j]$$

$$+ \sum_j a'(X')(v'_j) \otimes Y_j - \sum_i a'(Y')(u'_i) \otimes X_i. \quad (24)$$

This condition may be formulated even when the pullback Lie algebroid does not exist. We thus arrive at the following definition.

Definition 4.3.1 Let A' and A be Lie algebroids on bases M' and M respectively. A *morphism of Lie algebroids* from A' to A is a vector bundle morphism (21) such that the anchor preservation condition (22) and the bracket condition (24) hold.   □

There are now a substantial number of checks to carry out in order to confirm the correctness of this definition. First we verify that the RHS of (24) is well defined.

Fix $X' \in \Gamma(A')$ and fix a decomposition $\varphi^!(X') = \sum u'_i \otimes X_i$. Define a function of $v' \in C^\infty(M')$ and $Y \in \Gamma(A)$ by

$$E(v', Y) = a'(X')(v') \otimes Y + \sum (u'_i v' \otimes [X_i, Y]).$$

This is $\mathbb{R}$–bilinear and it is straightforward to check that, for $w \in C^\infty(M)$,

$$E(v', wY) = E(v'(w \circ f), Y),$$

So E defines a map $\widetilde{E} \colon C^\infty(M') \otimes \Gamma A \to C^\infty(M') \otimes \Gamma A$ with

$$\widetilde{E}\left(\sum v'_j \otimes Y_j\right) = \sum E(v'_j, Y_j).$$

In particular, $\sum_{i,j} u'_i v'_j \otimes [X_i, Y_j] + \sum_{i,j} a'(X')(v'_j) \otimes Y_j$ is independent of the choice of decomposition of Y' and hence so is the RHS of (24). Since the RHS of (24) is skew–symmetric, this completes the proof that it is well defined.

It is clear that in the base–preserving case, 4.3.1 reduces to 3.3.1. From the next result, there is a category $\mathscr{LA}$ of Lie algebroids.

Proposition 4.3.2 Let $\varphi \colon A' \to A$ be a morphism of Lie algebroids over $f \colon M' \to M$ and let $\psi \colon A'' \to A'$ be a morphism of Lie algebroids over $g \colon M'' \to M'$. Then $\psi \circ \varphi \colon A'' \to A$ is a morphism of Lie algebroids over $g \circ f \colon M'' \to M$.

Proof The anchor condition is immediate. For $X'' \in \Gamma(A'')$ take a ψ-decomposition $\psi^!(X'') = \sum u_i'' \otimes X_i'$ and then take a φ-decomposition of each X_i'. This produces a $\psi \circ \varphi$-decomposition of X'', and from here it is straightforward to complete the proof. $\qquad\square$

Proposition 4.3.3 *For any smooth map $f \colon M' \to M$, the tangent map $T(f) \colon TM' \to TM$ is a morphism of Lie algebroids.*

Proof Only the bracket condition needs proof. Take $X', Y' \in \mathfrak{X}(M)$ with $T(f)$–decompositions

$$T(f)^!(X') = \sum u_i' \otimes X_i, \qquad T(f)^!(Y') = \sum v_j' \otimes Y_j.$$

These assert that for $r, s \in C^\infty(M)$,

$$X'(r \circ f) = \sum u_i'(X_i(r) \circ f), \qquad Y'(s \circ f) = \sum v_j'(Y_j(s) \circ f).$$

Hence

$$
\begin{aligned}
X'(Y'(s \circ f)) &= \sum X'(v_j')(Y_j(s) \circ f) + \sum v_j' X'(Y_j(s) \circ f) \\
&= \sum X'(v_j')(Y_j(s) \circ f) + \sum v_j' u_i'(X_i(Y_j(s)) \circ f).
\end{aligned}
$$

Using the similar equation for $Y'(X'(s \circ f))$, we get

$$T(f)^!([X', Y']) = \sum_{i,j} u_i' v_j' \otimes [X_i, Y_j] + \sum_j X'(v_j') \otimes Y_j - \sum_i Y'(u_i') \otimes X_i,$$

as required. $\qquad\square$

Given a morphism of Lie groupoids $F \colon G' \to G$, $f \colon M' \to M$, the induced map $A(F) \colon AG' \to AG$, $f \colon M' \to M$ was defined in §3.5, p. 124.

Proposition 4.3.4 *For any morphism of Lie groupoids $F \colon G' \to G$, $f \colon M' \to M$, the induced map $A(F) \colon AG' \to AG$, $f \colon M' \to M$, is a morphism of Lie algebroids.*

Proof We already noted in §3.5 that $A(F)$ is anchor preserving. Take $X' \in \Gamma AG'$ with

$$A(F)^!(X') = \sum u_i' \otimes X_i.$$

where $u_i' \in C^\infty(M')$ and $X_i \in \Gamma AG$. This can be rewritten in the form $A(F) \circ X' = \sum u_i'(X_i \circ f)$. Recall that the right–invariant vector fields $\overrightarrow{X_i}$ and $\overrightarrow{X'}$ which correspond to X_i and X' are the pullback sections

defined by X_i and X' with respect to the vector bundle morphism
$\mathscr{R}\colon T^{\alpha}G \to AG$, $\beta\colon G \to M$ of 3.5.3 and the corresponding map for
G'. It follows that

$$T(F) \circ \overrightarrow{X'} = \sum (u'_i \circ \beta')(\overrightarrow{X_i} \circ F).$$

Similarly for $Y' \in \Gamma AG'$ with $A(F)^!(Y') = \sum v'_j \otimes Y_j$, we get

$$T(F) \circ \overrightarrow{Y'} = \sum (v'_j \circ \beta')(\overrightarrow{Y_j} \circ F).$$

Applying 4.3.3 we now get

$$T(F) \circ [\overrightarrow{X'}, \overrightarrow{Y'}] = \sum (u'_i \circ \beta')(v'_j \circ \beta')([\overrightarrow{X_i}, \overrightarrow{Y_j}] \circ F)$$
$$+ \sum \overrightarrow{X'}(v'_j \circ \beta')(\overrightarrow{Y_j} \circ F) - \sum \overrightarrow{Y'}(u'_i \circ \beta')(\overrightarrow{X_i} \circ F).$$

This is an equation for maps defined on G' with values in TG. Restrict-
ing it to the identity elements of G' and remembering that $\overrightarrow{X'}(v' \circ \beta') =$
$a'(X')(v') \circ \beta'$, we obtain

$$A(F) \circ [X', Y'] = \sum u'_i v'_j ([X_i, Y_j] \circ f)$$
$$+ \sum a'(X')(v'_j)(Y_j \circ f) - \sum a'(Y')(u'_i)(X_i \circ f),$$

from which the equation for $A(F)^!([X', Y'])$ follows. $\square$

As in the base–preserving case, we refer to $A(\varphi)$ as the *induced morphism of Lie algebroids*.

Examples 4.3.5 For any Lie group G, the map $TG \to \mathfrak{g}$ defined by
right–translation is a morphism of Lie algebroids, induced by the groupoid morphism $G \times G \to G$, $(g, h) \mapsto gh^{-1}$, from the pair groupoid to the
group. More generally, for any Lie groupoid $G \rightrightarrows M$, the map $\mathscr{R}$ of
(34) on p. 121 is a Lie algebroid morphism over $\beta\colon G \to M$. Similarly,
for any principal bundle $P(M, G, \pi)$, the projection from TP to $\frac{TP}{G}$ is
a Lie algebroid morphism over π. $\boxtimes$

The proof of the following is now trivial, but the result — that $f^{!!}A$
is the pullback of (18) in the category $\mathscr{L}\mathscr{A}$ — is important.

Theorem 4.3.6 *Let A' and A be Lie algebroids on bases M' and M
respectively and let $f\colon M' \to M$ be such that the pullback Lie algebroid
$f^{!!}A$ exists. Consider a morphism of vector bundles $\varphi\colon A' \to A$ over f
for which $a \circ \varphi = T(f) \circ a'$.*

Then (φ, f) *is a morphism of Lie algebroids if and only if the map* $\varphi^{!!} \colon A' \to f^{!!}A$ *of* (23) *is a morphism of Lie algebroids over* M'.

In particular, when the pullback $f^{!!}A$ exists, the canonical projection $f^{!!} \colon f^{!!}A \to A$ is a morphism of Lie algebroids over f.

It is easy to see directly that the canonical projection of an action Lie algebroid $A \lessdot f \to A, (X, m') \mapsto X$, is a morphism. The following reformulation of this is worth noting.

Proposition 4.3.7 *Let a Lie algebroid A on base M act on a surjective submersion $f \colon M' \to M$. Then the action Lie algebroid embeds into the pullback Lie algebroid by*

$$A \lessdot f \to f^{!!}A, \quad (X, m') \mapsto X^{\dagger}(m') \oplus X.$$

Proof This is a base–preserving morphism of vector bundles over M', and the preservation of the anchors is clear. For bracket preservation it is sufficient to consider sections of the form $1 \otimes X$, by use of the Leibniz condition on both sides. The image of $1 \otimes X$ is $X^{\dagger} \oplus X$ and the result follows from comparison of (6) with (24). $\qquad\square$

In the case where f is a covering, the map $A \lessdot f \to f^{!!}A$ is an isomorphism.

There are many cases in which Definition 4.3.1 can be simplified.

Proposition 4.3.8 *Let* (21) *be a vector bundle morphism of Lie algebroids, and suppose φ is a fibrewise surjection. Then if $a \circ \varphi = T(f) \circ a'$ and if for any $X', Y' \in \Gamma A'$, $X, Y \in \Gamma A$, we have*

$$X' \sim X, \; Y' \sim Y \implies [X', Y'] \sim [X, Y],$$

then (φ, f) is a morphism of Lie algebroids.

Proof Note first that $X' \in \Gamma A'$ and $X \in \Gamma A$ are φ–related if and only if $\varphi^{!} \circ X' = X^{!}$. Now consider any $X' \in \Gamma A'$ and write $\varphi^{!}(X') = \sum u'_i \otimes X_i$. Using a right inverse of $\varphi^{!}$, there is for each i a section X'_i of A' such that $X'_i \sim X_i$. Now $X' = \sum u'_i X'_i + Z'$ where $Z' \sim 0$. Similarly write

$Y' = \sum v'_j Y'_j + W'$. Now

$$[X', Y'] = \sum_{i,j} \{u'_i v'_j [X'_i, Y'_j] + u'_i a'(X'_i)(v'_j)Y'_j - v'_j a'(Y'_j)(u'_i)X'_i\}$$

$$+ \sum_i \{u'_i [X'_i, W'] - a'(W')(u'_i)X'_i\} + \sum_j \{v'_j[Z', Y'_j] + a'(Z')(v'_j)Y'_j\}$$

$$+ [Z', W'].$$

From the projectability relations assumed, applying $\varphi^!$ to the first of these eight terms gives $\sum_{i,j} u'_i v'_j \otimes [X_i, Y_j]$. Collecting together the second and the second last terms gives

$$\sum_j a' \left(\sum_i u'_i X'_i + Z' \right) (v'_j)Y'_j = \sum_j a'(X')(v'_j)Y'_j.$$

Proceeding in a similar fashion, (24) follows. $\qquad\square$

If f is a surjective submersion, the hypothesis on φ is that (φ, f) is a fibration, as defined in 4.4.1.

The key to this result is that $\Gamma A'$ is generated over $C^\infty(M')$ by the projectable sections. This case arises very frequently in practice, but it is nonetheless easy to construct examples in which the only projectable section of the domain is the zero section. For example, let A' and A be the trivial line bundles over $\mathbb{R}^2$ and $\mathbb{R}$, respectively, and define $\varphi((x, y), t) = (x, yt)$.

Proposition 4.3.8 continues to hold if φ is of constant rank.

Example 4.3.9 The canonical projections and injections in a direct product (see §4.2) are morphisms. It follows that the direct product is indeed the direct product in the category $\mathcal{LA}$. $\qquad\boxtimes$

The proof of the following is a straightforward diagrammatic manipulation.

Proposition 4.3.10 *Given Lie groupoids $G^i \rightrightarrows M^i$ for $i = 1, 2$, the Lie algebroid of the direct product $G^1 \times G^2$ is isomorphic to the direct product Lie algebroid $AG^1 \times AG^2$ under the map formed from the two induced morphisms $A(G^1 \times G^2) \to AG^1$ and $A(G^1 \times G^2) \to AG^2$.*

Proposition 4.3.11 *Let $G \rightrightarrows M$ be a Lie groupoid and $f: M' \to M$ a smooth map satisfying the hypotheses of 2.3.1, so that the pullback Lie* *groupoid $f^{\Downarrow}G$ exists. Then the pullback Lie algebroid $f^{!!}(AG)$ exists and is canonically isomorphic to the Lie algebroid of $f^{\Downarrow}G$.*

Proof Apply the Lie functor to $F = f^{\Downarrow}: f^{\Downarrow}G \to G$. By 4.3.4 we can apply 4.3.6 and obtain a base–preserving morphism $A(F)^{!!}: A(f^{\Downarrow}G) \to f^{!!}(AG)$. By a simple dimension count in the fibres, this is an isomorphism. $\qquad\square$

Use of connections

As in §4.1, one may use auxiliary Koszul connections to express the morphism condition. The proof of the following is straightforward.

Proposition 4.3.12 *Let A and A' be Lie algebroids on M and M' and let $\varphi: A' \to A$ be a vector bundle morphism over $f: M' \to M$ with $a \circ \varphi = T(f) \circ a'$. Then φ is a morphism of Lie algebroids if and only if, for some (and hence any) Koszul connection ∇ in A we have, for all $X', Y' \in \Gamma A'$,*

$$\varphi^{!}([X', Y']) = \overline{\nabla}_{a'X'}(\varphi^{!}(Y')) - \overline{\nabla}_{a'Y'}(\varphi^{!}(X')) - \overline{T}_{\nabla}(\varphi^{!}(X'), \varphi^{!}(Y')). \tag{25}$$

Alternatively, given that (φ, f) satisfies the anchor condition, suppose that there exist connections ∇' and ∇ in A' and A, with torsions T' and T such that

$$\varphi^{!}(\nabla'_{a'X'}(Y')) = \overline{\nabla}_{a'X'}(\varphi^{!}(Y')), \quad \varphi^{!}(T'(X', Y')) = \overline{T}(\varphi^{!}(X'), \varphi^{!}(Y')),$$

for all $X', Y' \in \Gamma A'$; then (φ, f) is a Lie algebroid morphism.

The RHS of (25) can also be used to express the bracket in an abstract pullback Lie algebroid.

Example 4.3.13 Let $\mathfrak{g}$ be a Lie algebra and M a manifold and consider when a vector bundle map $\varphi: TM \to \mathfrak{g}$ is a Lie algebroid morphism. Since the anchor on $\mathfrak{g}$ is zero, only the bracket condition need be considered. The trivial bundle $\mathfrak{g} \to \{\cdot\}$ has a single connection ∇, and the pullback of this to the vector bundle $f^{!}\mathfrak{g} = M \times \mathfrak{g}$ is the standard flat connection given by the Lie derivative (here f is the constant map). Thus in this case (25) becomes

$$\varphi^{!}([X', Y']) = L_{X'}(\varphi^{!}(Y')) - L_{Y'}(\varphi^{!}(X')) + [\varphi^{!}(X'), \varphi^{!}(Y')]. \tag{26}$$

Now $\varphi^!\colon TM \to M \times \mathfrak{g}$ is simply the 1–form corresponding to φ and (26) is simply the Maurer–Cartan equation

$$\delta\varphi + [\varphi, \varphi] = 0.$$

Thus a linear map $TM \to \mathfrak{g}$ is a Lie algebroid morphism if and only if it is a Maurer–Cartan form.

This result also follows by using 4.3.6 and 3.3.3 — φ is a morphism of Lie algebroids if and only if $\varphi^{!!}\colon TM \to f^{!!}\mathfrak{g} = TM \oplus (M \times \mathfrak{g})$ is a base–preserving morphism of (trivial) Lie algebroids.

In turn this description may be applied to morphisms of any transitive Lie algebroids. Assume that A' and A are transitive Lie algebroids on bases M' and M and that $\varphi\colon A' \to A$, $f\colon M' \to M$ is a vector bundle morphism which satisfies the anchor condition. Then the pull-back $f^{!!}A$ exists and is transitive, and φ is a morphism if and only if $\varphi^{!!}\colon A' \to f^{!!}A$ is a base–preserving morphism of Lie algebroids. A local description of base–preserving morphisms of transitive Lie algebroids is given in Chapter 6.   ☒

Lie subalgebroids

Suppose that A and A' are Lie algebroids on M and M' and that $\varphi\colon A' \to A$, $f\colon M' \to M$ is a morphism of vector bundles with f an injective immersion onto a closed embedded submanifold and φ a fibrewise injection. Then, regarding f and φ as inclusions, the anchor and bracket conditions that (φ, f) be a Lie algebroid morphism take the form of the conditions in the following definition.

Definition 4.3.14 Let A be a Lie algebroid on M and let $M' \subseteq M$ be a closed embedded submanifold. A *Lie subalgebroid of A over M'* is a vector subbundle $A' \to M'$ of $A|_{M'} \to M'$ such that:

- the anchor $a\colon A \to TM$ restricts to $A' \to TM'$;
- if $X, Y \in \Gamma A$ have $X|_{M'}$, $Y|_{M'} \in \Gamma A'$ then $[X, Y]|_{M'} \in \Gamma A'$ also;
- if $X, Y \in \Gamma A$ have $X|_{M'} = 0$ and $Y|_{M'} \in \Gamma A'$, then $[X, Y]|_{M'} = 0$.

                                                                    □

When these conditions are satisfied, there is a unique Lie algebroid structure on A' such that the inclusions $A' \to A$, $M' \to M$, constitute a morphism of Lie algebroids. The anchor of this structure is the restriction of a to $A' \to TM'$ and the bracket is given by

$$[x, y] = [X, Y]|_{M'},$$

where $X|_{M'} = x$, $Y|_{M'} = y$.

If $A \to M$ and $M' \subseteq M$ are given, and $A|_{M'} \cap (a|_{M'})^{-1}(TM')$ exists as a vector bundle, it will be a Lie subalgebroid of A over M', denoted $A_{M'}^{TM'}$, and called the *Lie algebroid restriction of A to M'*. Every Lie subalgebroid of A over M' will then be a wide Lie subalgebroid of $A_{M'}^{TM'}$.

If A is transitive, then $A_{M'}^{TM'}$ exists for all closed embedded submanifolds M'.

Examples 4.3.15 For any Lie algebroid $A \to M$, and any $m \in M$, the kernel L_m of $a_m \colon A_m \to T_m M$ is a Lie subalgebroid of A over $\{m\}$; indeed it is the Lie algebroid restriction of A to $\{m\}$.

For any closed embedded submanifold $N \subseteq M$, the restricted Lie algebroid of TM to N is TN. ⊠

Comorphisms

If one ignores the vector bundle structure underlying a Lie algebroid, and considers only the modules of sections over the algebras of functions, there is an alternative notion of morphism which is very natural.

Definition 4.3.16 Let $A' \to M'$ and $A \to M$ be Lie algebroids, and let $f \colon M' \to M$ be a smooth map. A *comorphism of Lie algebroids over f from A' to A*, is a vector bundle morphism $\gamma \colon f^! A \to A'$ such that the induced map $\overline{\gamma} \colon \Gamma A \to \Gamma A'$, $X \mapsto \varphi \circ (1 \otimes X)$ and $f^* \colon C^\infty(M) \to C^\infty(M')$ satisfy the conditions:

(i) $f^*(a(X)(u)) = a'(\overline{\gamma}(X))(f^*(u))$ for all $u \in C^\infty(M)$, $X \in \Gamma A$;
(ii) $\overline{\gamma}([X, Y]) = [\overline{\gamma}(X), \overline{\gamma}(Y)]$ for all $X, Y \in \Gamma A$.

□

Note that γ and f point in opposite directions.

This is a natural concept, and includes in particular the maps of cotangent Lie algebroids induced by Poisson maps; see §10.1.

Example 4.3.17 If $\varphi \colon A' \to A$, $f \colon M' \to M$ is an action morphism, then $(\varphi^!)^{-1} \colon f^! A \to A'$ is a comorphism over f. ⊠

There are thus two concepts of morphism for Lie algebroids, and accordingly two categories. However, any comorphism can be factored into an action and a base preserving morphism as follows.

Proposition 4.3.18 *Let* $\gamma\colon f^!A \to A'$ *be a comorphism over* $f\colon M' \to M$. *Then* A *acts on* f *by* $X^\dagger = a'(\overline{\gamma}(X))$, *and* γ *is then a morphism* $A \lessdot f \to A'$ *of Lie algebroids over* M'.

Proof That $X^\dagger$ projects under f to $a(X)$ follows directly from (i) and that $[X^\dagger, Y^\dagger] = [X, Y]^\dagger$ follows from (ii). Additivity is clear and the module condition follows from observing that $\overline{\gamma}(uX) = (u \circ f)\overline{\gamma}(X)$. So $X \mapsto X^\dagger$ is an action.

Now the map of sections induced by $\gamma\colon f^!A \to A'$ is precisely the map $u' \otimes X \mapsto u'\overline{\gamma}(X)$, and it is clear that this preserves the anchors and the brackets. $\qquad\Box$

4.4 General quotients and fibrations

The concept of Lie algebroid morphism is a very general one. It includes, as we have seen, the right–translation map $TG \to \mathfrak{g}$, $X_g \mapsto T(R_{g^{-1}})(X_g)$ of a Lie group G, and the map $TP \to \frac{TP}{G}$, $X_u \mapsto \langle X_u \rangle$ of a principal bundle $P(M, G)$. In this section we show that the treatment of morphisms such as these follows a broadly similar pattern to the groupoid case (§2.4). For Lie algebroids we also need the full force of the descent construction done for vector bundles in §2.1.

We define fibrations of Lie algebroids by direct analogy with 2.4.3.

Definition 4.4.1 *Let* $\varphi\colon A \to A'$, $f\colon M \to M'$ *be a morphism of Lie algebroids. Then* (φ, f) *is a* fibration *if both* f *and* $\varphi^!\colon A \to f^!A'$ *are surjective submersions.* $\qquad\Box$

That is, a morphism (φ, f) of Lie algebroids is a fibration if f is a surjective submersion and φ is a fibrewise surjection. Such a map has a kernel system $\mathscr{K} = (K, R(f), \theta)$ in the sense of §2.1 and we need only delineate the bracket properties of $\mathscr{K}$.

Firstly, if $X', Y' \in \Gamma K$ then $\varphi \circ X' = \varphi \circ Y' = 0$ and it follows from the definition of morphism that $\varphi \circ [X', Y'] = 0$; thus K is a Lie subalgebroid of A'. Secondly, from $a \circ \varphi = T(f) \circ a'$ it follows that a' maps K into T^fM'.

From 2.1.8 we know that a section X' of A' is φ–projectable if and only if it is θ–stable. So if $X' \in \Gamma K$, $Y' \in \Gamma A'$, and Y' is θ–stable, then $\varphi \circ X' = 0$ and $\varphi \circ Y' = Y \circ f$ for some $Y \in \Gamma A$ and so $\varphi \circ [X', Y'] = [0, Y] \circ f = 0$. Thus $[X', Y'] \in \Gamma K$. These four properties now constitute the following definition.

Definition 4.4.2 Let A' be a Lie algebroid on M' with anchor a'. An *ideal system* of A' is a triple $\mathscr{J} = (J, R, \theta)$ where J is a wide Lie subalgebroid of A', where $R = R(f)$ is a closed, embedded, wide, Lie subgroupoid of the pair groupoid $M' \times M'$ corresponding to a surjective submersion $f \colon M' \to M$, and where θ is a linear action of $R(f)$ on the vector bundle $A'/J \to M'$, such that:

(i) if $X', Y' \in \Gamma A'$ are θ–stable, then $[X', Y']$ is θ–stable;

(ii) if $X' \in \Gamma J$, $Y' \in \Gamma A'$, and Y' is θ–stable, then $[X', Y'] \in \Gamma J$;

(iii) the anchor $a' \colon A' \to TM'$ maps J into $T^f M'$;

(iv) the induced map $A'/J \to TM'/T^f M'$ is $R(f)$–equivariant with respect to θ and the canonical action θ_0 of $R(f)$ on $TM'/T^f M'$.   □

By the canonical action θ_0 we mean the action transported from the pullback bundle by the isomorphism $TM'/T^f M' \cong f^! TM$. In fact $(T^f M', R(f), \theta_0)$ is itself an ideal system of TM'. Proposition 2.1.4 ensures that the fourth condition above is satisfied by the kernel system of a fibration.

Note that if Y' in the second condition is replaced by $u'Y'$ for an arbitrary function u' then $[X', u'Y']$ will generally not be in ΓJ.

Theorem 4.4.3 *Let $A' \to M'$ be a Lie algebroid, and consider an ideal system $\mathscr{J} = (J, R(f), \theta)$ for A', where $f \colon M' \to M$ is a surjective submersion. Then there is a unique Lie algebroid structure on the quotient vector bundle $A = A'/\mathscr{J} \to M$ such that the natural map $\natural \colon A' \to A$, $f \colon M' \to M$ is a morphism of Lie algebroids with kernel system $\mathscr{J}$.*

If $\psi \colon A' \to A''$, $g \colon M' \to M''$, is a morphism of Lie algebroids which annuls $\mathscr{J}$ in the sense of 2.1.6, then the induced morphism of vector bundles $\Psi \colon A'/\mathscr{J} \to A''$, $h \colon M \to M''$, is a morphism of Lie algebroids with respect to this structure.

Proof It follows from conditions (iii) and (iv) of 4.4.2, and from 2.1.4, that there is a unique vector bundle morphism $a \colon A \to TM$ over M such that $a \circ \natural = T(f) \circ a'$. This will be the anchor of A.

Given $X, Y \in \Gamma A$ there exist $X', Y' \in \Gamma A'$ such that $X' \sim X$ and

$Y' \sim Y$ under $\natural$. In particular X', Y' are θ–stable and so, by condition (i), $[X, Y]$ is independent of the choice of X', Y'. It is now easy to verify that this bracket and anchor make A a Lie algebroid on base M. That $\natural \colon A' \to A$, $f \colon M' \to M$, is a Lie algebroid morphism follows easily from 4.3.8 and it is clear that $(\natural, f)$ is a fibration with kernel system $\mathcal{J}$. The uniqueness of the Lie algebroid structure on A is also clear.

Given (ψ, g) we obtain (Ψ, h) from 2.1.6 with $\Psi \circ \natural = \psi$. For the anchors, we have

$$(a'' \circ \Psi) \circ \natural = a'' \circ \psi = T(g) \circ a' = T(h) \circ T(f) \circ a' = T(h) \circ a \circ \natural$$

and since $\natural$ is surjective, it follows that $a'' \circ \Psi = T(h) \circ a$. For the brackets, take $X, Y \in \Gamma A$ and suppose their Ψ–decompositions are

$$\Psi \circ X = \sum u_i(X_i'' \circ h), \qquad \Psi \circ Y = \sum v_j(Y_j'' \circ h).$$

Choose $X', Y' \in \Gamma A'$ such that $\natural \circ X' = X \circ f$, $\natural \circ Y' = Y \circ f$. Then

$$\natural \circ [X', Y'] = [X, Y] \circ f \tag{27}$$

by definition of $[X, Y]$. On the other hand,

$$\psi \circ X' = \Psi \circ X \circ f = \sum u_i'(X_i'' \circ g), \qquad \psi \circ Y' = \Psi \circ Y \circ f = \sum v_j'(Y_j'' \circ g),$$

where $u_i' = u_i \circ f$, $v_j' = v_j \circ f$. Since (ψ, g) is a morphism of Lie algebroids, this implies that

$$\psi \circ [X', Y'] = \sum u_i' v_j'([X_i'', Y_j''] \circ g) +$$
$$\sum a'(X')(v_j')(Y_j'' \circ g) - \sum a'(Y')(u_i')(X_i'' \circ g).$$

Using (27), this can be rewritten

$$\Psi \circ [X, Y] \circ f = \Big\{ \sum u_i v_j([X_i'', Y_j''] \circ h) +$$
$$\sum a(X)(v_j)(Y_j'' \circ h) - \sum a(Y)(u_i)(X_i'' \circ h) \Big\} \circ f$$

and since f is surjective, it may be cancelled off and we have the bracket condition for (Ψ, h), which is therefore a morphism of Lie algebroids. $\square$

The Lie algebroid A on M just constructed is the *quotient Lie algebroid of A' over the ideal system $\mathcal{J}$*. The following case of 4.4.3 is the case used most often.

Corollary 4.4.4 *Let* $\varphi\colon A' \to A$, $f\colon M' \to M$, *be a fibration of Lie algebroids with kernel system* $\mathscr{J} = (J, R, \theta)$. *Then* φ *induces an isomorphism* $A'/\mathscr{J} \cong A$.

Thus fibrations of Lie algebroids, like fibrations of Lie groupoids, are completely characterized by their ideal systems. As in the groupoid case, fibrations of Lie algebroids are closed under composition, and form the largest subcategory of $\mathscr{LA}$, the maps of which can be determined by data entirely on the domain Lie algebroid.

Example 4.4.5 In the case of the right–translation map $TG \to \mathfrak{g}$ for a Lie group G, the standard kernel K is the zero bundle $G \times 0$, the kernel pair $R(f)$ is the whole of $G \times G$, and the action is $\theta(h, g)(X_g) = T(R_{g^{-1}h})(X_g)$. $\boxtimes$

Proposition 4.4.6 *Let* $F\colon G' \to G$, $f\colon M' \to M$, *be a fibration of Lie groupoids. Then* $A(F)\colon AG' \to AG$, $f\colon M' \to M$, *is a fibration of Lie algebroids.*

Proof It is only necessary to prove that AF is fibrewise a surjection, and this follows because the restrictions of F to the α–fibres are surjective submersions. $\square$

We now need to exhibit the relations between the kernel system of a fibration of Lie groupoids and that of the induced fibration of Lie algebroids. Before doing this, we need the groupoid version of the well–known process of linearizing a group action at a fixed point.

Consider a Lie groupoid $H \rightrightarrows M$ acting smoothly on a surjective submersion $q\colon B \to M$ by $\theta\colon H \times_M B \to B$. Assume that q has a global section σ which is stable under the action in the sense that $h\sigma(\alpha h) = \sigma(\beta h)$ for all $h \in H$.

Applying the tangent functor to $H * B \to B$ we get an action $T(\theta)$ of the tangent groupoid $TH \rightrightarrows TM$ on $T(q)\colon TB \to TM$. In particular, elements of TH of the form 0_h act only on the vertical vectors of TB. Writing $B_* = \sigma^! T^q B$ for the pullback, there is an evident restriction of $TH \times_{TM} TB \to TB$ to

$$H \times_M B_* \to B_*$$

which we denote θ_{lin} and call the *linearization of* θ *along* σ.

Suppose now that $F\colon G' \to G$, $f\colon M' \to M$, is a fibration of Lie groupoids with kernel system $\mathscr{N} = (N, R(f), \theta)$. Applying the Lie functor, it is trivial that $AF\colon AG' \to AG$, $f\colon M' \to M$, is a fibration of Lie algebroids, and that the kernel, in the usual sense, of AF is AN.

The action θ of $R(f)$ on $\alpha'\colon G' \overset{\cdot}{\cdot\;} N \to M'$ may now be linearized along the section $1'$, by virtue of condition (N2) of 2.4.7. We thereby have an action θ_{lin} of $R(f)$ on the pullback of $T^\alpha(G' \overset{\cdot}{\cdot\;} N)$ along $1'$. Now G' and N are both Lie groupoids on M' and their Lie algebroids are the pullbacks (as vector bundles) of $T^\alpha G'$ and $T^\alpha N$ along the identity section; it is thus easy to see that $(G' \overset{\cdot}{\cdot\;} N)_*$ is canonically isomorphic to AG'/AN. We denote the action θ_{lin} transferred to AG'/AN by θ_A. The remaining steps of the proof of the following theorem are now routine.

Theorem 4.4.7 *Let $F\colon G' \to G$, $f\colon M' \to M$, be a fibration of Lie groupoids with kernel system $\mathscr{N} = (N, R(f), \theta)$. Then $AF\colon AG' \to AG$, $f\colon M' \to M$, is a fibration of Lie algebroids with kernel system $\mathscr{J} = (AN, R(f), \theta_A)$, as described above.*

Regarding θ_A as a groupoid morphism $R(f) \to \Phi(AG'/AN)$ and applying the Lie functor, one gets a representation of $T^f M'$ on the vector bundle AG'/AN which is in effect a partial flat connection in AG'/AN.

Finally in this section we consider the uniform case.

Definition 4.4.8 (i) A fibration of Lie algebroids $\varphi\colon A' \to A$ over $f\colon M' \to M$, is *uniform* if $\varphi^{!!}\colon A' \to f^{!!}A$ is a surjective submersion.

(ii) An ideal system $\mathscr{J} = (J, R(f), \theta)$ for a Lie algebroid A' is *uniform* if the restriction of the anchor $a'\colon J \to T^f M'$ is surjective. □

Proposition 4.4.9 *Let $\varphi\colon A' \to A$, $f\colon M' \to M$ be a fibration of Lie algebroids. Then (φ, f) is uniform if and only its ideal system $\mathscr{J} = (J, R(f), \theta)$ is uniform.*

Proof This is a simple exercise in chasing elements. Assume that $\mathscr{J}$ is uniform. Take $x' \oplus X \in f^{!!}A$. Since (φ, f) is a fibration, there exists $X' \in A'$ with $\varphi(X') = X$. Write $y' = a'(X')$. Now

$$T(f)(y') = a(\varphi(X')) = a(X) = T(f)(x')$$

so $x' - y' \in T^f M'$. Hence there exists $Z' \in J$ with $a'(Z') = x' - y'$. Now $\varphi^{!!}(X' + Z') = x' \oplus X$. The converse is similar. □

In the case of Lie groupoids, a uniform normal subgroupoid system $(N, R(f), \theta)$ is entirely determined by N. For Lie algebroids the corresponding result is not quite true, since $T^f M'$ need not determine $R(f)$. Nonetheless quotients over uniform ideal systems are easier to handle than the general case.

The proof of the following result is similar to that of 4.4.6.

Proposition 4.4.10 *Let* $F\colon G' \to G$, $f\colon M' \to M$, *be a uniform fibration of Lie groupoids. Then* $AF\colon AG' \to AG$, $f\colon M' \to M$, *is a uniform fibration of Lie algebroids.*

4.5 General semidirect products

Rather than differentiate the definition 2.5.1 of a general action of one groupoid on another, we define split fibrations of Lie algebroids by analogy with split fibrations of groupoids 2.5.2 and deduce the definition of general action of Lie algebroids from it.

Definition 4.5.1 Let $\varphi\colon A' \to A$, $f\colon M' \to M$ be a fibration of Lie algebroids. A *splitting* of φ consists of an action of A on $f\colon M' \to M$, and a morphism of Lie algebroids $s\colon A \ltimes f \to A'$, defined on the resulting action Lie algebroid, such that $\varphi^{!} \circ s = \mathrm{id}_A$. The fibration is said to be *split* if it admits a splitting. □

Now suppose that φ is a split fibration with splitting s and with action $X \mapsto X^{\dagger}, \Gamma A \to \Gamma T M'$. Denote the conventional kernel of φ by K, and the restriction of a' to K by k. Given $X \in \Gamma A$ and $V \in \Gamma K$, define $\rho(X)(V) \in \Gamma K$ by

$$\rho(X)(V) = [s(1 \otimes X), V].$$

Then the map $\rho\colon \Gamma A \times \Gamma K \to \Gamma K$ is $\mathbb{R}$-bilinear, and has the following properties:

(i) $\rho(X)(u'V) = u'\rho(X)(V) + X^{\dagger}(u')V$;
(ii) $\rho(X)([U, V]) = [\rho(X)(U), V] + [U, \rho(X)(V)]$;
(iii) $k(\rho(X)(V))(u') = X^{\dagger}(k(V)(u')) - k(V)(X^{\dagger}(u'))$;
(iv) $\rho([X, Y])(V) = \rho(X)(\rho(Y)(V)) - \rho(Y)(\rho(X)(V))$;
(v) $\rho(uX)(V) = (u \circ f)\rho(X)(V)$;

where $X, Y \in \Gamma A$, $U, V \in \Gamma K$, $u' \in C^{\infty}(M')$, $u \in C^{\infty}(M)$.

We take these conditions as defining a generalized action of A on K via f.

Definition 4.5.2 Let A be a Lie algebroid on M, let $f\colon M' \to M$ be a surjective submersion, and let K be a Lie algebroid on M'. Assume that $T(f) \circ k = 0\colon K \to TM$. Then a *derivative representation of A on K via f*, or a *generalized action of A on K via f*, consists of an action of A on f together with an $\mathbb{R}$-bilinear map $\rho\colon \Gamma A \times \Gamma K \to \Gamma K$ satisfying conditions (i)–(v) above. $\square$

The following characterization extends 4.1.8.

Proposition 4.5.3 *Let A be a Lie algebroid on M and let E be a vector bundle on M'. Let $f\colon M' \to M$ be a smooth map.*

Let $\rho\colon \Gamma A \times \Gamma E \to \Gamma E$ and $X \mapsto X^{\dagger}$ constitute a derivative representation of A on E via f. Then the map $\bar{\rho}\colon A \lessdot f \to \mathfrak{D}(E)$ defined in terms of sections by $u' \otimes X \mapsto u'\rho(X)$, is a representation of $A \lessdot f$ on E.

Conversely, let $X \mapsto X^{\dagger}$ be an action of A on f and let $\rho\colon A \lessdot f \to \mathfrak{D}(E)$ be a representation of $A \lessdot f$ on E. Then the map which sends $X \in \Gamma A$ to $\rho(\overline{X})$, where $\overline{X} \in \Gamma(f^{!}A)$ is the pullback section, is a derivative representation of A on E via f.

Proof It is clear from condition (v) above that $u' \otimes X \mapsto u'\rho(X)$ is well defined and so defines a vector bundle map $A \lessdot f \to \mathfrak{D}(E)$. Condition (i) implies that $\bar{\rho}$ is anchor preserving and (iv) that it is bracket preserving. The converse is equally straightfoward. $\square$

Theorem 4.5.4 *Let A be a Lie algebroid on M, let $f\colon M' \to M$ be a surjective submersion, and let K be a Lie algebroid on M'. Suppose given a generalized action ρ of A on K. Then the vector bundle $f^{!}A \oplus K$ has a Lie algebroid structure on M', denoted $A \sqcap_{\rho} K$, with anchor*

$$a'\left(\sum(u'_i \otimes X_i) \oplus V\right) = \sum u'_i X_i^{\dagger} + k(V)$$

and bracket

$$\left[\sum(u'_i \otimes X_i) \oplus U, \sum(v'_j \otimes Y_j) \oplus V\right] =$$
$$\left\{\sum u'_i v'_j \otimes [X_i, Y_j] + \sum u'_i X_i^{\dagger}(v'_j) \otimes Y_j - \sum v'_j Y_j^{\dagger}(u'_i) \otimes X_i + \right.$$
$$\left.\sum k(U)(v'_j) \otimes Y_j - \sum k(V)(u'_i) \otimes X_i\right\}$$
$$\oplus \left\{[U, V] + \sum u'_i \rho(X_i)(V) - \sum v'_j \rho(Y_j)(U)\right\}$$

where $X_i, Y_j \in \Gamma A$, $u'_i, v'_j \in C^\infty(M')$, $U, V \in \Gamma K$. *The natural map* $A \sqsubset_\rho K \to A$ *is a split fibration with kernel* K.

The proof is long but straightforward and is omitted.

This structure $A \sqsubset_\rho K$ is the *generalized semidirect product Lie algebroid of* A *with* K *via* ρ. It of course includes the standard semidirect product of Lie algebras, as well as the action Lie algebroid construction described in §4.1.

Theorem 4.5.5 *Let* $G \rightrightarrows M$ *and* $W \rightrightarrows M'$ *be Lie groupoids and let* G *act smoothly on* W *via a surjective submersion* $f \colon M' \to M$. *Denote by* $\pi \colon G \sqsubset W \to G$ *the corresponding split fibration of 2.5.3. Then the fibration* $A(\pi) \colon A(G \sqsubset W) \to AG$ *is split with respect to the infinitesimal action of* AG *on* f *induced by the given action of* G *on* f. *The splitting induces a derivative representation of* AG *on* AW *via* f *with respect to which there is a canonical isomorphism*

$$A(G \sqsubset W) \cong AG \sqsubset AW.$$

Proof That $A(\pi)$ is a fibration was proved in 4.4.6. From 4.1.6 we know that $A(G \lessdot f)$ is canonically isomorphic to $AG \lessdot f$. Denote the canonical groupoid splitting by $\sigma_0 \colon G \lessdot f \to G \sqsubset W$. Now although σ_0 is a morphism of Lie groupoids, $\pi^! \colon G \sqsubset W \to G \lessdot f$ is not. Nonetheless, because it preserves the α–fibres and the identity elements it induces a map, abusively denoted $A(\pi^!)$, which is a morphism of vector bundles $A(G \sqsubset W) \to A(G \lessdot f)$. From $\pi^! \circ \sigma_0 = \mathrm{id}$ it follows as usual that $A(\pi^!) \circ A(\sigma_0) = \mathrm{id}$ and hence that $A(\sigma_0)$ is a splitting of the Lie algebroid fibration $A(\pi)$. The result now follows. $\qquad \square$

PBG–Lie algebroids

Finally in this section we deal with the infinitesimal version of the concept of PBG–Lie groupoid which was introduced in §2.5. Until the end of the section, let $P(M, G, p)$ be a fixed principal bundle.

Definition 4.5.6 A *PBG–Lie algebroid* on $P(M, G)$ is a transitive Lie algebroid A on base P together with a right action $(X, g) \mapsto Xg = \widetilde{R}_g(X)$ of G on A such that each $\widetilde{R}_g \colon A \to A$ is a Lie algebroid automorphism over $R_g \colon P \to P$. $\qquad \square$

The conditions of the definition are, firstly, that G act on A by vector bundle automorphisms and, secondly, that the anchor is equivariant with respect to the action of G on TP obtained by lifting the action of G on P. For the bracket condition, first extend the notation $\widetilde{R}_g$ to the map $\Gamma A \to \Gamma A$ defined by

$$(\widetilde{R}_g(X))(u) = \widetilde{R}_g(X(ug^{-1}))$$

for $u \in P$, $X \in \Gamma A$. Then the third condition is that, for $g \in G$ and $X, Y \in \Gamma A$,

$$[\widetilde{R}_g(X), \widetilde{R}_g(Y)] = \widetilde{R}_g([X, Y]).$$

Theorem 4.5.7 *Let A be a PBG–Lie algebroid on $P(M, G)$. Then the quotient manifold A/G exists and inherits a quotient structure of transitive Lie algebroid from A; further, it is an extension*

$$\frac{L}{G} \rightarrowtail \frac{A}{G} \longrightarrow\!\!\!\!\!\rightarrow \frac{TP}{G} = A\Omega$$

of the Lie algebroid of the gauge groupoid of $P(M, G)$ by the quotient LAB L/G.

Proof The proof that the manifold A/G exists follows very much as in 2.5.5. Denote the projection $A \to P$ by q, and write

$$\Gamma' = \{(X, Xg) \mid X \in A, \ g \in G\};$$

we must show that Γ' is a closed submanifold of $A \times A$. Now $\Gamma' \subseteq (q \times q)^{-1}(\Gamma)$ where $\Gamma = \{(u, ug) \mid u \in P, \ g \in G\}$. Since Γ is a closed submanifold of $P \times P$, and $q \times q$ is a surjective submersion, it suffices to prove that Γ' is a closed submanifold of $(q \times q)^{-1}(\Gamma)$. Define

$$f\colon (q \times q)^{-1}(\Gamma) \to A, \qquad (X, Y) \mapsto Xg - Y$$

where $q(Y) = q(X)g$. From the local triviality of A, it easily follows that f is a surjective submersion. The preimage of the zero section under q is Γ', and this shows that Γ' is a closed submanifold. Denote the quotient projection $A \to A/G$ by $\natural$.

The vector bundle structure of A quotients to A/G in a straightforward fashion. Since the anchor $a\colon A \to TP$ is G–equivariant, it quotients to a vector bundle morphism $A/G \to TP/G$ which is again a surjective submersion; denote this by π, and define $b = \widetilde{a} \circ \pi$ where $\widetilde{a}$ is the anchor of TP/G.

For the bracket structure of A/G, note first that $\Gamma(A/G)$ can be identified with the $C^\infty(M)$ module of G–equivariant sections of A as in §3.1. Since the bracket on ΓA restricts to the G–equivariant sections by assumption, this bracket transfers to $\Gamma(A/G)$. It is now straightforward to check that this makes A/G a Lie algebroid on M with anchor b, and $\natural\colon A \to A/G$ a Lie algebroid morphism over p. That π is a Lie algebroid morphism with kernel L/G is easily checked. □

Conversely, suppose we are given an extension of Lie algebroids

$$K \rightarrowtail B \overset{\pi}{\longrightarrow\!\!\!\!\!\gg} \frac{TP}{G} = A\Omega.$$

Define an action of B on P by $X \mapsto \overrightarrow{\pi(X)}$, where the arrow denotes the right–invariant vector field on P corresponding to the section of TP/G, and let $A = B \ltimes P$. Then it is straightforward to see that A has a PBG–Lie algebroid structure given by the natural action on the pullback bundle, and that $A/G \cong B$. The adjoint bundle of A is $p^! K$, the pullback LAB of K to P.

The next result follows from the preceding, and naturality arguments.

Proposition 4.5.8 *Let Υ be a PBG–Lie groupoid on $P(M,G)$. Then the Lie algebroid $A(\Upsilon/G)$ of the quotient groupoid is naturally isomorphic to the quotient Lie algebroid $(A\Upsilon)/G$.*

A result on the integrability of PBG–Lie algebroids is given in 8.3.3.

4.6 Classes of morphism

In this chapter, we have developed four main classes of Lie algebroid morphism in clear analogy with the classes of Lie groupoid morphism described in §2.6. Again, we give a brief summary here.

Consider a morphism of Lie algebroids $\varphi\colon A' \longrightarrow A$, $f\colon M' \longrightarrow M$, and assume, as before, that f is a surjective submersion. Form the pullback manifold $f^! A$ and factorize φ as follows

$$
\begin{array}{ccccc}
 & \overset{\varphi^!}{\longrightarrow} & & \longrightarrow & \\
A' & & f^! A & & A \\
\Big\downarrow{\scriptstyle q'} & & \Big\downarrow & & \Big\downarrow{\scriptstyle q} \\
M' & =\!=\!= & M' & \longrightarrow & M \\
 & & & \underset{f}{} &
\end{array}
\qquad (28)
$$

where $\varphi^!\colon A' \to f^!A$ is the induced map $X' \mapsto (q'(X'), \varphi(X'))$. Then:

- (φ, f) is a *fibration* if $\varphi^!$ is a surjective submersion;
- (φ, f) is an *action morphism* if $\varphi^!$ is a diffeomorphism.

Since f is a surjective submersion we can also form the pullback Lie algebroid $f^{!!}A$ and factorize φ as follows:

$$
\begin{array}{ccccc}
& \varphi^{!!} & & & \\
A' & \longrightarrow & f^{!!}A & \longrightarrow & A \\
a' \downarrow & & \downarrow & & \downarrow a \qquad (29)\\
TM' & =\!=\!= & TM' & \longrightarrow & RM \\
& & T(f) & &
\end{array}
$$

Recall that the induced map $\varphi^{!!}\colon A' \to f^{!!}A$ is $X' \mapsto (a'(X'), \varphi(X'))$. Now:

- (φ, f) is a *uniform fibration* if $\varphi^{!!}$ is a surjective submersion;
- (φ, f) is an *inductor* if $\varphi^{!!}$ is a diffeomorphism.

These four classes of maps characterize the basic algebraic notions of action, quotient and pullback.

4.7 Notes

§4.1. The notion of action of a Lie algebroid on a smooth map was first clearly defined in [Higgins and Mackenzie, 1990a], following earlier indications in [Mackenzie, 1988a, 1987c] which were themselves based upon Weinstein's calculation of the Lie algebroid structure of $A(G \ltimes M)$ for an action of a Lie group G on a manifold M (see the Notes to Chapter 3). Most of §4.1, including the general abstract equivalence between infinitesimal actions and action Lie algebroids and Theorem 4.1.6 comes from [Higgins and Mackenzie, 1990a].

There is a purely algebraic version for Lie pseudoalgebras of the equivalence between infinitesimal actions and action Lie algebras (see [Mackenzie, 1995]); this, at least in the case where the acting Lie pseudoalgebra is a Lie algebra, has been long known in ring theory as a *crossed product* associated with a *change of rings* procedure. See [Fel'dman, 1982] and [Malliavin, 1988].

The derivative representations of 4.1.8 were studied in [Kosmann, 1976], [Kosmann–Schwarzbach, 1980]. The latter paper, in particular, gives a thorough account in these terms, with substantial applications, of many of the constructions which are treated here by means of Lie algebroid actions. See [Kosmann–Schwarzbach and Mackenzie, 2002] for more detail on the relationship between these approaches.

The use of connections to simplify bracket formulas like (6) and (24) was first given in [Higgins and Mackenzie, 1990a]. The notion of torsion (14) for connections in the vector bundles underlying Lie algebroids is due to Pradines [1967b].

§4.2. The treatment of direct products and pullbacks given here follows [Higgins and Mackenzie, 1990a].

§4.3. The general notion of morphism of Lie algebroids comes from [Higgins and Mackenzie, 1990a], following an incomplete account in [Mackenzie, 1987c]. Almeida and Kumpera [1981] had earlier formulated essentially the same notion using localization in place of partitions of unity.

Definition 4.3.1 is effective in dealing with general constructions, such as elsewhere in this chapter, but can be very hard to work with in some applications. The alternative Poisson formulation dealt with in Chapter 10 is then often effective. In the author's work on double structures, it mostly turns out that one can use 4.3.6.

The case of Maurer–Cartan forms in Example 4.3.13 was observed independently by Xu [1995].

A strikingly different formulation was introduced by Vaĭntrob [1997]. The bracket structure on a Lie algebroid gives rise to a cochain complex (see §7.1) and, regarding the cochain complex as the ring of smooth functions on a super–manifold, the coboundary operator may be viewed as a vector field upon it; in the terminology used by [Vaĭntrob, 1997], as a *homological vector field*. It is then possible to define morphisms of Lie algebroids by adapting the notion of projectability of vector fields to the super case.

For an interval $I \subseteq \mathbb{R}$, morphisms of Lie algebroids $TI \to A$ and $TI^2 \to A$ into an arbitrary Lie algebroid A have arisen in work on σ–models; see [Schaller and Strobl, 1994], [Cattaneo and Felder, 2001] and [Crainic and Fernandes, 2003]. For convenience here, take $I = \mathbb{R}$. For a Lie algebroid morphism $\varphi \colon TI \to A$ over a curve $f \colon I \to M$, the bracket condition is vacuous and any vector bundle map $\varphi \colon TI \to A$ such that $a \circ \varphi = T(f)$ will be a Lie algebroid morphism. Regarding TI as $I \times \mathbb{R}$, write $F(t) = \varphi(t, 1)$; then any f and any $F \colon I \to A$ with $a \circ F = \frac{df}{dt}$ determine a Lie algebroid morphism φ.

Comorphisms The treatment of comorphisms follows [Higgins and Mackenzie, 1993]. Concepts of comorphism may be defined for arbitrary modules and for arbitrary vector bundles, in such a way that there is a duality between the category of vector bundle morphisms and the category of vector bundle comorphisms and, for finitely generated and projective modules, between the category of module morphisms and the category of module comorphisms. These extend the usual dualities, which are only defined for morphisms which preserve the base manifold, or base algebra. Further, the section functor from (the category of all base–preserving morphisms of) vector bundles to (the category of all base–preserving morphisms of) modules, extends to give a section functor from the category of all vector bundle morphisms to the category of all module comorphisms. (Dually, there is another section functor, from the category of all vector bundle comorphisms to the category of all module morphisms.) This extended section functor restricts to give a section functor from the category of all Lie algebroid morphisms to the category of all Lie pseudoalgebra comorphisms. This is done in detail in [Higgins and Mackenzie, 1993].

Comorphisms of vector bundles go back at least as far as [Bourbaki, 1982, 7.2.6] (see also [Guillemin and Sternberg, 1990]).

What we have called comorphisms of Lie pseudoalgebras seem to have first been defined by Huebschmann [1990]. In the context of the modules of sections, this appears as a very natural notion of morphism. It is worth pointing out (see [Higgins and Mackenzie, 1993]) that the notion of morphism of Lie algebroids (4.3.1) may be formulated for abstract Lie pseudoalgebras; this may well be worth further development.

There is a corresponding concept of comorphism for Lie groupoids [Higgins and Mackenzie, 1993] which is a particular case of the notion of Morita equivalence (see the Notes to Chapter 2). However a general concept of Morita equivalence for abstract Lie algebroids is lacking.

§4.4. Most of §4.4 has been taken directly from [Higgins and Mackenzie, 1990a, §4].

The ideas behind §4.4 (and §2.1) are closely related to Libermann's concept of 'fibre parallelism' (also called 'almost parallelism') [Libermann, 1973, 1974]. If $f\colon M' \to M$ is a surjective submersion, a *fibre parallelism for* f is a vector bundle $T_{\mathrm{red}}M'$ on base M, together with an isomorphism of vector bundles $TM' \cong f^!T_{\mathrm{red}}M'$ over M'; equivalently, using §2.1, a fibre parallelism for f is a linear action of $R(f)$ on TM'. In [Libermann, 1974], a fibre parallelism is called *pseudo–integrable* if the bracket of invariant vector fields on M' is invariant, and is called *projectable* if every projectable vector field on M' is f–projectable; it is noted [Libermann, 1974, p22] that for a fibre parallelism that is pseudo–integrable and projectable, the vector bundle $T_{\mathrm{red}}M'$ has a natural Lie algebroid structure on base M. This is effectively the case of 4.4.3 in which $A' = TM'$ and $K = M' \times 0$.

Since the base map of a quotient morphism corresponds to a simple foliation, and an ideal system may be regarded as a generalization of a simple foliation, it is reasonable to ask whether it is possible to develop a useful concept of ideal — and corresponding quotient — which would include arbitrary (non–singular) foliations. This would presumably require a concept of Lie algebroid with a generalized manifold as base.

§4.5. Most of §4.5 has been taken directly from [Higgins and Mackenzie, 1990a, §3].

An interesting alternative viewpoint to this general notion of action has been given by Moerdijk and Mrčun [2002].

The concept of PBG–Lie algebroids comes from [Mackenzie, 1987b]; for Theorem 4.5.7, see [Androulidakis, arXiv:0307282].

PART TWO:
THE TRANSITIVE THEORY

5

Infinitesimal
connection theory

Connection theory is inextricably bound up with the theory of transitive
Lie algebroids. Accounts of the general theory of connections (that is
to say, not merely the case of connections in vector bundles) are tra-
ditionally presented in terms of principal bundles, and can equally well
be presented in terms of locally trivial Lie groupoids. (Indeed the Lie
groupoid formulation is often clearer since there is no need to choose
arbitrary reference points, and holonomy groups, for example, are consequently well defined at each point, rather than only up to conjugacy.)
However a very great deal of general connection theory — that part
which we call the infinitesimal theory — naturally resides within the
context of abstract transitive Lie algebroids. It both clarifies the exposition and strengthens the results to maintain a separation between those
parts of the theory which require principal bundles (or Lie groupoids)
and those which do not.

On the other hand, there is a large family of results for transitive Lie
algebroids which are most naturally proved by using connection theory.
These are dealt with in §6.5.

This chapter and the following one are self–contained but make no
attempt to motivate the notion of connection or describe any of its many
applications. A reader who has never met the notion of connection
before will need to supplement these chapters with other accounts. Some
suggestions are included in the Notes to this chapter.

We begin in §5.1 by listing the main properties of the (right) Darboux
derivative which to a group–valued map $M \to G$ assigns a Maurer–
Cartan form in $\Omega^1(M, \mathfrak{g})$. These formulas are in fact local forms of the
basic properties of general connections.

In §5.2 we present the global infinitesimal aspects of connection theory. These are essentially algebraic and formal, but for that reason it

is valuable to see them stripped of the elaborate garb usual in principal bundle theory. The corresponding local formulation of the infinitesimal theory is given in §5.4. The long §5.3 is concerned with the translation between the principal bundle language and that of Lie algebroids; most readers may postpone this section until the need for it is felt.

5.1 The Darboux derivative

Let G be a Lie group and M a manifold, and let $f\colon M \to G$ be a smooth map. Then the *Darboux derivative* $\Delta(f)\colon TM \to M \times \mathfrak{g}$ of f is the $\mathfrak{g}$-valued 1–form on M defined by

$$\Delta(f)(X_x) = T(R_{f(x)^{-1}})(T(f)(X_x)).$$

Alternatively, $\Delta(f)$ is the pullback $f^*\theta$ of the right Maurer–Cartan form θ on G, and so $\omega = \Delta(f)$ satisfies the Maurer–Cartan equation

$$\delta(\omega) + [\omega, \omega] = 0. \tag{1}$$

with respect to the (right) bracket in $\mathfrak{g}$.

If $G = V$ is a vector space, then $\Delta(f)(X)\colon M \to V$ will be identified with the Lie derivative $X(f)$.

The product rule is

$$\Delta(f_1 f_2) = \Delta(f_1) + \mathrm{Ad}(f_1)(\Delta(f_2)) \tag{2}$$

where $f_1, f_2\colon M \to G$ are two maps. Here and elsewhere the symbol $\mathrm{Ad}(f_1)(\Delta(f_2))$ denotes the map

$$X \mapsto \mathrm{Ad}(f_1(x))(\Delta(f_2)(X)), \qquad X \in T_x(M).$$

From the product rule it follows that

$$\Delta(f^{-1}) = -\mathrm{Ad}(f^{-1})(\Delta(f)) \tag{3}$$

where f^{-1} denotes the pointwise inverse $x \mapsto f(x)^{-1}$.

If $f\colon M \to G$ and $s\colon M \to G$ are smooth maps, denote by $I_s(f)$ the map $x \mapsto s(x)f(x)s(x)^{-1}$. From the product rule it follows that

$$\Delta(I_s(f)) = \mathrm{Ad}(s)\{\mathrm{Ad}(f)(\Delta(s^{-1})) + \Delta(f) - \Delta(s^{-1})\} \tag{4}$$

In particular,

$$\Delta(R_g \circ f) = \Delta(f), \qquad \Delta(L_g \circ f) = \mathrm{Ad}g(\Delta(f)) \tag{5}$$

and

$$\Delta(I_g \circ f) = \mathrm{Ad}g(\Delta(f)) \tag{6}$$

where $g \in G$ is fixed, and $f\colon M \to G$ is a smooth map.

Suppose that V and W are vector spaces and that $f\colon M \to V$ and $\varphi\colon M \to \mathrm{Hom}(V, W)$ are smooth maps. Then it is easy to see that

$$X(\varphi(f)) = X(\varphi)(f) + \varphi(X(f)), \qquad X \in \mathfrak{X}(M) \tag{7}$$

Here, once again, $\varphi(f)$ denotes pointwise evaluation $x \mapsto \varphi(x)(f(x))$. If $W = V$ and φ takes values in $GL(V)$, then (7) can be rewritten as

$$X(\varphi(f)) = -\Delta(\varphi)(X)(\varphi(f)) + \varphi(X(f)), \qquad X \in \mathfrak{X}(M). \tag{8}$$

The reader is urged to check this formula directly. The minus sign and the double appearance of φ in the first term on the right–hand side arise from the identification of $T_I(GL(V))$ with $\mathfrak{gl}(V)$ via right–invariant vector fields.

It is easily verified that if $\varphi\colon G \to H$ is a morphism of Lie groups and $f\colon M \to G$ is a smooth map, then

$$\Delta(\varphi \circ f) = \varphi_* \circ \Delta(f) \tag{9}$$

In particular,

$$\Delta(\mathrm{Ad} \circ f) = \mathrm{ad} \circ \Delta(f). \tag{10}$$

This formula may be rewritten as

$$\Delta(\mathrm{Ad} \circ f)(X)(V) = [\Delta(f)(X), V] \tag{11}$$

where $X \in \mathfrak{X}(M)$ and $V\colon M \to \mathfrak{g}$, and the bracket is taken pointwise.

Proposition 5.1.1 *Let G and H be Lie groups and let M be a manifold. Let $\varphi\colon \mathfrak{g} \to \mathfrak{h}$ be a morphism of Lie algebras and let $f\colon M \to H$ be a smooth map. Define $\widetilde{\varphi}\colon M \to \mathrm{Hom}(\mathfrak{g}, \mathfrak{h})$ by $\widetilde{\varphi}(x) = \mathrm{Ad}(f(x)) \circ \varphi$. Then $\widetilde{\varphi}(x)$ is a Lie algebra morphism for $x \in M$ and*

$$X(\widetilde{\varphi}(V)) = \widetilde{\varphi}(X(V)) - [\Delta(f)(X), \widetilde{\varphi}(V)]$$

for $X \in \mathfrak{X}(M)$, $V\colon M \to \mathfrak{g}$.

Proof Write $\widetilde{\varphi}(V)$ as $(\mathrm{Ad} \circ f)(\varphi(V))$ where $\mathrm{Ad} \circ f\colon M \to \mathrm{Aut}(\mathfrak{h})$ and $\varphi(V)\colon M \to \mathfrak{h}$, and apply (8). This gives,

$$X(\widetilde{\varphi}(V)) = -\Delta(\mathrm{Ad} \circ f)(X)(\widetilde{\varphi}(V)) + (\mathrm{Ad} \circ f)(X(\varphi(V))).$$

Applying (11) to the first term, and (7) to the expression $X(\varphi(V))$, this becomes,

$$X(\widetilde{\varphi}(V)) = -[\Delta(f)(X), \widetilde{\varphi}(V)] + (\mathrm{Ad} \circ f)(\varphi(X(V))).$$

Now

$$(\mathrm{Ad} \circ f)(\varphi(X(V))) = \tilde{\varphi}(X(V))$$

by definition.   □

In the case where $\mathfrak{g} = \mathfrak{h}$ and $\varphi \in \mathrm{Aut}(\mathfrak{g})$ this equation can, by (8), be written more simply as

$$\Delta(\tilde{\varphi}) = \mathrm{ad} \circ \Delta(f)$$

where Δ on the left–hand side is with respect to the group $\mathrm{Aut}(\mathfrak{g})$.

The following result is the simplest interesting case of the integrability of Lie algebroid morphisms.

Theorem 5.1.2 *Let $\omega \in \Omega^1(M, \mathfrak{g})$ be a Maurer–Cartan form, where M is a simply–connected manifold and G is a Lie group. Given $x_0 \in M$, there is a unique smooth map $f \colon M \to G$ with $\Delta(f) = \omega$ and such that $f(x_0) = 1$.*

Proof Define a distribution D on $M \times G$ by

$$D_{(x,g)} = \{X + T(R_g)(\omega(X)) \mid X \in T_x(M)\}.$$

The projection $D_{(x,g)} \to T_x(M)$ is bijective and so D has constant rank. For exch $X \in \mathfrak{X}(M)$ write $\overline{X}$ for the pullback to $M \times G$. For any function $\varphi \colon M \to \mathfrak{g}$ write $\tilde{\varphi}$ for the map $M \times G \to TG$, $(x,g) \mapsto T(R_g)(\varphi(x))$. Then the sections of D are generated by those of the form $\overline{X} + \omega(\overline{X})$ for $X \in \mathfrak{X}(M)$. Now it is straightforward to verify that

$$[\overline{X}, \overline{Y}] = \overline{[X,Y]}, \qquad [\overline{X}, \tilde{\varphi}] = \widetilde{X(\varphi)}, \qquad [\tilde{\varphi}, \tilde{\psi}] = \widetilde{[\varphi, \psi]}.$$

In the last of these $[\varphi, \psi]$ is the pointwise bracket of $\mathfrak{g}$–valued maps; the proof in this case is best handled by use of a basis in $\mathfrak{g}$.

It now follows quickly from these that D is involutive. Denote by L the leaf through $(x_0, 1)$ and by p the restriction to $L \to M$ of the projection $M \times G \to M$. Then $T(p)$ is the restriction to $D|_L$ of the fibrewise bijection $D \to TM$ and so p is étale. Since M is simply–connected, p is a diffeomorphism, and we define $f \colon M \to G$ to be the composite of p^{-1} with the projection $L \to G$. This map f has $\Delta(f) = \omega$.   □

5.2 Infinitesimal connections and curvature

We begin with the case of vector bundles.

Definition 5.2.1 A *connection* in a vector bundle (E, q, M) is a map

$$\nabla \colon \, \mathfrak{X}(M) \times \Gamma E \to \Gamma E, \qquad (X, \mu) \mapsto \nabla_X(\mu),$$

which is $\mathbb{R}$-bilinear and satisfies the two identities

$$\nabla_{fX}(\mu) = f \nabla_X(\mu), \qquad \nabla_X(f\mu) = f \nabla_X(\mu) + X(f)\mu,$$

for all $X \in \mathfrak{X}(M)$, $\mu \in \Gamma E$, $f \in C^\infty(M)$. □

We will call this a *Koszul connection* when it is necessary to distinguish it from other formulations.

Example 5.2.2 It is clear that in a trivial bundle $E = M \times V$ the Lie derivatives define a connection

$$\nabla_X^0(\mu) = \mathfrak{L}_X(\mu), \qquad X \in \mathfrak{X}(M), \ \mu \colon M \to V,$$

which we will variously call the *standard flat connection*, the *trivial connection*, or the *Lie derivative connection*. ⊠

It follows that, given a trivializing atlas of a vector bundle, a connection may be defined locally over each trivializing open subset of the base by transport of standard flat connections; these, however, do not generally coincide on overlaps. The concept of Koszul connection trades off uniqueness and the property of flatness in order to obtain a global means of differentiating sections.

That every vector bundle admits a connection will follow from 5.2.7. For the rest of this section we consider a given connection ∇ in a given vector bundle E.

Definition 5.2.3 The *curvature* of ∇ is the map $\overline{R}_\nabla \colon TM \oplus TM \to \mathrm{End}(E)$ defined on sections by

$$\overline{R}_\nabla(X, Y)(\mu) = \nabla_{[X,Y]}(\mu) - \nabla_X(\nabla_Y(\mu)) + \nabla_Y(\nabla_X(\mu)). \qquad (12)$$

The connection ∇ is *flat* if $\overline{R}$ is identically zero. □

It is an easy exercise to check that the RHS is $C^\infty(M)$-linear in X, Y and μ.

It is clear that ∇_X, for each $X \in \mathfrak{X}(M)$, is a derivative endomorphism of ΓE. Indeed we can identify ∇ with a vector bundle morphism $\gamma \colon TM \to \mathfrak{D}(E)$ by $\gamma(X)(\mu) = \nabla_X(\mu)$.

Denote the linear vector field corresponding to ∇_X by (C_X, X). From these vector fields we construct a map

$$\mathsf{C}\colon E \times_M TM \to TE, \qquad (e, X) \mapsto \mathsf{C}_X(e).$$

From the definition it follows that C is right–inverse to

$$(p_E, T(q))\colon TE \to E \times_M TM.$$

Furthermore, since each $\nabla_X(\mu)$ is linear in μ and ∇_X is linear in X, it is a right splitting (as defined on p. 187) of both the sequences (19) and (20) of Chapter 3. This proves the first half of the next result, and the converse is straightforward.

Proposition 5.2.4 *There is a bijective correspondence between connections in the vector bundle E and maps $\mathsf{C}\colon E \times_M TM \to TE$ which are simultaneous splittings of sequences (19) and (20) on p. 112.*

*   *   *   *   *

We observed that a Koszul connection in a vector bundle E can be regarded as a map $TM \to \mathfrak{D}(E)$. The definition of the general notion of connection is a simple extension of this.

Definition 5.2.5 Let A be a transitive Lie algebroid over M with anchor a. A *connection* in A, or a *Lie algebroid connection* if confusion is possible, is a morphism of vector bundles $\gamma\colon TM \to A$ over M such that $a \circ \gamma = \mathrm{id}_{TM}$. A *connection reform* in A is a morphism of vector bundles $\omega\colon A \to L$ over M such that $\omega \circ j = \mathrm{id}_L$.

Let Ω be a Lie groupoid on M. An *infinitesimal connection* in Ω is a connection in the Lie algebroid $A\Omega$. A *connection reform* in Ω is a connection reform in $A\Omega$. □

The following general result about vector bundles is worth stating in full. It may be proved by a partition of unity argument, or by introducing a metric.

Proposition 5.2.6 *Consider a short exact sequence of vector bundles over M*

$$E' \overset{\iota}{>\!\!\longrightarrow} E \overset{\pi}{\longrightarrow\!\!\!\gg} E'', \tag{13}$$

that is, an assembly of three vector bundles E', E, E'' over M and two base preserving morphisms $\iota\colon E' >\!\!\longrightarrow E$ and $\pi\colon E \longrightarrow\!\!\!\gg E''$ such that

$$E'_m \overset{\iota_m}{>\!\!\longrightarrow} E_m \overset{\pi_m}{\longrightarrow\!\!\!\gg} E''_m$$

is a short exact sequence of vector spaces, for each $m \in M$. Then there is a right inverse $\gamma \colon E'' \to E$ to π and a left inverse $\lambda \colon E \to E'$ to ι. Furthermore, given any right–inverse γ or any left–inverse ι, the other can be chosen uniquely so that

$$\iota \circ \lambda + \gamma \circ \pi = \mathrm{id}_E.$$

In this case

$$E'' \overset{\gamma}{\rightarrowtail} E \overset{\lambda}{\twoheadrightarrow} E'$$

is then also a short exact sequence of vector bundles over M. We call γ a right–splitting of (13) *and λ* a left–splitting of (13).

It now follows immediately that there is a bijective correspondence between connections γ and connection reforms ω, such that

$$j \circ \omega + \gamma \circ a = \mathrm{id}_A. \tag{14}$$

The following fundamental result is also an immediate consequence.

Corollary 5.2.7 *Every transitive Lie algebroid admits a connection.*

Lemma 5.2.8 *Let $L \overset{j}{\rightarrowtail} A \overset{a}{\twoheadrightarrow} TM$ be a transitive Lie algebroid with a connection γ. Then the map*

$$\mathfrak{X}(M) \times \mathfrak{X}(M) \to \Gamma L, \qquad (X, Y) \mapsto \gamma[X, Y] - [\gamma X, \gamma Y]$$

is bilinear over $C^\infty(M)$.

The proof is a simple calculation. The lemma justifies the following definition.

Definition 5.2.9 The *curvature* of a connection γ in A is the alternating vector bundle morphism $\overline{R}_\gamma \colon TM \oplus TM \to L$ defined by

$$j(\overline{R}_\gamma(X, Y)) = \gamma[X, Y] - [\gamma X, \gamma Y]$$

for $X, Y \in \mathfrak{X}(M)$. A connection is *flat* if is has zero curvature. □

Remark 5.2.10 The notation $L \overset{j}{\rightarrowtail} A \overset{a}{\twoheadrightarrow} TM$ in 5.2.8 emphasizes that the curvature $\overline{R}_\gamma$ depends on L and j as well as on γ. Already in §3.2, for a principal bundle $P(M, G, \pi)$ we replaced $\frac{T^\pi P}{G} = \ker(\pi_*)$ by the adjoint bundle. In general we can replace $L = \ker(a)$ by any isomorphic LAB L'. If $\varphi \colon L \to L'$ is the isomorphism, then $\overline{R}'_\gamma = \varphi \circ \overline{R}_\gamma$.

If L is abelian, then any $t \in \mathbb{R}$, $t \neq 0$, defines a rescaling of curvature forms.

When the Lie algebroid is the Atiyah sequence of a principal bundle, connections and their curvatures can be expressed directly in terms of the principal bundle. For this see the next section. We now consider some basic examples.

Example 5.2.11 Let $TM \oplus (M \times \mathfrak{g})$ be a trivial Lie algebroid. Then $TM \to TM \oplus (M \times \mathfrak{g})$, $X \mapsto X \oplus 0$ is a connection, called the *standard flat connection* in $TM \oplus (M \times \mathfrak{g})$ and denoted by γ^0.

An arbitrary connection in $TM \oplus (M \times \mathfrak{g})$ is of the form $X \mapsto X \oplus \omega(X)$ where $\omega : TM \to M \times \mathfrak{g}$ is a $\mathfrak{g}$–valued 1–form on M. The corresponding connection reform is $X \oplus V \mapsto V - \omega(X)$. The curvature of this connection is $-(\delta\omega + [\omega, \omega]) \in \Omega^2(M, \mathfrak{g})$.   ⊠

Example 5.2.12 Let E be a vector bundle on M. We have already noted that a Lie algebroid connection in $\mathfrak{D}(E)$ is equivalent to a Koszul connection in E. In practice we will often allow ∇ to denote the map $TM \to \mathfrak{D}(E)$ as well as using it in the usual way. The curvature of ∇ as a Lie algebroid connection is precisely $\overline{R}_\nabla : TM \oplus TM \to \operatorname{End}(E)$ as defined in (12).

If $\langle\ ,\ \rangle$ is a Riemannian structure in E then a connection in the Lie algebroid $\mathfrak{D}_{\mathfrak{so}}(E)$ of 3.6.11 is a Koszul connection ∇ in E such that

$$\langle \nabla_X(\mu), \nu \rangle + \langle \mu, \nabla_X(\nu) \rangle = X(\langle \mu, \nu \rangle), \quad X \in \mathfrak{X}(M),\ \mu, \nu \in \Gamma E.$$

It follows from the calculation of the adjoint bundle in 3.6.11 that the curvature of such a connection takes values in $\mathfrak{so}(E)$; that is,

$$\langle \overline{R}_\nabla(X, Y)(\mu), \nu \rangle + \langle \mu, \overline{R}_\nabla(X, Y)(\nu) \rangle = 0$$

for all $X, Y \in \mathfrak{X}(M)$, $\mu, \nu \in \Gamma E$. Such a connection is a *Riemannian connection* or *metric connection*; 3.6.11 and 5.2.7 show that such a connection always exists.

Now let L be an LAB on M. Then a connection in the Lie algebroid $\mathfrak{D}_{\operatorname{Der}}(L)$ of 3.6.11 is a Koszul connection ∇ in the vector bundle L such that

$$\nabla_X([V, W]) = [\nabla_X(V), W] + [V, \nabla_X(W)], \quad X \in \mathfrak{X}(M),\ V, W \in \Gamma L.$$

It follows from the calculation of the adjoint bundle in 3.6.11 that the curvature of such a connection takes values in $\operatorname{Der}(L)$; that is,

$$\overline{R}_\nabla(X, Y)([V, W]) = [\overline{R}_\nabla(X, Y)(V), W] + [V, \overline{R}_\nabla(X, Y)(W)],$$

for $X \in \mathfrak{X}(M)$, $V, W \in \Gamma L$. We call a connection in $\mathfrak{D}_{\mathrm{Der}}(L)$ a *Lie connection* in L. Again 3.6.11 and 5.2.7 show that such connections always exist. ⊠

We next need the concept of a produced connection.

Definition 5.2.13 Let $\varphi \colon A \to A'$ be a morphism of transitive Lie algebroids over M and let γ be a connection in A. Then $\gamma' = \varphi \circ \gamma$ is called the *produced connection* in A'. □

Clearly, then

$$\overline{R}_{\gamma'} = \varphi^+ \circ \overline{R}_\gamma. \tag{15}$$

Example 5.2.14 Let E be a vector bundle on M, and let ∇ be a connection in E. Then ∇ induces connections in the vector bundles $\mathrm{Hom}^n(E; M \times \mathbb{R})$, $\mathrm{Hom}^n(E; E)$ and in the various tensor, exterior and symmetric algebra bundles built over E through the representations 3.6.10. For example, the produced connection $\widetilde{\nabla}$ in $\mathrm{Hom}^n(E; M \times \mathbb{R})$ is

$$\widetilde{\nabla}_X(\varphi)(\mu_1, \ldots, \mu_n) = X(\varphi(\mu_1, \ldots, \mu_n)) - \sum_{i=1}^{n} \varphi(\mu_1, \ldots, \nabla_X(\mu_i), \ldots \mu_n).$$

Similarly, if E' is a second vector bundle on M and ∇' a connection in E' then the representation of $\mathfrak{D}(E) \oplus_{TM} \mathfrak{D}(E')$ on $\mathrm{Hom}(E; E')$ given in 3.6.10(iii) induces the connection $\widetilde{\nabla}$ in $\mathrm{Hom}(E; E')$ given by

$$\widetilde{\nabla}_X(\varphi)(\mu) = \nabla'_X(\varphi(\mu)) - \varphi(\nabla_X(\mu)).$$

⊠

Definition 5.2.15 Let $\varphi \colon E \to E'$ be a morphism of vector bundles over M and let ∇ and ∇' be connections in E and E' respectively. Then φ *maps* ∇ *to* ∇' if $\varphi(\nabla_X(\mu)) = \nabla'_X(\varphi(\mu))$ for all $X \in \Gamma TM$ and all $\mu \in \Gamma E$. □

If φ in 5.2.15 is an isomorphism, then it induces an isomorphism of Lie groupoids $\widetilde{\varphi} \colon \Phi(E) \to \Phi(E')$, $\xi \mapsto \varphi_{\beta\xi} \circ \xi \circ (\varphi_{\alpha\xi})^{-1}$. This differentiates to $\widetilde{\varphi}_* \colon \mathfrak{D}(E) \to \mathfrak{D}(E')$ where $\widetilde{\varphi}_*(D)(\mu) = (\varphi \circ D \circ \varphi^{-1})(\mu)$ for $D \in \Gamma \mathfrak{D}(E)$ and $\mu \in \Gamma E$ (see §3.6). Now, given a connection ∇ in E, the produced connection $\widetilde{\varphi}_*(\nabla)$ is the unique connection in E' to which φ maps ∇.

The following case of a produced connection will be used repeatedly.

Definition 5.2.16 Let $L \overset{j}{>\!\!\!-\!\!\!-\!\!\!>} A \overset{a}{-\!\!\!-\!\!\!\gg} TM$ be a transitive Lie algebroid on M and let $\gamma \colon TM \to A$ be a connection. Then the produced connection ad $\circ \, \gamma \colon TM \to \mathfrak{D}_{\mathrm{Der}} L$ in the adjoint bundle L will be denoted ∇^γ and called the *adjoint connection* of γ.   □

Example 5.2.17 If ∇ is a connection in a vector bundle E then the adjoint connection in $\mathrm{End}(E)$ is the connection

$$\nabla^\nabla_X(\varphi)(\mu) = \nabla_X(\varphi(\mu)) - \varphi(\nabla_X(\mu)).$$

⊠

Proposition 5.2.18 Let $L \overset{j}{>\!\!\!-\!\!\!-\!\!\!>} A \overset{a}{-\!\!\!-\!\!\!\gg} TM$ be a transitive Lie algebroid on M and let $\gamma \colon TM \to A$ be a connection. Then:

 (i) ∇^γ is a Lie connection in L;
 (ii) if L is abelian then ∇^γ is independent of γ — that is, there is a single adjoint connection in L — and it has curvature zero.

Proof (i) is a simple consequence of the Jacobi identity.

(ii) Let $\gamma' = \gamma + j \circ \ell$ be a second connection in A. Then $j(\nabla^{\gamma'}_X(V)) = [\gamma'(X), j(V)] = [\gamma(X), j(V)] + j[\ell(X), V] = [\gamma(X), j(V)] = j(\nabla^\gamma_X(V))$. By (15), the curvature of ∇^γ is $\mathrm{ad}^+ \circ \overline{R}_\gamma$, and $\mathrm{ad}^+ \colon L \to \mathrm{Der}(L)$ is zero if L is abelian.   □

The next result is known as Bianchi's (second) identity. It will be of central importance in Chapter 7. The symbol $\mathfrak{S}$ denotes the cyclic sum over X, Y, Z.

Proposition 5.2.19 Let $L \overset{j}{>\!\!\!-\!\!\!-\!\!\!>} A \overset{a}{-\!\!\!-\!\!\!\gg} TM$ be a transitive Lie algebroid on M and let $\gamma \colon TM \to A$ be a connection. Then

$$\mathfrak{S}\{\nabla^\gamma_X(\overline{R}_\gamma(Y, Z)) - \overline{R}_\gamma([X, Y], Z)\} = 0 \tag{16}$$

for all $X, Y, Z \in \mathfrak{X}(M)$.

Proof Apply j to the LHS. The result is

$$\mathfrak{S}\{[\gamma X, \gamma[Y, Z] - [\gamma Y, \gamma Z]] - \gamma[[X, Y], Z] + [\gamma[X, Y], \gamma Z]\}$$
$$= \mathfrak{S}\{[\gamma X, \gamma[Y, Z]] - [\gamma X, [\gamma Y, \gamma Z]] - \gamma[[X, Y], Z] + [\gamma[X, Y], \gamma Z]\}.$$

The first and the last terms cancel, once $\mathfrak{S}$ is applied. The second and the third terms vanish by the Jacobi identity.   □

The remainder of this section may be omitted at a first reading. It demonstrates how the algebraic formalism of covariant derivatives — a less commonly used part of standard connection theory — is valid for abstract transitive Lie algebroids. This material will not be used in the remainder of the chapter, but will be relevant in Chapter 7.

Let A be a transitive Lie algebroid on M and let $\gamma \colon TM \to A$ be a connection.

For $n \geqslant 0$ denote the vector bundle $\mathrm{Alt}^n(TM;L)$ by $C^n(TM,L)$; thus elements of $\Gamma C^n(TM;L)$ are alternating n–forms on M with values in the vector bundle L. Treat $C^0(TM;L)$ as L itself.

Define differential operators $\nabla^\gamma \colon \Gamma C^n(TM,L) \to \Gamma C^{n+1}(TM,L)$ by

$$\nabla^\gamma(f)(X_1,\ldots,X_{n+1}) = \sum_{r=1}^{n+1}(-1)^{r+1}\nabla^\gamma_{X_r}(f(X_1,\ldots,\widehat{X_r},\ldots,X_{n+1}))$$
$$+ \sum_{r<s}(-1)^{r+s}f([X_r,X_s],X_1,\ldots,\widehat{X_r},\ldots,\widehat{X_s},\ldots,X_{n+1}). \quad (17)$$

For $V \in \Gamma L = \Gamma C^0(TM,L)$, the covariant derivative $X \mapsto \nabla^\gamma(V)(X)$ is the adjoint connection $X \mapsto \nabla^\gamma_X(V)$ itself. The ∇^γ constitute the *(exterior) covariant derivative* associated with the connection.

Note that the Bianchi identity (16) can now be written $\nabla^\gamma(\overline{R}_\gamma) = 0$.

Proposition 5.2.20 *Let γ' be a second connection in A and let ℓ be the map $TM \to L$ with $\gamma' = \gamma + j \circ \ell$. Then*

$$\overline{R}_{\gamma'} - \overline{R}_\gamma = -\{\nabla^\gamma(\ell) + [\ell,\ell]\}.$$

Proof For $X,Y \in \Gamma TM$, expanding out $R_{\gamma'}(X,Y) - R_\gamma(X,Y)$ gives

$$\gamma'[X,Y] - \gamma[X,Y] - [\gamma'X,\gamma'Y] + [\gamma X,\gamma Y]$$
$$= j\ell[X,Y] - [\gamma X,j\ell Y] - [j\ell X,\gamma Y] - [j\ell X,j\ell Y]$$
$$= j\{\ell[X,Y] - \nabla^\gamma_X(\ell Y) + \nabla^\gamma_Y(\ell X) - [\ell X,\ell Y]\},$$

which gives the result. $\qquad\square$

In particular, if two connections have the same curvature, then their difference map $\ell \colon TM \to A$ satisfies $\nabla^\gamma(\ell) + [\ell,\ell] = 0$, which is a Maurer–Cartan equation with respect to the adjoint connection ∇^γ.

We also need a covariant derivative for forms on A. With $C^n(A, L) =$ $\text{Alt}^n(A; L)$, define $D^\gamma \colon \Gamma C^n(A, L) \to \Gamma C^{n+1}(A, L)$ by

$$D^\gamma(f)(X_1, \ldots, X_{n+1}) = \sum_{r=1}^{n+1} (-1)^{r+1} \nabla^\gamma_{aX_r} (f(X_1, \ldots, \widehat{X_r}, \ldots, X_{n+1}))$$

$$+ \sum_{r<s} (-1)^{r+s} f([X_r, X_s], X_1, \ldots, \widehat{X_r}, \ldots, \widehat{X_s}, \ldots, X_{n+1}) \quad (18)$$

Proposition 5.2.21 *Let* $\omega \colon A \to L$ *be the connection reform corresponding to* γ. *Then for* $X, Y \in \Gamma A$,

$$\overline{R}_\gamma(aX, aY) = (D^\gamma(\omega) + [\omega, \omega])(X, Y).$$

Proof Using $j \circ \omega + \gamma \circ a = \text{id}$, we expand $[X, Y]$ and get

$$\begin{aligned}
[X, Y] &= [j\omega X, j\omega Y] + [j\omega X, \gamma a Y] + [\gamma a X, j\omega Y] + [\gamma a X, \gamma a Y] \\
&= j[\omega X, \omega Y] - j\nabla^\gamma_{ax}(\omega X) + j\nabla^\gamma_{aY}(\omega Y) + [\gamma a X, \gamma a Y]
\end{aligned}$$

so

$$[X, Y] - [\gamma a X, \gamma a Y] = j\{[\omega X, \omega Y] - \nabla^\gamma_{aY}(\omega X) + \nabla^\gamma_{aX}(\omega Y)\}$$

and therefore

$$\begin{aligned}
R_\gamma(aX, aY) &= \gamma[aX, aY] - [\gamma a X, \gamma a Y] \\
&= \gamma a[X, Y] - [\gamma a X, \gamma a Y] \\
&= [X, Y] - j\omega[X, Y] - [\gamma a X, \gamma a Y] \\
&= j\{-\omega[X, Y] + [\omega X, \omega Y] - \nabla^\gamma_{aY}(\omega X) + \nabla^\gamma_{aX}(\omega Y)\},
\end{aligned}$$

whence the result. □

In particular, if γ is flat then ω satisfies an equation of Maurer–Cartan type.

If $\rho \colon A \to \mathfrak{D}(E)$ is any representation of A on a vector bundle E, then exterior covariant derivatives can be defined in both $C^*(TM, E)$ and $C^*(A, E)$, using the produced connection $\rho \circ \gamma$ in place of ∇^γ in equations (17) and (18).

In the connection theory of Lie algebroids it is usually easier to work with connections $\gamma \colon TM \to A$ rather than with connection reforms $\omega \colon A \to L$. A connection γ is anchor–preserving ($a \circ \gamma = \text{id}$) whereas ω is not, and for this reason γ fits into the algebraic formalism of Lie algebroids better than does ω. Note also that the definition of curvature in terms of ω would need to be via 5.2.21 and would present curvature as the failure of ω to be a Maurer–Cartan form; the simple definition 5.2.9 for the curvature of γ is only meaningful because γ is anchor–preserving.

On the other hand, it is desirable to have the formalism of connection reforms available, firstly to relate the global Lie algebroid formalism to the standard theory, and secondly because the connection reform formalism is needed in Chapter 7.

5.3 Infinitesimal connections and curvature in principal bundles

This section is concerned with the translation between the classical language of connections in principal bundles and the definitions of §5.2. The classical treatment is significantly lengthier than the Lie algebroid formulation.

Consider a principal bundle $P(M, G, \pi)$ with Atiyah sequence

$$\frac{P \times \mathfrak{g}}{G} \;\overset{j}{\rightarrowtail}\; \frac{TP}{G} \;\overset{\pi_*}{\longrightarrow\!\!\!\!\!\rightarrow}\; TM.$$

A connection $\gamma \colon TM \to \frac{TP}{G}$ must be injective, so its image I is a vector subbundle of $\frac{TP}{G}$. The preimage $\mathscr{H} = \natural^{-1}(I)$ under $\natural \colon TP \to \frac{TP}{G}$ is therefore a vector subbundle of TP; that is, it is a distribution on P. Since $\natural$ collapses orbits, we have $\mathscr{H}_{ug} = T(R_g)(\mathscr{H}_u)$ for all $u \in P$ and $g \in G$. Lastly, since γ is right–inverse to π_*, it follows that $\frac{TP}{G} = \frac{T^\pi P}{G} \oplus I$, and hence that $TP = T^\pi P \oplus \mathscr{H}$.

Definition 5.3.1 Let $P(M, G, \pi)$ be a principal bundle. A distribution $\mathscr{H}$ on P is *horizontal* if it is complementary to the vertical subbundle $T^\pi P$; that is, if $TP = T^\pi P \oplus \mathscr{H}$. It is *invariant* if $\mathscr{H}_{ug} = T(R_g)(\mathscr{H}_u)$ for all $u \in P$ and $g \in G$. □

We have proved half of the following result.

Proposition 5.3.2 *Let $P(M, G, \pi)$ be a principal bundle. There is a bijective correspondence between connections γ in $P(M, G)$ and invariant horizontal distributions $\mathscr{H}$ on P, given by $\mathscr{H} = \natural^{-1}(\operatorname{im} \gamma)$.*

Proof It remains to show that $\mathscr{H}$ determines γ. The G–invariance implies that the action of G on TP restricts to an action of G on $\mathscr{H}$

and it is straightforward to show that $\mathscr{H}$ admits equivariant charts; in the notation of 3.2.1, the chart

$$\mathscr{U} \times \mathbb{R}^n \to \mathscr{H}_\mathscr{U} \qquad (u, \mathbf{t}) \mapsto h\psi(u, \mathbf{t}, 0),$$

where hX is the horizontal component of $X \in T_u(P)$, is equivariant. So the decomposition $TP = T^\pi P \oplus \mathscr{H}$ quotients to the decomposition $\frac{TP}{G} = \frac{T^\pi P}{G} \oplus \frac{\mathscr{H}}{G}$. Now $\pi_* \colon \frac{TP}{G} \to TM$ is surjective and its kernel is $\frac{T^\pi P}{G}$, so the restriction $\pi_* \colon \frac{\mathscr{H}}{G} \to TM$ is an isomorphism of vector bundles over M. We define $\gamma \colon TM \to \frac{\mathscr{H}}{G} \subseteq \frac{TP}{G}$ to be its inverse. $\square$

Now consider a connection reform $\omega \colon \frac{TP}{G} \to \frac{P \times \mathfrak{g}}{G}$. Define $\overrightarrow{\omega} \colon TP \to P \times \mathfrak{g}$ by $\overrightarrow{\omega}_u(X) = (\natural_u^{P \times \mathfrak{g}})^{-1}(\omega(\langle X \rangle))$, where $\natural^{P \times \mathfrak{g}}$ is the projection $P \times \mathfrak{g} \to \frac{P \times \mathfrak{g}}{G}$. This $\overrightarrow{\omega}$ is smooth because $\natural^{P \times \mathfrak{g}} \circ \overrightarrow{\omega} = \omega \circ \natural^{TP}$, and $\natural^{P \times \mathfrak{g}}$ is a submersion. From the definitions of the two $\natural$ maps, it follows that $w = \overrightarrow{\omega}$ is a connection form in the sense of the following definition.

Definition 5.3.3 Let $P(M, G, \pi)$ be a principal bundle. A *connection form* in $P(M, G)$ is a $\mathfrak{g}$–valued 1–form $w \colon TP \to P \times \mathfrak{g}$ such that

$$w_{ug}(T(R_g)_u X) = \mathrm{Ad}(g^{-1})(w_u(X))$$

for all $X \in T_u(P)$, $u \in P$, $g \in G$, and such that

$$w(A^*) = A,$$

for all $A \in \mathfrak{g}$, where A^* is the fundamental vector field induced by the action of G on P. $\square$

Proposition 5.3.4 *Let $P(M, G, \pi)$ be a principal bundle. There is a bijective correspondence between connection reforms ω in $P(M, G)$ and connection forms $w \in \Omega^1(P, \mathfrak{g})$, given by $\natural^{P \times \mathfrak{g}} \circ w = \omega \circ \natural^{TP}$,*

Proof It remains to show that w determines ω. The first of the conditions in 5.3.3 states that w, regarded as a map $TP \to P \times \mathfrak{g}$, preserves the actions of G on TP and on $P \times \mathfrak{g}$, and it therefore quotients to a map $w^{/G} \colon \frac{TP}{G} \to \frac{P \times \mathfrak{g}}{G}$. The second condition now implies that $w^{/G} \circ j = \mathrm{id}$. $\square$

*   *   *   *   *

We now proceed to the various formulations of curvature. First recall from 3.3.16 that if $E = \frac{P \times V}{G}$ is a vector bundle associated to $P(M, G)$ via a representation $g \mapsto (v \mapsto gv)$ of G on a vector space V, then there is an associated Lie algebroid representation ρ of $\frac{TP}{G}$ on E. The following result is now immediate.

Proposition 5.3.5 *If γ is a connection in $P(M,G)$ and E is an associated vector bundle, then*

$$\nabla^\gamma_X(\mu) = \rho(\gamma X)(\mu), \quad X \in \mathcal{X}(M), \ \mu \in \Gamma E$$

defines a linear connection ∇^γ in E, called the connection in E associated to or induced by γ.

In the case of the adjoint bundle $\frac{P \times \mathfrak{g}}{G}$, there is an alternative formula.

Proposition 5.3.6 *The action of $\frac{TP}{G}$ on the associated bundle $\frac{P \times \mathfrak{g}}{G}$ is given by*

$$j(X(V)) = [X, j(V)], \quad X \in \Gamma(\tfrac{TP}{G}), \ V \in \Gamma(\tfrac{P \times \mathfrak{g}}{G}).$$

Proof Let $\{\varphi_t\}$ be a saturated local flow for $\overline{X}$ as in 3.2.7. Then $\varphi_t \circ m_{\varphi_{-t}}(u) = m_u$ for all $u \in P$, and a modification of part of the proof of 3.2.5 shows that

$$[\overline{X}, \overline{j(V)}](u) = T_1(m_u)(\overline{X}(\tilde{V})(u)),$$

for all $u \in P$. But $T_1(m_u)(\overline{X}(\tilde{V})(u)) = \overline{j(X(V))}(u)$ and so the result follows. $\square$

Corollary 5.3.7 *If γ is a connection in $P(M,G)$, then the induced connection ∇^γ in $\frac{P \times \mathfrak{g}}{G}$ is given by*

$$j(\nabla^\gamma_X(V)) = [\gamma X, jV], \quad X \in \mathcal{X}(M), \ V \in \Gamma(\tfrac{P \times \mathfrak{g}}{G}).$$

This connection is the *adjoint connection* of γ.

We now proceed to study curvature. The following is a restatement of 5.2.9.

Definition 5.3.8 Let $\gamma \colon TM \to \frac{TP}{G}$ be a connection in $P(M,G)$. The *curvature* of γ is the skew–symmetric vector bundle map

$$\overline{R}_\gamma \colon TM \oplus TM \to \frac{P \times \mathfrak{g}}{G}$$

defined by $j(\overline{R}_\gamma(X,Y)) = \gamma[X,Y] - [\gamma X, \gamma Y]$. $\square$

To prove that this is indeed the classical curvature form in disguise requires some preparation. First recall some traditional terminology:

Definition 5.3.9 Let ρ be a representation of G on a vector space V.

A form $\varphi \in \Omega^r(P, V)$ is called *equivariant* or *pseudotensorial of type* (ρ, V) if $R_g^*(\varphi) = \rho(g^{-1}) \circ \varphi$ for all $g \in G$. The set of equivariant r–forms on P with values in V is denoted $\Omega^r(P, V)^G$.

A form $\varphi \in \Omega^r(P, V)$ is called *horizontal* if, at any given point $u \in P$, we have $\varphi(X_1, \ldots, X_r)(u) = 0$ whenever one or more of the $X_i(u)$ is vertical.

A form $\varphi \in \Omega^r(P, V)$ is called *basic* or *tensorial of type* (ρ, V) if it is both equivariant and horizontal. The set of basic r–forms on P with values in V is denoted $\Omega^r_{\text{basic}}(P, V)$. $\qquad\square$

Note that the concept of a horizontal form does not depend on the presence of a connection.

The constructions which relate connection forms and connection re-forms can be extended to arbitrary degree.

Proposition 5.3.10 (i) *There exists a bijective correspondence between equivariant r–forms $\varphi \in \Omega^r(P, V)^G$ and skew–symmetric vector bundle morphisms $\underline{\varphi}$: $\oplus^r\left(\frac{TP}{G}\right) \to \frac{P \times V}{G}$. A corresponding pair $\varphi, \underline{\varphi}$ are related by the diagram*

$$
\begin{array}{ccc}
\oplus^r TP & \xrightarrow{\ \varphi\ } & P \times V \\
{\scriptstyle \oplus^r\natural}\downarrow & & \downarrow{\scriptstyle \natural^{P \times V}} \\
\oplus^r\left(\frac{TP}{G}\right) & \xrightarrow{\ \underline{\varphi}\ } & \frac{P \times V}{G}.
\end{array}
\qquad (19)
$$

(ii) *There exists a bijective correspondence between basic r–forms φ in $\Omega^r_{\text{basic}}(P, V)$ and skew–symmetric vector bundle morphisms $\underline{\underline{\varphi}}$ from $\oplus^r TM$ to $\frac{P \times V}{G}$. A corresponding pair $\varphi, \underline{\underline{\varphi}}$, are related by the diagram*

$$
\begin{array}{ccc}
\oplus^r TP & \xrightarrow{\ \varphi\ } & P \times V \\
{\scriptstyle \natural^{\oplus^r}}\downarrow & & \downarrow{\scriptstyle \natural^{P \times V}} \\
\oplus^r\left(\frac{TP}{G}\right) & \xrightarrow{\ \underline{\varphi}\ } & \frac{P \times V}{G} \\
{\scriptstyle \oplus^r\pi}\downarrow & & \parallel \\
\oplus^r TM & \xrightarrow{\ \underline{\underline{\varphi}}\ } & \frac{P \times V}{G}.
\end{array}
\qquad (20)
$$

Proof Using 3.2.1(i) it is straightforward to show that the action of G on the bundle $\oplus^r TP \to P$ by

$$(X_1 \oplus \ldots \oplus X_r)g = X_1 g \oplus \ldots \oplus X_r g$$

satisfies the conditions of 3.1.1 and that the vector bundle morphism $\oplus^r\natural$: $\oplus^r TP \to \oplus^r\left(\frac{TP}{G}\right)$ quotients to an isomorphism $\frac{\oplus^r TP}{G} \cong \oplus^r\left(\frac{TP}{G}\right)$.

Given $\varphi \in \Omega^r(P, V)^G$, regarded as $\varphi \colon \oplus^r TP \to P \times V$, the equivariance of φ implies that it quotients, using 3.1.2(ii), to a vector bundle morphism $\varphi^{/G} \colon \frac{\oplus^r TP}{G} \to \frac{P \times V}{G}$. We let $\underline{\varphi}$ be the equivalent morphism $\oplus^r(\frac{TP}{G}) \to \frac{P \times V}{G}$. Clearly $\underline{\varphi}$ inherits skew–symmetry from φ and satisfies (19).

Conversely, let $\underline{\varphi} \colon \oplus^r(\frac{TP}{G}) \to \frac{P \times V}{G}$ be a given skew–symmetric vector bundle morphism, and consider $\underline{\varphi} \circ \oplus^r \natural \colon \oplus^r TP \to \frac{P \times V}{G}$. Since $\natural^{P \times V}$ is a pullback over π, there is a unique vector bundle morphism $\oplus^r TP \to P \times V$, denoted φ, such that $\natural^{P \times V} \circ \varphi = \underline{\varphi} \circ \oplus^r \natural$. Since $\natural^{P \times V}$ is fibrewise an isomorphism, φ inherits skew–symmetry from $\underline{\varphi}$.

It is straightforward to check that these constructions are mutual inverses.

(ii) Let γ be any connection in $P(M, G)$. If $\varphi \in \Omega^r(P, V)^G$ is horizontal, then $\underline{\varphi}(X_1, \ldots, X_r)(x)$ vanishes whenever one or more of the arguments $X_i(x)$ is in $\mathrm{im}(j) = \frac{T^\pi P}{G}$. Therefore the vector bundle morphism $\underline{\underline{\varphi}} = \underline{\varphi} \circ \oplus^r \gamma \colon \oplus^r TM \to \frac{P \times V}{G}$ does not depend on the choice of γ. Clearly $\underline{\underline{\varphi}}$ is skew–symmetric, since $\underline{\varphi}$ is, and $\underline{\underline{\varphi}} \circ \oplus^r \pi_* = \underline{\varphi}$ since each $X_i - \gamma \pi_*(X_i) = j\omega(X_i)$ is in $\frac{T^\pi P}{G}$, where ω is the connection form corresponding to γ.

Conversely, let $\underline{\underline{\varphi}} \colon \oplus^r TM \to \frac{P \times V}{G}$ be a given skew–symmetric vector bundle morphism and consider $\underline{\underline{\varphi}} \circ \oplus^r \pi_* \colon \oplus^r(\frac{TP}{G}) \to \frac{P \times V}{G}$. This is certainly a skew–symmetric vector bundle morphism and therefore induces, by (i), an equivariant form $\varphi \in \Omega^r(P, V)^G$ which is horizontal since $(\underline{\underline{\varphi}} \circ \oplus^r \pi_*)(X_1, \ldots, X_r)(x)$ vanishes whenever one or more of the arguments $X_i(x)$ is in $\frac{T^\pi P}{G} = \ker \pi_*$.

Again, it is straightforward to check that these constructions are mutual inverses. □

Anticipating Chapter 7, we denote by $C^r(\frac{TP}{G}, \frac{P \times V}{G})$ the vector bundle $\mathrm{Alt}^r(\frac{TP}{G}, \frac{P \times V}{G})$ whose fibres are the alternating r–multilinear maps $\oplus^r(\frac{TP}{G})\big|_x \to \frac{P \times V}{G}\big|_x$ for $x \in M$. Likewise denote by $C^r(TM, \frac{P \times V}{G})$ the vector bundle $\mathrm{Alt}^r(TM, \frac{P \times V}{G})$. Then $\Gamma C^r(\frac{TP}{G}, \frac{P \times V}{G})$ is naturally isomorphic to the $C^\infty(M)$–module of alternating bundle morphisms $\oplus^r(\frac{TP}{G}) \to \frac{P \times V}{G}$, and it is trivial to check that the correspondence of 5.3.10(i) becomes a $C^\infty(M)$–isomorphism of $\Gamma C^r(\frac{TP}{G}, \frac{P \times V}{G})$ with $\Omega^r(P, V)^G$, where the latter has module structure $f\varphi = (f \circ \pi)\varphi$. In a similar way $\Gamma C^r(TM, \frac{P \times V}{G})$ is isomorphic as a $C^\infty(M)$–module to $\Omega^r_{\mathrm{basic}}(P, V)$.

It is well–known, and easy to check directly, that the graded module

$\Omega^*(P,V)^G$ is closed under the exterior derivative δ. It follows that δ can be transferred to $\Gamma C^*(\frac{TP}{G}, \frac{P\times V}{G})$:

Proposition 5.3.11 *The diagram*

$$
\begin{array}{ccc}
\Omega^r(P,V)^G & \xrightarrow{\ \delta\ } & \Omega^{r+1}(P,V)^G \\
{\scriptstyle\cong}\Big\downarrow & & {\scriptstyle\cong}\Big\downarrow \\
\Gamma C^r\left(\frac{TP}{G}, \frac{P\times V}{G}\right) & \xrightarrow{\ d\ } & \Gamma C^{r+1}\left(\frac{TP}{G}, \frac{P\times V}{G}\right)
\end{array}
$$

commutes, where d is the Lie algebroid coboundary of 7.1.1, and the vertical arrows are the isomorphisms of 5.3.10(i).

Proof First note that, for any $\varphi \in \Omega^r(P,V)^G$, where $X_i \in \Gamma(\frac{TP}{G})$, (19) implies that $\varphi(\overline{X}_1, \ldots, \overline{X}_r) = \widetilde{\varphi(X_1, \ldots, X_r)}$ as functions $P \to V$. Now, for $X_i \in \Gamma(\frac{TP}{G})$, $x \in M$, and any $u \in p^{-1}(x)$,

$$
\begin{aligned}
(\underline{d\varphi})(X_1, \ldots, X_{r+1})(x) &= \langle u, d\varphi(\overline{X}_1, \ldots, \overline{X}_{r+1})(u) \rangle \\
&= \langle u, \sum (-1)^{i+1} \overline{X}_i(\varphi(\overline{X}_1, \ldots \hat{\ldots}, \overline{X}_{r+1}))(u) \\
&\quad + \sum (-1)^{i+j} \varphi([\overline{X}_i, \overline{X}_j], \overline{X}_1, \ldots \hat{\ldots} \hat{\ldots}, \overline{X}_{r+1})(u) \rangle \\
&= \langle u, \sum (-1)^{i+1} X_i(\widetilde{\varphi(X_1, \ldots \hat{\ldots}, X_{r+1}))} \ \widetilde{}\ (u) \\
&\quad + \sum (-1)^{i+j} \widetilde{\varphi([X_i, X_j], X_1, \ldots \hat{\ldots} \hat{\ldots}, X_{r+1})} \ \widetilde{}\ (u) \rangle \\
&= \sum (-1)^{i+1} X_i(\varphi(X_i, \ldots \hat{\ldots}, X_{r+1}))(x) \\
&\quad + \sum (-1)^{i+j} \varphi([X_i, X_j], X_1, \ldots \hat{\ldots} \hat{\ldots}, X_{r+1})(x),
\end{aligned}
$$

where the circumflexes refer to the arguments corresponding to the dummy indices. This completes the proof. $\qquad\square$

Now suppose that we have a connection γ in $P(M,G)$. Denote the corresponding projection $TP \to Q \subseteq TP$ by h and let $h^*\colon \Omega^r(P,V) \to \Omega^r(P,V)$ be the map dual to h, that is,

$$
h^*(\varphi)(X_1, \ldots, X_r) = \varphi(hX_1, \ldots, hX_r), \quad X_i \in \Gamma TP.
$$

Clearly h^* maps $\Omega^r(P,V)^G$ into $\Omega^r_{\text{basic}}(P,V)$.

Lemma 5.3.12 *The diagram*

$$
\begin{array}{ccc}
\Omega^r(P,V)^G & \xrightarrow{\ h^*\ } & \Omega^r_{\text{basic}}(P,V) \\
{\scriptstyle\cong}\Big\downarrow & & {\scriptstyle\cong}\Big\downarrow \\
\Gamma C^r\left(\frac{TP}{G}, \frac{P\times V}{G}\right) & \xrightarrow{\ \gamma^*\ } & \Gamma C^r\left(TM, \frac{P\times V}{G}\right)
\end{array}
$$

commutes, where the vertical arrows are the isomorphisms of 5.3.10, *and* γ^* *is the map* $\varphi \mapsto \varphi \circ \gamma^r$.

Proof Take $\varphi \in \Omega^r(P,V)^G$; we must prove that $\underline{h^*(\varphi)} = \underline{\gamma^*(\varphi)}$. For $X_i \in \Gamma TM$, $x \in M$, and any $u \in p^{-1}(x)$,

$$
\begin{aligned}
\underline{h^*(\varphi)}(X_1,\ldots,X_r)(x) &= \underline{h^*(\varphi)}(\gamma X_1,\ldots,\gamma X_r)(x) \\
&= \langle u, h^*(\varphi)(\overline{\gamma X_1},\ldots,\overline{\gamma X_r})(u)) \rangle \\
&= \langle u, \varphi(h(\overline{\gamma X_1}(u)),\ldots,h(\overline{\gamma X_r}(u))) \rangle.
\end{aligned}
$$

Now for any $X \in \Gamma TM$, we have $\gamma X \in \Gamma(\frac{Q}{G})$ (see the proof of 5.3.2) so $\overline{\gamma X}(u) \in Q_u$ and so $h(\overline{\gamma X}(u)) = \overline{\gamma X}(u)$. The expression therefore reduces to

$$
\langle u, \varphi(\overline{\gamma X_1},\ldots,\overline{\gamma X_r})(u) \rangle = \underline{\varphi}(\gamma X_1,\ldots,\gamma X_r)(x)
$$

and the result follows. $\qquad\square$

The classical definition of the curvature of a connection in terms of the invariant horizontal distribution Q is in terms of the operators

$$
\nabla^Q = h^* \circ \delta \colon \Omega^r(P,V)^G \to \Omega^{r+1}_{\text{basic}}(P,V).
$$

Putting 5.3.11 and 5.3.12 together we have:

Proposition 5.3.13 *Let* γ *be a connection with invariant horizontal distribution* Q. *Then*

$$
\begin{array}{ccc}
\Omega^r(P,V)^G & \xrightarrow{\nabla^Q} & \Omega^{r+1}_{\text{basic}}(P,V) \\
\cong \big\downarrow & & \cong \big\downarrow \\
\Gamma C^r\left(\frac{TP}{G}, \frac{P\times V}{G}\right) & \xrightarrow{\mathscr{D}^\gamma} & \Gamma C^{r+1}\left(\frac{TP}{G}, \frac{P\times V}{G}\right)
\end{array}
$$

commutes, where the vertical arrows are the isomorphisms of 5.3.10, *and* $\mathscr{D}^\gamma$ *is the map* $\varphi \to \gamma^*(d\varphi)$.

We can now show that the curvature $\overline{R}_\gamma \in \Gamma C^2(\frac{TP}{G}, \frac{P\times \mathfrak{g}}{G})$ defined in 5.3.8 does indeed correspond to the curvature form $\Omega = \nabla^Q(\omega) \in \Omega^2_{\text{basic}}(P,\mathfrak{g})$ as classically defined.

Proposition 5.3.14 *Let* $\omega \in \Omega^1(P,\mathfrak{g})^G$ *be a connection form and let* $\underline{\omega} \colon \frac{TP}{G} \to \frac{P\times\mathfrak{g}}{G}$ *be the corresponding connection reform in the Atiyah sequence of* $P(M,G)$. *Then*

$$
\underline{\Omega} = \underline{\nabla^Q(\omega)} = \mathscr{D}^\gamma(\underline{\omega}) = \overline{R}_\gamma,
$$

where γ is the connection corresponding to $\underline{\omega}$.

Proof It has just been proved that $\nabla^Q(\omega) = \mathscr{D}^\gamma(\underline{\omega}) = \gamma^*(d\underline{\omega})$, and $\Omega = \nabla^Q(\omega)$ by definition. So it remains to prove that $\overline{R}_\gamma = \gamma^*(d\underline{\omega})$. For $X, Y \in \Gamma T M$,

$$
\begin{aligned}
d\underline{\omega}(\gamma X, \gamma Y) &= \gamma X(\underline{\omega}(\gamma X)) - \gamma Y(\underline{\omega}(\gamma X)) - \underline{\omega}([\gamma X, \gamma Y]) \\
&= -\underline{\omega}([\gamma X, \gamma Y]) \qquad \text{since } \underline{\omega} \circ \gamma = 0 \\
&= \underline{\omega}(\gamma([X, Y]) - [\gamma X, \gamma Y]) \\
&= (\underline{\omega} \circ j)(\overline{R}_\gamma(X, Y)) \\
&= \overline{R}_\gamma(X, Y) \qquad \text{since } \underline{\omega} \circ j = \text{id.}
\end{aligned}
$$

$\square$

The following equation is a useful expression for curvature in terms of the connection reform.

Proposition 5.3.15 *If γ is a connection in $P(M, G)$ and ω the corresponding connection reform, then*

$$
(\pi_*)^* \overline{R}_\gamma = d\omega - [\omega, \omega]
$$

Proof For $X, Y \in \Gamma(\frac{TP}{G})$, expanding out $j(d\omega - [\omega, \omega])(X, Y)$ we get

$$
\begin{aligned}
j(X(\omega Y)) - j(Y(\omega X)) - j\omega[X, Y] - [j\omega X, j\omega Y] \\
= [X, j\omega Y] + [j\omega X, Y] - j\omega[X, Y] - [j\omega X, j\omega Y] \\
= [\gamma \pi_* X, j\omega Y] + [X - \gamma \pi_* X, Y] - j\omega[X, Y] \\
= [\gamma \pi_* X, -\gamma \pi_* Y] + [X, Y] - j\omega[X, Y] \\
= j(R_\gamma(\pi_* X, \pi_* Y))
\end{aligned}
$$

where we used $j \circ \omega + \gamma \circ \pi_* = \text{id}$ repeatedly, and 5.3.6. The result follows. $\square$

Proposition 5.3.15 does not possess the importance in the Atiyah sequence/Lie algebroid formulation of connection theory that the corresponding equation $\Omega = d\omega + [\omega, \omega]$, known as the structure equation, does in the traditional treatment. The reason for this is straightforward: the traditional definition of curvature, $\Omega = h^*(d\omega)$, is difficult to work with in both theoretical and practical calculations and the structure equation is the usual means by which curvature is calculated. The Lie algebroid curvature $\overline{R}_\gamma$, on the other hand, is very easy to work with for a very wide range of theoretical purposes, and it can easily be localized to a family of local 2–forms in $\Omega^2(U_i, \mathfrak{g})$, $U_i \subseteq M$, for computational work (see §5.4). There is thus no need for an alternative formula.

We conclude this section with two examples of working with the Atiyah sequence/Lie algebroid formulation in 'theoretical' problems.

Example 5.3.16 Consider a principal bundle $P(M, H, \pi)$ on which a Lie group G acts to the left; that is, there are actions $G \times P \to P$ and $G \times M \to M$ with respect to which π is equivariant, and such that $g(uh) = (gu)h$ for all $g \in G$, $u \in P$, $h \in H$. Denote $u \to gu$, $P \to P$ by L_g and $x \to gx$, $M \to M$ by λ_g.

Then G acts on the Atiyah sequence of $P(M, H)$ in the following way: G acts on $\frac{TP}{H}$ by $g\langle X_u \rangle = \langle T(L_g)_u X_u \rangle$, on TM by $gX_x = T(\lambda_g)_x(X_x)$, and on $\frac{P \times \mathfrak{h}}{H}$ by $g\langle u, X \rangle = \langle gu, X \rangle$. It is easy to check that the projection of each of these vector bundles is equivariant, and that j and π_* are equivariant. For $g \in G$, $(\lambda_g)_*\colon \Gamma TM \to \Gamma TM$ denotes the induced map of vector fields, $(\lambda_g)_*(X)(x) = T(\lambda_g)_{g^{-1}x}(X_{g^{-1}x}) = g(X(g^{-1}x))$, and $(L_g)_*\colon \Gamma TP \to \Gamma TP$ denotes the corresponding map of vector fields on P. Also denote by $(L_g)_*$ the maps $\Gamma(\frac{TP}{H}) \to \Gamma(\frac{TP}{H})$ and $\Gamma(\frac{P \times \mathfrak{h}}{H}) \to \Gamma(\frac{P \times \mathfrak{h}}{H})$ defined by $(L_g)_*(\overline{X})(x) = \overline{g(X(g^{-1}x))}$; it then easily follows that $\overline{(L_g)_* X} = (L_g)_*(\overline{X})$ for $X \in \Gamma(\frac{TP}{H})$. We thus have:

$$(\lambda_g)_*([X, Y]) = [(\lambda_g)_*(X), (\lambda_g)_*(Y)], \qquad \text{for all } X, Y \in \Gamma TM. \quad (21)$$

From the corresponding result for $(L_g)_*\colon \Gamma TP \to \Gamma TP$ we get

$$(L_g)_*([X, Y]) = [(L_g)_*(X), (L_g)_*(Y)], \qquad \text{for all } X, Y \in \Gamma(\frac{TP}{H}) \quad (22)$$

and thus we obtain the corresponding result for $(L_g)_*\colon \Gamma(\frac{P \times \mathfrak{h}}{H}) \to \Gamma(\frac{P \times \mathfrak{h}}{H})$.

A connection (form) $\omega \in \Omega^1(P, \mathfrak{h})$ is often defined to be *G–invariant* if $(L_g)^*\omega = \omega$, for all $g \in G$. We define a connection $\gamma\colon TM \to \frac{TP}{H}$ to be *G–equivariant* if it is equivariant with respect to the actions of G on TM and $\frac{TP}{H}$. These two definitions are easily seen to be equivalent. We wish to show that if γ is G–equivariant, then $\overline{R}_\gamma$ is also; that is

$$\overline{R}_\gamma(gX_x, gY_x) = g\overline{R}_\gamma(X_x, Y_x), \quad (23)$$

for all $g \in G$, $X_x, Y_x \in T_x(M)$, $x \in M$.

To prove (23), let X, Y be vector fields on M with the given values at the chosen $x \in M$. Then $(\lambda_g)_*(X)(gx) = gX_x$ and likewise for Y, so

$$j \circ \overline{R}_\gamma(gX_x, gY_x) = j \circ \overline{R}_\gamma((\lambda_g)_*X, (\lambda_g)_*Y)(gx)$$
$$= (\gamma[(\lambda_g)_*X, (\lambda_g)_*Y] - [\gamma \circ (\lambda_g)_*X, \gamma \circ (\lambda_g)_*Y])(gx).$$

Now the G–equivariance of γ implies that $\gamma \circ (\lambda_g)_* = (L_g)_* \circ \gamma$ so, using this and (21), (22), the above becomes

$$(L_g)_*(\gamma[X, Y] - [\gamma X, \gamma Y])(gx) = j((L_g)_*(\overline{R}_\gamma(X, Y)))(gx))$$

and hence $\overline{R}_\gamma(gX_x, gY_x) = g(\overline{R}_\gamma(X, Y)(x)) = g\overline{R}_\gamma(X_x, Y_x)$, as required.

Thus the curvature of a G–equivariant connection in $P(M, H)$ is determined by its values over any one $x \in M$.

The principal example of such an action on a principal bundle is the action of a Lie group G on a homogeneous bundle $G(G/H, H)$. In this case the Atiyah sequence is isomorphic to

$$\frac{G \times \mathfrak{h}}{H} \xrightarrow{j_1} \frac{G \times \mathfrak{g}}{H} \xrightarrow{a_1} \frac{G \times \mathfrak{g}/\mathfrak{h}}{H} \qquad (24)$$

where j_1 and a_1 are induced by the corresponding maps in the exact sequence $\mathfrak{h} \xrightarrow{\subseteq} \mathfrak{g} \xrightarrow{T_1(\pi)} \mathfrak{g}/\mathfrak{h}$ as in 3.2.9. It is easy to verify that if G acts on each of the bundles in (24) by $g_1\langle g_2, X\rangle = \langle g_1 g_2, X\rangle$, where here X can be in $\mathfrak{h}, \mathfrak{g}$ or $\mathfrak{g}/\mathfrak{h}$, then the isomorphism of (24) onto the Atiyah sequence of $G(G/H, H)$ described in 3.2.9 is G–equivariant. Thus a G–equivariant connection in $G(G/H, H)$ can be identified with a G–equivariant map $\frac{G \times \mathfrak{g}/\mathfrak{h}}{H} \to \frac{G \times \mathfrak{g}}{H}$ which is right–inverse to a_1. We now need the following result.

Lemma 5.3.17 *Let $\frac{G \times V}{H}$ and $\frac{G \times V'}{H}$ be two vector bundles associated to $G(G/H, H)$ via actions ρ, ρ' of H on V, V'. Then every G–equivariant map $\varphi: \frac{G \times V}{H} \to \frac{G \times V'}{H}$ is of the form*

$$\varphi(\langle g, v\rangle) = \langle g, \varphi_1(v)\rangle, \qquad (25)$$

for some H–equivariant map $\varphi_1: V \to V'$, and every H–equivariant map $\varphi_1: V \to V'$ defines a G–equivariant map φ by (25).

Proof Let $\varphi: \frac{G \times V}{H} \to \frac{G \times V'}{H}$ be G–equivariant; that is, $\varphi(g_1\langle g_2, v\rangle) = g_1\varphi(\langle g_2, v\rangle)$ for all $g_1, g_2 \in G, v \in V$. Define $\varphi_1: V \to V'$ by $\varphi(\langle 1, v\rangle) = \langle 1, \varphi_1(v)\rangle$. Then $\varphi(\langle g, v\rangle) = \varphi(g\langle 1, v\rangle) = g\langle 1, \varphi_1(v)\rangle = \langle g, \varphi_1(v)\rangle$, which establishes (25). That φ_1 is H–equivariant follows from

$$\langle 1, \rho'(h)\varphi_1(v)\rangle = \langle h, \varphi_1(v)\rangle = \varphi(\langle h, v\rangle) = \varphi(\langle 1, \rho(h)v\rangle) = \langle 1, \varphi_1(\rho(h)v)\rangle.$$

The converse is straightforward. □

This lemma is of course a part of the well–known result that the category of G–vector bundles and G–equivariant morphisms over G/H is isomorphic to the category of H–vector spaces and H–equivariant maps.

It follows that G–equivariant connections in $G(G/H, H)$ are in bijective correspondence with maps $\gamma_1 : \mathfrak{g}/\mathfrak{h} \to \mathfrak{g}$ which are right–inverse to $T_1(\pi)$ and H–equivariant, that is $\gamma_1(\mathrm{Ad}hX + \mathfrak{h}) = \mathrm{Ad}h\gamma_1(X + \mathfrak{h})$, for all $h \in H$, $X \in \mathfrak{g}$. By chasing around the diagram

$$
\begin{array}{ccc}
\frac{TG}{H} & \xleftarrow{\quad \gamma \quad} & T(G/H) \\
\cong \big\uparrow & & \cong \big\uparrow \\
\frac{G \times \mathfrak{g}}{H} & \xleftarrow{\frac{\mathrm{id} \times \gamma_1}{H}} & \frac{G \times \mathfrak{g}/\mathfrak{h}}{H}
\end{array}
$$

it can be seen that the connection γ corresponding to γ_1 is given by

$$\gamma(\langle T_H(\lambda_g)(X + \mathfrak{h}) \rangle) = \langle T_1(L_g)(\gamma_1(X + \mathfrak{h})) \rangle.$$

It is also easy to check that $\omega_1 : \mathfrak{g} \to \mathfrak{h}$, the left split map corresponding to $\gamma_1 : \mathfrak{g}/\mathfrak{h} \to \mathfrak{g}$, is the restriction to $\mathfrak{g} \to \mathfrak{h}$ of the connection form $\omega : TG \to G \times \mathfrak{h}$ corresponding to γ.

With these preliminaries established, we can calculate the curvature of a G–equivariant connection γ over the coset $H \in G/H$. Take $X + \mathfrak{h}$, $Y + \mathfrak{h} \in \mathfrak{g}/\mathfrak{h}$ and write $\xi = \gamma_1(X + \mathfrak{h})$, $\eta = \gamma_1(Y + \mathfrak{h})$. Let $\overleftarrow{\xi}, \overleftarrow{\eta}$ denote the left–invariant vector fields on G corresponding to ξ, η; then $\langle \overleftarrow{\xi}(1) \rangle = \gamma(\langle X + \mathfrak{h} \rangle)$ and similarly for η and Y. Now

$$
\begin{aligned}
\overline{R}_\gamma(X + \mathfrak{h}, Y + \mathfrak{h}) &= d\underline{\omega}(\gamma(X + \mathfrak{h}), \gamma(Y + \mathfrak{h})) && \text{by 5.3.14} \\
&= \delta\omega(\overleftarrow{\xi}(1), \overleftarrow{\eta}(1)) && \text{by 5.3.11} \\
&= \delta\omega(\overleftarrow{\xi}, \overleftarrow{\eta}(1)) \\
&= -\omega([\overleftarrow{\xi}, \overleftarrow{\eta}])(1) \text{ since } \omega(\overleftarrow{\xi}), \omega(\overleftarrow{\eta}) \text{ are constant} \\
&= -\omega_1([\xi, \eta]_L) \\
&= \omega_1([\xi, \eta]_R) \\
&= \omega_1([\gamma_1(X + \mathfrak{h}, \gamma_1(Y + \mathfrak{h})])
\end{aligned}
$$

and this, together with (23), completely determines $\overline{R}_\gamma$. If $\mathfrak{h}$ is an ideal of $\mathfrak{g}$ the last expression can be written as $-\overline{R}_{\gamma_1}(X + \mathfrak{h}, Y + \mathfrak{h})$, as in 5.3.14, but in general $\gamma_1([X + \mathfrak{h}, Y + \mathfrak{h}])$ has no meaning. ⊠

Example 5.3.18 Let $\varphi(\mathrm{id}_M, f) : P(M, G) \to Q(M, H)$ be a morphism of principal bundles over a common base, and consider the induced morphism of their Atiyah sequences (see 3.2.8)

$$
\begin{array}{ccccc}
\frac{P\times\mathfrak{g}}{G} & \rightarrowtail^{\ j\ } & \frac{TP}{G} & \xrightarrow{\ \pi_*\ }\!\!\!\!\!\rightarrow & TM \\
\varphi^+_* \downarrow & & \varphi_* \downarrow & & \| \\
\frac{Q\times\mathfrak{h}}{H} & \rightarrowtail^{\ j'\ } & \frac{TQ}{H} & \xrightarrow{\ \pi'_*\ }\!\!\!\!\!\rightarrow & TM.
\end{array}
$$

Let γ be a connection in $P(M,G)$ with corresponding connection form $\omega \in \Omega^1(P,\mathfrak{g})$, and connection reform $\underline{\omega} \colon \frac{TP}{G} \to \frac{P\times\mathfrak{g}}{G}$. In the abstract Lie algebroid context (5.2.13) the produced connection $\gamma' \colon TM \to \frac{TQ}{H}$ in $Q(M,H)$ was defined by $\gamma' = \varphi_* \circ \gamma$. The traditional description of an 'induced connection' is considerably less simple; we obtain it now for purposes of comparison.

The induced connection form $\omega' \in \Omega^1(Q,\mathfrak{h})$ is characterized by the condition $\varphi^*\omega' = f_* \circ \omega$ and in fact only the values of ω' on the image in TQ of $T(\varphi)$ are given in the standard treatments. Using the Atiyah sequence/Lie algebroid formulation, we can quickly derive the general formula for ω'.

Let ω' and $\underline{\omega}'$ now denote the connection form and connection reform corresponding to γ'. Then $j'\underline{\omega}' = \mathrm{id} - \gamma'p'_* = \mathrm{id} - \varphi_*\gamma p'_*$. Using this, and the fact that any connection γ in any $P(M,G)$ and its connection form $\omega \in \Omega^1(P,\mathfrak{g})$ are related by

$$
\gamma(X_x) = \langle Z_u - \omega_u(Z)^* |_u \rangle,
$$

where Z is any element of $T_u(P)$ with $T_u(\pi)(Z) = X \in T_x(M)$, it is straightforward to establish that the connection form $\omega' \in \Omega^1(Q,\mathfrak{h})$ corresponding to γ' is given by

$$
\omega'(Y_v) = B + \mathrm{Ad}h^{-1}f_*(\omega_u(X)), \qquad Y \in T_v(Q) \tag{26}
$$

where

u is any element of P with $\pi(u) = \pi'(v)$,
X is any element of $T_u(P)$ with $T_u(\pi)(X) = T_v(\pi')(Y)$,
h is the element of H for which $\varphi(u)h = v$, and
$B \in \mathfrak{h}$ is determined by $B^*|_v = Y - T(R_h)T_u(\varphi)(X)$.

It is straightforward, if tedious, to check that this ω' is well defined and is a connection form in $Q(M,H)$. If $Y_v = T_u(\varphi)(X)$ for some $X \in T_u(P)$ then we may use this u and X in (26) and take $h = 1$ so that we get $\omega'(T_u(\varphi)X) = f_*(\omega_u(X))$. This confirms that ω' is indeed the induced connection (form) in the traditional sense.

This example shows that (26), although it concerns connection forms in the traditional sense, is most easily derived using the Atiyah sequence/Lie algebroid formulation.

One may derive a similar formula to (26) for expressing the curvature form Ω' of ω' in terms of the curvature form Ω of ω. However in the Lie algebroid language we need only note that $\overline{R}_{\gamma'} = \varphi_*^+ \circ \overline{R}_\gamma$ (see (15)). Here $\varphi_*^+ : \frac{P \times \mathfrak{g}}{G} \to \frac{Q \times \mathfrak{h}}{H}$ is the map $\langle u, X \rangle \mapsto \langle \varphi(u), f_*(X) \rangle$.

The definition $\gamma' = \varphi_* \circ \gamma$ and the resulting equation $\overline{R}_{\gamma'} = \varphi_*^+ \circ \overline{R}_\gamma$ are considerably simpler than both (26) and the corresponding equation for Ω' in terms of Ω, yet $\gamma' = \varphi_* \circ \gamma$ and $\overline{R}_{\gamma'} = \varphi_*^+ \circ \overline{R}_\gamma$ contain more information than the standard $\varphi^* \omega' = f_* \circ \omega$, $\varphi^* \Omega' = f_* \circ \Omega$. ☒

5.4 Local descriptions

We will prove in Chapter 8 that every transitive Lie algebroid is locally trivial in the sense defined in 5.4.1 below. In order to give the local formulas of this section without unnecessary restrictions, we introduce the concept here. The main development of the notion is in Chapter 8.

Consider a transitive Lie algebroid $L \rightarrowtail A \twoheadrightarrow TM$. Let $U \subseteq M$ be open and suppose that there is a Lie algebra $\mathfrak{g}$ and an isomorphism of Lie algebroids

$$S : TU \oplus (U \times \mathfrak{g}) \to A_U$$

from the trivial Lie algebroid on U with Lie algebra $\mathfrak{g}$ to the restriction A_U of A to U. The map of the adjoint bundles $S^+ : U \times \mathfrak{g} \to L_U$ is a chart for L; denote it by ψ. The map $X \mapsto S(X \oplus 0)$ is a flat connection in A_U; denote it by Θ. For any $X \in \mathfrak{X}(U)$ and $V : U \to \mathfrak{g}$, expanding out the equation

$$[S(X \oplus 0), S(0 \oplus V)] = S[X \oplus 0, 0 \oplus V]$$

gives

$$[\Theta(X), \psi(V)] = \psi(X(V)). \tag{27}$$

Conversely, suppose given an open $U \subseteq M$, a flat connection $\Theta : TU \to A_U$ and a bracket–preserving vector bundle chart $\psi : U \times \mathfrak{g} \to L_U$ such that (27) holds. Then $S : TU \oplus (U \times \mathfrak{g}) \to A_U$, $S(X \oplus V) = \Theta(X) + \psi(V)$, is an isomorphism of Lie algebroids.

Definition 5.4.1 A transitive Lie algebroid $L \rightarrowtail A \twoheadrightarrow TM$ is *locally trivial* if there is an open cover $\{U_i\}$ of the base with respect to which there is both a family of local flat connections $\Theta^i \colon TU_i \to A_{U_i}$ and an atlas of bracket–preserving trivializations $\{\psi_i \colon U_i \times \mathfrak{g} \to L_{U_i}\}$ for the adjoint bundle L, such that (27) holds for all i. Such a collection $\{U_i, \psi_i, \Theta^i\}$ is called a *Lie algebroid atlas* for A. □

It is clear that the adjoint bundle L of a locally trivial Lie algebroid is an LAB, and that $\{\psi_i\}$ is an LAB atlas. Denote by a_{ij} the transition functions for $\{\psi_i\}$. Thus $a_{ij} \colon U_{ij} \to \mathrm{Aut}(\mathfrak{g})$, where $\mathrm{Aut}(\mathfrak{g})$ is the group of Lie algebra automorphisms.

For $U_{ij} \neq \emptyset$ we define a $\mathfrak{g}$–valued 1–form χ_{ij} on U_{ij} by

$$\Theta^j = \Theta^i + j \circ \ell_{ij}, \qquad \chi_{ij} = \psi_i^{-1} \circ \ell_{ij}.$$

Proposition 5.4.2 $\chi_{ij} \in \Omega^1(U_{ij}, \mathfrak{g})$ *is a Maurer–Cartan form.*

Proof Since Θ^i and Θ^j are both bracket–preserving, it follows by a manipulation that

$$\nabla^{\Theta^i}(\ell^{ij}) + [\ell_{ij}, \ell_{ij}] = 0,$$

where ∇^{Θ^i} is the adjoint connection in L_{U_i} induced by the flat Lie algebroid connection Θ^i. Now the compatibility equation (27) may be written

$$\nabla^{\Theta^i}_X(\psi_i(v)) = \psi_i(X(v)), \qquad (28)$$

in which form it shows that ψ_i maps the standard flat connection in $U_i \times \mathfrak{g}$ to ∇^{Θ^i}.

Since $\ell_{ij} = \psi_i \circ \chi_{ij}$, the Maurer–Cartan equation for ψ_{ij} now follows. □

The forms χ_{ij} are the *transition forms* for the Lie algebroid atlas $\{U_i, \psi_i, \Theta^i\}$. We return to the general study of these systems in Chapter 8.

For the present consider a locally trivial Lie groupoid Ω on base M. Let $\{\sigma_i \colon U_i \to \Omega_m\}$ be a section–atlas for Ω, and write $G = \Omega_m^m$. Then

$$\Sigma_i \colon U_i \times G \times U_i \to \Omega_{U_i}^{U_i}, \qquad (y, g, x) \mapsto \sigma_i(y) g \sigma_i(x)^{-1},$$

is an isomorphism of Lie groupoids from the trivial Lie groupoid to the restriction of Ω to U_i. Applying the Lie functor, we obtain an isomorphism of Lie algebroids

$$A(\Sigma_i) \colon TU_i \oplus (U_i \times \mathfrak{g}) \to A\Omega|_{U_i}$$

which we take as S_i. Together with a few more manipulations, this proves:

Theorem 5.4.3 *Let Ω be a locally trivial Lie groupoid on M. Then $A\Omega$ is a locally trivial Lie algebroid. Given a section-atlas $\{\sigma_i \colon U_i \to \Omega_m\}$, define*

$$\theta_i \colon U_i \times U_i \to \Omega_{U_i}^{U_i}, \quad \theta_i(y, x) = \sigma_i(y)\sigma_i(x)^{-1},$$

and let $\psi_i \colon U_i \times \mathfrak{g} \to L\Omega|_{U_i}$ be defined by $\psi_{i,x} = (I_{\sigma_i(x)})_*$, the Lie algebra isomorphism induced by the inner automorphism $I_{\sigma_i(x)} \colon G \to \Omega_x^x$.

Then $\{U_i, (\theta_i)_*, \psi_i\}$ is a Lie algebroid atlas for $A\Omega$, where $(\theta_i)_*$ is the Lie algebroid morphism induced by the Lie groupoid morphism θ_i.

Now consider an arbitrary connection γ in a locally trivial Lie algebroid $L \rightarrowtail A \twoheadrightarrow TM$. Continue the notations $\Theta_i, \psi_i, \chi_{ij}, a_{ij}$ used above.

For each i, define $\omega_i \colon TU_i \to U_i \times \mathfrak{g}$ by

$$\psi_i(\omega_i) = \gamma|_{U_i} - \Theta_i. \tag{29}$$

The $\omega_i \in \Omega^1(U_i, \mathfrak{g})$ are the *local connection forms* for γ with respect to the given system of local data. Clearly, on a U_{ij} which is non-void,

$$\omega_i = a_{ij}(\omega_j) + \chi_{ij} \tag{30}$$

and, conversely, if $\{\omega_i \in \Omega^1(U_i, \mathfrak{g})\}$ is a family of forms which satisfy (30), then they define by (29) a connection γ in A. Using (29) and (27), the next result is easily proved.

Proposition 5.4.4 *With the above notation, the adjoint connection ∇^γ is given locally by*

$$\nabla_X^\gamma(\psi_i(V)) = \psi_i\{X(V) + [\omega_i(X), V]\}$$

where $X \in \mathfrak{X}(U_i)$ and $V \colon U_i \to \mathfrak{g}$.

Next define $R_\gamma^i \in \Omega^2(U_i, \mathfrak{g})$ by

$$R_\gamma^i = \psi_i^{-1} \circ \overline{R}_\gamma.$$

The R_γ^i are the *local curvature forms* for γ with respect to the given system of local data. Clearly $R_\gamma^j = a_{ji}(R_\gamma^i)$ when $U_{ij} \neq \emptyset$. Again, the next result follows easily from (29) and (27).

Proposition 5.4.5 *Continuing the above notation,*

$$R^i_\gamma = -(\delta\omega_i + [\omega_i, \omega_i]).$$

We conclude the section with two illustrations of the use of these methods in familiar situations.

Proposition 5.4.6 *Let Ω be a locally trivial Lie groupoid on M.*

(i) *If $\{\sigma_i\}$ is a section–atlas for Ω for which the cocycle s_{ij} consists of constant maps, then*

$$\gamma|_{U_i} = (\theta_i)_*$$

is a well–defined and flat infinitesimal connection in Ω.

(ii) *If γ is a flat infinitesimal connection in Ω then there is a section–atlas $\{\sigma_i\}$ for Ω for which the cocycle consists of constant maps and for which*

$$\gamma|_{U_i} = (\theta_i)_*$$

on all U_i.

Proof (i) Since s_{ij} is constant, it follows that $\chi_{ij} = 0$ and so (30) admits the solution $\omega_i = 0$ for all i.

(ii) Let $\{\sigma_i\colon U_i \to \Omega_m\}$ be a section–atlas in which each U_i is connected and simply–connected. Since γ is flat, the ω_i themselves are Maurer–Cartan forms and so there exist maps $f_i\colon U_i \to G$ such that

$$\Delta(f_i) = \omega_i.$$

Define $\tau_i\colon U_i \to \Omega_m$ by $\tau_i = \sigma_i f_i$. It is easy to verify that $\Delta(t_{ij}) = 0$, where t_{ij} is the cocycle for τ_i, and so t_{ij} is constant.

Denote by θ'_i the local morphism induced by τ_i. Then, using

$$(\theta'_i)_*(X) = T(R_{\tau_i(x)^{-1}})T(\tau_i)(X),$$

for $X \in T(U_i)_x$, it is easy to verify that

$$(\theta'_i)_* = (\theta_i)_* + \psi_i(\Delta(f_i)).$$

Hence $(\theta'_i)_* = \gamma|_{U_i}$ as claimed.   □

The following version of 5.4.6 is used several times in the sequel.

Proposition 5.4.7 *Let $M \times G \times M$ be a trivial Lie groupoid with M connected and simply–connected. Then for any flat connection γ, there is an automorphism φ of $M \times G \times M$ over M such that $\varphi_* \circ \gamma^0 = \gamma$, where γ^0 is the standard flat connection.*

Proof The connection γ is of the form $X \mapsto X \oplus \omega(X)$ where $\omega \in \Omega^1(M, \mathfrak{g})$ is a Maurer–Cartan form (see 1.4.4). Let $f \colon M \to G$ be such that $\Delta(f) = \omega$; then $\varphi(y, g, x) = (y, f(y)gf(x)^{-1}, x)$ has the required property. □

Proposition 5.4.8 *Let M be a manifold and let $\langle\,,\,\rangle$ be a Riemannian structure in the vector bundle TM. Let ∇ be a Riemannian connection such that*

$$\nabla_X(Y) - \nabla_Y(X) - [X, Y] = 0, \qquad X, Y \in \Gamma TM \qquad (31)$$

Then if ∇ is flat, M is locally isometric to Euclidean space.

Proof Let $\varphi \colon \mathbb{R}^n \to U$ be a chart for M. Pull the Riemannian structure and the connection on $TM|_U$ back to $T(\mathbb{R}^n)$. Continue to use the notations $\langle\,,\,\rangle$ and ∇. By 1.7.9 there is a neighbourhood W of $0 \in \mathbb{R}^n$, which we may assume to be connected and simply–connected, and a decomposing section $\sigma \colon W \to \Phi_{\mathscr{O}}(T\mathbb{R}^n)_0$. This σ defines an automorphism of the vector bundle TW which maps the given Riemannian structure to another, still denoted $\langle\,,\,\rangle$, for which $\Phi_{\mathscr{O}}(TW)$ is $W \times \mathscr{O}(n) \times W$; furthermore, the value of σ at 0 can be chosen so that $\langle\,,\,\rangle_0$ is the standard metric on $\mathbb{R}^n$. We transport ∇ under this automorphism also; ∇ is still a Riemannian connection in TW and still satisfies (31), and still is flat.

By 5.4.7 there is a map $f \colon W \to \mathscr{O}(n)$ such that $F \colon W \times \mathbb{R}^n \to W \times \mathbb{R}^n, (x, X) \longmapsto (x, f(x)(X))$ maps the standard flat connection ∇^0 to ∇; that is,

$$\nabla_X(Y) = F(X(F^{-1}(Y)));$$

furthermore, we can also require that $f(0) = I \in \mathscr{O}(n)$. Let $\{\frac{\partial}{\partial x_i}\}$ be the standard vector fields on $\mathbb{R}^n$ and define $X_i = F\left(\frac{\partial}{\partial x_i}\right)$. Then for any vector field X,

$$\nabla_X(X_i) = F(X(\tfrac{\partial}{\partial x_i})) = 0 \qquad (32)$$

since $\frac{\partial}{\partial x_i}$ is constant as a map $W \to \mathbb{R}^n$. Hence from (31) it follows that $[X_i, X_j] = 0$ for all i, j and so there is a local coordinate system $\{y_l, \ldots, y_n\}$ around 0 in W such that

$$\frac{\partial}{\partial y_i} = X_i$$

for all i.

Now in this coordinate system the metric is canonical, for

$$X(\langle \tfrac{\partial}{\partial y_i}, \tfrac{\partial}{\partial y_j} \rangle) = \langle \nabla_X(X_i), X_j \rangle + \langle X_i, \nabla_X(X_j) \rangle = 0$$

by (32) and so $\langle \tfrac{\partial}{\partial y_i}, \tfrac{\partial}{\partial y_j} \rangle$ is constant and we arranged the value at 0 to be δ_{ij}. $\square$

5.5 Notes

Connection theory. For readers who have not previously met connection theory, here are a few comments on other accounts:

The fundamental source for the general theory of connections in principal bundles is the account in the magisterial volumes of Kobayashi and Nomizu [1963, 1969]. About a decade later, the second of the three volumes of [Greub, Halperin, and Vanstone, 1973] provided a usefully different perspective on parts of this material. Neither of these works is particularly easy for the beginner.

For most of the 20th century the two main motivations for studying connection theory came from Riemannian geometry and from characteristic class theory, and to these was added in the 1970s gauge theory. However a reader who simply wishes to see some concrete examples of the use of connection theory would be better advised to consult books for which generality is not a main concern. For connections as covariant derivatives on surfaces, see, for example, [do Carmo, 1976, Chap. 4] and [Spivak, 1979, Vol. II].

§5.1 This section follows Appendix B of [Mackenzie, 1987a]. The value of the Darboux derivative as a working technique could be more widely appreciated; see [Malliavin, 1972].

§5.2 Most of the material in this section has been taken from [Mackenzie, 1987a, III§5], with some substantial reorganization.

The formulation of connections in vector bundles in terms of the two vector bundle structures on TE is several decades old; see, for example, [Dieudonné, 1972] or [Besse, 1978] for accounts which are still modern in spirit. As this formulation indicates, connection theory in vector bundles developed for some years independently of the general principal bundle theory.

The definition of curvature 5.2.3 has the opposite sign to that used in many accounts. Nonetheless, this is consistent with the general definition in 5.2.9 which in turn corresponds exactly to the definition of curvature as a Lie algebra valued 2–form on a principal bundle (see §5.3). The choice of signs in 5.2.3 has been used by Milnor [1963] and others.

What are here called connection reforms were called *back–connections* in [Mackenzie, 1987a]. They are a reformulation of the standard notion of connection 1–form.

Proposition 5.2.6 may be found in [Dieudonné, 1972].

In the connection theory of principal bundles, the definition of a connection as an invariant horizontal distribution is usually given pre–eminence, perhaps because it is easier to visualize than the associated connection form. Nonetheless it is easier to compute with the connection form than with the associated distribution, and most computations are done in terms of forms, either global or local.

§5.3 This section has been taken from Appendix A, §3, and elsewhere in [Mackenzie, 1987a], with light revision.

This section is addressed to readers who know the standard treatment of Kobayashi and Nomizu. Readers who wish to compare the merits of the two approaches may find the following comments useful:

That the definition 5.3.5 of an associated connection in a vector bundle coincides with the classical definition of the induced connection in an associated bundle follows from [Kobayashi and Nomizu, 1963, III 1.3].

The terminology 5.3.9 comes from [Kobayashi and Nomizu, 1963, § II.5] and [Greub et al., 1973, §3.15, §6.6].

The notion of equivariance for connections in 5.3.16 is a standard one: see for example [Greub et al., 1973, §6.28]. All of this example should be compared with the traditional account in [Greub et al., 1973, §6.30, §6.31].

For a traditional account of the 'induced connection' referred to in 5.3.18, see [Kobayashi and Nomizu, 1963, § II.6].

§5.4 This section has been taken from [Mackenzie, 1987a, III§5], with a little re-organization. The concept of a Lie algebroid atlas (under the name *system of local data*) was new in [Mackenzie, 1987a]. The notion of local triviality for Lie algebroids is an expository device introduced here; it is comparable to the standard practice in bundle theory of building local triviality into the definition and then proving that a bundle is trivializable over any contractible open subset.

Proposition 5.4.8 is what [Spivak, 1979, Volume II] calls the 'Test Case' for evaluating techniques in this area of differential geometry; he gives seven different proofs.

6

Path connections
and Lie theory

The previous chapter was concerned with connections in abstract Lie algebroids, both as global maps and localized with respect to open covers of the base. The present chapter deals with the process by which an infinitesimal connection in the Lie algebroid of a locally trivial Lie groupoid may be integrated to a law of path lifting; that is, to a *path connection*.

Preparatory to this, in §6.1 we construct the monodromy groupoid of a locally trivial and α–connected Lie groupoid. This plays the role which for Lie groups is played by the universal covering of a connected Lie group. Indeed in terms of a principal bundle with connected total space P, the monodromy bundle has total space the universal cover of P (and the same base space).

In §6.2 we prove versions of the First and Second integrability theorems of Lie for locally trivial Lie groupoids and transitive Lie algebroids. The proofs are reasonably straightforward extensions of the methods known for Lie groups and Lie algebras; we will see, however, that they cover a range of phenomena not within the scope of Lie group theory.

§6.3 formalizes the concept of path connection in a locally trivial Lie groupoid Ω and establishes the correspondence between path connections in Ω and infinitesimal connections in the Lie algebroid $A\Omega$. The path lifting associated with an infinitesimal connection is usually treated as a subsidiary concept — given an infinitesimal connection, there is an associated path connection. Our purpose in treating this concept separately is to emphasize the distinction between the infinitesimal aspect of connection theory, which takes place for abstract transitive Lie algebroids, and those parts of connection theory — the concept of path lifting and holonomy — which require the Lie algebroid to be realized as the Lie algebroid of a specific Lie groupoid. The situation is exactly parallel

212

to that existing with Lie groups and Lie algebras: the one transitive Lie algebroid may arise from several distinct Lie groupoids, which are only locally isomorphic, and although the curvature, for example, depends only on the Lie algebroid, the holonomy, and its associated concepts, depend on the connectivity properties of the Lie groupoid.

§6.4 treats the relationship between the action of an infinitesimal connection and the action of its holonomy groupoid. This material is essential to what follows in Chapters 7 and 8. In 6.4.4 we prove a very general result concerning structures on vector bundles defined by tensor fields: Theorem 1.6.23 may be reformulated to state that such a structure is locally trivial if and only if the structures defined on the fibres of the vector bundle are pairwise isomorphic; in 6.4.4 we prove that this is so if and only if the bundle admits a connection compatible with the structure. From this it follows, for example, that a morphism of vector bundles $\varphi\colon E^1 \to E^2$ over a base M is of locally constant rank if and only if it maps some connection ∇^1 in E^1 to a connection ∇^2 in E^2.

In 6.4.18 to 6.4.21 we give a strong, Lie groupoid form of the Ambrose–Singer theorem. 6.4.18 and 6.4.19 give an abstract construction of the Lie algebroid of the holonomy groupoid of a connection which may be easily seen to hold in any transitive Lie algebroid.

6.4.23 is a connection–theoretic analysis of morphisms of trivial Lie algebroids over a fixed base. In 1.2.5 it is pointed out that a morphism of trivial groupoids $\mu\colon M \times G \times M \to M \times H \times M$ can be constructed from any morphism $F = \varphi_m^m\colon G \to H$ and any map $\theta\colon M \to H$; however 6.4.23 shows that an arbitrary Maurer–Cartan form $\omega = \Delta(\theta) \in \Omega^1(M, \mathfrak{h})$ and Lie algebra morphism $\mathfrak{g} \to \mathfrak{h}$ need to satisfy a further compatibility condition in order to define a morphism of trivial Lie algebroids $TM \oplus (M \times \mathfrak{g}) \to TM \oplus (M \times \mathfrak{h})$. This difference in behaviour turns out to be typical. Using 6.4.23, we obtain a second proof of the local integrability of morphisms of transitive Lie algebroids.

§6.5 is concerned with abstract transitive Lie algebroids. For the Lie algebroid of a locally trivial Lie groupoid Ω, the fact that $\mathscr{I}\Omega$ is an LGB implies immediately that the adjoint bundle of $A\Omega$ is an LAB; likewise, for a morphism of Lie algebroids which is induced by a morphism F of locally trival Lie groupoids, the local constancy of F implies that of the induced morphism. In §6.5 we apply the results of the preceding sections to show that these results, and various refinements of them, hold for abstract transitive Lie algebroids and morphisms between them.

6.1 The monodromy groupoid of a locally trivial Lie groupoid

Consider first a principal bundle $P(M, G, p)$ with P connected. Denote elements of the fundamental groupoid $\Pi(P)$ by $\langle \nu \rangle$ where $\nu \colon [0, 1] \to P$ is a path in P. The action of G on P lifts to an action on $\Pi(P)$ by $\langle \nu \rangle g = \langle R_g \circ \nu \rangle$. This is a smooth action by Lie groupoid automorphisms over the given action $P \times G \to P$. In the terminology of 2.5.4, $\Pi(P)$ is a PBG–groupoid over $P(M, G)$.

Definition 6.1.1 The locally trivial Lie groupoid $\frac{\Pi(P)}{G} \rightrightarrows M$, denoted $\mathcal{M}(P(M, G))$ or $\mathcal{M}_P$, is the *monodromy groupoid* of $P(M, G)$. $\quad\square$

For reference we restate from §2.5 the structure on $\mathcal{M}_P$. For a path ν in P we denote the corresponding element of $\mathcal{M}_P$ by $\|\nu\|$. The source and target maps are given by

$$\beta(\|\nu\|) = p(\nu(1)), \qquad \alpha(\|\nu\|) = p(\nu(0))$$

and for $\|\nu\|$, $\|\nu'\|$ with $\nu(0) = \nu'(1)g$ we define

$$\|\nu\|\,\|\nu'\| = \|\nu(R_g \circ \nu')\|.$$

From §2.5 we know that $\mathcal{M}_P$ is a locally trivial Lie groupoid and that the natural projection $\natural \colon \Pi(P) \to \mathcal{M}_P$ is an action morphism. Furthermore, the short exact sequence $\mathscr{I}\Pi(P) \rightarrowtail \Pi(P) \twoheadrightarrow P \times P$ quotients to an extension of Lie groupoids

$$\frac{\mathscr{I}\Pi(P)}{G} \rightarrowtail \mathcal{M}_P = \frac{\Pi(P)}{G} \overset{\mathfrak{m}}{\twoheadrightarrow} \frac{P \times P}{G}. \tag{1}$$

We call $\mathfrak{m}$ the *monodromy projection*. Since $\natural$ is an action morphism, the α–fibres of $\mathcal{M}_P$ can be identified with the α–fibres of $\Pi(P)$; that is, with the universal cover of P. Given $u_0 \in P$, each class in the vertex group of $\mathcal{M}_P$ at $m_0 = p(u_0)$ contains a unique homotopy class of paths ν with $\nu(0) = u_0$, and can be represented by the homotopy class in $\Pi(P)$ of ν. Since $p(\nu(1)) = m_0$ also, there is a unique $g \in G$ such that $\nu(1) = u_0 g$; this g is $\mathfrak{m}(\|\nu\|)$. The extension of principal bundles corresponding to (1) is therefore

$$\pi_1 P \rightarrowtail \widetilde{P}(M, H) \overset{\mathfrak{m}}{\twoheadrightarrow} P(M, G)$$

where we write H for a typical representative of the vertex groups of $\mathcal{M}_P$. The bundle $\widetilde{P}(M, H)$ is the *monodromy bundle* of $P(M, G)$. The group H is itself an extension

$$\pi_1 P \rightarrowtail H \overset{\mathfrak{m}}{\twoheadrightarrow} G.$$

but is not determined by G and $\pi_1 P$ alone. Consideration of the long exact homotopy sequence of these bundles gives the following result.

Proposition 6.1.2 *Let $P(M,G)$ be a principal bundle with P connected, and let $\widetilde{P}(M,H)$ be the monodromy bundle. Then:*

(i) *$\pi_0 H \cong \pi_1 M$ under the boundary morphism of the long exact homotopy sequence of $\widetilde{P}(M,H)$;*

(ii) *$\pi_1 H = \ker(\pi_1 G \to \pi_1 P)$.*

Example 6.1.3 For G a connected Lie group and H a closed subgroup, let $K = c^{-1}(H)$ where $c\colon \widetilde{G} \to G$ is the covering projection. Then K is a closed subgroup of $\widetilde{G}$, the spaces $\widetilde{G}/K$ and G/H are equivariantly diffeomorphic, and $\widetilde{G}(\widetilde{G}/K, K)$ is the monodromy bundle of $G(G/H, H)$.

In particular, the Hopf bundle $SO(3)(S^2, SO(2))$ has monodromy bundle $SU(2)(S^2, U(1))$.

Next, let the universal covering $\varphi\colon SU(2) \times SU(2) \to SO(4)$ be realized as the map $(p,q) \mapsto (h \mapsto phq^{-1})$ where p, q are unit quarternions and h is any quaternion. Let $A \in SO(3)$ act on $SO(4)$ as $\left[\begin{smallmatrix} 1 & 0 \\ 0 & A \end{smallmatrix}\right]$; then the monodromy bundle of $SO(4)(S^3, SO(3))$ is the trivial bundle $SU(2) \times SU(2)(S^3, SU(2))$, where the group acts as the diagonal subgroup.

Lastly, consider $SO(4)(SO(4)/T^2, T^2)$ where T^2 is the maximal torus

$$\{\left[\begin{smallmatrix} A & 0 \\ 0 & B \end{smallmatrix}\right] \mid A, B \in SO(2)\}$$

in $SO(4)$. The homogeneous space $SO(4)/T^2$ is the Grassmannian $\widetilde{G}_{4,2}$ of oriented 2–planes in $\mathbb{R}^4$, and a calculation shows that $\varphi^{-1}(T^2)$ is the maximal torus

$$K = \{((e^{i\theta}, 0), (e^{i\theta'}, 0)) \mid \theta, \theta' \in \mathbb{R}\}$$

where the notation (z, w), for $z, w \in \mathbb{C}$ with $|z|^2 + |w|^2 = 1$, denotes the element $\left[\begin{smallmatrix} z & w \\ -\overline{w} & \overline{z} \end{smallmatrix}\right]$ of $SU(2)$.

Now the action of K on $SU(2) \times SU(2)$ is the cartesian square of the standard action of $U(1)$ on $SU(2)$, so the monodromy bundle of $SO(4)(\widetilde{G}_{4,2}, T^2)$ is $SU(2) \times SU(2)(S^2 \times S^2, U(1) \times U(1))$. Note that the restriction of $\varphi\colon SU(2) \times SU(2) \to SO(4)$ to $K \to T^2$ is

$$((e^{i\theta}, 0), (e^{i\theta'}, 0)) \mapsto \begin{bmatrix} R_{\theta - \theta'} & 0 \\ 0 & R_{\theta + \theta'} \end{bmatrix}$$

where R_α is the rotation matrix $\left[\begin{smallmatrix} \cos\alpha & -\sin\alpha \\ \sin\alpha & \cos\alpha \end{smallmatrix}\right]$. $\boxtimes$

Now consider a locally trivial and α–connected Lie groupoid Ω on M. Given any $m \in M$ we can construct the monodromy groupoid of the vertex bundle at m, and clearly these monodromy groupoids will be isomorphic. As with the universal cover of a space, the universal properties are normally more important than details of the construction. However, the following gives a description which is independent of a choice of base point.

Let $\mathscr{P}^\alpha = \mathscr{P}^\alpha(\Omega)$ be the set of continuous and piecewise–smooth paths $\nu: I \to \Omega$ (where $I = [0,1]$) for which $\alpha \circ \nu: I \to M$ is constant; elements of P^α are called α–*paths* in Ω. Let $P_M^\alpha = P_M^\alpha(\Omega)$ be the subset of α–paths which commence at an identity of Ω; every $\nu \in \mathscr{P}^\alpha$ is of the form $R_\xi \circ \nu'$ where $\nu' \in P_M^\alpha$ and $\xi = \nu(0)$. Define $\nu, \nu' \in P^\alpha$ to be α–*homotopic*, written $\nu \overset{\alpha}{\sim} \nu'$, if $\nu(0) = \nu'(0)$, $\nu(1) = \nu'(1)$, and there is a continuous and piecewise–smooth homotopy $H: I \times I \to \Omega$ such that $H(0,-) = \nu$, $H(1,-) = \nu'$, and such that $H(s,0)$ and $H(s,1)$ are constant with respect to $s \in I$, and $H(s,-) \in \mathscr{P}^\alpha$ for all $s \in I$. Such a map H is called an α–*homotopy from* ν *to* ν'. The α–homotopy class containing $\nu \in P^\alpha$ is written $\langle \nu \rangle$.

Define $\mathcal{M}\Omega$ to be the set $\{\langle \nu \rangle \mid \nu \in P_M^\alpha\}$ with the following groupoid structure: the projections $\overline{\alpha}, \overline{\beta}: \mathcal{M}\Omega \to M$ are $\overline{\alpha}(\langle \nu \rangle) = \alpha\nu(0)$ and $\overline{\beta}(\langle \nu \rangle) = \beta\nu(1)$; consequently if $\overline{\alpha}(\langle \nu' \rangle) = \overline{\beta}(\langle \nu \rangle)$, then $\nu'(0)\nu(1)$ is defined and so is the standard concatenation $(R_{\nu(1)} \circ \nu')\nu$; we define $\langle \nu' \rangle \langle \nu \rangle$ to be $\langle (R_{\nu(1)} \circ \nu')\nu \rangle$. It is straightforward to verify that this product is well defined and makes $\mathcal{M}\Omega$ into a groupoid on M. The identities are $\overline{1}_x = \langle \kappa_{1_x} \rangle$ and the inverse of $\langle \nu \rangle$ is $\langle R_\xi \circ \nu^\leftarrow \rangle$ where $\xi = \nu(1)^{-1}$ and $\nu^\leftarrow$ is the reverse of ν.

Definition 6.1.4 Let $\Omega \rightrightarrows M$ be a locally trivial and α–connected Lie groupoid. The *monodromy groupoid* $\mathcal{M}\Omega$ of Ω is the groupoid just constructed with the Lie structure induced from its identification with the monodromy groupoid of any vertex bundle. $\square$

Since the projection $\mathsf{m}: \mathcal{M}\Omega \to \Omega$ is a surjective submersion with fibrewise–discrete kernel, the next result is immediate.

Proposition 6.1.5 *The projection* $\mathsf{m}: \mathcal{M}\Omega \to \Omega$ *induces an isomorphism of Lie algebroids* $A(\mathsf{m}): A(\mathcal{M}\Omega) \to A\Omega$.

We now treat the problem of globalizing a local morphism of Lie groupoids.

Definition 6.1.6 Let Ω and Ω' be Lie groupoids on bases M and M' respectively. A *local morphism* of Lie groupoids, denoted $\varphi\colon \Omega \multimap \Omega'$, consists of a smooth map $\varphi\colon \mathscr{U} \to \Omega'$ defined on an open neighbourhood $\mathscr{U}$ of the base of Ω, together with a smooth map $\varphi_\circ\colon M \to M'$, such that $\alpha' \circ \varphi = \varphi_\circ \circ \alpha$, $\beta' \circ \varphi = \varphi_\circ \circ \beta$, $\varphi \circ 1 = 1' \circ \varphi_\circ$, and such that

(i) $\varphi(\eta\xi) = \varphi(\eta)\varphi(\xi)$ whenever $\alpha\eta = \beta\xi$ and each of $\xi, \eta, \eta\xi$ is in $\mathscr{U}$; and

(ii) $\varphi(\xi^{-1}) = \varphi(\xi)^{-1}$ whenever both of ξ and ξ^{-1} are in $\mathscr{U}$.

Two local morphisms $\varphi, \psi\colon \Omega \multimap \Omega'$ are *germ–equivalent* if the maps φ and ψ are equal on an open neighbourhood of the base of Ω.

A local morphism $\varphi\colon \Omega \multimap \Omega'$, $\varphi_\circ\colon M \to M'$ is a *local isomorphism* if there exists a local morphism $\varphi'\colon \Omega' \multimap \Omega$ such that $\varphi \circ \varphi'$ and $\varphi' \circ \varphi$ are germ–equivalent to $\mathrm{id}_{\Omega'}$ and id_Ω. $\qquad\square$

Example 6.1.7 Suppose that Ω is a locally trivial Lie groupoid with a section–atlas $\{\sigma_i\colon U_i \to \Omega_m\}$ which has the property that each transition function $s_{ij}\colon U_{ij} \to G = \Omega_m^m$ is constant. Let $\mathscr{U}$ be the open neighbourhood $\bigcup_i (U_i \times U_i)$ of the base in $M \times M$ and define $\theta\colon \mathscr{U} \to \Omega$ by $\theta(y, x) = \sigma_i(y)\sigma_i(x)^{-1}$ whenever $(y, x) \in U_i \times U_i$. It is easy to see that because the transition functions $\{s_{ij}\}$ are constant, θ is well defined, and so gives a local morphism $M \times M \multimap \Omega$ over M.

Conversely, let Ω be a locally trivial Lie groupoid on a manifold M and suppose there exists a local morphism $\theta\colon M \times M \multimap \Omega$ over M. Choose $m \in M$ and an open cover $\{U_i\}$ of M such that $\bigcup_i (U_i \times U_i)$ is contained in the domain of θ. In each U_i choose an x_i and for each i choose some $\xi_i \in \Omega_m^{x_i}$. Define $\sigma_i\colon U_i \to \Omega_m$ by $\sigma_i(x) = \theta(x, x_i)\xi_i$. Then the transition functions for $\{\sigma_i\}$ are constant, and the local morphism induced by $\{\sigma_i\}$ is θ.

In particular there is a local right–inverse τ to the anchor χ of the fundamental groupoid $\Pi(M)$. $\qquad\boxtimes$

Proposition 6.1.8 *Let Ω be a locally trivial Lie groupoid on a connected base M and let $\theta\colon M \times M \multimap \Omega$ be a local morphism. Then there is a unique morphism of Lie groupoids $F\colon \Pi(M) \to \Omega$ such that $F \circ \tau = \theta$.*

Proof Take a simple open cover $\{U_i\}$ of M such that $\bigcup_i (U_i \times U_i)$ is inside the domain of θ.

Consider a path ν in M. Choose points $0 = t_0 < t_1 < \cdots < t_n = 1$ such that, for the corresponding points $x_r = \nu(t_r)$ on the curve, we have

$(x_r, x_{r-1}) \in U_{i_r} \times U_{i_r}$ for some i_r for $n \geqslant r \geqslant 1$. Define

$$F(\nu) = \prod_{r=1}^{n} \theta_{i_r}(x_r, x_{r-1}),$$

where θ_i is the restriction of θ to $U_i \times U_i$.

Note first that if some x_r is moved to x_r' satisfying the same conditions, then the fact that θ_{r+1} and θ_r coincide on (x_r, x_r') means that the product is unchanged.

To show that F is well defined on homotopy classes, we need some terminology. If the restriction of ν to some subinterval is of the form $\mu^{\leftarrow}\mu$, call the path obtained by deleting $\mu^{\leftarrow}\mu$ a *revision* of ν; equally, call the path obtained by inserting a path of the form $\mu^{\leftarrow}\mu$ into ν a *revision* of ν. Next, say that ν and ν', with the same endpoints, are *equally good* if either can be obtained from the other by a finite number of revisions. Lastly, a *lasso* (for the cover $\{U_i\}$) is a loop of the form $\mu^{-1}\ell\mu$ where ℓ is a loop lying entirely in one U_i. We now need a standard lemma.

Lemma 6.1.9 *For any loop in M which is homotopic to zero, there is a finite product of lassos which is equally good.*

Proof Denote the loop by ν and let $h\colon I \times I \to M$ be the homotopy with $h(t,0) = \nu(t)$, $h(t,1) = x$, and $h(0,s) = h(1,s) = x$ for all $t, s \in I$. Subdivide $I \times I$ into n^2 equal closed squares so that the image of each under h lies in some U_i. For each $1 \leqslant r, s \leqslant n$ define a path $\lambda(r,s)$ in $I \times I$ to be the sequence of straight line segments as illustrated in Figure 6.1 (the separation between the initial two and final two segments is for clarity only).

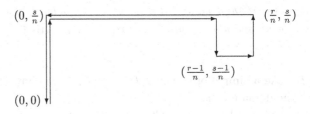

Fig. 6.1.

Now write $\nu(r,s) = \nu \circ \lambda(r,s)$ and define

$$\nu' = \nu(n,n)\nu(n-1,n)\cdots\nu(1,n)\nu(n,n-1)\cdots\nu(2,1)\nu(1,1).$$

This ν' is a product of lassos and is equally good as ν. $\qquad\square$

We resume the proof of 6.1.8. For a loop ℓ which is entirely within some U_i it is clear that $F(\ell)$ is an identity of Ω. It follows that F of any lasso is an identity of Ω and hence, from the lemma, that F maps all null–homotopic loops to identities. It thus quotients to a well defined morphism of groupoids $\Pi(M) \to \Omega$, also denoted F. Now for any x, y in a single U_i and any path ν from x to y within U_i, it is clear that $F(\nu) = \theta(y,x)$. This shows that $F \circ \tau = \theta$ and, by 1.3.6, that F is smooth. $\qquad\square$

In particular, if M is simply–connected, then θ extends uniquely to all of $M \times M$.

Theorem 6.1.10 *Let Ω and Ω' be locally trivial Lie groupoids on the same base M, and let $\theta \colon \Omega \dashrightarrow \Omega'$ be a local morphism over M. Then if Ω is α–connected and α–simply connected, θ extends to a global morphism $\Omega \to \Omega'$.*

Proof Let $p \colon P \to M$ be the restriction of β to an α–fibre of Ω. Then the pullback groupoid $p^{!!}\Omega$ is $P \times P$ and the pullback of θ is a G–equivariant local morphism

$$\widetilde{\theta} \colon P \times P \dashrightarrow p^{!!}\Omega'.$$

Using 6.1.12 below, it is straightforward to give a G–equivariant version of the proof of 6.1.8. The required global morphism is then obtained by quotienting over the action of G. $\qquad\square$

Definition 6.1.11 Let $P(M,G)$ be a principal bundle. A *G–simple cover* of P is a simple cover $\{U_i\}$ of P such that given i and given $g \in G$ there is a j such that $R_g(U_i) \subseteq U_j$. $\qquad\square$

Proposition 6.1.12 *Let $P(M,G)$ be a smooth principal bundle. Then $P(M,G)$ admits G–simple open covers.*

Proof Firstly note that P admits a G–invariant Riemannian metric since the vector bundle $\frac{TP}{G} \to M$ admits a fibre metric and any such metric can be pulled back to a G–invariant Riemannian metric on P via the fibrewise isomorphism $TP \to \frac{TP}{G}$.

Let $\{U_i\}$ be the set of all open subsets of P such that any two points lying in U_i can be joined by exactly one geodesic in U_i. It is standard that $\{U_i\}$ covers P and since the metric is invariant under the right action of G, it follows that $\{U_i\}$ is stable under G in the sense of 6.1.11. By construction, $\{U_i\}$ is simple. $\qquad\square$

Proposition 6.1.13 *Let Ω be a locally trivial and α–connected Lie groupoid. Then* m: $\mathscr{M}\Omega \to \Omega$ *has a local right–inverse.*

Proof Choose $m \in M$ and write $P = \Omega_m$, $G = \Omega_m^m$, $Q = \mathscr{M}\Omega|_m = \widetilde{P}$, $H = \mathscr{M}\Omega|_m^m$. Let $\{U_i\}$ be a G–simple cover of P. For each i choose $\xi_i \in U_i$, a path ν_i from 1_m to ξ_i within P and a function θ_i which to $\xi \in U_i$ assigns a path in U_i from ξ_i to ξ. Define $\sigma_i \colon U_i \to Q$ by $\sigma_i(\xi) = \langle \theta_i(\xi)\nu_i \rangle$ the homotopy class of the concatenation of ν_i followed by $\theta_i(\xi)$. We will show that $\varphi_i \colon \frac{U_i \times U_i}{G} \to \frac{Q \times Q}{H}$ by $\langle \xi', \xi \rangle \mapsto \langle \sigma_i(\xi'), \sigma_i(\xi) \rangle$ is a well–defined local morphism $\frac{P \times P}{G} \multimap \frac{Q \times Q}{H}$.

Consider $\sigma_j(\xi g)$ where $\xi \in U_i$, $g \in G$ and $R_g(U_i) \subseteq U_j$. Let

$$\lambda = R_g(\nu_i^{\leftarrow}\theta_i(\xi)^{\leftarrow})\theta_j(\xi g)\nu_j;$$

λ is a path in P from 1_m to g and thus defines an element $\langle \lambda \rangle$ of H. Clearly $\sigma_j(\xi g) = \sigma_i(\xi)\langle \lambda \rangle$. Let λ' denote the element obtained similarly from ξ'. To show that $\langle \lambda' \rangle = \langle \lambda \rangle$, it suffices to show that

$$R_g(\theta_i(\xi)^{\leftarrow})\theta_j(\xi g) \sim R_g(\theta_i(\xi')^{\leftarrow})\theta_j(\xi' g);$$

where both sides are paths from ξ_j to $\xi_i g$, as in Figure 6.2. Since

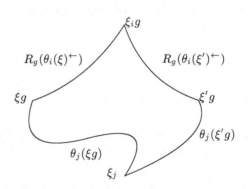

Fig. 6.2.

$R_g(U_i) \subseteq U_j$, it follows that all four paths in Figure 6.2 lie in U_j and so the deformation can be carried out.

Considering the σ_i as local sections of the bundle $Q(P, \pi_1 P)$, it is clear that their transition functions are constant, and the remainder of the proof follows as in 6.1.7. $\qquad\square$

Corollary 6.1.14 *Let Ω be a locally trivial α–connected Lie groupoid. Then $\mathscr{M}\Omega$ is locally isomorphic to Ω under* $\mathfrak{m}$.

As in the case of Lie groups, it follows that two α–connected, locally trivial Lie groupoids are locally isomorphic if and only if their monodromy groupoids are isomorphic.

6.2 Lie subalgebroids and morphisms

This section treats the correspondence between α–connected reductions of a locally trivial Lie groupoid and reductions of its Lie algebroid, and the correspondence between local morphisms of locally trivial Lie groupoids and morphisms of their Lie algebroids.

Theorem 6.2.1 *Let Ω be a locally trivial Lie groupoid on a connected base M and let $L' \!>\!\!-\!\!-\!\!> A' \longrightarrow\!\!\!\gg TM$ be a reduction of its Lie algebroid $L\Omega \!>\!\!-\!\!-\!\!> A\Omega \longrightarrow\!\!\!\gg TM$. Then there is a unique α–connected locally trivial Lie subgroupoid Φ of Ω such that $A\Phi = A'$.*

Proof The proof is modelled on the proof for Lie groups and we deal only with the features that are new.

Denote by Δ the inverse image bundle $\beta^* A'$ on Ω. Since $T^\alpha \Omega$ is a pullback of $A\Omega$ (see (34) on p. 121), it follows that there is a unique injective vector bundle morphism $\Delta \xrightarrow{i} T^\alpha \Omega$ over Ω such that

$$
\begin{array}{ccc}
T^\alpha \Omega & \xrightarrow{\;\;\mathscr{R}\;\;} & A\Omega \\
{\scriptstyle i}\big\uparrow & & \big\uparrow{\scriptstyle \subseteq} \\
\Delta & \xrightarrow{\hspace{2cm}} & A'
\end{array}
$$

commutes. Since $\Delta = \beta^* A'$ it follows that $\Gamma\Delta = C^\infty(\Omega) \otimes_{C^\infty(M)} \Gamma A'$ and using this one can mimic the standard proof for groups and show that Δ is an involutive distribution on Ω.

For each $x \in M$, let Δ_x be the restriction of Δ to Ω_x; thus Δ_x is an involutive distribution on Ω_x. Let Φ_x be the (connected) integral manifold of Δ which contains 1_x and let Φ be the union of the *sets*

$\Phi_x, x \in M$. For any $x, y \in M$ and $\xi \in \Omega_x^y$ it follows from the definition of Δ that $T(R_\xi^{-1})(\Delta_x) = \Delta_y$, and therefore if $\xi \in \Phi_x^y$ it follows that $R_\xi^{-1}(\Phi_y) = \Phi_x$. Thus Φ is a subgroupoid of Ω in the set sense.

Write α', β' for the restrictions of α, β to Φ. Since $a': A' \to TM$ is surjective by assumption, the restriction $\beta_x': \Phi_x \to M$, for $x \in M$, is a submersion, because $T_\xi(\beta_x')$, for $\xi \in \Phi_x$, is the composite

$$T_\xi(\Phi_x) = \Delta_\xi \overset{\cong}{\to} A'_{\beta\xi} \overset{a'_\xi}{\to} T_{\beta\xi}(M).$$

Hence each $\mathrm{im}(\beta_x')$, $x \in M$, is open in M. Since Φ is a groupoid in the set sense, the $\mathrm{im}(\beta_x')$, $x \in M$, partition M and since M is connected it follows that each β_x' is onto M. Thus Φ is transitive.

We now give Φ a smooth structure. Choose $m \in M$ and write $H = \Phi_m^m$. Since β_m' is a surjective submersion, H is a closed embedded sub–manifold of Φ_m. Also, β_m' has a family of local right–inverses $\sigma_i: U_i \to \Phi_m$, where the U_i cover M. Let

$$\Sigma_i^j: U_j \times H \times U_i \to \Phi_{U_i}^{U_j}$$

be the bijections defined using $\{\sigma_i\}$; to show that the overlap maps are smooth, it suffices to show that the transition functions $s_{ij}: U_{ij} \to H$ for $\{\sigma_i\}$ are smooth. Now, again using the fact that β_m' is a surjective submersion,

$$\Phi_m * \Phi_m = \{(\eta, \xi) \in \Phi_m \times \Phi_m \mid \beta'(\eta) = \beta'(\xi)\}$$

is a submanifold of $\Phi_m \times \Phi_m$, and hence of $\Omega_m \times \Omega_m$. Consider the restriction of

$$\delta': \Omega_m * \Omega_m \to \Omega_m, \qquad (\eta, \xi) \to \eta^{-1}\xi$$

to $\Phi_m * \Phi_m$. Since Φ_m is a leaf of the distribution Δ_m on Ω_m, it follows that $\Phi_m * \Phi_m \to \Omega_m$ is smooth as a map into Φ_m, and since H is an embedded submanifold of Φ_m, it follows that $\Phi_m * \Phi_m \to \Phi_m$ is smooth as a map into H. Hence $s_{ij}: U_{ij} \to H$, $x \mapsto \sigma_i(x)^{-1}\sigma_j(x)$ is smooth, as required. Also, it follows that $H \times H \to H$, $(h, h') \mapsto h^{-1}h'$ is smooth and so H is a Lie subgroup of Ω_m^m. The maps Σ_i^j now define a smooth structure on Φ with respect to which it is a Lie groupoid on M. Because the σ_i are smooth into Φ_m with its leaf differentiable structure, it follows that each $\Phi_x, x \in M$, inherits from Φ the differentiable structure it was originally given by Δ, and because the inclusion $\Phi \subseteq \Omega$ is locally (with respect to the Σ_i^j) represented by $\mathrm{id}_{U_j} \times (H \subseteq \Omega_m^m) \times \mathrm{id}_{U_i}$, it follows that Φ is a Lie subgroupoid of Ω.

The inclusion $\Phi \subseteq \Omega$ induces an injective Lie algebroid morphism $A\Phi \to A\Omega$ over M; since $A\Phi|_x = T_{1_x}(\Phi_x) = \Delta_{1_x} = A'_{x'}$ for all $x \in M$, it follows that $A\Phi \to A\Omega$ is a Lie algebroid isomorphism onto A'.

Suppose now that $\psi \colon \Psi \to \Omega$ is a Lie subgroupoid of Ω and that $\psi_*(A\Psi) = A'$. From the diagram

$$
\begin{array}{ccccc}
T^\alpha(\Psi) & & \Delta & \subseteq & T^\alpha(\Omega) \\
{\scriptstyle\mathscr{R}}\downarrow & & \downarrow & & {\scriptstyle\mathscr{R}}\downarrow \\
A\Psi & \xrightarrow{\ \psi_\star\ } & A' & \subseteq & A\Omega
\end{array}
$$

in which each vertical arrow is a pullback, it follows that $T^\alpha(\psi)$ is onto Δ and hence for each $x \in M$, $\psi_x(\Psi_x)$ is an integral manifold for Δ through 1_x. So there exists a smooth map $\varphi_x \colon \Psi_x \to \Phi_x$ such that the

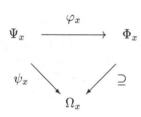

triangle commutes. It is easy to see that $\varphi_y \circ R_\xi = R_{\varphi_x(\xi)} \circ \varphi_x$ for each $\xi \in \Psi_x^y$, $x,y \in M$ and therefore $\varphi = \bigcup_x \varphi_x$ is a morphism of set groupoids $\Psi \to \Phi$ over M. Since Ψ and Φ are Lie and φ_x is smooth, it follows that φ is smooth. From the triangle it follows that φ_x is an injective immersion and so φ is an injective immersion.

Lastly, from $\psi_*(A\Psi) = A' = A\Phi$, it follows that Ψ and Φ have the same dimension, so φ is étale, and therefore, by 3.5.16, φ is onto Φ. This completes the proof of 6.2.1. $\qquad\square$

Corollary 6.2.2 *Let $F \colon \Omega \to \Omega'$, $f \colon M \to M'$ be a morphism of locally trivial Lie groupoids and let Φ be a reduction of Ω'. Then if F takes values in Φ, it is smooth as a map $\Omega \to \Phi$.*

Proof First assume that Ω is α–connected. Let Ψ be the α–identity component subgroupoid of Φ. Then for each $x \in M$, the submanifold $\Psi_{f(x)}$ of Ω' is a leaf of the foliation on Ω' defined by $A\Phi$. So $F_x \colon \Omega_x \to \Psi_{f(x)}$ is smooth. Hence, by 1.3.6(i), $F \colon \Omega \to \Psi$ is smooth; since Ψ is open in Φ it follows that $F \colon \Omega \to \Phi$ is smooth.

The case where Ω is not α–connected now follows from 1.5.2 and 1.3.6(ii). $\qquad\square$

* * * * *

We come now to the correspondence between local morphisms of Lie groupoids and morphisms of their Lie algebroids.

It is easy to extend the construction of the Lie functor to the case of smooth local morphisms $F\colon G \multimap G'$, $f\colon M \to M'$ of Lie groupoids. Denote the domain of F by $\mathscr{U}$. Then $T(F)\colon TG|_{\mathscr{U}} \to TG'$ restricts to $T^\alpha(F)\colon T^\alpha G|_{\mathscr{U}} \to T^\alpha G'$ and one can form the composition

$$
\begin{array}{ccccc}
AG & \longrightarrow & T^\alpha G|_{\mathscr{U}} & \overset{T^\alpha(F)}{\longrightarrow} & T^\alpha G' \\
\downarrow & & \downarrow & & \downarrow \\
M & \overset{1}{\longrightarrow} & \mathscr{U} & \overset{F}{\longrightarrow} & G'
\end{array}
\tag{2}
$$

and proceed as in §3.5. Note that now we have that $\overrightarrow{X}|_{\mathscr{U}} \overset{F}{\sim} \overrightarrow{X'}$ if and only if $F_* \circ X = X' \circ f$. However 3.5.10 holds without change. The following uniqueness result can now be proved in the most general setting.

Proposition 6.2.3 *Let $F_1, F_2\colon G \multimap G'$ be local morphisms of Lie gro-upoids over the same base map $f\colon M \to M'$. If $A(F_1) = A(F_2)\colon AG \to AG'$, then F_1 and F_2 are germ–equivalent. If F_1 and F_2 are global morphisms and G is α–connected then $A(F_1) = A(F_2)$ implies $F_1 = F_2$.*

Proof Let $\mathscr{U}$ be the intersection of the domains of F_1 and F_2. From $A(F_1) = A(F_2)$ it follows immediately that $T^\alpha(F_1)$ and $T^\alpha(F_2)$ coincide on $T^\alpha G|_{\mathscr{U}}$. Now the diagram

$$
\begin{array}{ccccc}
T_g(G_{\alpha g}) & \rightarrowtail & T_g(G) & \overset{T(\alpha)}{\twoheadrightarrow} & T_{\alpha g}(M) \\
{\scriptstyle T_g^\alpha(F)} \downarrow & & {\scriptstyle T_g(F)} \downarrow & & \downarrow {\scriptstyle T_{\alpha g}(f)} \\
T_{F(g)}(G'_{f(\alpha g)}) & \rightarrowtail & T_{F(g)}(G') & \overset{T(\alpha')}{\twoheadrightarrow} & T_{f(\alpha g)}(M')
\end{array}
$$

is valid for F_1 and F_2, so it follows that $T(F_1)$ and $T(F_2)$ coincide on $T(G)_{\mathscr{U}}$.

Now F_1 and F_2 are two maps $G \multimap G'$ which coincide on the closed embedded submanifold 1_M of G and whose tangent maps coincide on an open neighbourhood of 1_M; it follows that F_1 and F_2 themselves coincide on an open neighbourhood of 1_M.

The second assertion follows from 1.5.8. $\qquad\square$

We now return to locally trivial Lie groupoids.

Theorem 6.2.4 *Let Ω and Ω' be locally trivial Lie groupoids on base*

M, and assume that Ω is α–connected and α–simply connected. Then if $\varphi\colon A\Omega \to A\Omega'$ is a morphism of Lie algebroids over M, there is a morphism $F\colon \Omega \to \Omega'$ of Lie groupoids over M such that $F_* = \varphi$.

Proof Define $\overline{\varphi}\colon A\Omega \to A\Omega \oplus_{TM} A\Omega'$ by $X \mapsto X \oplus \varphi(X)$. Then $\overline{\varphi}$ is an injective vector bundle morphism over M, and so its image, $\mathrm{im}(\overline{\varphi})$, is a vector subbundle of $A\Omega \oplus_{TM} A\Omega'$. It is easy to see that $\mathrm{im}(\overline{\varphi})$ is a transitive Lie subalgebroid of $A\Omega \oplus_{TM} A\Omega'$. Therefore, by 6.2.1, there is a unique α–connected Lie subgroupoid Φ of $\Omega \times_{M \times M} \Omega'$ such that $A\Phi = \mathrm{im}(\overline{\varphi})$. Let π denote the restriction of the projection $\Omega \times_{M \times M} \Omega' \to \Omega$ to Φ. Then $\pi_*\colon A\Phi \to A\Omega$ is $X \oplus \varphi(X) \mapsto X$ and is evidently an isomorphism of Lie algebroids over M.

From 6.2.5 below it follows that each $\pi_x\colon \Phi_x \to \Omega_x$ is a covering and since Φ and Ω are α–connected and Ω is α–simply connected, it follows that each π_x is a diffeomorphism. So, by 3.5.15, π itself is a diffeomorphism and it is easy to check now that $F = \pi_2 \circ \pi^{-1}\colon \Omega \to \Omega'$ has the required properties. $\square$

Proposition 6.2.5 *Let $F\colon \Omega \to \Omega'$ be a morphism of locally trivial Lie groupoids over M. If $F_*\colon A\Omega \to A\Omega'$ is an isomorphism and Ω and Ω' are α–connected, then each $F_m\colon \Omega_m \to \Omega'_m$ is a covering. In particular, F is a surjective submersion.*

Proof Use the notation $P = \Omega_m$, $G = \Omega_m^m$, $Q = \Omega'_m$, $H = \Omega'^m_m$ and $K = \ker(F_m^m)$.

From 3.5.15 it follows that F is étale and so $F(\Omega)$ is open in Ω'. By 1.5.8, it follows that F is onto, and so $P(Q, K, \varphi_m^m)$ is a principal bundle. Next, $F_*\colon A\Omega \to A\Omega'$ is an isomorphism and so, by 3.3.12, $(F_m^m)_*\colon \mathfrak{g} \to \mathfrak{h}$ is an isomorphism and so $K \leqslant G$ is discrete. Since P and Q are connected, it follows that F_m is a covering. $\square$

Theorem 6.2.6 *Let Ω and Ω' be locally trivial Lie groupoids on M and let $\varphi\colon A\Omega \to A\Omega'$ be a morphism of Lie algebroids over M. Then there is a local morphism $F\colon \Omega \dashrightarrow \Omega'$ of Lie groupoids over M, such that $F_* = \varphi$.*

Proof It is no loss of generality to assume that Ω is α–connected. Let $\psi\colon \mathscr{M}\Omega \to \Omega$ be the projection of the monodromy groupoid, and apply 6.2.4 to $\varphi \circ \psi_*\colon A\mathscr{M}\Omega \to A\Omega'$. There is then a morphism $f\colon \mathscr{M}\Omega \to \Omega'$ over M such that $f_* = \varphi \circ \psi_*$. Now, by 6.1.13, ψ has a local right–inverse morphism $\chi\colon \Omega \multimap M\Omega$, and $f \circ \chi\colon \Omega \multimap \Omega'$ is now a local morphism over M with $(f \circ \chi)_* = \varphi \circ \psi_* \circ \chi_* = \varphi$. $\qquad\square$

Corollary 6.2.7 *Let* $F\colon \Omega \multimap \Omega'$ *be a local morphism of locally trivial Lie groupoids over* M. *Then* F *is a local isomorphism if and only if* $F_*\colon A\Omega \to A\Omega'$ *is an isomorphism.*

Proof Follows from 6.2.6 and 6.2.3. $\qquad\square$

Remark 6.2.8 Consider a locally trivial Lie groupoid Ω on a connected base M. A flat infinitesimal connection $\gamma\colon TM \to A\Omega$ is a morphism of Lie algebroids and so, by 6.2.4, integrates to a local morphism of Lie groupoids $\theta\colon M \times M \multimap \Omega$. By 6.1.7, local morphisms $M \times M \multimap \Omega$ are equivalent to section–atlases whose transition functions are constant, and this argument therefore gives an alternative proof of 5.4.6(ii).

Since $A\Pi(M) \cong TM$, a flat infinitesimal connection γ may also be integrated to a global morphism $\Pi(M) \to \Omega$ which we call the *holonomy morphism* of γ. Conversely, any Lie groupoid morphism $\Pi(M) \to \Omega$ differentiates to a flat infinitesimal connection in Ω. If M is simply–connected, then $\Pi(M) = M \times M$ and since morphisms $M \times M \to \Omega$ correspond to global sections of a vertex bundle, we have the following:

Proposition 6.2.9 *Let* Ω *be a locally trivial Lie groupoid on a simply–connected base* M. *If* Ω *admits a flat infinitesimal connection, then it is trivializable.*

A further local description of morphisms is given at the end of §6.4.

Example 6.2.10 Let Ω be an α–connected locally trivial Lie groupoid on M and let $\rho\colon A\Omega \to \mathfrak{D}(E)$ be a representation of $A\Omega$ in a vector bundle E on M. Then, by 6.2.4, there is a representation $\mathsf{P}\colon \mathscr{M}\Omega \to \Phi(E)$ of the monodromy groupoid of Ω in E with $\mathsf{P}_* = \rho$. In the case where Ω is $M \times M$, the monodromy groupoid of Ω is $\Pi(M)$ and a representation of $\Pi(M)$ in E is precisely a local system of coefficients on M with values in E. In analogy with this case we call a representation $\mathscr{M}\Omega \to \Phi(E)$ a *local system of coefficients for* Ω *with values in* E.

Note in particular that a flat connection in a vector bundle E on M constitutes a local system of coefficients with values in E. $\qquad\boxtimes$

It is a simple matter to extend 6.2.4 to base–changing morphisms. Let $\Omega \rightrightarrows M$ and $\Omega' \rightrightarrows M'$ be locally trivial Lie groupoids, with Ω

α–connected and α–simply connected. Let $\varphi\colon A\Omega \to A\Omega'$ be a morphism of Lie algebroids over $f\colon M \to M'$. Since Ω' is locally trivial, $f^{\Downarrow}\Omega' \rightrightarrows M$ exists and $A(f^{\Downarrow}\Omega') \cong f^{!!}(A\Omega')$; see 4.3.11. By 4.3.6 the map $\varphi^{!!}\colon A\Omega \to f^{!!}(A\Omega')$ is a morphism of Lie algebroids over M and so, by 6.2.4, integrates to $\widetilde{F}\colon \Omega \to f^{\Downarrow}\Omega'$. Now, with the notation of 2.3.1, composing with $f^{\Downarrow}\colon f^{\Downarrow}\Omega' \to \Omega'$ gives $F\colon \Omega \to \Omega'$ with $A(F) = \varphi$. This proves:

Theorem 6.2.11 *Let* $\Omega \rightrightarrows M$ *and* $\Omega' \rightrightarrows M'$ *be locally trivial Lie groupoids, and assume that* Ω *is* α–*connected and* α–*simply connected. Let* $\varphi\colon A\Omega \to A\Omega'$ *be a morphism of Lie algebroids over* $f\colon M \to M'$. *Then there is a morphism* $F\colon \Omega \to \Omega'$ *of Lie groupoids over* f *such that* $A(F) = \varphi$.

*   *   *   *   *

Two items in the Lie theory of Lie groups and Lie algebras which do not generalize to locally trivial Lie groupoids and transitive Lie algebroids are the correspondence between connected normal subgroups and ideals, and the result that a connected Lie group with abelian Lie algebra is abelian.

Concepts of ideal for Lie algebroids are treated in §4.4 and §6.5, but here we need only assume that a transitive Lie algebroid is an ideal of itself. Then the α–connected Lie subgroupoid of Ω corresponding to $A\Omega$ itself is the identity component subgroupoid Ψ of Ω and 1.5.5 shows that Ψ need not be normal in Ω.

If Ω is an α–connected locally trivial Lie groupoid with abelian Lie algebroid then the adjoint bundle $L\Omega$ is abelian and so the identity components of the vertex groups of Ω are abelian. If the base M is simply–connected then the vertex groups of Ω must be connected (by the long exact homotopy sequence for the vertex bundles), and so Ω is abelian. However if M is not simply–connected, then the vertex groups of Ω need not be abelian, and $I\colon \Omega \to \Phi(\mathscr{I}\Omega)$ need not quotient to $M \times M \to \Phi(\mathscr{I}\Omega)$. It is however true that if $A\Omega$ is abelian, then $\mathrm{Ad}\colon \Omega \to \Phi_{\mathrm{Aut}}(L\Omega)$ factorizes as $\mathrm{Ad} = h \circ \widetilde{\chi}$ where $h\colon \Pi(M) \to \Phi_{\mathrm{Aut}}(L\Omega)$ is the holonomy morphism of the flat connection defined by ad and $\widetilde{\chi}\colon \Omega \to \Pi(M)$ integrates the anchor $A \to TM$.

6.3 Path connections

When a transitive Lie algebroid A is the Lie algebroid of a (necessarily locally trivial) Lie groupoid $\Omega \rightrightarrows M$, a connection in A induces a means of lifting paths in M to paths in Ω; the abstract formulation of this process is the concept of path connection. Whereas flat infinitesimal connections can be integrated to local morphisms as in 6.2.8 or 5.4.6, the concept of path connection is what is needed in order to formulate the data of an infinitesimal connection solely in terms of the groupoid.

Consider a locally trivial Lie groupoid Ω on a base M; we assume M to be connected. We use the notations $\mathscr{P}^\alpha(\Omega)$ and $\mathscr{P}^\alpha_M(\Omega)$ of §6.1, although we do not assume that Ω is α–connected. Similarly we denote by $\mathscr{P}(M)$ the set of all continuous and piecewise smooth paths $\nu\colon I \to M$.

Definition 6.3.1 A *path connection* in Ω is a map $\mathscr{P}(M) \to \mathscr{P}^\alpha_M(\Omega)$, denoted Γ and usually written $\nu \mapsto \overline{\nu}$, satisfying the following conditions:

(i) $\overline{\nu}(0) = 1_{\nu(0)}$ and $\beta \circ \overline{\nu} = \nu$;

(ii) if $\varphi\colon [0,1] \to [a,b] \subseteq [0,1]$ is a diffeomorphism then

$$\overline{\nu \circ \varphi} = R_{\overline{\nu}(\varphi(0))^{-1}} \circ (\overline{\nu} \circ \varphi);$$

(iii) if $\nu \in \mathscr{P}(M)$ is smooth at $t_0 \in I$ then $\overline{\nu}$ is also smooth at t_0;

(iv) if $\nu_1, \nu_2 \in \mathscr{P}(M)$ have $\frac{d\nu_1}{dt}(t_0) = \frac{d\nu_2}{dt}(t_0)$ for some $t_0 \in I$, then

$$\frac{d\overline{\nu}_1}{dt}(t_0) = \frac{d\overline{\nu}_2}{dt}(t_0);$$

(v) if $\nu_1, \nu_2, \nu_3 \in \mathscr{P}(M)$ are such that $\frac{d\nu_1}{dt}(t_0) + \frac{d\nu_2}{dt}(t_0) = \frac{d\nu_3}{dt}(t_0)$ for some $t_0 \in I$, then

$$\frac{d\overline{\nu}_1}{dt}(t_0) = \frac{d\overline{\nu}_2}{dt}(t_0) + \frac{d\overline{\nu}_3}{dt}(t_0).$$

$\square$

From (i) it follows that $\overline{\nu}$ is a path in $\Omega_{\nu(0)}$ with $\overline{\nu}(t) \in \Omega^{\nu(t)}_{\nu(0)}$ for all t. We call $\overline{\nu}$ the Γ–*lift* of ν.

(ii) is the *reparametrization condition*. It is geometrically natural inasmuch as it guarantees that the images of paths in M can be meaningfully lifted to the images of α–paths in Ω. It also allows the definition of $\overline{\nu}$ to be extended to open paths of the form $\nu\colon (-\varepsilon, \varepsilon) \to M$, or $\nu\colon \mathbb{R} \to M$, and so to lift local 1–parameter groups of local transformations on M.

Lemma 6.3.2 *Let Γ be a path connection in Ω and let $\nu\colon \mathbb{R} \to M$ be continuous and piecewise smooth. Then there is a unique continuous and piecewise smooth $\bar{\nu}\colon \mathbb{R} \to \Omega$, with $\alpha(\bar{\nu}(t)) = \nu(0)$ for all t, $\beta \circ \bar{\nu} = \nu$, $\bar{\nu}(0) = 1_{\nu(0)}$ and such that for every closed interval $[a, b] \subseteq \mathbb{R}$, the restriction $\bar{\nu}|_{[a,b]}$ is $R_{\bar{\nu}(a)} \circ \Gamma(\nu|_{[a,b]})$, both restrictions being reparametrized by the same diffeomorphism $[0, 1] \to [a, b]$.*

Proof The path $\bar{\nu}$ is most conveniently defined by lifting a suitably reparametrised $\nu|_{[n,n+1]}$ for each $n \in \mathbb{Z}$, and right–translating the results, so that the relevant endpoints match. The uniqueness result is then easy to see. $\square$

We also refer to $\bar{\nu}$, for $\nu\colon \mathbb{R} \to M$ or $\nu\colon (-\varepsilon, \varepsilon) \to M$, as the Γ–*lift* of ν.

The reparametrisation condition also guarantees that Γ preserves the algebraic operations on the sets of paths. The numbering of the following three results continues the numbering in 6.3.1.

Proposition 6.3.3 *Let Γ be a path connection in Ω. Then:*

(vi) $\bar{\kappa}_x = \kappa_{1_x}$ *(where κ_p denotes the path constant at p).*

(vii) $\overline{\nu^{\leftarrow}} = R_{\bar{\nu}(1)^{-1}} \circ (\bar{\nu})^{\leftarrow}$ *(where $\nu^{\leftarrow}$ denotes the reversal, $\nu^{\leftarrow}(t) = \nu(1 - t)$, of ν).*

(viii) $\overline{\nu'\nu} = (R_{\nu(1)} \circ \bar{\nu'})\bar{\nu}$ *(where juxtaposition denotes the usual concatenation of paths).*

Proof For (vi), take $t \in (0, 1]$ and define $\rho_t\colon [0, 1] \to [0, t]$ by $s \mapsto st$. Then

$$\bar{\kappa}_x(t) = (\bar{\kappa}_x \circ \rho_t)(1) = (R_{\bar{\kappa}_x(0)} \circ (\overline{\kappa_x \circ \rho_t}))(1) = \overline{\kappa_x \circ \rho_t}(1)\bar{\kappa}_x(0)$$

and since $\bar{\kappa}_x(0) = 1_x$ and $\kappa_x \circ \rho_s = \kappa_x$, this is just $\kappa_x(1)$. So $\bar{\kappa}_x(1) = \bar{\kappa}_x(t)$ for all $t > 0$, and since $\bar{\kappa}_x$ is continuous, it follows that $\bar{\kappa}_x(1) = \bar{\kappa}_x(0)$ also.

For (vii), use (ii) with $\varphi(t) = 1 - t$. The proof of (viii) is similar. $\square$

Returning to 6.3.1, we refer to (iii) and (iv) as the *tangency conditions* and to (v) as the *additivity condition*.

Proposition 6.3.4 *Let $\Gamma\colon \nu \mapsto \bar{\nu}$ be a path connection in Ω. Then:*

(ix) *if $\nu_1, \nu_2 \in \mathscr{P}(M)$ are such that $\frac{d\nu_1}{dt}(t_0) = k\frac{d\nu_2}{dt}(t_0)$ for some $t_0 \in I$ and $k \in \mathbb{R}$, then $\frac{d\bar{\nu_1}}{dt}(t_0) = k\frac{d\bar{\nu_2}}{dt}(t_0)$;*

(x) *if* $\varphi_t\colon U \times (-\varepsilon, \varepsilon) \to M$ *is a local 1–parameter group of local diffeomorphisms on* M, *then the map* $\overline{\varphi}_t\colon \Omega^U \times (-\varepsilon, \varepsilon) \to \Omega$ *constructed in the proof is a local 1–parameter group of local diffeomorphisms on* Ω, *and*

$$\beta \circ \overline{\varphi}_t = \varphi_t \circ \beta$$

for all $t \in (-\varepsilon, \varepsilon)$.

Proof (ix) follows from the reparametrization condition on Γ.

We prove (x) in the case that the flow is global, for ease of notation. Clearly $\overline{\varphi}_t(\xi) \in \Omega_{\alpha\xi}^{\varphi_t(\beta\xi)}$, and this establishes the last equation. Since $\varphi_t\colon M \to M$ is smooth, and β is a surjective submersion, it follows that $\overline{\varphi}_t$ is smooth; once we have established the group property, it will follow that $\overline{\varphi}_t$ is a diffeomorphism. Likewise, to prove that $\overline{\varphi}\colon \mathbb{R} \times \Omega \to \Omega$ is smooth, observe that

$$
\begin{array}{ccc}
\mathbb{R} \times \Omega & \xrightarrow{\overline{\varphi}} & \Omega \\
{\scriptstyle \mathrm{id} \times \beta} \downarrow & & \downarrow {\scriptstyle \beta} \\
\mathbb{R} \times M & \xrightarrow{\varphi} & M
\end{array}
$$

commutes.

It remains to prove the group property. Given $\xi \in \Omega$ and $s, t \in \mathbb{R}$, consider the curves $t \mapsto \overline{\varphi}_{t+s}(\xi)$ and $t \mapsto \overline{\varphi}_t(\overline{\varphi}_s(\xi))$. Clearly both project under β to the curve $t \mapsto \varphi_{t+s}(\xi) = \varphi_t(\varphi_s(\xi))$ in M and both have value $\overline{\varphi}_s(\xi)$ at $t = 0$. It is easily checked that both are Γ–lifts of the corresponding curve in M. So, by uniqueness, $\overline{\varphi}_{t+s}(\xi) = \overline{\varphi}_t(\overline{\varphi}_s(\xi))$. $\qquad\square$

Theorem 6.3.5 *There is a bijective correspondence between path connections* $\Gamma\colon \nu \mapsto \overline{\nu}$ *in* Ω *and infinitesimal connections* $\gamma\colon TM \to A\Omega$ *in* $A\Omega$, *such that a corresponding* Γ *and* γ *are related by*

$$\frac{d}{dt}\overline{\nu}(t_0) = T(R_{\overline{\nu}(t_0)})\left(\gamma\left(\frac{d}{dt}\nu(t_0)\right)\right), \qquad \nu \in \mathscr{P}(M), \ t_0 \in I. \quad (3)$$

Proof Suppose given a path connection Γ. For $X \in T_x(M)$ take any $\nu \in \mathscr{P}(M)$ with $\nu(t_0) = x$ and $\frac{d\nu}{dt}(t_0) = X$ for some $t_0 \in I$, and define

$$\gamma(X) = T(R_{\overline{\nu}(t_0)^{-1}})\left(\frac{d}{dt}\overline{\nu}(t_0)\right).$$

Since $\alpha \circ \overline{\nu}$ is constant, the RHS is defined, and lies in $T_{1_x}(\Omega_x)$. By the tangency conditions, $\gamma(X)$ is well defined with respect to the choice of ν. By (v) and (ix), $\gamma\colon T_x(M) \to A\Omega|_x$ is $\mathbb{R}$–linear.

Now let X be a vector field on M. We prove that $\gamma(X)$ is a smooth section of $A\Omega$. Let $\varphi_t\colon U \times (-\varepsilon, \varepsilon) \to M$ be a local flow for X and let $\overline{\varphi}_t\colon \Omega^U \times (-\varepsilon, \varepsilon) \to \Omega$ be the Γ–lift of $\{\varphi_t\}$. Let X^* be the (local) vector field on Ω derived from $\{\overline{\varphi}_t\}$. Then, from the definition of $\overline{\varphi}_t$ in 6.3.4, it is clear that X^* is right–invariant. From the definition of γ above, it is clear that X^* is the (local) vector field on Ω associated to $\gamma(X)$; in the notation of 3.5.9, $X^* = \mathbf{X}^* = \gamma(X)|_U$. Hence, by 3.5.9, $\gamma(X)$ is smooth. Therefore $\gamma\colon TM \to A\Omega$ is smooth.

Conversely, suppose given an infinitesimal connection $\gamma\colon TM \to A\Omega$, and a path $\nu \in \mathscr{P}(M)$. Let $\sigma\colon U \to \Omega_m$ be a local decomposing section of Ω with $\nu(0) \in U$, and let $\Sigma\colon U \times G \times U \to \Omega_U^U$ be the corresponding chart. Let $t_1 > 0$ be such that $\nu([0, t_1]) \subseteq U$ and ν is smooth on $[0, t_1]$. Write $\overline{\nu}$ on $[0, t_1]$ as $\Sigma \circ (\nu, a, \nu(0))$, where $a\colon [0, t_1] \to G$ has $a(0) = 1$. Then (3) becomes

$$\frac{da}{dt} = T(R_{a(t)}) \left(\omega \left(\frac{d\nu}{dt} \right) \right), \tag{4}$$

where $\omega \in \Omega^1(U, \mathfrak{g})$ is the local connection form of γ with respect to σ. In terms of the right–derivative of a this can be rewritten as

$$\Delta(a) = \nu^*\omega \tag{5}$$

where $\nu^*\omega \in \Omega^1([0, t_1], \mathfrak{g})$ is the pullback of ω. Now $\nu^*\omega$ is a Maurer–Cartan form, since $[0, t_1]$ is 1–dimensional, and so there is a unique smooth solution a to (5) on $[0, t_1]$ with $a(0) = 1$. Since ν has only a finite number of points where it is not smooth, and since $\nu(I)$ is covered by a finite number of domains of decomposing sections $\sigma_i\colon U_i \to \Omega_m$, this process yields a curve $\overline{\nu} \in \mathscr{P}_M^\alpha(\Omega)$ satisfying (3) and with properties (i), (iii), (iv) and (v). The remaining property, (ii), is easily seen from the form of (4) and the uniquenss of its solutions. $\qquad\square$

Corollary 6.3.6 (of the proof) *Let* $\gamma\colon TM \to A\Omega$ *be an infinitesimal connection in* Ω *and let* Γ *be the corresponding path connection. Then, for all* $X \in \mathfrak{X}(M)$,

$$\text{Exp } t\gamma(X)(x) = \Gamma(\varphi, x)(t)$$

where $\{\varphi_t\}$ *is a local flow for* X *near* x, *and* $\Gamma(\varphi, x)\colon \mathbb{R} \circ\!\!\to \Omega_x$ *is the lift of* $t \mapsto \varphi_t(x)$.

Proof This follows from the definition of γ in terms of Γ in the proof and from the construction of $\overline{\varphi}$ in 6.3.4.   □

Theorem 6.3.5 establishes that the relationship between path connections and infinitesimal connections is comparable to that between morphisms of Lie groupoids and morphisms of Lie algebroids. The failure of an infinitesimal connection to be a morphism is measured by its curvature; we turn now to the corresponding global concept for path connections.

Definition 6.3.7 For $\nu \in \mathscr{P}(M)$, the element $\widehat{\nu} = \overline{\nu}(1) \in \Omega$ is the *holonomy* of ν.   □

The next result is an immediate corollary of 6.3.3.

Proposition 6.3.8 *Let Γ be a path connection in Ω. Then*

$$(\text{vi}') \quad \widehat{\kappa_x} = 1_x, \qquad (\text{vii}') \quad \widehat{\nu^{\leftarrow}} = (\widehat{\nu})^{-1}, \qquad (\text{viii}') \quad \widehat{\nu'\nu} = \widehat{\nu'}\widehat{\nu}.$$

Definition 6.3.9 Let Γ be a path connection in Ω. Then

$$\Theta = \Theta(\Gamma) = \{\widehat{\nu} \mid \nu \in \mathscr{P}(M)\}$$

is the *holonomy subgroupoid* of Γ. The vertex group Θ_x^x at $x \in M$ is the *holonomy group* of Γ at x. The vertex bundle at $x \in M$ is the *holonomy bundle* of Γ at x.   □

From 6.3.8 it is clear that Θ is a wide subgroupoid of Ω; since M is path–connected, Θ is transitive.

Proposition 6.3.10 *Let Γ be a path connection in Ω. Then for each $\nu \in \mathscr{P}(M)$, the Γ–lift $\overline{\nu}$ lies entirely in Θ and, in particular, Θ is an α–connected subset.*

Proof For each $t \in (0,1]$ consider $\nu_t = \nu \circ \rho_t$ where ρ_t is $s \mapsto st$ as in the proof of 6.3.3(iii). Then $\widehat{\nu_t} = \overline{\nu}(t)$ and this establishes $\overline{\nu}(t) \in \Theta$.   □

This is also, of course, true for any open path $\nu\colon (-\varepsilon, \varepsilon) \to M$ or $\nu\colon \mathbb{R} \to M$.

Example 6.3.11 The pair groupoid $M \times M$ admits a single path connection, namely $\Gamma(\nu)(t) = (\nu(t), \nu(0))$.

Slightly less trivially, the fundamental groupoid $\Pi(M)$ admits a single path connection, namely $\Gamma(\nu)(t) = \langle \nu \circ \rho_t \rangle$, where $\rho_t \colon [0,1] \to [0,t]$ is the reparametrization used in 6.3.3.

In both cases the holonomy subgroupoid is the whole groupoid. ⊠

Definition 6.3.12 Let Ω and Ω' be locally trivial Lie groupoids on M. Let $\varphi \colon \Omega \to \Omega'$ be a morphism of Lie groupoids over M, and let Γ be a path connection in Ω.

Then $\varphi \circ \Gamma$ denotes the map $\nu \mapsto \varphi \circ \Gamma(\nu)$, which is easily seen to be a path connection in Ω', and is called the *produced path connection* in Ω'. □

The proof of the following is immediate.

Proposition 6.3.13 *Let* $\varphi \colon \Omega \to \Omega'$ *be a morphism of locally trivial Lie groupoids over* M. *Let* Γ *be a path connection in* Ω *and let* $\Gamma' = \varphi \circ \Gamma$ *be the produced path connection. Then* $\varphi(\Theta) = \Theta'$, *where* Θ *and* Θ' *are the holonomy subgroupoids for* Γ *and* Γ'.

Proposition 6.3.14 *Let* $F \colon \Omega \to \Omega'$ *be a morphism of locally trivial Lie groupoids over* M, *let* γ *be an infinitesimal connection in* Ω, *and write* Γ *for the corresponding path connection. Write* γ' *for the produced connection,* $\gamma' = F_* \circ \gamma$. *Then the path connection* Γ' *associated to* γ' *is the produced path connection,* $\Gamma' = F \circ \Gamma$, *and* $F(\Theta) = \Theta'$.

Proof Take $\nu \in \mathscr{P}(M)$. Then, since $\bar{\nu}$ satisfies equation (3) for γ, it immediately follows that $F \circ \bar{\nu}$ satisfies equation (3) for $F_* \circ \gamma$. Thus $\Gamma' = F \circ \Gamma$ and $F(\Theta) = \Theta'$. □

Corollary 6.3.15 *If* Ω' *is a reduction of* Ω *and* $\gamma \colon TM \to A\Omega$ *takes values in* $A\Omega'$, *then* $\Theta \leqslant \Omega'$.

Proof This is a special case of 6.3.14. □

Example 6.3.16 Let Ω be an α–connected locally trivial Lie groupoid on M, and let Γ be a path connection in Ω. Then there is a unique path connection $\tilde{\Gamma}$ in the monodromy groupoid $\mathscr{M}\Omega$ such that the monodromy projection $\psi \colon \mathscr{M}\Omega \to \Omega$ maps $\tilde{\Gamma}$ to Γ.

Namely, given $\nu \in \mathscr{P}(M)$, define $\bar{\bar{\nu}} = \tilde{\Gamma}(\nu)$ by $\bar{\bar{\nu}}(t) = \langle \bar{\nu} \circ \rho_t \rangle = \langle \overline{\nu \circ \rho_t} \rangle$ where, again, ρ_t is $[0,1] \to [0,t]$, $s \mapsto st$.

The uniqueness of $\widetilde{\Gamma}$ follows from the uniqueness of lifts across the universal covering projections $\mathcal{M}\Omega|_x \to \Omega_x$.

If Ω and Ω' are locally trivial Lie groupoids on M and $F\colon \Omega \rightsquigarrow \Omega'$ is a local morphism, then a path connection Γ in Ω induces an infinitesimal connection γ, which induces a produced infinitesimal connection γ' in Ω', which may be integrated to a path connection $\widetilde{\Gamma}$ in $\mathcal{M}\Omega$. In this case $F(\Theta) \subseteq \Theta'$, but equality need not hold.   ⊠

Example 6.3.17 Let (E, q, M) be a vector bundle on a connected manifold M, and let Γ be a path connection in $\Phi(E)$. Then, for $\nu \in \mathscr{P}(M)$ and $t \in I$, $\overline{\nu}(t)$ is an isomorphism $E_{\nu(0)} \to E_{\nu(t)}$, generally known as *parallel translation along ν*.

From 6.3.6 and 3.6.6, it follows that

$$(\nabla_X(\mu))(x) = -\left.\frac{d}{dt}\overline{\varphi}_{\bullet}(x)(\mu(\varphi_{-t}(x)))\right|_0 . \tag{6}$$

Here φ_t is a local flow for X around x and $\overline{\varphi}_{\bullet}(x)$ is the lift of the path $\varphi_t(x)$ through x.   ⊠

Example 6.3.18 Let $P(M, G, \pi)$ be a principal bundle with gauge groupoid $\Omega = \frac{P \times P}{G}$. Let Γ be a path connection in Ω. For any given $\nu \in \mathscr{P}(M)$, the lift $\overline{\nu}$ starts at the identity $1_{\nu(0)}$ which can be written as $\langle u, u \rangle$ for any in $u \in \pi^{-1}(\nu(0))$. Fix such a u. Since $\alpha\overline{\nu}(t) = \nu(0)$ for all t, each $\overline{\nu}(t)$ can be written as $\langle \nu^\star(t), u \rangle$ with $\nu^\star(t)$ uniquely determined by $\overline{\nu}(t)$ and u. From $\beta\overline{\nu}(t) = \nu(t)$, it follows that $\pi(\nu^\star(t)) = \nu(t)$. Clearly $\nu^\star(0) = u$. We call $\nu^\star \in \mathscr{P}(P)$ the Γ–*lift* of ν starting at u, and will denote it by $\Gamma(\nu; u)$ when it is necessary to indicate the dependence of $\nu^\star$ on u. This Γ is a map $\mathrm{ev}_0^\star P \to \mathscr{P}(P)$, where $\mathrm{ev}_0^\star P$ is the pull–back bundle of $P(M, G)$ over $\mathrm{ev}_0\colon \mathscr{P}(M) \to M$, $\nu \mapsto \nu(0)$.

If, instead of $u \in \pi^{-1}(\nu(0))$, a second choice $u' \in \pi^{-1}(\nu(0))$ is made then $u' = ug$ for some $g \in G$ and from $\langle \nu^\star(t), u \rangle = \langle \nu^\star(t)g, ug \rangle$ it follows that $\Gamma(\nu; ug) = R_g \circ \Gamma(\nu; u)$.

Now consider the holonomy subgroupoid Θ of Γ. The holonomy group Θ_x^x at $x \in M$ consists of the set of all $\widehat{\ell}$ where ℓ is a loop in M at x. Choose $u \in \pi^{-1}(x)$ as reference point for all ℓ at x; then $\overline{\ell}(0) = \langle u, u \rangle$ and $\widehat{\ell} = \overline{\ell}(1) = \langle \ell^\star(1), u \rangle$, where $\ell^\star = \Gamma(\ell; u)$. Since $\pi\ell^\star(1) = \ell(1) = x$, there is a unique $g \in G$ such that $\ell^\star(1) = ug$ and this g, which is the holonomy of ℓ with reference point $u \in \pi^{-1}(x)$, corresponds to $\widehat{\ell} \in \Theta_x^x$ under the isomorphism of 1.3.5(i), using u as reference point. Similarly it may be seen that Θ_x corresponds, under the isomorphism of 1.3.5(i), with $u \in \pi^{-1}(x)$ as reference point, to the holonomy bundle of $P(M, G)$ through u. ⊠

Return to a general locally trivial Lie groupoid Ω on M. Let $\mathfrak{L}(M) \subseteq \mathscr{P}(M)$ denote the (piecewise smooth) loops in M and for $x \in M$ let $\mathfrak{L}_x(M) \subseteq \mathfrak{L}(M)$ denote the loops in M based at x. Denote by $\mathfrak{L}_x^0(M)$ the set of loops based at x which are (piecewise smoothly) contractible to the constant loop at x.

Then the holonomy group of Γ at x is

$$\Theta_x^x = \{\widehat{\ell} \mid \ell \in \mathfrak{L}_x(M)\};$$

denote this temporarily by H_x. Likewise write

$$H_x^0 = \{\widehat{\ell} \mid \ell \in \mathfrak{L}_x^0(M)\};$$

this is the *restricted holonomy group* of Γ at x. Since the concatenation of contractible loops is contractible, and the conjugate of a contractible loop by any loop is contractible, it follows that H_x^0 is normal in H_x. Furthermore, the map $\mathfrak{L}_x(M) \to H_x$ quotients to a morphism

$$\pi_1(M, x) \to H_x/H_x^0$$

which is surjective.

Theorem 6.3.19 *Let Γ be a path connection in Ω. Then the holonomy groupoid Θ of Γ has a natural structure of Lie subgroupoid of Ω.*

Proof The main work of the proof is to prove that the holonomy groups are Lie groups. First we prove that H_x^0 is a path–connected subgroup of H_x for $x \in M$. By the same argument as in the proof of 6.1.8, it is sufficient to show that, for a loop ℓ based at x and lying within a single U_i, the contraction of ℓ to κ_x gives a path from $\widehat{\ell}$ to 1_x. The set U_i may be chosen to be within a coordinate patch and the question is then local.

Now a path connected subgroup of a Lie group is a Lie subgroup. To show that H_x itself is a Lie group, we must show that H_x/H_x^0 is countable. This follows from the facts that H_x/H_x^0 is a quotient of $\pi_1(M, x)$ and that, since M is second–countable, $\pi_1(M, x)$ is countable.

Finally, consider an open set $U \subseteq M$ which is the domain of a decomposing section for Ω and the image of a chart for M. Fix $x_0 \in U$ and for any $x \in U$ let ν_x be the path from x_0 to x that is the image of a straight line under the chart. Define $\sigma'(x) = \widehat{\nu_x}$. Then $\sigma' : U \to \Omega_{x_0}$

is smooth and takes values in Θ_{x_0}. Given an atlas $\{U_i\}$ of such sets and choosing paths from x_0 to chosen points in each U_i, we obtain smooth functions $\sigma_i\colon U_i \to \Omega_{x_0}$ that take values in Θ_{x_0}. It follows that the transition functions are smooth into $\Omega_{x_0}^{x_0}$ and take values in H_{x_0}; since Lie subgroups are always quasi–regular, it follows that the transition functions are smooth into H_{x_0}. Thus Θ can be given a smooth structure from the bijections

$$U_j \times H_{x_0} \times U_i \to \Theta_{U_i}^{U_j}, \qquad (y, h, x) \mapsto \sigma_j(y)h\sigma_i(x)^{-1},$$

and thus Θ is a Lie subgroupoid of Ω. $\qquad\qquad\square$

6.4 Parallel sections and stabilizer subgroupoids

Given a vector bundle E and a Koszul connection ∇, a section μ of E is parallel if all covariant derivatives $\nabla_X(\mu)$ are zero. This simple geometric notion generalizes the notion of a constant tensor field on a Euclidean space. However, it also arises naturally in the relationship between a stablizer subgroupoid and its Lie algebroid, and as such the notion of parallel section pervades the development of the abstract theory of transitive Lie algebroids.

In the course of this section we also calculate the Lie algebroid of a holonomy groupoid. The results we obtain are used throughout Chapter 7.

Throughout this section, except where noted, we assume that base manifolds are connected.

Definition 6.4.1 Let E be a vector bundle on M, and let ∇ be a connection in E. A section μ of E satisfying $\nabla(\mu) = 0$ is said to be *parallel* with respect to ∇. $\qquad\qquad\square$

Theorem 6.4.2 *Let E be a vector bundle on M, and let ∇ be a connection in E. Let $\Theta = \Theta(\nabla) \leqslant \Phi(E)$ be the holonomy groupoid of ∇. Write*

$$(\Gamma E)^{\nabla} = \{\mu \in \Gamma E \mid \nabla(\mu) = 0\}$$

and recall that

$$(\Gamma E)^{\Theta} = \{\mu \in \Gamma E \mid \xi\mu(\alpha\xi) = \mu(\beta\xi), \text{for all } \xi \in \Theta\}.$$

Then

$$(\Gamma E)^{\nabla} = (\Gamma E)^{\Theta}.$$

Equivalently, $\nabla(\mu) = 0$ if and only if $\Theta(\nabla)$ is contained in the stablizer subgroupoid of $\Phi(E)$ at μ.

Proof ($\supseteq$) Take $\mu \in (\Gamma E)^{\Theta}$. Then μ is Θ–deformable, and therefore is also $\Phi(E)$–deformable. Hence, by 1.6.23, the stabilizer subgroupoid $\Sigma = \Phi(E)\{\mu\}$ is a closed reduction of $\Phi(E)$. Now $\Theta \leqslant \Sigma$ by assumption and since, using 3.6.9,

$$\Gamma A \Sigma = \{D \in \Gamma \mathfrak{D}(E) \mid D(\mu) = 0\},$$

we have $D(\mu) = 0$ for all $D \in \Gamma A \Theta$. But Θ is the holonomy groupoid for ∇ and therefore, by 6.3.6, $\nabla_X \in \Gamma A \Theta$ for all $X \in \mathfrak{X}(M)$. Hence $\nabla(\mu) = 0$.

($\subseteq$) Assume $\nabla(\mu) = 0$. We are to prove that $\xi\mu(\alpha\xi) = \mu(\beta\xi)$ for all $\xi \in \Theta$. Since Θ is α–connected it is sufficient (by 1.5.8) to establish the equation for $\xi \in \Theta_U^U$, where U is the domain of a decomposing section $U \to \Phi(E)_m$ for E.

So we can assume that E is a trivial vector bundle $U \times V$. Now $\nabla \colon TU \to TU \oplus (U \times \mathfrak{gl}(V))$ has the form $\nabla(X) = X \oplus \omega(X)$, where $\omega \in \Omega^1(U, \mathfrak{gl}(V))$ is the local connection form of ∇ with respect to $x \mapsto (x, \mathrm{id}_V, x)$. As in the proof of 6.3.5, the lift of any $\nu \colon I \to U$ is $t \mapsto (\nu(t), a(t), \nu(0))$, where $\nabla(a) = \nu^*\omega$ and $a(0) = \mathrm{id}_V$.

We need to show that $a(t)\mu(\nu(0)) = \mu(\nu(t))$ for all $t \in I$, where μ is regarded as a map $U \to V$. We show that $\frac{d}{dt}(a(t)^{-1}\mu(\nu(t)))$ is identically zero.

Write $f = \mu \circ \nu \colon I \to V$. From (8) on p. 183 we obtain

$$\frac{d}{dt}(a^{-1}(f)) = -\Delta(a^{-1})\left(\frac{d}{dt}(a^{-1}(f))\right) + a^{-1}\left(\frac{d}{dt}f\right)$$

and from (3) on p. 182 we have

$$\Delta(a^{-1}) = -\mathrm{Ad}(a^{-1})(\Delta(a)).$$

Putting these together we get

$$\begin{aligned}
\frac{d}{dt}(a^{-1}(f)) &= \mathrm{Ad}(a^{-1})\left(\Delta(a)\left(\frac{d}{dt}(a^{-1}(f))\right)\right) + a^{-1}\left(\frac{d}{dt}f\right) \\
&= a^{-1}\left(\Delta(a)\left(\frac{d}{dt}f\right)\right) + a^{-1}\left(\frac{d}{dt}f\right) \\
&= a^{-1}\left(\omega\left(\frac{d\nu}{dt}\right)(\mu) + \frac{d\nu}{dt}(\mu)\right).
\end{aligned}$$

Now the hypothesis $\nabla(\mu) = 0$ is exactly that $X(\mu) + \omega(X)(\mu) = 0$ for all $X \in T(U)$, and so, putting $X = \frac{d\nu}{dt}$, the result follows.   □

Applying 1.7.4 to $\Theta * E \to E$ we have:

Corollary 6.4.3 *Continue the notation of 6.4.2 and let $m \in M$ be any reference point. There is an isomorphism of vector spaces*

$$(\Gamma E)^\nabla \to V^H, \quad \mu \mapsto \mu(m)$$

where $V = E_m$ and $H = \Theta_m^m$.

Theorem 6.4.4 *Let Ω be a locally trivial Lie groupoid on M and let $\rho \colon \Omega * E \to E$ be a smooth linear action of Ω on a vector bundle E. For $\mu \in \Gamma E$, the following four conditions are equivalent:*

(i) *μ is Ω–deformable;*
(ii) *the stabilizer groupoid Σ of μ is a Lie subgroupoid of Ω;*
(iii) *Ω possesses a section–atlas $\{\sigma_i \colon U_i \to \Omega_m\}$ such that $\rho(\sigma_i(x)^{-1})\mu(x)$ is a constant map $U_i \to E_m$;*
(iv) *Ω possesses an infinitesimal connection γ such that $(\rho_* \circ \gamma)(\mu) = 0$.*

Proof (i) $\implies$ (ii) is 1.6.23; (ii) $\implies$ (iii) is immediate; (iii) $\implies$ (i) follows from the connectivity of M.

(ii) $\implies$ (iv) follows from 3.6.9.

(iv) $\implies$ (i) Let Θ be the holonomy groupoid of γ, and let $\Theta' \leqslant \Phi(E)$ be the holonomy groupoid of $\rho_* \circ \gamma$. Then $\rho(\Theta) = \Theta'$ by 6.3.14. Now $(\rho_* \circ \gamma)(\mu) = 0$ so, by 6.4.2, $\mu \in (\Gamma E)^{\Theta'}$ and now $\Theta' = \rho(\Theta)$ shows that μ is Θ–deformable. In particular, μ is Ω–deformable.   □

In (iii), the various constant maps $U_i \to E_m$ may be chosen to have the same value.

Theorem 6.4.5 *Let L be a vector bundle on M and let $[\ ,\ ]$ be a field of Lie algebra brackets on L. Then the following three conditions are equivalent:*

(i) *The fibres of L are pairwise isomorphic as Lie algebras;*
(ii) *L admits a connection ∇ such that*

$$\nabla_X([V,W]) = [\nabla_X(V), W] + [V, \nabla_X(W)]$$

for all $X \in \mathfrak{X}(M)$ and $V, W \in \Gamma L$;

(iii) *L is an LAB.*

Proof Let ρ: $\Phi(L) * \mathrm{Alt}^2(L; L) \to \mathrm{Alt}^2(L; L)$ denote the action 1.7.7(ii). Then (i) is the condition that [,] is $\Phi(L)$–deformable, and (iii) is the condition that $\Phi(L)$ admits a section–atlas $\{\sigma_i\}$ such that the corresponding charts for $\mathrm{Alt}^2(L; L)$ induced via ρ map [,] $\in \Gamma \mathrm{Alt}^2(L; L)$ to constant maps $U_i \to \mathrm{Alt}^2(L_m; L_m)$. So (i) and (iii) are equivalent by the equivalence (i) $\iff$ (iii) of 6.4.4.

From 3.6.10(ii) it follows that ρ_*: $\mathfrak{D}(L) \to \mathfrak{D}(\mathrm{Alt}^2(L; L))$ is

$$\rho_*(D)(\varphi)(V, W) = D(\varphi(V, W)) - \varphi(D(V), W) - \varphi(V, D(W)).$$

Therefore (ii) is the condition that L admits a connection ∇ such that $(\rho_* \circ \nabla)([,]) = 0$. Hence (i) $\iff$ (ii) follows from the equivalence of (i) and (iv) in 6.4.4. $\qquad\square$

Recall from 5.2.1 that a connection in L satisfying (ii) is called a Lie connection in L.

Theorem 6.4.6 *Let E and E' be vector bundles on M and consider a morphism φ: $E \to E'$ over M. The following three conditions are equivalent:*

(i) *$x \mapsto \mathrm{rk}(\varphi_x)$ is constant;*

(ii) *there exist connections ∇ in E and ∇' in E' such that $\nabla'_X(\varphi(\mu)) = \varphi(\nabla_X(\mu))$ for all $\mu \in \Gamma E$ and $X \in \mathfrak{X}(M)$;*

(iii) *there exist atlases $\{\psi_i$: $U_i \times V \to E_{U_i}\}$ and $\{\psi'_i$: $U_i \times V' \to E'_{U_i}\}$ for E and E' such that each φ: $E_{U_i} \to E'_{U_i}$ is of the form $\varphi(x, v) = (x, f_i(v))$, where f_i: $V \to V'$ is a linear map depending only on i.*

Proof This follows from 6.4.4 in the same way as does 6.4.5, using now 1.7.7(iii), 3.6.10(iii) and the lemma which follows. $\qquad\square$

Lemma 6.4.7 *Let φ_1: $V \to V'$ and φ_2: $W \to W'$ be morphisms of vector spaces with $\dim V = \dim W$, $\dim V' = \dim W'$ and $\mathrm{rk}(\varphi_1) = \mathrm{rk}(\varphi_2)$. Then there are isomorphisms α: $V \to W$, α': $V' \to W'$ such that $\alpha' \circ \varphi_1 = \varphi_2 \circ \alpha$.*

In order to ease reference to 6.4.6, we use the following terminology:

Definition 6.4.8 Let M be any manifold, not necessarily connected, and let $\varphi \colon E \to E'$ be a morphism of vector bundles over M. Then φ is of *locally constant rank* if $x \mapsto \mathrm{rk}(\varphi_x)$, $M \to \mathbb{Z}$, is locally constant.

Secondly, φ is a *locally constant morphism* if it satisfies condition (iii) of 6.4.6. $\qquad\qquad\square$

Then 6.4.6 may be paraphrased for morphisms $\varphi \colon E \to E'$ of vector bundles over any base as follows: φ is of locally constant rank if and only if it is locally constant and if and only if there are connections ∇, ∇' in E, E' such that φ maps ∇ to ∇'. In 6.4.6 itself, where the base is connected, the maps f_i may be arranged to be identical.

We will also need the following LAB version of 6.4.6.

Theorem 6.4.9 *Let L and L' be LABs on M and let $\varphi \colon L \to L'$ be a morphism of LABs over M. Then the following three conditions are equivalent:*

(i) *for each x and y in M, there are Lie algebra isomorphisms $\alpha \colon L_x \to L_y$ and $\alpha' \colon L'_x \to L'_y$ such that $\varphi_y \circ \alpha = \alpha' \circ \varphi_x$;*

(ii) *L and L' possess Lie connections ∇ and ∇' such that, for all $V \in \Gamma L$ and $X \in \mathfrak{X}(M)$, we have $\varphi(\nabla_X(V)) = \nabla'_X(\varphi(V))$;*

(iii) *there exist LAB atlases $\{\psi_i \colon U_i \times \mathfrak{g} \to L_{U_i}\}$ and $\{\psi'_i \colon U_i \times \mathfrak{g}' \to L'_{U_i}\}$ for L and L' such that each $\varphi \colon L_{U_i} \to L'_{U_i}$ is of the form $\varphi(x, A) = (x, f_i(A))$, where $f_i \colon \mathfrak{g} \to \mathfrak{g}'$ is a Lie algebra morphism depending only on i.*

The proof is similar to that of 6.4.6. We refer to a morphism of LABs over any (not necessarily connected) base M which satisfies (iii) of 6.4.9 as a *locally constant morphism of LABs*. When the base is connected, f_i may be chosen to be independent of i as well.

Further applications of 6.4.4 are made in §6.5.

The following result is similar in spirit to 6.4.5, 6.4.6 and 6.4.9.

Theorem 6.4.10 *Let E be a vector bundle on M, and let E^1 and E^2 be vector subbundles of E. Then $E^1 \cap E^2$ is a vector subbundle of E if there is a connection ∇ in E such that $\nabla(\Gamma E^1) \subseteq \Gamma E^1$ and $\nabla(\Gamma E^2) \subseteq \Gamma E^2$.*

Here $\nabla(\Gamma E') \subseteq \Gamma E'$ is an abbreviation for the condition

$$\mu' \in \Gamma E', \ X \in \mathfrak{X}(M) \implies \nabla_X(\mu') \in \Gamma E'.$$

Though 6.4.10 is related to 6.4.4, it requires a more circuitous proof and will only be completed after 6.4.16.

Proposition 6.4.11 *Let E be a vector bundle on M and let E' be a subbundle. Let $\rho\colon \Omega * E \to E$ be a smooth linear action of a Lie groupoid Ω on E. Denote by Ω' the subgroupoid $\{\xi \in \Omega \mid \xi(E'_{\alpha\xi}) = E'_{\beta\xi}\}$ of Ω.*

Then if Ω' is a transitive subgroupoid of Ω, it is a closed embedded reduction of Ω.

Proof Let q be the rank of E' and let $G_q(E) \to M$ be the fibre bundle with fibres, for $x \in M$, which are the Grassmannians $G_q(E_x)$ of q–dimensional subspaces of E_x, and charts induced from the charts of E in the natural fashion. Then E' is a smooth Ω–deformable section of $G_q(E)$. Now apply 1.6.23. □

Notation 6.4.12 *Let E be a vector bundle on M and E' a vector subbundle of E. Then $\Phi(E, E')$ denotes the Lie groupoid*

$$\{\varphi \in \Phi(E) \mid \varphi(E'_{\alpha\varphi}) = E'_{\beta\varphi}\}.$$

□

Proposition 6.4.13 *With the notation of 6.4.12, the isomorphism $\mathscr{D}$ from $A\Phi(E)$ to $\mathfrak{D}(E)$ of 3.6.6 maps $\Gamma A\Phi(E, E')$ isomorphically onto*

$$\{D \in \Gamma\mathfrak{D}(E) \mid D(\Gamma E') \subseteq \Gamma E'\}.$$

We complete the proof of 6.4.13 after 6.4.15. Although 6.4.13 resembles 3.6.9, the method there cannot be used in a general fibre bundle and we are obliged to give a different proof.

Lemma 6.4.14 *E admits a connection ∇ such that $\nabla(\Gamma E') \subseteq \Gamma E'$.*

Proof Let $\langle\,,\,\rangle$ be a Riemannian structure on E, and let E'' be the orthogonal complement to E' in E. Let ∇' and ∇'' be connections in E' and E'' and define ∇ in $E = E' \oplus E''$ by $\nabla_X(\mu' \oplus \mu'') = \nabla'_X(\mu') \oplus \nabla''_X(\mu'')$. □

Proposition 6.4.15 *There is a transitive Lie subalgebroid $\mathfrak{D}(E, E')$ of $\mathfrak{D}(E)$ which has the property*

$$\Gamma\mathfrak{D}(E, E') = \{D \in \Gamma\mathfrak{D}(E) \mid D(\Gamma E') \subseteq \Gamma E'\}.$$

Proof Let $\operatorname{End}(E, E')$ be the sub LAB of $\operatorname{End}(E)$ defined by

$$\Gamma\operatorname{End}(E, E') = \{\varphi \in \Gamma\operatorname{End}(E) \mid \varphi(\Gamma E') \subseteq \Gamma E'\};$$

it is easy to prove (see 3.3.9 and the subsequent discussion) that a unique such LAB exists. Let ∇ be a connection in E such that $\nabla(\Gamma E') \subseteq \Gamma E'$ and define

$$i\colon TM \oplus \operatorname{End}(E, E') \to \mathfrak{D}(E)$$

by $i(X \oplus \varphi) = \nabla_X + \varphi$. Then i is an injection since $\operatorname{End}(E, E') \to \operatorname{End}(E)$ is an injection, and so $\operatorname{im}(i)$ is a vector subbundle of $\mathfrak{D}(E)$. It is easily verified that $[\nabla_X, \nabla_Y] = \nabla_X \circ \nabla_Y - \nabla_Y \circ \nabla_X$ maps $\Gamma E'$ into $\Gamma E'$ for all $X, Y \in \mathfrak{X}(M)$, and that $[\nabla_X, \psi] = \nabla_X \circ \psi - \psi \circ \nabla_X$ does likewise, for $X \in \mathfrak{X}(M)$ and $\psi \in \Gamma\operatorname{End}(E, E')$. It now follows that $\operatorname{im}(i)$ is closed under the bracket on $\mathfrak{D}(E)$, and hence $\operatorname{im}(i)$ is a reduction of $\mathfrak{D}(E)$.

If $\widetilde{\nabla}$ is a second connection in E such that $\widetilde{\nabla}(\Gamma E') \subseteq \Gamma E'$, then $\widetilde{\nabla} - \nabla$ takes values in $\operatorname{End}(E, E')$ and so it is easily seen that $\operatorname{im}(\widetilde{i}) = \operatorname{im}(i)$. Denote this common image by $\mathfrak{D}(E, E')$; if $D \in \Gamma\mathfrak{D}(E)$ has the property that $D(\Gamma E') \subseteq \Gamma E'$ then $D = \nabla_{q(D)} + (D - \nabla_{q(D)})$ shows that $D \in \Gamma\mathfrak{D}(E, E')$. $\square$

A similar construction may be carried out with any suitable family of connections on E.

Finally we need an elementary lemma.

Lemma 6.4.16 *Let V be a vector space and V' a subspace. Let*

$$GL(V, V') = \{A \in GL(V) \mid A(V') = V'\}$$

and let $\mathfrak{gl}(V, V') = \{X \in \mathfrak{gl}(V) \mid X(V') \subseteq V'\}$. *Then* $\mathfrak{gl}(V, V')$ *is the Lie algebra of* $GL(V, V')$.

Proof Take $X \in \mathfrak{gl}(V, V')$. Then $X^n \in \mathfrak{gl}(V, V')$ for all integers $n \geqslant 0$ and therefore, since V' is closed, $\operatorname{Exp}tX$ maps V' into V' for all t. Thus $\operatorname{Exp}tX \in GL(V, V')$. $\square$

Proof of 6.4.13: If $X \in \Gamma A\Phi(E, E')$ then $\operatorname{Exp}tX$ takes values in $\Phi(E, E')$ for all t and so, by the definition of $\mathscr{D}$ in 3.6.6, $\mathscr{D}(X) \in \Gamma\mathfrak{D}(E, E')$. Thus $\mathscr{D}$ maps $A\Phi(E, E')$ into $\mathfrak{D}(E, E')$.

To prove that $\mathscr{D}(A\Phi(E, E')) = \mathfrak{D}(E, E')$, it suffices (by 3.3.12) to prove that $\mathscr{D}^+(L\Phi(E, E')) = \operatorname{End}(E, E')$. Fibrewise, $\mathscr{D}^+$ is

$$T_I(GL(V, V')) \to \mathfrak{gl}(V, V'),$$

and is an isomorphism by 6.4.16. $\qquad\square$

Proof of 6.4.10: Let $\Theta \leqslant \Phi(E)$ be the holonomy groupoid of ∇. Now $\nabla_X \in \Gamma \mathfrak{D}(E, E^1) = \Gamma A\Phi(E, E^1)$ so, by 6.3.15, $\Theta \leqslant \Phi(E, E^1)$. Similarly $\Theta \leqslant \Phi(E, E^2)$. So every element ξ of Θ maps $E^1_{\alpha\xi} \cap E^2_{\alpha\xi}$ to $E^1_{\beta\xi} \cap E^2_{\beta\xi}$.

Let $\{\varphi_i \colon U_i \to \Theta_b\}$ be a section-atlas for Θ. Then the associated charts $\psi_i \colon U_i \times E_b \to E_{U_i}$, $(x, v) \mapsto \sigma_i(x)(v)$, restrict to charts for $E^1 \cap E^2$. $\qquad\square$

We also need an LAB version of 6.4.10; the proof is exactly analogous.

Theorem 6.4.17 *Let L be an LAB on M, and let L^1 and L^2 be sub LABs of L. Then $L^1 \cap L^2$ is a sub–LAB of L if there is a Lie connection ∇ in L such that $\nabla(\Gamma L^1) \subseteq \nabla(\Gamma L^1)$ and $\nabla(\Gamma L^2) \subseteq \Gamma L^2$.*

We can now calculate the Lie algebroid of the holonomy groupoid of a connection. First we show that, given an infinitesimal connection γ in a Lie groupoid Ω, there is a least reduction, denoted $(A\Omega)^\gamma$, of $A\Omega$ which 'contains' γ. As we will see in §6.5, this construction may be carried out in any transitive Lie algebroid. It then follows immediately that this reduction $(A\Omega)^\gamma$ is the Lie algebroid of the holonomy groupoid of γ.

Until we reach 6.4.20, let Ω be a locally trivial Lie groupoid on M and let γ be an infinitesimal connection in Ω.

Proposition 6.4.18 *Let L' be a sub–LAB of $L\Omega$ such that*

(i) $\overline{R}_\gamma(X, Y) \in L'$ *for all $X, Y \in TM$, and*
(ii) $\nabla^\gamma(\Gamma L') \subseteq \Gamma L'$.

Then there is a reduction $A' \leqslant A\Omega$ defined by

$$\Gamma A' = \{X \in \Gamma A \mid X - \gamma a(X) \in \Gamma L'\}$$

which has L' as adjoint bundle and is such that $\gamma(X) \in A'$ for all $X \in TM$.

Proof Define $\varphi \colon TM \oplus L' \to A\Omega$ by $\varphi(X \oplus V') = \gamma(X) + V'$. Then $\text{im}(\varphi) = A'$ and, applying the 5–lemma of category theory (see 3.3.12) to the diagram in Figure 6.3, it follows that A' is a vector subbundle of $A\Omega$.

$$L' \rightarrowtail TM \oplus L' \longrightarrow\!\!\!\!\!\rightarrow TM$$

$$\subseteq \downarrow \qquad\qquad \downarrow \qquad\qquad \|$$

$$L\Omega \rightarrowtail A\Omega \longrightarrow\!\!\!\!\!\rightarrow TM$$

Fig. 6.3.

Clearly $\gamma(X) \in A'$ for $X \in TM$ and so the restriction of a to A' is surjective. Clearly $\ker(a|_{A'}) = L'$.

To prove that $\Gamma A'$ is closed under the bracket on ΓA, take $X, Y \in \Gamma A'$ and write $X = \gamma aX + V'$, $Y = \gamma aY + W'$, where $V', W' \in \Gamma L'$. Then

$$[X, Y] = \gamma[aX, aY] - \overline{R}_\gamma(aX, aY) + \nabla_{aX}^\gamma(W') - \nabla_{aY}^\gamma(V') + [V', W']$$

and the last four terms are in $\Gamma L'$ by (i) and (ii). $\qquad\square$

Proposition 6.4.19 *There is a least sub–LAB, denoted $(L\Omega)^\gamma$, of $L\Omega$ which has the properties* (i) *and* (ii) *of 6.4.18.*

Proof It suffices to prove that if L^1 and L^2 each satisfy (i) and (ii), then $L^1 \cap L^2$ does also. The only point that is not clear is that $L^1 \cap L^2$ is a sub–LAB, and since ∇^γ is a Lie connection this is established by 6.4.17. $\qquad\square$

The corresponding reduction of $A\Omega$ is denoted by $(A\Omega)^\gamma$ and called the *γ–curvature reduction* of $A\Omega$.

Theorem 6.4.20 *Let Ω be a locally trivial Lie groupoid on M and let $\gamma\colon TM \to A\Omega$ be an infinitesimal connection. Denote the associated path connection by Γ and the holonomy groupoid of Γ by Θ.*
Then $A\Theta = (A\Omega)^\gamma$.

Proof By 6.3.6, γ takes values in $A\Theta$. Hence $L\Theta$ satisfies the conditions of 6.4.18 and therefore $L\Theta \geqslant (L\Omega)^\gamma$ and $A\Theta \geqslant (A\Omega)^\gamma$. On the other hand, γ takes values in $(A\Omega)^\gamma$ and so, by 6.3.15, $A\Theta \leqslant (A\Omega)^\gamma$. $\qquad\square$

Since Θ is α–connected, it is determined by $(A\Omega)^\gamma$ via 6.2.1. Theorem 6.4.20 is a strong form of the classical Ambrose–Singer Theorem. We indicate how the standard formulation may be obtained from 6.4.20.

Example 6.4.21 Let $P(M, G)$ be a principal bundle and consider a connection 1–form $\omega \in \Omega^1(P, \mathfrak{g})$ and its curvature 2–form $\Omega \in \Omega^2(P, \mathfrak{g})$. (For the relationship between ω and $\gamma \colon TM \to \frac{TP}{G}$ and between Ω and $\overline{R}_\gamma \colon TM \oplus TM \to \frac{P \times \mathfrak{g}}{G}$, see §5.3.) Let $\mathfrak{h}$ be the Lie subalgebra of $\mathfrak{g}$ generated by $\{\Omega(X, Y) \mid X, Y \in T(P)\}$. Since $\mathrm{Ad}_g \Omega(X, Y) = \Omega(T(R_{g^{-1}})X, T(R_{g^{-1}})Y)$, for $g \in G$, it follows that $\mathfrak{h}$ is stable under $\mathrm{Ad}G$.

Let $\{\sigma_i \colon U_i \to P\}$ be a section–atlas for P and let $\psi_i \colon U_i \times \mathfrak{g} \to \frac{P \times \mathfrak{g}}{G}\big|_{U_i}$ be the associated charts for $\frac{P \times \mathfrak{g}}{G}$. Since the transition functions $\psi_i^{-1} \psi_j$ take values in $\mathrm{Ad}G \leqslant \mathrm{Aut}(\mathfrak{g})$, it follows that $\mathfrak{h}$ translates into a well–defined sub–LAB K of $\frac{P \times \mathfrak{g}}{G}$.

From 5.3.14 it follows that $\overline{R}_\gamma(X, Y) \in K$ for all $X, Y \in TM$. To show that condition (ii) of 6.4.18 holds, note that $\nabla_X^\gamma(\psi_i(V)) = \psi_i(X(V) + [\omega_i(X), V])$, where $V \colon U_i \to \mathfrak{g}$, and the $\omega_i \in \Omega^1(U_i, \mathfrak{g})$ are the local connection forms of γ with respect to $\{\sigma_i\}$ (see 5.4.4). From this it follows easily that $\nabla^\gamma(\Gamma K) \subseteq \Gamma K$.

Hence $(L\Omega)^\gamma \leqslant K$. Now $\mathfrak{h}$ is the least Lie subalgebra of $\mathfrak{g}$ which contains all the values of Ω so, by following through the relationships between $\overline{R}_\gamma$ and Ω, and between K and $\mathfrak{h}$, it follows that any sub–LAB of $L\Omega$ which satisfies (i) of 6.4.18, also contains K. Hence $(L\Omega)^\gamma \geqslant K$.

Note that $\mathfrak{h}$ is an ideal of $\mathfrak{g}$, since it is stable under $\mathrm{Ad}G$. ⊠

The account of flat connections given in 6.2.8 can now be made more precise.

Proposition 6.4.22 *Let Ω be a locally trivial Lie groupoid on M and let $\gamma \colon TM \to A\Omega$ be a flat connection. Then the holonomy groupoid Θ of γ is a quotient of $\Pi(M)$.*

Proof Since $\overline{R}_\gamma = 0$ we can take $L' = M \times \{0\}$ in 6.4.18. So $L\Theta = (L\Omega)^\gamma = M \times \{0\}$ and γ is an isomorphism $TM \to A\Theta$. Now, as in 6.2.8, γ integrates to a morphism $h^\gamma \colon \Pi(M) \to \Theta$. Since both $\Pi(M)$ and Θ are α–connected, it follows from 6.2.5 that h^γ is a surjective submersion and fibrewise a covering. □

Thus the locally constant transition functions found in 6.2.8 form a cocycle for Θ, and the path connection for γ in Ω is the image under h^γ of the unique path connection in $\Pi(M)$.

In the case of flat connections in a vector bundle E, the holonomy morphism h^γ gives a smooth action of $\Pi(M)$ on E.

We close this section with a more detailed analysis of morphisms of transitive Lie algebroids, based on 6.4.4—6.4.9.

Example 6.4.23 Let $\varphi\colon TM\oplus(M\times\mathfrak{g})\to TM\oplus(M\times\mathfrak{h})$ be a morphism of trivial Lie algebroids. By 3.3.3, φ is $\varphi(X\oplus V)=X\oplus(\omega(X)+\varphi^+(V))$ where $\omega\in\Omega^1(M,\mathfrak{h})$ is a Maurer–Cartan form, $\varphi^+\colon M\times\mathfrak{g}\to M\times\mathfrak{h}$ is the induced morphism of LABs, and φ^+ and ω satisfy the compatibility equation

$$X(\varphi^+(V))-\varphi^+(X(V))+[\omega(X),\varphi^+(V)]=0. \tag{7}$$

Let ∇^0 be the standard flat connection in $M\times\mathfrak{g}$, and define a connection ∇^1 in $M\times\mathfrak{h}$ by

$$\nabla^1_X(W)=X(W)+[\omega(X),W].$$

Since ω is a Maurer–Cartan form, ∇^1 is flat, and it is easily seen to be Lie. In terms of 6.4.9, (7) now asserts that $\varphi^+(\nabla^0_X(V))=\nabla^1_X(\varphi^+(V))$, for $V\colon M\to\mathfrak{g}$ and $X\in\mathfrak{X}(M)$. Hence φ^+ is a locally constant morphism of LABs.

Further, ∇^0 and ∇^1 together induce, by 3.6.10(iii), a connection ∇ in $\mathrm{Hom}(M\times\mathfrak{g},M\times\mathfrak{h})=M\times\mathrm{Hom}(\mathfrak{g},\mathfrak{h})$, the vector bundle whose sections are the vector bundle morphisms $M\times\mathfrak{g}\to M\times\mathfrak{h}$. This ∇ is

$$\begin{aligned}\nabla^X(\psi)(V)&=\nabla^1_X(\psi(V))-\psi(\nabla^0_\zeta V))\\&=X(\psi(V))-\psi(X(V))+[\omega(X),\psi(V)].\end{aligned}$$

Since ∇^0 and ∇^1 are both flat, ∇ is flat.

(7) now asserts that $\nabla(\varphi^+)=0$ and, applying 6.4.3, φ^+ is determined by its restriction to any single fibre, $f=\varphi^+|_m\colon\mathfrak{g}\to\mathfrak{h}$. By analogy with the situation for morphisms of trivial groupoids, one would expect this. However 6.4.3 also shows that f cannot be an arbitrary Lie algebra morphism, but must lie in $\mathrm{Hom}(\mathfrak{g},\mathfrak{h})^H$, where H is the holonomy group of ∇. Since ∇ is induced from ∇^0 and ∇^1 via 3.6.10(iii), it follows (from 6.3.14) that H is the direct product $H^0\times H^1$ of the holonomy groups of ∇^0 and ∇^1 and acts on $\mathrm{Hom}(\mathfrak{g},\mathfrak{h})$ by

$$(\alpha^0,\alpha^1)(f)=\alpha^1\circ f\circ(\alpha^0)^{-1}.$$

Since ∇^0 has trivial holonomy, we conclude that f lies in $\mathrm{Hom}(\mathfrak{g},\mathfrak{h}^{\pi_1 M})$ where $\mathfrak{h}^{\pi_1 M}$ is the vector space of elements of $\mathfrak{h}$ invariant under the ∇^1–holonomy action of $\pi_1 M$ (see 6.4.22). A direct calculation with (4) shows that $\pi_1 M$ must act by elements of $\mathrm{Int}(\mathfrak{h})$.

This is a real restriction; there are certainly Lie algebras with nontrivial discrete groups of inner automorphisms; for example the 3–dimensional Lie algebra with $[e_1, e_2] = e_3$, $[e_2, e_3] = [e_3, e_1] = 0$.

Thus if $\pi_1 M \neq 0$, if $\omega \neq 0$, if $\mathfrak{h}$ is not abelian, and if the action of $\pi_1 M$ on $\mathfrak{h}$ induced by ω is nontrivial, then there will be Lie algebra morphisms $\mathfrak{g} \to \mathfrak{h}$ which are not the restriction of a Lie algebroid morphism which induces ω.   ⊠

This analysis also yields an alternative proof of 6.2.6.

Proposition 6.4.24 *Let M be a simply–connected manifold and let $\varphi \colon TM \oplus (M \times \mathfrak{g}) \to TM \oplus (M \times \mathfrak{h})$ be a morphism of trivial Lie algebroids over M. If G and H are Lie groups with Lie algebras $\mathfrak{g}$ and $\mathfrak{h}$, then there is a local morphism of trivial Lie groupoids*

$$F \colon M \times G \times M \looparrowright M \times H \times M$$

such that $F_* = \varphi$, and F is unique up to germ–equivalence.

Proof For convenience we assume that G is connected and simply–connected; the general case is only notationally more complicated.

With φ^+ and ω as in 6.4.23, choose $m \in M$ and define $\theta \colon M \to H$ to be the solution to $\Delta(\theta) = \omega$, $\theta(m) = 1$, and $f \colon G \to H$ to be the Lie group morphism with $f_* = \varphi^+|_m$. Define $F \colon M \times G \times M \to M \times H \times M$ by $F(y, g, x) = (y, \theta(y)f(g)\theta(x)^{-1}, x)$. Then, by 3.5.13, F_* is $X \oplus V \mapsto X \oplus \{\Delta(\theta)(X) + \mathrm{Ad}(\theta)f_*(V)\}$; thus the Maurer–Cartan form for F_* is $\Delta(\theta) = \omega$ and $F_*^+ = \mathrm{Ad}(\theta)f_*$. We need to show that $F_*^+ = \varphi^+$.

Since both F_*^+ and φ^+ are LAB morphisms compatible with ω, they are each parallel with respect to ∇. Also, $F_*^+|_m = (F_m^m)_* = f_* = \varphi^+|_m$, since $\theta(m) = 1$. So, as in 6.4.23, it follows that $F_*^+ = \varphi^+$.

That $F_*^+ = \mathrm{Ad}(\theta)f_*$ is parallel with respect to ∇ also follows from 5.1.1.

For uniqueness, let $F' \colon M \times G \times M \to M \times H \times M$ be any other morphism with $F_*' = \varphi$. Then, by 1.2.5,

$$F'(y, g, x) = (y, \theta'(y)f'(g)\theta'(x)^{-1}, x)$$

where $\theta'(m) = 1$. It follows that $\theta' = \theta$ and $f' = f$.   □

Theorem 6.4.25 *Let Ω and Ω' be locally trivial Lie groupoids on M and let $\varphi \colon A\Omega \to A\Omega'$ be a morphism of Lie algebroids over M. Then there is a local morphism $\mu \colon \Omega \looparrowright \Omega'$ of Lie groupoids over M such that $\mu_* = \varphi$, and μ is unique up to germ–equivalence.*

Proof Let $\{U_i\}$ be a simple cover of M, and let $\{\sigma_i\colon U_i \to \Omega_m\}$ and $\{\sigma_i'\colon U_i \to \Omega_m'\}$ be section–atlases for Ω and Ω' over $\{U_i\}$. Let Σ_i denote the isomorphism

$$U_i \times G \times U_i \to \Omega_{U_i}^{U_i}, \quad (y,g,x) \mapsto (y, \sigma_i(y)g\sigma_i(x)^{-1}, x)$$

and $(\Sigma_i)_*\colon TU_i \oplus (U_i \times \mathfrak{g}) \to A\Omega|_{U_i}$ its derivative. Let $S_{ij} = \Sigma_i^{-1} \circ \Sigma_j$ for $U_{ij} \neq \emptyset$. Similarly with Σ_i' and S_{ij}'.

Define $\varphi^i = (\Sigma_i')_* \circ \varphi \circ (\Sigma_i^{-1})_*$; by 6.4.24, φ^i integrates to a well–defined local morphism $\mu^i\colon U_i \times G \times U_i \rightarrowtail U_i \times G' \times U_i$. In order to show that the $\Sigma_i' \circ \mu^i \circ \Sigma_i^{-1}$ stick together into a well–defined local morphism $\Omega \rightarrowtail \Omega'$, it is sufficient to prove that $(S_{ij}')^{-1} \circ \mu^i \circ S_{ij} = \mu^j$ whenever $U_{ij} \neq \emptyset$. By the uniqueness result in 6.4.24, it suffices to prove that $(S_{ij}')_*^{-1} \circ \mu_*^i \circ (S_{ij})_* = \mu_*^j$, and this follows from the definition of φ^i, φ^j.

Likewise, the uniqueness statement follows from the uniqueness result in 6.4.24. $\qquad\square$

6.5 The abstract theory of transitive Lie algebroids

This section uses the results of §6.4 to prove several algebraic results about transitive Lie algebroids. In 6.5.1 we prove that the adjoint bundle of a transitive Lie algebroid is an LAB, and in 6.5.3 we prove that if $\varphi\colon A \to A'$ is a base–preserving morphism of transitive Lie algebroids then $\varphi^+\colon L \to L'$ is a locally constant morphism of LABs; it then follows that such morphisms have well–defined kernels and images. In 6.5.12 we prove that if ρ is a representation of a transitive Lie algebroid A on a vector bundle E, then E^L is a flat vector bundle with a natural flat connection, and in 6.5.16 we prove that, under the same hypotheses, $(\Gamma E)^A$ is naturally isomorphic to $(V^{\mathfrak{g}})^{\pi_1 M}$, where V and $\mathfrak{g}$ are the fibre types of E and L and $\pi_1 M$ acts via the holonomy of the natural flat connection in E^L. These results are fundamental to the cohomology theory developed in the next chapter, and to any development of the algebraic theory of transitive Lie algebroids.

Theorem 6.5.1 *Let* $L \overset{j}{>\!\!-\!\!-\!\!>} A \overset{a}{-\!\!-\!\!\twoheadrightarrow} TM$ *be a transitive Lie algebroid on base* M. *Then* L *is an LAB with respect to the bracket structure on* ΓL *induced from the bracket on* ΓA.

Proof Recall from 3.3.15 the adjoint representation $\mathrm{ad}\colon A \to \mathfrak{D}(L)$ of A on the vector bundle L. Let γ be a connection in A and consider the produced connection $\nabla = \text{ad} \circ \gamma\colon TM \to \mathfrak{D}(L)$ in the vector bundle L. A calculation with the Jacobi identity for ΓA shows that ∇ satisfies condition (ii) of 6.4.5 with respect to the field of Lie algebra brackets on L which is induced from the bracket on ΓA, as for (16) on p. 106. Hence, by 6.4.5, L is an LAB. □

The following result is an immediate consequence of the Jacobi identity. It is only by virtue of 6.5.1, however, that it is possible to formulate it.

Proposition 6.5.2 *Let* $L \!>\!\!-\!\!\!\longrightarrow\! A \longrightarrow\!\!\!\gg TM$ *be a transitive Lie algebroid on* M. *Then the adjoint representation of* A *on* L, $\text{ad}\colon A \to \mathfrak{D}(L)$, *takes values in* $\mathfrak{D}_{\mathrm{Der}}\, L$.

Recall from 5.2.16 that for a connection γ in A, the produced connection $\text{ad} \circ \gamma\colon TM \to \mathfrak{D}_{\mathrm{Der}}\, L$ is the *adjoint connection* of γ, and is denoted ∇^γ. It is a Lie connection in L.

From now on we will call L the *adjoint LAB* of A. This needs to be distinguished from the expression 'adjoint LAB of L', which refers to $\text{ad}(L)$.

The following theorem is a Lie algebroid analogue of 2.2.5.

Theorem 6.5.3 *Let* $\varphi\colon A \to A'$ *be a morphism of transitive Lie algebroids over* M. *Then* $\varphi^+\colon L \to L'$ *is a locally constant morphism of LABs and, as a morphism of vector bundles over* M, $\varphi\colon A \to A'$ *is of locally constant rank.*

Proof Let γ be a connection in A and let $\gamma' = \varphi \circ \gamma$ be the produced connection in A'. Let ∇^γ and $\nabla^{\gamma'}$ be the corresponding adjoint connections in L and L'. Then it is easily checked that $\varphi^+\colon L \to L'$ maps ∇^γ to $\nabla^{\gamma'}$, that is, φ^+ satisfies (ii) of 6.4.9. Now 6.4.9 establishes that φ^+ is a locally constant morphism of LABs.

That φ itself is of locally constant rank follows from applying the 5–lemma to φ^+ and φ, as in the proof of 3.3.12. □

Thanks to 6.5.3, there is a significant algebraic theory of transitive Lie algebroids. We start by giving a rapid account of quotients for transitive Lie algebroids over a fixed base; this is consistent with, but independent of, the general treatment in §4.4.

Definition 6.5.4 Let $\varphi\colon A \to A'$ be a morphism of transitive Lie algebroids over M. Then the *kernel* of φ, denoted $\text{ker}(\varphi)$, is the subbundle

$\ker(\varphi^+)$ of L. The *image* of φ, denoted $\operatorname{im}(\varphi)$, is the transitive Lie algebroid $\operatorname{im}(\varphi^+) \rightarrowtail \operatorname{im}(\varphi) \longrightarrow\!\!\!\!\!\gg TM$. $\qquad\square$

Proposition 6.5.5 *Let* $\varphi\colon A \to A'$ *be a morphism of transitive Lie algebroids over* M. *Then* $\ker(\varphi)$ *is a sub–LAB of* L.

Proof Let γ be a connection in A and ∇^γ the corresponding adjoint connection in L. Write $K = \ker(\varphi)$. Then $\nabla^\gamma(\Gamma K) \subseteq \Gamma K$; for if $V \in \Gamma K$ and $X \in \mathfrak{X}(M)$, then $\varphi^+(\nabla^\gamma_X(V)) = \varphi([\gamma(X), V]) = [\varphi\gamma(X), \varphi^+(V)] = 0$. So K is a vector bundle with a field of Lie algebra brackets which admits a connection — the restriction of ∇^γ — that satisfies 6.4.5(ii). $\qquad\square$

We leave to the reader the proof that $\operatorname{im}(\varphi)$ is actually a reduction of A', as implied in 6.5.4.

Definition 6.5.6 Let $L \rightarrowtail A \longrightarrow\!\!\!\!\!\gg TM$ be a transitive Lie algebroid on M. An *ideal* of A is a sub–LAB K of L such that

$$X \in \Gamma A, \ V \in \Gamma K \quad\Longrightarrow\quad [X, V] \in \Gamma K$$

We denote that fact that K is an ideal of A by $K \trianglelefteq A$.

An *ideal reduction* of A is a reduction $L' \rightarrowtail A' \longrightarrow\!\!\!\!\!\gg TM$ of A such that L' is an ideal of A. $\qquad\square$

Clearly the kernel of a morphism of transitive Lie algebroids over M is an ideal of its domain Lie algebroid. Other examples of ideals of a transitive Lie algebroid $L \rightarrowtail A \longrightarrow\!\!\!\!\!\gg TM$ are ZL and $[L, L]$.

Example 6.5.7 If $\mathfrak{g}'$ is an ideal of a Lie algebra $\mathfrak{g}$ then $M \times \mathfrak{g}'$ is an ideal of the trivial Lie algebroid $A = TM \oplus (M \times \mathfrak{g})$, and $A' = TM \oplus (M \times \mathfrak{g}')$ is an ideal reduction of A. But note that, for $X \in \Gamma A$ and $Y' \in \Gamma A'$, it is not necessarily true that $[X, Y'] \in \Gamma A'$. $\qquad\boxtimes$

Proposition 6.5.8 *Let* $L \overset{j}{\rightarrowtail} A \overset{a}{\longrightarrow\!\!\!\!\!\gg} TM$ *be a transitive Lie algebroid on* M *and let* L' *be an ideal of* A. *Let* $\overline{A}$ *and* $\overline{L}$ *be the quotient vector bundles* $A/j(L')$ *and* L/L', *and let* $\overline{a}\colon \overline{A} \to TM$ *and* $\overline{j}\colon \overline{L} \to \overline{A}$ *be the vector bundle morphisms induced by* a *and* j. *Define a bracket on* $\Gamma(\overline{A})$ *by*

$$[X + \Gamma L', Y + \Gamma L'] = [X, Y] + \Gamma L'$$

for $X, Y \in \Gamma A$. *Then* $\overline{L} \overset{\overline{j}}{>\!\!-\!\!\!\!\longrightarrow} \overline{A} \overset{\overline{a}}{-\!\!\!\twoheadrightarrow} TM$ *is a transitive Lie algebroid on* M *and the natural projection* $\natural \colon A \longrightarrow\!\!\!\!\twoheadrightarrow \overline{A}$, $X \mapsto X + L'$, *is a surjective submersion of Lie algebroids over* M, *and has kernel* L'.

If $\varphi \colon A \to A'$ *is any surjective submersion of transitive Lie algebroids over* M, *and* $K \trianglelefteq A$ *is its kernel, then there is a unique isomorphism* $\overline{\varphi} \colon A/j(K) \to A'$ *of Lie algebroids over* M *such that* $\varphi = \overline{\varphi} \circ \natural$.

The proof is straightforward. This $\overline{A} = A/j(L')$ is the *quotient transitive Lie algebroid* of A over the ideal L'. We usually denote $A/j(L')$ by A/L'.

Proposition 6.5.9 *Let* $\varphi \colon \Omega \longrightarrow\!\!\!\!\twoheadrightarrow \Omega'$ *be a surjective submersive morphism of locally trivial Lie groupoids over* M, *and let* N *denote its kernel. Then* $AN = N_*$ *is an ideal of* $A\Omega$ *and* $\varphi_* \colon A\Omega \to A\Omega'$ *induces an isomorphism* $\overline{\varphi_*} \colon A\Omega/N_* \cong A\Omega'$.

Proof This follows by putting together 2.2.7, 3.5.17 and 6.5.8 above. $\qquad\square$

The following remark extends 6.4.23.

Remark 6.5.10 Let $\varphi \colon A \to A'$ be a morphism of transitive Lie algebroids over M. As in 6.5.3, let γ be a connection in A, let $\gamma' = \varphi \circ \gamma$, and let ∇^γ and $\nabla^{\gamma'}$ be the corresponding adjoint connections. Then the condition that φ^+ maps ∇^γ to $\nabla^{\gamma'}$ is equivalent to $\widetilde{\nabla}(\varphi^+) = 0$, where $\widetilde{\nabla}$ is the connection in $\mathrm{Hom}(L, L')$ induced from ∇^γ and $\nabla^{\gamma'}$.

The equation $\widetilde{\nabla}(\varphi^+) = 0$ may be roughly paraphrased as the statement that the rate–of–change of $\varphi^+ \colon M \to \mathrm{Hom}(L, L')$ is zero in every direction within M; the morphisms $\varphi_x^+ \colon L_x \to L_x'$ are in this sense constant with respect to x. When φ is equal to $f_* \colon A\Omega \to A\Omega'$ for a morphism of Lie groupoids $f \colon \Omega \to \Omega'$ over M, we have $\varphi_y^+ = \mathrm{Ad}(\varphi(\xi))^{-1} \circ \varphi_x^+ \circ \mathrm{Ad}(\xi)$, for every $\xi \in \Omega_x^y$ (this follows from $f_y^y = I_{\varphi(\xi)}^{-1} \circ f_x^x \circ I_\xi$). The condition $\varphi^+(\nabla^\gamma) = \nabla^{\gamma'}$ is an infinitesimal version of this equation. It is remarkable that the structure of a transitive Lie algebroid is sufficiently tight to impose this local constancy on φ^+.

In §8.2 we will give a second proof of 6.5.3, which sheds further light on the structure of φ.

The following generalization of 6.5.3 is proved by the same method.

Theorem 6.5.11 *Let A^1 and A^2 be transitive Lie algebroids on M, let $\varphi\colon A^1 \to A^2$ be a morphism of Lie algebroids over M, let ρ^1 and ρ^2 be representations of A^1 and A^2 on vector bundles E^1 and E^2, and let $\psi\colon E^1 \to E^2$ be a φ–equivariant morphism of vector bundles over M, as defined in 3.3.13. Then ψ is of locally constant rank.*

Theorem 6.5.3 is actually a special case of 6.5.11. To see this, note that if $\varphi\colon A \to A'$ is a morphism of transitive Lie algebroids over M, then there is the representation ad of A on L and the representation $X \mapsto (V' \mapsto \mathrm{ad}(\varphi(X))(V'))$ of A on L', and φ^+ is id_A–equivariant with respect to them.

For representations of transitive Lie algebroids, a refinement of 6.5.3 is necessary.

Theorem 6.5.12 *Let A be a transitive Lie algebroid on M, and let $\rho\colon A \to \mathfrak{D}(E)$ be a representation of A on a vector bundle E. Then there exist an LAB atlas $\{\psi_i\colon U_i \times \mathfrak{g} \to L_{U_i}\}$ for L and an atlas $\{\varphi_i\colon U_i \times V \to E_{U_i}\}$ for E, and representations $f_i\colon \mathfrak{g} \to \mathfrak{gl}(V)$ of $\mathfrak{g}$ on V, such that*

$$
\begin{array}{ccc}
L_{U_i} & \xrightarrow{\;\rho^+\;} & \mathrm{End}(E)_{U_i} \\[4pt]
{\scriptstyle \psi_i}\uparrow & & \uparrow{\scriptstyle \overline{\varphi}_i} \\[4pt]
U_i \times \mathfrak{g} & \xrightarrow{\;\mathrm{id}\times f_i\;} & U_i \times \mathfrak{gl}(V)
\end{array}
$$

commutes, where $\{\overline{\varphi}_i\}$ is the atlas for $\mathrm{End}(E)$ induced from the atlas $\{\varphi_i\}$ for E as in the discussion preceding 3.3.10.

Proof Let γ be a connection in A, let ∇^γ be the adjoint connection in L, let $\rho \circ \gamma$ be the produced connection in E, and let $\overline{\nabla}$ be the connection in $\mathrm{End}(E)$ induced by $\rho \circ \gamma$.

Let $\Omega = \Phi_{\mathrm{Aut}}(L) \times_{M \times M} \Phi(E)$. By a double application of 1.7.7(iii), one obtains a canonical action of Ω on $\mathrm{Hom}(L, \mathrm{End}(E))$. Then $A\Omega = \mathfrak{D}_{\mathrm{Der}}(L) \oplus_{TM} \mathfrak{D}(E)$; let $\widetilde{\nabla}$ be the connection in $A\Omega$ defined by ∇^γ and $\rho \circ \gamma$. Then it is easy to check that $\widetilde{\nabla}(\rho^+) = 0$.

Now apply 6.4.4 to $\Omega, \overline{\nabla}$ and $\rho^+ \in \Gamma\mathrm{Hom}(L, \mathrm{End}(E))$. It is easy to see that a section–atlas for Ω, with respect to which ρ^+ is locally constant, is composed of an LAB atlas $\{\psi_i\}$ and a vector bundle atlas $\{\varphi_i\}$ with the required property. $\qquad\square$

When M is connected, a single representation f of $\mathfrak{g}$ on V may be used for all i.

Corollary 6.5.13 *With the assumptions of 6.5.12, let* $E^L|_x$, *for* $x \in$ *M, denote*

$$E^{L_x}_x = \{u \in E_x \mid \rho^+_x(W)(u) = 0, \text{ for all } W \in L_x\}.$$

Then $E^L = \bigcup_{x \in M} E^L|_x$ *is a vector subbundle of* E.

Proof Firstly note that, for $W \in \mathfrak{g}$ and $u \in V$,

$$\rho^+(\psi_i(x, W))(\varphi_i(x, u)) = \varphi_i(x, f_i(W)(u)).$$

From this it follows that each $\varphi_{i,x} \colon V \to E_x$ restricts to $V^{\mathfrak{g}} \to E^{L_x}_x$. □

Continue the notation of 6.5.12. The representation ρ of A on E restricts to a representation, denoted $\bar{\rho}$, of A on E^L, since we have

$$\rho^+(W)(\rho(X)(\mu)) = [\rho^+(W), \rho(X)](\mu) + \rho(X)(\rho^+(W)(\mu))$$

for all $W \in \Gamma L$, $X \in \mathfrak{X}(M)$, $\mu \in \Gamma E$, and so if $\mu \in \Gamma(E^L)$, then the second term obviously vanishes, and the first term vanishes because $[W, X]$ is in (the image in ΓA of) ΓL.

For $\bar{\rho} \colon A \to \mathfrak{D}(E^L)$, the representation $(\bar{\rho})^+ \colon L \to \mathrm{End}(E^L)$ is of course zero.

Let γ be any connection in A, and consider the produced connection $\bar{\rho} \circ \gamma$ in E^L. Since $\overline{R}_{\bar{\rho} \circ \gamma} = (\bar{\rho})^+ \circ R_\gamma$ and since $(\bar{\rho})^+ \colon L \to \mathrm{End}(E^L)$ is zero, $\bar{\rho} \circ \gamma$ is flat. Furthermore, if γ' is a second connection in A, say $\gamma' = \gamma + j \circ \ell$ where $\ell \colon TM \to L$, then for $\mu \in \Gamma(E^L)$ and $X \in \mathfrak{X}(M)$,

$$\begin{aligned}
(\bar{\rho} \circ \gamma')(X)(\mu) &= (\bar{\rho} \circ \gamma)(X)(\mu) + (\bar{\rho})^+(\ell(X))(\mu) \\
&= (\bar{\rho} \circ \gamma)(X)(\mu).
\end{aligned}$$

Thus the representation ρ of A on E induces a unique flat connection in E^L, which we will denote by ∇^ρ. We summarize all this for reference.

Proposition 6.5.14 *Let* $L \rightarrowtail A \twoheadrightarrow TM$ *be a transitive Lie algebroid, and let* ρ *be a representation of* A *on a vector bundle* E. *Then* ρ *restricts to a representation of* A *on* E^L, *denoted* $\bar{\rho}$, *and* $\bar{\rho}$ *maps every connection in* A *to a single flat connection,* ∇^ρ, *in* E^L.

This phenomenon has a simple precursor in a purely algebraic setting: if $N \rightarrowtail G \twoheadrightarrow Q$ is an exact sequence of (set) groups, then every representation of G on a vector space V induces a representation of Q on V^N.

Definition 6.5.15 Let A be a Lie algebroid on M and let ρ be a representation of A on a vector bundle E. Then

$$(\Gamma E)^A = \{\mu \in \Gamma E \mid \rho(X)(\mu) = 0, \quad \text{for all } X \in \Gamma A\}$$

is called the *space of A–parallel sections* of E. $\qquad\square$

If A is totally intransitive, then $(\Gamma E)^A$ is a $C^\infty(M)$–submodule of ΓE. Otherwise $(\Gamma E)^A$ is merely an $\mathbb{R}$–vector subspace of ΓE. Even if A is an LAB, $(\Gamma E)^A$ need not correspond to a vector subbundle of E.

Theorem 6.5.16 *Let A be a transitive Lie algebroid on a connected base M, and let ρ be a representation of A on a vector bundle E. Choose $m \in M$ and write $\mathfrak{g} = L_m$, $V = E_m$, and $\pi_1 M = \pi_1(M, m)$. Then the evaluation map $\Gamma E \to V$ restricts to an isomorphism of vector spaces*

$$(\Gamma E)^A \overset{\cong}{\to} (V^{\mathfrak{g}})^{\pi_1 M}$$

where $\pi_1 M$ acts on $V^{\mathfrak{g}}$ via the holonomy morphism of ∇^ρ.

Proof Clearly $(\Gamma E)^A = (\Gamma(E^L))^{\nabla^\rho}$, in the notation of 6.4.2. The result now follows from 6.4.3 and 6.4.22. $\qquad\square$

By way of comparison, if ρ is a representation of a Lie groupoid Ω on a vector bundle E, then $E^{\mathscr{I}\Omega}$ is a trivializable subbundle of E, isomorphic to $M \times V^G$ as in 1.7.6, and $(\Gamma E)^\Omega$ is isomorphic to V^G (see 1.7.4). One may say that the sections of $E^{\mathscr{I}\Omega}$ invariant under the action of $M \times M$ are the constant sections, that is, they are the elements of V^G.

Theorem 6.5.16 may be regarded as the calculation of the Lie algebroid cohomology of A with coefficients in E, at degree zero. In the next chapter we will extend this to arbitrary degrees.

6.6 Notes

§6.1 This section covers the same ground as [Mackenzie, 1987a, II§6], but has been revised to make use of the notion of PBG–groupoid from [Mackenzie, 1987b, 1988b].

Given an arbitrary α–connected Lie (or topological) groupoid G, Pradines [1967b] stated the existence of an α–simply connected Lie (resp. topological) groupoid $\mathscr{M}G$, the monodromy groupoid of G, with an α–étale morphism $\mathscr{M}G \to G$. That the α–fibres of $\mathscr{M}G$ must be universal covers of the α–fibres of G, and that the set groupoid $\mathscr{M}G$ could be constructed from α–paths as is done here, was always clear; the problem was to deduce a global smooth structure on $\mathscr{M}G$ from the smooth structures on its α–fibres. This problem is closely related to the construction of the holonomy groupoid of a foliation, and led to the general scheme set out in [Pradines, 1966] for holonomy of α–structured groupoids.

The treatment in [Mackenzie, 1987a] was specific to the locally trivial case, as is that here, since in the locally trivial case the whole monodromy groupoid is determined by its vertex structure. Despite various endeavours over the years, a wholly satisfactory treatment of the monodromy groupoid for a general Lie groupoid has only emerged very recently in the work of Crainic and Moerdijk [2001].

One would think that the notion of monodromy principal bundle would have been standard in the literature before [Mackenzie, 1987a], but the only intimation of it that I have ever seen is in [Kamber and Tondeur, 1971, 6.3].

The PBG–groupoid structure of the fundamental groupoid of a principal bundle generalizes the observation of Brown and Spencer [1976b] that the fundamental groupoid of a connected topological group has a double groupoid structure of a special type. Much later Brown and I [1992] showed that the fundamental groupoid of a general α–connected Lie groupoid has the structure of a double groupoid and that the core of this double groupoid is the monodromy groupoid. This gives an intrinsic construction of the monodromy groupoid of a groupoid, without reference to its vertex bundles.

Lemma 6.1.9 is from [Kobayashi and Nomizu, 1963, App. 7].

§6.2 The material of this section is from [Mackenzie, 1987a, III§6]. The proof of 6.2.1 is essentially a groupoid reformulation of the main part of the proof in [Kobayashi and Nomizu, 1963] of the Ambrose–Singer theorem; the gist of it was given by Bowshell [1971]. The assumption that A' is transitive is essential to the possibility of transferring the smooth structures on the Φ_x globally to the groupoid Φ. This is a similar problem to that mentioned in the note above to §6.1; indeed if $A\Omega = TM$ and A' is an involutive distribution on M, then one has again the problem of realizing the graph of a foliation by a Lie groupoid, and one is led again to considerations of the the holonomy groupoid.

Theorem 6.2.1 was originally announced by Pradines [1966, 1967b] in full generality for arbitrary Lie groupoids. The full apparatus of the holonomy groupoid construction also announced in [Pradines, 1966] was needed to obtain a global integration of general Lie subalgebroids. This is a much more complicated matter than the case we have treated and a satisfactory treatment has only recently been given by Crainic and Moerdijk [2001].

Morphisms. Again, Theorem 6.2.4 was originally announced by Pradines [1966, 1967b] in full generality for arbitrary Lie groupoids. Proofs of the general result have been given in [Mackenzie and Xu, 2000] and [Moerdijk and Mrčun, 2002].

The proof of Theorem 6.2.4 given here by replacing the morphism by its graph and using the integrability of subobjects is of course a completely standard technique, and was used by Almeida and Kumpera [1981]. (For the case of Lie groups I recommend [Varadarajan, 1974, 2.7.5].)

A different proof of 6.2.4 is given in 6.4.25.

For the notion of a local system of coefficients, see, for example, [Hu, 1959, IV.15].

A treatment of local morphisms in terms of principal bundles is given by Kubarski [1989].

§6.3. This section is a revision of II§7 and part of III§7 of [Mackenzie, 1987a].

The definition of a path connection with reparametrization in a groupoid as used here is from [Virsik, 1971]. Most of the geometric interest depends upon the reparametrization condition.

Treatments of what we call a path connection have been given, albeit in passing, in [Bishop and Crittenden, 1964, 5.2], [Singer and Thorpe, 1967, 7.1].

The proof of 6.3.19 is the standard proof as in [Kobayashi and Nomizu, 1963, II 7.1, II 4.2], adapted to the groupoid setting. A simple proof that a path connected subgroup of a Lie group is a Lie subgroup is given in [Kobayashi and Nomizu, 1963, App. 4].

§6.4. This section is a light revision of [Mackenzie, 1987a, III§7].

Theorem 6.4.4 includes Theorems 1 and 2 of [Greub et al., 1973, Chapter VIII]; the proof given here was new when it appeared in [Mackenzie, 1987a]. In the series of applications of 6.4.4 which follow it, the first, 6.4.5, may be deduced equally well from [Greub et al., 1973].

Though 6.4.20 is considerably stronger than the standard statement of the theorem of Ambrose and Singer, everything required to prove 6.4.20 is implicit in the standard proofs [Kobayashi and Nomizu, 1963, II 7.1, II 8.1]; what the language of Lie groupoids and Lie algebroids has provided is the means to formulate the full content of these results.

Proposition 6.4.24 and 6.4.25 are based on the analysis of morphisms given by Ngô Van Quê [1968].

Although part of the analysis in 6.4.23 relies on 6.2.6, it would be possible to develop connection theory sufficiently to prove 6.4.25 without making use of 6.2.3 to 6.2.10. Thus the connection–theoretic proof of 6.4.25 is independent of the results in §6.2 on the local integrability of morphisms of Lie algebroids.

Notice that the construction in 6.4.18 and 6.4.19 of the γ–curvature reduction of a Lie algebroid $A\Omega$ derived from a Lie groupoid Ω and corresponding to a connection γ in $A\Omega$, may be extended to abstract transitive Lie algebroids without difficulty. The only point in 6.4.18 or 6.4.19 where Ω is used is to ensure that $L\Omega$ is an LAB.

§6.5. This section has been taken from [Mackenzie, 1987a, IV§1] with light revision. The results were new when they appeared in [Mackenzie, 1987a].

7

Cohomology
and Schouten calculus
for Lie algebroids

Despite its presence in Part II, most of the material in this chapter is valid for arbitrary Lie algebroids.

§7.1 constructs the cohomology of an arbitrary Lie algebroid with coefficients in an arbitrary representation and gives interpretations in degrees 2, 1 and 0. We use the most straightforward definition, in terms of a standard resolution of de Rham, or Chevalley–Eilenberg, type. This definition, as well as being the simplest, is the closest to the geometric applications. In 7.1.5 and 7.1.6 we prove that when A is a transitive Lie algebroid with adjoint bundle L, the cohomology spaces $\mathcal{H}^\bullet(L, \rho^+, E)$ are the modules of sections of certain flat vector bundles $H^\bullet(L, \rho^+, E)$. This is a generalization to all degrees of 6.5.13 and is proved using a generalization of the calculus of differential forms on a manifold, and other results of Chapter 6. In 7.1.7 and 7.1.8 we calculate $\mathcal{H}^\bullet(A\Omega, \rho, E)$ for a locally trivial Lie groupoid Ω and any representation ρ of $A\Omega$, in terms of the equivariant de Rham cohomology of an associated principal bundle.

In the second part of §7.1 we interpret $\mathcal{H}^2(A, \rho, E)$ in terms of equivalence classes of operator extensions of A by E. This is a straightforward generalization of the corresponding extension theory of Lie algebras, but we have given at least sketch proofs of most results, since there is no readily available account of the Lie algebra theory in the detail which is required for the geometric applications here. In Lie algebra cohomology, as in other cohomology theories of algebraic type, there is little interest in specific cocycles or in specific transversals for extensions: one is there only interested in cohomological invariants. In Lie algebroid cohomology, however, transversals are infinitesimal connections (either in the Lie algebroid or in an associated structure) and cocycles are, in degree two, curvature forms, and, in degree three, the left–hand sides of Bianchi identities. Thus the focus of geometric interest is usually on specific transversals or cocycles. It is for this reason that the standard cochain complex is the best for our purposes.

In §7.1 we treat only extensions by abelian (totally intransitive) Lie algebroids; that is, by vector bundles. The general case, which includes the most important applications, is treated in §7.3. If $P(M,G)$ is a principal bundle with abelian structure group then

$$M \times \mathfrak{g} \cong \frac{P \times \mathfrak{g}}{G} \rightarrowtail \frac{TP}{G} \longrightarrow\!\!\!\!\!\rightarrow TM$$

is an extension of TM by the trivial vector bundle $M \times \mathfrak{g}$; if G is compact and M is simply–connected then the cohomology class of $\frac{TP}{G}$ in $\mathscr{H}^2(TM, M \times \mathfrak{g}) \cong H^2_{\mathrm{deRh}}(M, \mathfrak{g})$ is the sum of the Chern classes of the component $SO(2)$–bundles of P (see §7.4). In the case where G is not compact and $\pi_1 M$ is arbitrary, the class in $\mathscr{H}^2(TM, M \times \mathfrak{g}) \cong H^2_{\mathrm{deRh}}(\widetilde{M}, \mathfrak{g})^{\pi_1 M}$ defined by $\frac{TP}{G}$ may still be regarded as a characteristic class of $P(M,G)$.

Given a non–abelian extension $K \overset{\iota}{\rightarrowtail} A' \overset{\pi}{\longrightarrow\!\!\!\!\!\rightarrow} A$, each $X' \in \Gamma A'$ induces a bracket derivation $V \mapsto \iota^{-1}[X', \iota(V)]$ of K. In the abelian case these operators depend only on $\pi(X')$; in the non–abelian case, however, $X'' \in \Gamma A'$ with $\pi(X'') = \pi(X')$ will induce an operator which differs from that induced by X' by an element of $\Gamma \operatorname{ad}(K) \subseteq \Gamma \mathfrak{D}_{\mathrm{Der}} K$. The extension therefore induces, not a representation of A on K, but a morphism $\Xi \colon A \to \mathfrak{D}_{\mathrm{Der}} K/\operatorname{ad}(K)$ which we call a *coupling* of A with K. The construction of $\mathfrak{D}_{\mathrm{Der}} K/\operatorname{ad}(K)$, which we call the Lie algebroid of outer bracket derivations, and denote by $\operatorname{Out}\mathfrak{D}(K)$, is given in §7.2.

Given a coupling $\Xi \colon A \to \operatorname{Out}\mathfrak{D}(K)$, there is a natural representation, denoted ρ^Ξ, of A on ZK, and from 7.2.2 to 7.2.12 we are concerned to show that Ξ defines an element of $\mathscr{H}^3(A, \rho^\Xi, ZK)$, called the *obstruction class* of Ξ and denoted $\operatorname{Obs}(\Xi)$. This completes the necessary preparation for the classification theory of non–abelian extensions of Lie algebroids, and its application to the problem of constructing a connection with a prescribed curvature form, which is the subject of §7.3.

From 7.3.1 through to 7.3.6 we give the detailed construction of the coupling arising from a (nonabelian) extension and in 7.3.5 prove that for such a coupling the obstruction class is zero. From 7.3.7 through to 7.3.18 we establish the converse of 7.3.5. The main construction result is 7.3.7 which in 7.3.9 shows that a coupling Ξ with obstruction class zero arises from an extension. From 7.3.10 through to 7.3.18 we classify the extensions which induce a given coupling Ξ; the analysis shows that

$\mathscr{H}^2(A, \rho^\Xi, ZK)$ acts freely and transitively on the set of equivalence classes of Ξ–operator extensions of A by K. In 7.3.19 to 7.3.21 we address the question of semidirect (or flat) extensions, and note that not every coupling with zero obstruction class arises from a semidirect extension; on the other hand there may be several inequivalent semidirect extensions inducing a given coupling. The section closes with a brief application of 7.3.7 to the construction of produced Lie algebroids.

The technique used in §7.3 is algebraic extension theory of a standard type: it does however yield results of genuine geometric interest. From 7.3.7 we obtain a solution to the algebraic half of the problem: When is a 2–form the curvature of a connection? Namely it follows that an LAB–valued 2–form $R \in \Omega^2(M, L)$ on a manifold M is the curvature of a connection in a Lie algebroid if and only if there is a Lie connection ∇ in L such that $\overline{R}_\nabla = \mathrm{ad} \circ R$ and $\nabla(R) = 0$. In Chapter 8 we will answer the question of when the resulting Lie algebroid can be integrated to a Lie groupoid.

A second application of the results of §7.3 is also in Chapter 8 where the classification of nonabelian extensions is the key to the proof that a transitive Lie algebroid on a contractible base admits a flat connection.

For a transitive Lie algebroid $L \rightarrowtail A \twoheadrightarrow TM$ and an arbitrary representation of A on a vector bundle E we construct in §7.4 a natural spectral sequence $\mathscr{H}^s(TM, H^t(L, E)) \implies \mathscr{H}^n(A, E)$ which gives the cohomology of A in terms of those of TM and L. The construction follows closely that of the spectral sequence of an extension of Lie algebras, and is at the same time a generalization of the Leray–Serre spectral sequence of a principal bundle in de Rham cohomology.

We are concerned particularly with the relationship between extensions of a transitive Lie algebroid and extensions of its adjoint bundle. Given a principal bundle $P(M, G)$ and a surjective group morphism $H \twoheadrightarrow G$, there are well–known obstructions to lifting P to H. Given a transitive Lie algebroid $L \rightarrowtail A \twoheadrightarrow TM$ and a surjective morphism of LABs $L' \twoheadrightarrow L$, there is a further requirement of consistency along the fibres. Thus we study the images and kernels of the restriction and inflation maps: given an extension of transitive Lie algebroids $E \rightarrowtail A' \twoheadrightarrow A$, its *restriction* is the induced extension of the adjoint bundles, $E \rightarrowtail L' \twoheadrightarrow L$; the inflation map constructs certain extensions $E \rightarrowtail A' \twoheadrightarrow A$ from transitive Lie algebroids on M with adjoint bundle E^L. The images and kernels of these maps have natural expressions in terms of the spectral sequence.

The spectral sequence of a transitive Lie algebroid provides an abstraction and algebraization of the Leray–Serre spectral sequence of a principal bundle in de Rham cohomology. Because the construction of the Lie algebroid spectral sequence is algebraic, and because coefficients in general vector bundles are permitted, it is possible to apply techniques developed in the algebraic theory for extensions of discrete groups, or of Lie algebras, to the de Rham spectral sequence of a principal bundle.

Throughout §7.1—§7.4, one sees that the standard identities of infinitesimal connection theory arise naturally in cohomological terms.

Just as the calculus of differential forms extends to arbitrary Lie algebroids, so too does the classical Schouten calculus of multi–vector fields. In §7.5 we give the basic formulas, which will be needed in Chapter 10.

7.1 Cohomology and abelian extensions

Up to 7.1.4, let A be an arbitrary Lie algebroid on a base M. It is not assumed that A is transitive. Let $\rho\colon A \to \mathfrak{D}(E)$ be a representation of A on a vector bundle E.

Definition 7.1.1 The *standard complex* associated with the vector bundle E and the representation ρ of A is the sequence of vector bundles $C^n(A, E) = \mathrm{Alt}^n(A; E)$, $n \geqslant 0$, and the differential operators $d\colon \Gamma C^n(A, E) \to \Gamma C^{n+1}(A, E)$ defined by

$$df(X_1, \ldots, X_{n+1}) = \sum_{r=1}^{n+1} (-1)^{r+1} \rho(X_r)(f(X_1, \ldots, \widehat{X_r}, \ldots, X_{n+1}))$$

$$+ \sum_{r<s} (-1)^{r+s} f([X_r, X_s], X_1, \ldots, \widehat{X_r}, \ldots, \widehat{X_s}, \ldots, X_{n+1})$$

for $f \in \Gamma C^n(A, E)$ and $X_1, \ldots, X_{n+1} \in \Gamma A$. (The circumflex denotes omission of the argument indicated.)

The *cohomology spaces* $\mathscr{H}^n(A, \rho, E)$, or $\mathscr{H}^n(A, E)$, are the cohomology spaces of this complex, namely $\mathscr{H}^n(A, E) = \mathscr{Z}^n(A, E)/\mathscr{B}^n(A, E)$, where $\mathscr{Z}^n(A, E) = \ker d\colon \Gamma C^n(A, E) \to \Gamma C^{n+1}(A, E)$ and $\mathscr{B}^n(A, E) = \mathrm{im}\, d\colon \Gamma C^{n-1}(A, E) \to \Gamma C^n(A, E)$ for $n \geqslant 1$, $\mathscr{B}^0(A, E) = (0)$. □

It is routine to verify that $d^2 = 0$. When A is totally intransitive (so that it is a vector bundle together with a field of Lie algebra brackets), the operators d are $C^\infty(M)$–linear. This is easy to check. Hence in this case, the d induce vector bundle morphisms $C^n(A, E) \to C^{n+1}(A, E)$, for each n, also denoted d. It is not true, however, that in this case the d must be of locally constant rank, and so the images and kernels may not be subbundles. This is so even if A is an LAB. Examples are easy to construct, using the same device as in 3.3.6.

In all other cases the anchor is non–zero and the $d\colon \Gamma C^n(A, E) \to \Gamma C^{n+1}(A, E)$ are first–order differential operators, and do not induce morphisms of the underlying vector bundles. Thus the $\mathscr{H}^n(A, E)$ are quotients of infinite–dimensional real vector spaces, and are at this stage rather formless. However for a transitive Lie algebroid A, we will show in §7.4 that the $\mathscr{H}^n(A, E)$ are computable.

When M is a point and A is a finite–dimensional real Lie algebra, the $\mathscr{H}^n(A, E)$ reduce to the Chevalley–Eilenberg cohomology spaces. When A is the tangent bundle TM and ρ is the trivial representation of TM in a product vector bundle $M \times V$ however, $\mathscr{H}^n(TM, M \times V)$ is the real de Rham cohomology space $H^n_{\mathrm{deRh}}(M, V)$. In 7.1.7 we will calculate $\mathscr{H}^n(A\Omega, \rho_*, E)$ for a locally trivial Lie groupoid Ω and a representation ρ_* induced from a groupoid representation $\rho\colon \Omega \to \Phi(E)$.

Definition 7.1.2 Fix $X \in \Gamma A$.

(i) The *Lie derivative* $\mathfrak{L}_X\colon \Gamma C^n(A, E) \to \Gamma C^n(A, E)$ is defined by

$$\mathfrak{L}_X(f)(X_1, \ldots, X_n) = \rho(X)(f(X_1, \ldots, X_n))$$
$$- \sum_{r=1}^{n} f(X_1, \ldots, [X, X_r], \ldots, X_n).$$

Here $f \in \Gamma C^n(A, E)$ and $X_r \in \Gamma A$, $1 \leqslant r \leqslant n$.

(ii) The *interior multiplication* $\iota_X\colon \Gamma C^{n+1}(A, E) \to \Gamma C^n(A, E)$, for $n > 0$, is defined by

$$\iota_X(f)(X_1, \ldots, X_n) = f(X, X_1, \ldots, X_n),$$

for $f \in \Gamma C^n(A, E)$, $X_r \in \Gamma A$, $1 \leqslant r \leqslant n$. $\qquad \square$

The operators $\mathfrak{L}_X, \iota_X$ and d satisfy a set of formulas identical in form to those which hold in the calculus of vector–valued forms on a manifold. Those which are used in the sequel follow. The proofs are completely straightforward.

Proposition 7.1.3 (i) *For $X, Y \in \Gamma A$, $u \in C^\infty(M)$, $f \in \Gamma C^n(A, E)$,*

$$\iota_X(uf) = u\iota_X(f), \qquad \iota_{uX}(f) = u\iota_X(f) \qquad and \qquad \iota_X\iota_Y = -\iota_Y\iota_X,$$

(ii) *for $X \in \Gamma A$, $u \in C^\infty(M)$, $f \in \Gamma C^n(A, E)$,*

$$\mathfrak{L}_X(uf) = u\mathfrak{L}_X(f) + a(X)(u)f,$$

(iii) *for* $X, X_1, \ldots, X_n \in \Gamma A$, $u \in C^\infty(M)$, $f \in \Gamma C^n(A, E)$;

$$\mathcal{L}_{uX}(f)(X_1, \ldots, X_n) = u\mathcal{L}_X(f)(X_1, \ldots, X_r)$$

$$+ \sum_{r=1}^{n} (-1)^{r-1} a(X_r)(u) \iota_X(f)(X_1, \ldots, \widehat{X_r}, \ldots, X_n),$$

(iv) $\mathcal{L}_{[X,Y]} = \mathcal{L}_X \circ \mathcal{L}_Y - \mathcal{L}_Y \circ \mathcal{L}_X$ *for* $X, Y \in \Gamma A$,
(v) $\mathcal{L}_X = \iota_X \circ d + d \circ \iota_X$ *for* $X \in \Gamma A$,
(vi) $\mathcal{L}_X \circ d = d \circ \mathcal{L}_X$ *for* $X \in \Gamma A$,
(vii) $[\mathcal{L}_X, \iota_Y] = \mathcal{L}_X \circ \iota_Y - \iota_Y \circ \mathcal{L}_X = \iota_{[X,Y]}$ *for* $X, Y \in \Gamma A$.

With the definition of wedge product given in §7.4, (iii) can be written

$$\mathcal{L}_{uX}(f) = u\mathcal{L}_X(f) + du \wedge \iota_X(f), \tag{1}$$

where du is taken with respect to the representation a of A on $M \times \mathbb{R}$.

Proposition 7.1.4 *Let* $L \overset{j}{>\!\!-\!\!\!\longrightarrow} A \overset{a}{-\!\!\!\longrightarrow\!\!\!\!\gg} TM$ *be a transitive Lie algebroid on* M. *Then for* $X \in \Gamma A$ *and* $f \in \Gamma C^n(L, E)$, *the Lie derivative* $\mathcal{L}_X(f)$ *of* f *is in* $\Gamma C^n(L, E)$, *the map* $\mathcal{L}_X \colon \Gamma C^n(L, E) \to \Gamma C^n(L, E)$ *is in* $\Gamma \mathfrak{D}(C^n(L, E))$ *and* $X \mapsto \mathcal{L}_X$ *defines a representation of* A *on* $C^n(L, E)$.

Proof For the first statement, note that

$$\mathcal{L}_X(f)(V_1, \ldots, V_n) = \rho(X)(f(V_1, \ldots, V_n))$$

$$- \sum_{r=1}^{n} f(V_1, \ldots, \mathrm{ad}(X)(V_r), \ldots, V_n)$$

for $V_r \in \Gamma L$, $X \in \Gamma A$, $f \in \Gamma C^n(L, E)$. The remaining assertions follow from 7.1.3(ii), (iii), (iv). $\qquad\qquad\square$

In the cohomology theory of Lie algebras, $\mathcal{L}$ is a representation of a Lie algebra $\mathfrak{g}$ on the associated spaces $C^n(\mathfrak{g}, V)$. In the context of general Lie algebroids however, the map $\mathcal{L} \colon \Gamma A \to \Gamma \mathfrak{D}(C^n(A, E))$ is not $C^\infty(M)$–linear, and therefore cannot be said to be a representation. As in the case of ad$\colon X \mapsto (Y \mapsto [X, Y])$, $\Gamma A \to \Gamma \mathfrak{D}(A)$, which $\mathcal{L}$ of course generalizes, this lack of $C^\infty(M)$–linearity can be avoided by lifting $\mathcal{L}$ to the 1–jet prolongation of A. However the representation of A on $C^n(L, E)$ will suffice for our purposes.

Although the equation $\mathcal{L}_X = \iota_X \circ d + d \circ \iota_X$ is meaningless for the restricted $\mathcal{L}_X$ of 7.1.4, its consequence, $d \circ \mathcal{L}_X = \mathcal{L}_X \circ d$, continues to be valid. This follows trivially from 7.1.3(vi). From this formula we have the following crucial result.

Theorem 7.1.5 *Let* $L \overset{j}{>\!\!\!-\!\!\!-\!\!\!>} A \overset{a}{-\!\!\!-\!\!\!\gg} TM$ *be a transitive Lie algebroid on* M. *Then the coboundaries*

$$d^n \colon C^n(L, E) \to C^{n+1}(L, E)$$

are of locally constant rank, and consequently there are well–defined vector bundles $Z^n(L, \rho, E) = \ker d^n$, $B^n(L, \rho, E) = \operatorname{im} d^{n-1}$ *and*

$$H^n(L, \rho, E) = Z^n(L, \rho, E)/B^n(L, \rho, E)$$

such that $\Gamma H^n(L, \rho, E) = \mathcal{H}^n(L, \rho^+, E)$.

Furthermore, $\mathfrak{L} \colon A \to \mathfrak{D}(C^n(L, E))$ *induces a well–defined representation, also denoted* $\mathfrak{L}$, *of* A *on* $H^n(L, E)$.

Proof Let γ be a connection in A, and let γ^n be the connection $\mathfrak{L} \circ \gamma$ in $C^n(L, E)$. Then the equation $\mathfrak{L}_X \circ d^n = d^{n+1} \circ \mathfrak{L}_X$ for $X \in \Gamma A$ implies in particular that $\mathfrak{L}_{\gamma(X)}(d(f)) = d(\mathfrak{L}_{\gamma(X)}(f))$ for $X \in \Gamma TM, f \in \Gamma C^n(L, E)$; that is, $\gamma^{n+1}(X)(d(f)) = d(\gamma^n(X)(f))$. Thus d maps γ^n to γ^{n+1} and so, by 6.4.6, d is of locally constant rank.

That the representation $\mathfrak{L}$ of A on $C^n(L, E)$ induces a well–defined representation of A on $H^n(L, E)$ follows likewise from the equation $\mathfrak{L}_X \circ d = d \circ \mathfrak{L}_X, X \in \Gamma A$. $\qquad\square$

Theorem 7.1.6 *Continuing the notation of 7.1.5, for the representation* $\mathfrak{L}$ *of* A *on* $H^n(L, E)$ *we have*

$$H^n(L, E)^L = H^n(L, E).$$

In particular, $H^n(L, E)$ *is a flat vector bundle, and* $\mathfrak{L}$ *maps every connection in* A *to a single flat connection, denoted* $\nabla^{\rho, n}$, *in* $H^n(L, E)$.

Proof Applying 7.1.3 to the Lie algebroid L and the representation ρ^+ we have

$$\mathfrak{L}_V = d \circ \iota_V + \iota_V \circ d,$$

for $V \in \Gamma L$. Now take $f \in C^n(L, E)$ with $df = 0$. We have $\mathfrak{L}_V(f) = d(\iota_V(f)) + 0$ and so

$$\mathfrak{L}_V([f]) = [\mathfrak{L}_V(f)] = [d(\iota_V(f))] = 0.$$

Thus $[f] \in H^n(L, E)^L$.

The remaining statements follow from applying 6.5.14 to $\mathfrak{L}\colon A \to \mathfrak{D}\,(H^n(L,E))$. □

We hope that the proofs of 7.1.5 and 7.1.6 give the reader some amusement. Thanks to these results it will be possible, in §7.4, to consider the cohomology of TM with coefficients in $H^\bullet(L,E)$, and to relate the spaces $\mathscr{H}^\bullet(TM, H^\bullet(L,E))$ to the cohomology of A.

In the case of the transitive Lie algebroid of a locally trivial Lie groupoid, the Lie algebroid cohomology is the equivariant de Rham cohomology of the corresponding principal bundle.

Theorem 7.1.7 *Let Ω be a locally trivial Lie groupoid on M and let $\rho\colon \Omega \to \Phi(E)$ be a representation of Ω on a vector bundle E. Then there are natural isomorphisms*

$$\mathscr{H}^n(A\Omega, \rho_*, E) \cong H^n_{\mathrm{deRh}}(P,V)^G$$

where $P = \Omega_m, G = \Omega_m^m$ and $V = E_m$ for some chosen $m \in M$.

Proof This is an immediate consequence of 5.3.11. □

Corollary 7.1.8 *If Ω is an α–connected locally trivial Lie groupoid on M and $\rho\colon A\Omega \to \mathfrak{D}\,(E)$ is any representation of $A\Omega$ on a vector bundle E, then there are natural isomorphisms*

$$\mathscr{H}^n(A\Omega, \rho, E) \cong H^n_{\mathrm{deRh}}(\widetilde{P},V)^H$$

where H is the structure group of the monodromy bundle $\widetilde{P}(M,H)$ of $P(M,G)$.

In particular, for a flat vector bundle E on M, and a flat connection ∇ in E,

$$\mathscr{H}^n(TM, \nabla, E) \cong H^n_{\mathrm{deRh}}(\widetilde{M},V)^{\pi_1 M}, \qquad (2)$$

the equivariant de Rham cohomology of the universal cover of M, constructed from forms $\omega \in \Omega^*(\widetilde{M},V)$ which are equivariant with respect to the holonomy action of $\pi_1(M)$ on the fibre type V. In this way Lie algebroid cohomology may be regarded as a generalization of de Rham cohomology in which coefficients in local systems of vector spaces are permitted.

Thus in the case of the Lie algebroids of locally trivial Lie groupoids, the Lie algebroid cohomology is a known invariant, though one which has only been extensively studied in the case where the structure group is compact. One of the strengths of the Lie algebroid formulation, however, is that it is a cohomology theory of algebraic type, comparable to the cohomology theories of Lie algebras and of discrete groups, and that what is significant from the point of view of the algebraic cohomology theory is also significant geometrically. We will spend much of the rest of this chapter justifying this observation and developing its consequences. The first step in this process is the interpretation of $\mathscr{H}^2(A, E)$ in terms of equivalence classes of extensions of A by E, and for this we need the following general concept of curvature.

Proposition 7.1.9 *Let A and A' be Lie algebroids on M, not necessarily transitive, and let $\varphi\colon A \to A'$ be a morphism of vector bundles over M such that $a' \circ \varphi = a$. Then*

$$R_\varphi(X, Y) = \varphi([X, Y]) - [\varphi(X), \varphi(Y)] \tag{3}$$

defines a map $\Gamma A \times \Gamma A \to \Gamma A'$ which is alternating and $C^\infty(M)$-bilinear, and thus defines a section of $\mathrm{Alt}^2(A; A')$, called the curvature *of φ.*

For $X, Y, Z \in \Gamma A$ we have

$$\mathfrak{S}\{[\varphi(X), R_\varphi(Y, Z)] - R_\varphi([X, Y], Z)\} = 0 \tag{4}$$

where $\mathfrak{S}$ is the cyclic sum.

Proof In the first paragraph only the $C^\infty(M)$-bilinearity is not clear, and it follows by calculation. Formula (4) follows from the Jacobi identity in $\Gamma A'$. $\qquad\square$

The curvature map $\overline{R}_\gamma\colon TM \oplus TM \to L$ defined in 5.2.9 for a connection $\gamma\colon TM \to A$ in a transitive Lie algebroid is related to this R_γ by $R_\gamma = j \circ \overline{R}_\gamma$. Equation (4) generalizes (16) on p. 190, and may be called an abstract Bianchi identity. Vector bundle morphisms $\varphi\colon A \to A'$ with $a' \circ \varphi = a$ will be called *anchor–preserving maps*. The next result is a simple calculation.

Proposition 7.1.10 *Let $\varphi\colon A \to A'$ and $\psi\colon A' \to A''$ be anchor–preserving maps of Lie algebroids. Then*

$$R_{\psi \circ \varphi} = R_\psi \circ (\varphi \times \varphi) + \psi \circ R_\varphi.$$

We return now to the conventions made at the start of this section: A is a Lie algebroid on base M, not necessarily transitive, and ρ is a representation of A on a vector bundle M. In the following definition, we repeat from 3.3.19 the definition of an extension of Lie algebroids.

Definition 7.1.11 An *extension* of A by the vector bundle E is an exact sequence

$$E \rightarrowtail^{\iota} A' \twoheadrightarrow^{\pi} A \tag{5}$$

of Lie algebroids over M, where E is considered to be an abelian Lie algebroid.

A *transversal* in (5) is a vector bundle morphism $\chi \colon A \to A'$ such that $\pi \circ \chi = \mathrm{id}_A$. A *retroversal* in (5) is a vector bundle morphism $\lambda \colon A' \to E$ such that $\lambda \circ \iota = \mathrm{id}_E$.

A transversal is *flat* if it is a morphism of Lie algebroids; equivalently, if it has zero curvature. The extension (5) is *flat* if it has a flat transversal.   □

As with connections (see the discussion in §5.2) there is a bijective correspondence between transversals χ and retroversals λ, given by

$$\iota \circ \lambda + \chi \circ \pi = \mathrm{id}_{A'},$$

and a corresponding pair satisfy $\lambda \circ \chi = 0$. Since π is a surjective submersion and a morphism of vector bundles over M, transversals always exist. Any choice of transversal determines an isomorphism of vector bundles $A' \cong A \oplus E$.

From $\pi \circ \chi = \mathrm{id}_A$ it follows that $a' \circ \chi = a$, so a transversal is automatically anchor–preserving. From this and $a \circ \pi = a'$ it follows that $\mathrm{im}(a) = \mathrm{im}(a')$. Hence A' is transitive if and only if A is transitive, and A' is totally intransitive if and only if A is totally intransitive.

For transversals $\chi \colon A \to A'$ of (5), we will normally use as curvature the map $\overline{R}_\chi \colon A \oplus A \to E$ with $\iota \circ \overline{R}_\chi = R_\chi$. Note that $\overline{R}_\chi \in \Gamma C^2(A, E)$.

Recall from 3.3.20 that an extension such as (5) induces a representation $\rho^{A'}$ of A on E, which can be written as

$$\iota(\rho^{A'}(X)(\mu)) = [\chi(X), \iota(\mu)],$$

where $X \in \Gamma A$, $\mu \in \Gamma E$, for any transversal χ.

Definition 7.1.12 Given the Lie algebroid A and the action ρ of A on E, the extension (5) is an *operator extension* of A by E if $\rho^{A'} = \rho$.

Two operator extensions $E \rightarrowtail^{\iota_s} A^s \twoheadrightarrow^{\pi_s} A$, $s = 1, 2$ are *equivalent* if there is a morphism of Lie algebroids $\varphi\colon A^1 \to A^2$ over M (necessarily an isomorphism) such that $\varphi \circ \iota_1 = \iota_2$ and $\pi_2 \circ \varphi = \pi_1$. $\qquad\square$

The set of equivalence classes of operator extensions is denoted by $\mathscr{O}\mathrm{pext}(A, \rho, E)$, or by $\mathscr{O}\mathrm{pext}(A, E)$ if ρ is understood. There is a natural bijection between $\mathscr{H}^2(A, \rho, E)$ and $\mathscr{O}\mathrm{pext}(A, \rho, E)$ described in the following proposition. The proof is a straightforward series of calculations.

Proposition 7.1.13 (i) *Let $f \in \Gamma C^2(A, E)$ be a cocycle, that is, let*

$$\mathfrak{S}\{\rho(X)(f(Y, Z)) - f([X, Y], Z)\} = 0$$

for all $X, Y, Z \in \Gamma A$. Denote by A^f the vector bundle $A \oplus E$ equipped with the anchor $a^f(X \oplus \mu) = a(X)$, and the bracket

$$[X \oplus \mu, Y \oplus \nu] = [X, Y] \oplus \{\rho(X)(\nu) - \rho(Y)(\mu) - f(X, Y)\}$$

on ΓA^f. Then A^f is a Lie algebroid on M and, with respect to the natural maps $\iota_2\colon E \to A \oplus E$ and $\pi^1\colon A \oplus E \to A$,

$$E \xrightarrow{\;\iota_2\;} A^f \xrightarrow{\;\pi_1\;} A$$

is an operator extension of A by E. The transversal $\iota_1\colon A \to A^f$ has curvature $\overline{R}_{\iota_1} = f$.

(ii) *Conversely, let $E \xrightarrow{\;\iota\;} A' \xrightarrow{\;\pi\;} A$ be an operator extension of A by E and let χ be a transversal. Then $\overline{R}_\chi$ is a cocycle, that is, $d(\overline{R}_\chi) = 0$, with respect to the coboundary induced by ρ. Furthermore,*

$$\mathscr{E}_\chi\colon A^{\overline{R}_\chi} \to A', \quad X \oplus \mu \mapsto \chi(X) + \iota(\mu)$$

is an equivalence of $E \xrightarrow{\quad} A^{\overline{R}_\chi} \xrightarrow{\quad} A$ with $E \xrightarrow{\quad} A' \xrightarrow{\quad} A$, and $\mathscr{E}_\chi$ maps ι_1 to χ.

(iii) *Let $g \in \Gamma C^1(A, E)$; recall that*

$$dg(X, Y) = \rho(X)(g(Y)) - \rho(Y)(g(X)) - g([X, Y])$$

for $X, Y \in \Gamma A$. Given any $f \in \mathscr{Z}^2(A, E)$, the map

$$\varepsilon_g\colon A^f \to A^{f+dg}, \quad X \oplus \mu \mapsto X \oplus (\mu + g(X))$$

is an equivalence of extensions.

(iv) *Conversely, if $E \xrightarrow{\;\iota_s\;} A^s \xrightarrow{\;\pi_s\;} A$, $s = 1, 2$, are operator extensions, and $\varphi\colon A^1 \to A^2$ is an equivalence, then for any pair of transversals χ_1, χ_2 for A^1, A^2, there is a unique $g \in \Gamma C^1(A, E)$ such that*

$$\varphi = \mathscr{E}_{\chi_2} \circ \varepsilon_g \circ \mathscr{E}_{\chi_1}^{-1},$$

namely $g = \lambda_2 \circ \varphi \circ \chi_1$ *and then* $dg = f_2 - f_1$.

Indeed each cochain $g \in \Gamma C^1(A, E)$ *induces a permutation* $\chi \mapsto \chi^g$ *of the transversals in any operator extension* $E \overset{\iota}{>\!\!\!-\!\!\!-\!\!\!>} A' \overset{\pi}{-\!\!\!-\!\!\!\gg} A$. *If* χ *is a transversal, then* $\chi^g = \chi + \iota \circ g$ *is another, and* $\overline{R}_{\chi^g} = \overline{R}_\chi - dg$.

The linear structure on $\mathscr{H}^2(A, E)$ now transfers to $\mathscr{O}\mathrm{pext}(A, E)$. Recall that if A is totally intransitive, $\mathscr{H}^2(A, E)$ is a $C^\infty(M)$–module and is otherwise merely an $\mathbb{R}$–vector space. The module multiplication is given by the following construction.

Proposition 7.1.14 *Consider an operator extension (5) of A by E. Let $\widetilde{E}$ be a second vector bundle on M and $\widetilde{\rho}\colon A \to \mathfrak{D}(\widetilde{E})$ a representation of A on $\widetilde{E}$. Then if $\varphi\colon E \to \widetilde{E}$ is an A–equivariant morphism of vector bundles over M, there is a unique extension $\widetilde{E} >\!\!\!-\!\!\!-\!\!\!> \widetilde{A} -\!\!\!-\!\!\!\gg A$ of A by $\widetilde{E}$ which induces $\widetilde{\rho}$ and which is such that there is a morphism of Lie algebroids $\widetilde{\varphi}\colon A \to \widetilde{A}$ making*

$$
\begin{array}{ccccc}
E & >\!\!\!-\!\!\!-\!\!\!> & A' & -\!\!\!-\!\!\!\gg & A \\
{\scriptstyle\varphi}\downarrow & & {\scriptstyle\widetilde{\varphi}}\downarrow & & \| \\
\widetilde{E} & >\!\!\!-\!\!\!-\!\!\!> & \widetilde{A} & -\!\!\!-\!\!\!\gg & A
\end{array}
$$

commute.

Proof Choose a transversal $\chi\colon A \to A'$ and define a bracket on $\Gamma(A \oplus \widetilde{E})$ by

$$
[X \oplus \widetilde{\mu}, Y \oplus \widetilde{\mu}] = [X, Y] \oplus \{\widetilde{\rho}(X)(\widetilde{\nu}) - \widetilde{\rho}(Y)(\widetilde{\mu}) - \varphi(\overline{R}_\chi(X, Y))\}.
$$

With this bracket and with anchor $\widetilde{a}(X \oplus \widetilde{\mu}) = a(X)$, $A \oplus \widetilde{E}$ becomes a Lie algebroid on M; denote it by $\widetilde{A}$. It is easily checked that $\widetilde{A}$ is independent of the choice of χ, up to isomorphism, that

$$
\widetilde{E} \overset{\iota_2}{>\!\!\!-\!\!\!-\!\!\!>} \widetilde{A} \overset{\pi_1}{-\!\!\!-\!\!\!\gg} A \tag{6}
$$

is a $\widetilde{\rho}$–operator extension, and that $X' \mapsto \pi(X') \oplus \varphi\lambda(X')$ has the properties required of $\widetilde{\varphi}$. $\square$

The extension (6) is the *pushout extension* of (5) along φ. It is tedious, and unrewarding, to give a proof of 7.1.14 without using 7.1.13.

Now, for an arbitrary Lie algebroid A and an extension (5), the map $E \to E$, $\mu \mapsto k\mu$, where $k \in \mathbb{R}$ is a constant, is equivariant and the corresponding pushout is the scalar multiple of $[E >\!\!\!-\!\!\!-\!\!\!> A' -\!\!\!-\!\!\!\gg A] \in \mathscr{O}\mathrm{pext}(A, E)$ by k. If $A = L$ is totally intransitive, and $u \in C^\infty(M)$ is a function, then $E \to E$, $\mu \mapsto u(q\mu)\mu$, is equivariant, and the pushout extension similarly defines the $C^\infty(M)$–module structure.

To define the addition on $\mathcal{O}\text{pext}(A, E)$ the construction of pullback extensions is required. This construction can be given in greater generality than the construction of pushouts. Again, the proof is straightforward.

Proposition 7.1.15 *Let* $\varphi^1\colon A^1 \to A$ *and* $\varphi^2\colon A^2 \to A$ *be morphisms of Lie algebroids over* M *such that*

$$\text{im}\,\varphi_x^1 + \text{im}\,\varphi_x^2 = A_x$$

for all $x \in M$. *Then the pullback vector bundle*

$$\widetilde{A} = \{X_1 \oplus X_2 \in A^1 \oplus A^2 \mid \varphi^1(X_1) = \varphi^2(X^2)\}$$

is a Lie algebroid with respect to the anchor $\widetilde{a}(X_1 \oplus X_2) = a^1(X_1) = a^2(X_2)$, *and the bracket*

$$[X_1 \oplus X_2, Y_1 \oplus Y_2] = [X_1, X_2] \oplus [X_2, Y_2].$$

Further, the restrictions to $\widetilde{A}$ *of the projections* $\pi_1\colon A^1 \oplus A^2 \to A^1$ *and* $\pi_2\colon A^1 \oplus A^2 \to A^2$ *are Lie algebroid morphisms over* M, *and if* $\overline{A}$ *is any other Lie algebroid on* M *and* $\psi^1\colon \overline{A} \to A^1$, $\psi^2\colon \overline{A} \to A^2$ *are morphisms of Lie algebroids over* M *such that* $\varphi^1 \circ \psi^1 = \varphi^2 \circ \psi^2$, *then there is a unique Lie algebroid morphism* $\psi\colon \overline{A} \to \widetilde{A}$ *such that* $\pi_1 \circ \psi = \psi_1, \pi_2 \circ \psi = \psi_2$.

This $\widetilde{A}$ is a pullback Lie algebroid in the more general sense of §4.2, and may be denoted $A^1 \oplus_A A^2$. The direct sum Lie algebroid of 3.3.22 is a particular instance. The second paragraph of 7.1.15 embodies the pullback property. We will normally denote the maps $\widetilde{A} \to A^1$, $\widetilde{A} \to A^2$ by $\widetilde{\varphi}_2$, $\widetilde{\varphi}_1$ respectively.

Proposition 7.1.16 *Let* $E \overset{\iota}{\rightarrowtail} A' \overset{\pi}{\longrightarrow\!\!\!\!\!\rightarrow} A$ *be an extension of* A *by* E *and let* $\varphi\colon A'' \to A$ *be a morphism of Lie algebroids over* M. *Then*

$$E \overset{\widetilde{\iota}}{\rightarrowtail} \widetilde{A} \overset{\widetilde{\pi}}{\longrightarrow\!\!\!\!\!\rightarrow} A'',$$

where $\widetilde{A}$ *is the pullback of* π *and* φ, *is an extension of* A'' *by* E, *where*

$\tilde{\iota}$ *is* $\mu \mapsto \iota(\mu) \oplus 0$, *and, furthermore,*

$$
\begin{array}{ccccc}
E & \overset{\tilde{\iota}}{>\!\!\!-\!\!\!-\!\!\!>} & \tilde{A} & \overset{\tilde{\pi}}{-\!\!\!-\!\!\!\gg} & A'' \\
\| & & \tilde{\varphi} \downarrow & & \varphi \downarrow \\
E & \underset{\iota}{>\!\!\!-\!\!\!-\!\!\!>} & A' & \underset{\pi}{-\!\!\!-\!\!\!\gg} & A
\end{array}
$$

commutes.

The proof is a routine verification. This $E \overset{\tilde{\iota}}{>\!\!-\!\!-\!\!\!>} \tilde{A} \overset{\tilde{\pi}}{-\!\!-\!\!\!\gg} A''$ is the *pullback extension* of $E >\!\!-\!\!-\!\!\!> A' -\!\!-\!\!\!\gg A$ over φ.

The addition on $\mathcal{O}\text{pext}(A, E)$, known as the *Baer sum*, can now be defined: given operator extensions $E >\!\!-\!\!-\!\!\!> A^s -\!\!-\!\!\!\gg A$, $s = 1, 2$, first form $E \oplus E >\!\!-\!\!-\!\!\!> A^1 \oplus A^2 -\!\!-\!\!\!\gg A \oplus A$. If A is transitive, the direct sums $A \oplus A$ and $A^1 \oplus A^2$ are to be taken as the direct sum Lie algebroid $\oplus_{TM}$ of 3.3.22; if A is totally intransitive, the $\oplus$s are to be taken as ordinary vector bundle sums over M: we do not need any other case. Next take the pullback of this over the diagonal map $A \to A \oplus A$, and lastly take the pushout of the result over the sum map $E \oplus E \to E$. The details are left to the reader.

The bijection $\mathcal{H}^2(A, E) \longleftrightarrow \mathcal{O}\text{pext}(A, E)$ is now an isomorphism of vector spaces for any Lie algebroid A, and of $C^\infty(M)$–modules when A is totally intransitive.

The zero element of $\mathcal{H}^2(A, E)$ corresponds, of course, to the flat extensions of A by E. We refer to the extension A^0 constructed in 7.1.13(i) from the zero cocycle as the *semidirect product* of A by E and denote it by $A \ltimes_\rho E$, or by $A \ltimes E$ if ρ is understood.

Any transversal χ in $A \ltimes E$ has the form $\chi(X) = X \oplus g(X)$ for $g \in \Gamma C^1(A, E)$, and χ is flat if and only if $dg = 0$. Thus $\mathcal{H}^1(A, E)$ can be interpreted as the space of flat transversals in $A \ltimes E$ modulo those flat transversals of the form $\chi(X) = X \oplus \rho(X)(\mu)$ for some $\mu \in \Gamma E$.

More generally, recall from 7.1.13(iv) that in any operator extension $E \overset{\iota}{>\!\!-\!\!-\!\!\!>} A' \overset{\pi}{-\!\!-\!\!\!\gg} A$ of A by E, a cochain $g \in \Gamma C^1(A, E)$ induces a permutation of the transversals of A', namely $\chi \mapsto \chi^g = \chi + \iota \circ g$. From the formula $\overline{R}_{\chi^g} = \overline{R}_\chi - dg$ it follows that the cocycles $g \in \mathcal{Z}^1(A, E)$ are precisely those cochains whose permutations preserve the curvature of transversals. If g is a coboundary $d\mu$ for $\mu \in \Gamma E$, then

$$(\chi + \iota \circ d\mu)(X) = \chi(X) + \iota(\rho(X)(\mu)) = \chi(X) + [\chi(X), \iota(\mu)]$$

Thus we have:

Proposition 7.1.17 *Let $E \rightarrowtail^{\iota} A' \xrightarrow{\pi} A$ be an operator exten-sion of A by E. Then $\mathscr{H}^1(A, \rho, E)$ may be realized as the space of those automorphisms $A' \to A'$ of the form $X' \mapsto X' + \iota g(\pi X')$, where g is a vector bundle map $A \to E$, which preserve the curvature of transver-sals of A', modulo the space of automorphisms $A' \to A'$ of the form $X' \mapsto X' + [X', \iota(\mu)]$, where $\mu \in \Gamma E$.*

This result may be interpreted for connections in a principal bundle with abelian structure group.

Lastly we have the calculation of $\mathscr{H}^0(A, E)$.

Proposition 7.1.18

$$\mathscr{H}^0(A, E) = \{\mu \in \Gamma E \mid \rho(X)(\mu) = 0 \text{ for all } X \in \Gamma A\} = (\Gamma E)^A$$

If A is totally intransitive then $(\Gamma E)^A$ is a $C^\infty(M)$–submodule of ΓE. It need not correspond to a vector subbundle of E, since the rank of ρ need not be locally constant. This is so even if A is an LAB.

For A not totally intransitive, $(\Gamma E)^A$ is merely an $\mathbb{R}$–vector space. For A transitive, $(\Gamma E)^A$ was calculated in 6.5.16.

7.2 Couplings of Lie algebroids and $\mathscr{H}^3$

In order to consider in the next section extensions by a non–abelian ker-nel, we first need the Lie algebroid analogue of the outer automorphism group of a discrete group.

Definition 7.2.1 Let K be an LAB on M. Then the quotient Lie alg-ebroid

$$\text{Der}(K)/\text{ad}(K) \rightarrowtail^{\bar{\jmath}} \mathfrak{D}_{\text{Der}}\, K/\text{ad}(K) \xrightarrow{\bar{a}} TM$$

is denoted by

$$\text{Out}(K) \rightarrowtail^{\bar{\jmath}} \text{Out}\mathfrak{D}(K) \xrightarrow{\bar{q}} TM$$

and elements of $\Gamma \text{Out}\mathfrak{D}(K)$ are called *outer bracket derivations* on K.

$\square$

The quotient Lie algebroid construction used here is defined in §6.5. That $\text{ad}(K) = \text{im}(\text{ad}\colon K \to \text{Der}(K))$ is a sub–LAB of $\text{Der}(K)$ follows from 3.3.10. That $\text{ad}(K)$ is an ideal of $\text{Out}\mathfrak{D}(K)$ follows from the formula

$$[D, (j \circ \text{ad})(V)] = (j \circ \text{ad})(D(V)) \tag{7}$$

for $D \in \Gamma\, \mathfrak{D}_{\mathrm{Der}}\, K$ and $V \in \Gamma K$, which is proved by an easy manipulation of the Jacobi identity. We now have Figure 7.1 in which both rows and columns are exact.

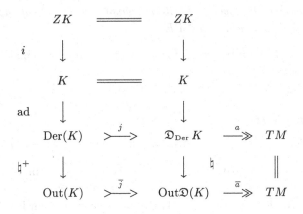

Fig. 7.1.

From now until 7.3.18 we consider a single Lie algebroid A on M. The anchor of A is denoted a^A. It is not assumed that A is transitive.

Definition 7.2.2 A *coupling* of A is an LAB K together with a morphism of Lie algebroids $\Xi\colon A \to \mathrm{Out}\mathfrak{D}(K)$; we also say that A and K are *coupled* by Ξ.   □

Every transitive Lie algebroid $L \rightarrowtail A \twoheadrightarrow TM$ induces a coupling of TM to L, namely $\natural \circ \nabla^\gamma$ for any connection $\gamma\colon TM \to A$.

Now fix a coupling Ξ of an LAB K to A until 7.2.13. Since $\natural$ is a surjective submersion, and a map of vector bundles over M, there are vector bundle morphisms $\nabla\colon A \to \mathfrak{D}_{\mathrm{Der}}\, K$, $X \mapsto \nabla_X$, such that $\natural \circ \nabla = \Xi$. We call ∇ a *Lie derivation law* covering Ξ. (See 7.2.8 for the formal definition.) Since $\bar{a} \circ \natural = a$ it follows that $a \circ \nabla = a^A$; that is, ∇ is an anchor–preserving map. Therefore the curvature of ∇ is a well–defined map $R_\nabla\colon A \oplus A \to \mathfrak{D}_{\mathrm{Der}}\, K$. Since $\natural \circ \nabla = \Xi$ is a morphism, it follows that $\natural \circ R_\nabla = 0$ and so R_∇ takes its values in $\mathrm{ad}(K) \subseteq \mathrm{Der}(K)$; we denote this map $A \oplus A \to \mathrm{ad}(K)$ by $\overline{R}_\nabla$. Now, as with Ξ above, there are alternating vector bundle morphisms $\Lambda\colon A \oplus A \to K$ such that $\mathrm{ad} \circ \Lambda = \overline{R}_\nabla$. This follows from momentarily considering $\overline{R}_\nabla$ to be defined on $\Lambda^2(A)$ and lifting it from $\Lambda^2(A) \to \mathrm{ad}(K)$ to $\Lambda^2(A) \to K$ across the surjective vector bundle morphism $\mathrm{ad} \colon K \to \mathrm{ad}(K)$. We call any alternating map $\Lambda \colon A \oplus A \to K$ with $\mathrm{ad} \circ \Lambda = \overline{R}_\nabla$ a *lift* of R_∇.

The coupling $\Xi \colon A \to \mathrm{Out}\mathfrak{D}(K)$ induces a representation of A on ZK, the centre of K. To see this, let ∇ be any Lie derivation law covering Ξ. Then for $X \in \Gamma A$, the operator $\nabla_X \colon \Gamma K \to \Gamma K$ restricts to $\Gamma ZK \to \Gamma ZK$, for if $Z \in \Gamma ZK$ and $\nabla \in \Gamma K$ then

$$[V, \nabla_X(Z)] = \nabla_X([V, Z]) - [\nabla_X(V), Z] = \nabla_X(0) - 0 = 0,$$

since Z is central. Write $\rho(X)$ for the restriction of ∇_X to $\Gamma ZK \to \Gamma ZK$. Then ρ defines a vector bundle map $A \to \mathfrak{D}_{\mathrm{Der}}\, ZK = \mathfrak{D}(ZK)$ which is easily seen to be a Lie algebroid morphism. If ∇' is a second Lie derivation law for Ξ then $\nabla'_X - \nabla_X$ is in $\Gamma(\mathrm{ad}(K))$, for all $X \in \Gamma A$, and therefore vanishes on ΓZK. Hence ρ is independent of the choice of ∇.

Definition 7.2.3 The representation $\rho \colon A \to \mathfrak{D}(ZK)$ just constructed is called the *central representation* of Ξ and is denoted ρ^Ξ.   □

Our concern now is to show that every coupling Ξ of A to K defines an element of $\mathscr{H}^3(A, \rho^\Xi, ZK)$. This will take us until 7.2.13.

Lemma 7.2.4 *Let ∇ be a Lie derivation law covering Ξ and let Λ be a lift of R_∇. Then for all $X, Y, Z \in \Gamma A$ the element*

$$\mathfrak{S}\{\nabla_X(\Lambda(Y, Z)) - \Lambda([X, Y], Z)\}$$

of ΓK lies in ΓZK.

Proof Apply $j \circ \mathrm{ad}$. We obtain, firstly,

$$j \circ \mathrm{ad}(\nabla_X(\Lambda(Y, Z))) = [\nabla_X, (j \circ \mathrm{ad})(\Lambda(Y, Z))]$$

by (7), and we have $j \circ \mathrm{ad} \circ \Lambda = j \circ \overline{R}_\nabla = R_\nabla$ by definition. So $j \circ \mathrm{ad}$ of the cyclic sum is

$$\mathfrak{S}\{[\nabla_X, R_\nabla(Y, Z)] - R_\nabla([X, Y], Z)\}$$

and this is zero by the general Bianchi identity (4).   □

Write $f(X, Y, Z)$ for the element in the lemma. It is easily checked that f is an alternating and $C^\infty(M)$–trilinear function of $X, Y, Z \in \Gamma A$, and it therefore defines an element of $\Gamma C^3(A, ZK)$, also denoted f.

Lemma 7.2.5 $df = 0$ *where* d *is the coboundary induced by* ρ.

Proof This is a long but straightforward calculation, and requires no ingenuity. □

So $f \in \mathscr{Z}^3(A, \rho, ZK)$. This f is the *obstruction cocycle*, or simply *obstruction*, defined by ∇ and Λ for the coupling Ξ. We may write $f = f(\nabla, \Lambda)$.

Lemma 7.2.6 *Fix a Lie derivation law* ∇ *covering* Ξ *and let* Λ *and* Λ' *be two lifts of* R_∇, *with corresponding obstruction cocycles* $f = f(\nabla, \Lambda)$ *and* $f' = f(\nabla, \Lambda')$. *Then* $\Lambda' - \Lambda = i \circ g$ *for some* $g \in \Gamma C^2(A, ZK)$ *and* $dg = f' - f$.

Proof Since $j \circ \mathrm{ad} \circ (\Lambda' - \Lambda) = R_\nabla - R_\nabla = 0$ there is a unique $g \colon A \oplus A \to ZK$ with $i \circ g = \Lambda' - \Lambda$. Since Λ' and Λ are alternating, it follows that g is also. Thus $g \in \Gamma C^2(A, ZK)$. Now, for $X, Y, Z \in \Gamma A$,

$$
\begin{aligned}
i(f'(X, Y, Z) - f(X, Y, Z)) &= \mathfrak{S}\{\nabla_X(i \circ g(y, Z)) - i \circ g([X, Y], Z)\} \\
&= i(\mathfrak{S}\{\rho(X)(g(Y, Z)) - g([X, Y], Z)\}) \\
&= i(dg(X, Y, Z)).
\end{aligned}
$$

□

We now need to show that the cohomology class of f is independent of the choice of ∇.

Proposition 7.2.7 *Let* ∇ *and* ∇' *be two Lie derivation laws covering* Ξ. *Then* $\nabla' = \nabla + j \circ \mathrm{ad} \circ \ell$ *for various maps* $\ell \colon A \to K$, *and*

$$
\overline{R}_{\nabla'} - \overline{R}_\nabla = -\mathrm{ad}\left(\nabla_X(\ell(Y)) - \nabla_Y(\ell(X)) - \ell[X, Y] + [\ell(X), \ell(Y)]\right).
$$

Proof The existence of ℓ follows as before. For $X, Y \in \Gamma A$ we have

$$
\begin{aligned}
(R_{\nabla'} - R_\nabla)(X, Y) &= \nabla'_{[X,Y]} - \nabla_{[X,Y]} - [\nabla'_X, \nabla'_Y] + [\nabla_X, \nabla_Y] \\
&= (j \circ \mathrm{ad} \circ \ell)[X, Y] - [(j \circ \mathrm{ad} \circ \ell)(X), \nabla_Y] \\
&\quad - [\nabla_X, (j \circ \mathrm{ad} \circ \ell)(Y)] \\
&\quad \qquad - [j \circ \mathrm{ad} \circ \ell)(X), (j \circ \mathrm{ad} \circ \ell)(Y)].
\end{aligned}
$$

Using (7), this becomes

$$
\begin{aligned}
(j \circ \mathrm{ad} \circ \ell)[X, Y] &+ (j \circ \mathrm{ad})(\nabla_Y(\ell(X))) \\
&- (j \circ \mathrm{ad})(\nabla_X(\ell(Y))) - (j \circ \mathrm{ad})[\ell(X), \ell(Y)],
\end{aligned}
$$

whence the result.   □

This can be expressed more succinctly by extending a piece of formalism from the traditional accounts of infinitesimal connection theory. This will take us until after 7.2.10.

Definition 7.2.8 Let A be any Lie algebroid and let K be any LAB on the same base. A *Lie derivation law* for A with coefficients in K is an anchor–preserving map $\nabla\colon A \to \mathfrak{D}_{\mathrm{Der}}\, K$.   □

Definition 7.2.9 Let ∇ be a Lie derivation law for A with coefficients in K. The *(exterior) covariant derivative* induced by ∇ is the sequence of operators $\nabla\colon \Gamma C^n(A, K) \to \Gamma C^{n+1}(A, K)$ defined by

$$\nabla(f)(X_1,\ldots,X_{n+1}) = \sum_{r=1}^{n+1}(-1)^{r+1}\nabla_{X_r}(f(X_1,\ldots,\widehat{X_r},\ldots,X_{n+1}))$$
$$+ \sum_{r<s}(-1)^{r+s}f([X_r,X_s],X_1,\ldots,\widehat{X_r},\ldots,\widehat{X_s},\ldots,X_{n+1}).$$

    □

The requirement that ∇ be anchor–preserving ensures that the RHS actually is $C^\infty(M)$–multilinear. There is a formula for $\nabla \circ \nabla$ in terms of the curvature of the Lie derivation law.

The equation in 7.2.7 can now be written

$$\overline{R}_{\nabla'} - \overline{R}_\nabla = -\mathrm{ad}(\nabla(\ell) + [\ell,\ell]).$$

The general result follows; the proof is formally identical to that of 7.2.7.

Proposition 7.2.10 Let $L' \rightarrowtail A' \twoheadrightarrow TM$ be a transitive Lie algebroid on M and let A be any Lie algebroid on M. Let $\varphi_1, \varphi_2\colon A \to A'$ be two anchor–preserving maps and write $\varphi_2 = \varphi_1 + j' \circ \ell$, where $\ell\colon A \to L'$. Then

$$\overline{R}_{\varphi_2} - \overline{R}_{\varphi_1} = -(\nabla(\ell) + [\ell,\ell])$$

where $\nabla\colon A \to \mathfrak{D}_{\mathrm{Der}}\, L'$ is the Lie derivation law defined by

$$j'(\nabla_X(V')) = [\varphi_1(X), j'(V')].$$

We return to the coupling Ξ. Notice that the definition of f can now be written $f = \nabla(\Lambda)$, and f thus measures the extent to which Λ satisfies the Bianchi identity with respect to ∇.

Proposition 7.2.11 *Let ∇ be a Lie derivation law covering Ξ and let Λ be a lift of R_∇. Let ∇' be a second Lie derivation law covering Ξ and write $\nabla' = \nabla + j \circ \mathrm{ad} \circ \ell$, where ℓ is a map $A \to K$. Then*

$$\Lambda' = \Lambda - (\nabla(\ell) + [\ell, \ell])$$

is a lift of $R_{\nabla'}$ and $f(\nabla', \Lambda') = f(\nabla, \Lambda)$.

Proof Certainly Λ' is alternating, and $\mathrm{ad} \circ \Lambda' = \overline{R}_\nabla - \mathrm{ad}(\nabla(\ell) + [\ell, \ell]) = \overline{R}_{\nabla'}$, by 7.2.7. It remains to show that $\nabla'(\Lambda') = \nabla(\Lambda)$.

Now ∇' is a linear operator, and so

$$\nabla'(\Lambda') - \nabla(\Lambda) = \nabla'(\Lambda) - \nabla'(\nabla(\ell) + [\ell, \ell]) - \nabla(\Lambda).$$

For $X, Y, Z \in \Gamma A$,

$$(\nabla'(\Lambda) - \nabla(\Lambda))(X, Y, Z) = \mathfrak{S}\{(j \circ \mathrm{ad} \circ \ell)(X)(\Lambda(Y, Z))\}$$
$$= \mathfrak{S}\{[\ell(X), \Lambda(Y, Z)]\}. \quad (8)$$

Similarly by expanding and regrouping, $\nabla'(\nabla(\ell) + [\ell, \ell])(X, Y, Z)$ becomes

$$\nabla(\nabla(\ell) + [\ell, \ell])(X, Y, Z) + \mathfrak{S}\{(j \circ \mathrm{ad} \circ \ell)(X)((\nabla)(\ell) + [\ell, \ell])(Y, Z))\}.$$

First calculate $\nabla(\nabla(\ell))$. By regrouping terms and using the Jacobi identity one obtains

$$\nabla(\nabla(\ell))(X, Y, Z) = -\mathfrak{S}\{R_\nabla(X, Y)(\ell(Z))\},$$

and since $j \circ \mathrm{ad} \circ \Lambda = R_\nabla$, this cancels with (8) above. Next, by expanding the cyclic sum and using $\nabla_X([V, W]) = [\nabla_X(V), W] + [V, \nabla_X(W)]$ repeatedly, one obtains:

$$\nabla([\ell, \ell])(X, Y, Z) = \mathfrak{S}[\nabla(\ell)(X, Y), \ell(Z)].$$

Lastly,

$$\mathfrak{S}\{(j \circ \mathrm{ad} \circ \ell)(X)((\nabla(\ell) + [\ell, \ell])(Y, Z))\} = \mathfrak{S}\{[\ell(X), \nabla(\ell)(Y, Z)]\}$$
$$+ \mathfrak{S}\{[\ell(X), [\ell(Y), \ell(Z)]]\}.$$

The second term of this vanishes, by the Jacobi identity, and the first cancels with $\nabla([\ell, \ell])(X, Y, Z)$.

Thus we obtain $\nabla'(\Lambda') - \nabla(\Lambda) = 0$, as desired. $\square$

Putting 7.2.6 and 7.2.11 together, we obtain:

Theorem 7.2.12 *Let A be a Lie algebroid on M and let K be an LAB on M. Let Ξ be a coupling of A with K. Then the cohomology class in $\mathscr{H}^3(A, \rho^\Xi, ZK)$ of $f(\nabla, \Lambda)$, where ∇ is a Lie derivation law covering Ξ, and Λ is a lift of R_∇, depends only on Ξ and is independent of the choice of ∇ and Λ.*

This class is called the *obstruction class* of the coupling Ξ, and will be denoted $\mathrm{Obs}(\Xi)$.

The following observation will be important later.

Proposition 7.2.13 *Let A, K and Ξ be as in 7.2.12. Let ∇ be any Lie derivation law covering Ξ, and let f' be any cocycle in $\mathrm{Obs}(\Xi)$. Then there is a lift Λ' of R_∇ such that $f(\nabla, \Lambda') = f'$.*

Proof Let Λ be any lift of R_∇, and write $f = f(\nabla, \Lambda)$. Then f and f' are cohomologous; choose any $g \in \Gamma C^2(A, ZK)$ such that $f' = f + dg$. Define $\Lambda' = \Lambda + i \circ g$; then $\mathrm{ad} \circ \Lambda' = \mathrm{ad} \circ \Lambda = \overline{R}_\nabla$ and 7.2.6 shows that $f' = f(\nabla, \Lambda')$. □

7.3 Non–abelian extensions of Lie algebroids

We now describe the coupling associated to a general (not necessarily abelian) extension of Lie algebroids, and its obstruction class. Up to 7.3.6, let A be a fixed Lie algebroid on M and let K be an LAB on M.

Definition 7.3.1 An *extension* of A by K is an exact sequence of Lie algebroids over M

$$K \overset{\iota}{>\!\!\!-\!\!\!-\!\!\!>} A' \overset{\pi}{-\!\!\!-\!\!\!\gg} A, \qquad (9)$$

as defined in 3.3.19.

Two extensions of A by K, $K \overset{\iota_r}{>\!\!\!-\!\!\!-\!\!\!>} A^r \overset{\pi}{-\!\!\!-\!\!\!\gg} A$, $r = 1, 2$, are *equivalent* if there is a morphism of Lie algebroids over M, $\varphi\colon A^1 \to A^2$, necessarily an isomorphism, such that $\varphi \circ \iota_1 = \iota_2$ and $\pi_2 \circ \varphi = \pi_1$. □

We define the concepts of transversal, retroversal, flat transversal and flat extension exactly as for extensions by vector bundles. The discussion following 7.1.11 also applies, except that an extension (9) by a general LAB need not induce a representation of A on K. Instead the representation formula defines a Lie derivation law.

The example of an extension of Lie algebroids which is of most importance to us is the extension $L >\!\!\!-\!\!\!-\!\!\!> A -\!\!\!-\!\!\!\gg TM$ associated with a transitive Lie algebroid. Applied to this example, all the results concerning general extensions include results for infinitesimal connections.

Fix an extension (9) of A by K until 7.3.6.

Proposition 7.3.2 *Let* $\chi: A \to A'$ *be a transversal in* (9). *Then*

$$\iota(\nabla_X^\chi(V)) = [\chi(X), \iota(V)]$$

for $\chi \in \Gamma A$, $V \in \Gamma K$, *defines a Lie derivation law for* A *with coefficients in* K.

Proof It need only be checked that $\nabla^\chi: A \to \mathfrak{D}_{\mathrm{Der}} K$ is anchor-preserving, and this follows from $a^{A'} \circ \chi = a^A$. $\square$

Lemma 7.3.3 *With the notation of* 7.3.2,

$$\overline{R}_{\nabla\chi} = \mathrm{ad} \circ \overline{R}_\chi.$$

Proof For $X, Y \in \Gamma A$ and $V \in \Gamma K$,

$$
\begin{aligned}
&\iota(R_{\nabla\chi}(X,Y)) \\
&= \quad [\chi[X,Y], \iota(V)] - [\chi(X), [\chi(Y), \iota(V)]] + [\chi(Y), [\chi(X), \iota(V)]] \\
&= \quad [\chi[X,Y], \iota(V)] + [\iota(V), [\chi(X), \chi(Y)]] \\
&\qquad\qquad (\text{ by the Jacobi identity in } \Gamma A') \\
&= \quad [R_\chi(X,Y), \iota(V)]
\end{aligned}
$$

and the result follows. $\square$

Hence the composition $A \xrightarrow{\nabla^\chi} \mathfrak{D}_{\mathrm{Der}} K \xrightarrow{\natural} \mathrm{Out}\mathfrak{D}(K)$ is a morphism of Lie algebroids. If $\chi': A \to A'$ is a second transversal then $\chi' = \chi + \iota \circ \ell$ for some map $\ell: A \to K$ and $\nabla^{\chi'} = \nabla^\chi + \mathrm{ad} \circ \ell$. Hence $\natural \circ \nabla^{\chi'} = \natural \circ \nabla^\chi$.

Definition 7.3.4 The coupling $\natural \circ \nabla^\chi: A \to \mathfrak{D}_{\mathrm{Der}} K$, where χ is any transversal in the extension $K \stackrel{\iota}{>\!\!\!-\!\!\!-\!\!\!\!\longrightarrow} A' \stackrel{\pi}{-\!\!\!-\!\!\!\twoheadrightarrow} A$, is the *coupling of* A *with* K *induced by the extension.* $\square$

Note that the choice of a transversal χ determines both a Lie derivation law ∇^χ covering the coupling induced by the extension, and (by 7.3.3) a lift $\overline{R}_\chi$ of $R_{\nabla\chi}$. Thus each χ determines an obstruction cocycle.

Proposition 7.3.5 *For every transversal* χ,

$$f(\nabla^\chi, \overline{R}_\chi) = 0$$

Proof We know that $f(X, Y, Z) = \mathfrak{S}\{\nabla^\chi_X(\overline{R}_\chi(Y, Z)) - \overline{R}_\chi([X, Y], Z))\}$ and, applying ι to the RHS, it becomes

$$\mathfrak{S}\{[\chi(X), R_\chi(Y, Z)] - R_\chi([X, Y], Z)\}$$

which is zero by the general Bianchi identity (4). □

This equation, $f(\nabla^\chi, \overline{R}_\chi) = 0$, is called the *Bianchi identity* for χ.

It follows in particular from 7.3.5 that the obstruction cohomology class for the coupling detemined by an extension is zero.

The next result shows that every Lie derivation law covering the coupling determined by an extension, arises from a transversal.

Proposition 7.3.6 *Let* ∇ *be any Lie derivation law covering the coupling determined by* $K \overset{\iota}{>\!\!-\!\!\longrightarrow} A' \overset{\pi}{-\!\!\!\longrightarrow\!\!\!\gg} A$. *Then there is a transversal* $\chi\colon A \to A'$ *such that* $\nabla^\chi = \nabla$.

Proof Let χ' by any transversal. Since $\flat \circ \nabla = \flat \circ \nabla^{\chi'}$, there is a map $\overline{\ell}\colon A \to \operatorname{ad}(K)$ such that $\nabla = \nabla^{\chi'} + j \circ \overline{\ell}$. Since $\operatorname{ad}\colon K \to \operatorname{ad}(K)$ is a surjective submersion of vector bundles, $\overline{\ell}$ can be lifted to $\ell\colon A \to K$. Define $\chi = \chi' + \iota \circ \ell$. Certainly χ is a transversal, and for all $X \in \Gamma A$ and $V \in \Gamma K$,

$$
\begin{aligned}
\iota(\nabla^\chi_X(V)) &= [\chi'(X), \iota(V)] + \iota[\ell(X), V] \\
&= i((\nabla^{\chi'}_X(V)) + (j \circ \operatorname{ad} \circ \ell)(X)(V)) \\
&= \iota(\nabla_X(V)),
\end{aligned}
$$

as required. □

Thus the nonzero elements of the obstruction class arise from having chosen a Lie derivation law ∇^χ and having then failed to choose the natural lift, namely $\overline{R}_\chi$, of R_{∇^χ}; see 7.2.6. On the other hand, it is not true that if Λ is a lift of $R^{\chi'}_\nabla$ for which $f(\nabla^\chi, \Lambda) = 0$, then $\Lambda = \overline{R}_\chi$; there are usually, for example, many closed non–zero two–forms on a manifold.

$$* \quad * \quad * \quad * \quad *$$

We arrive now at the construction and enumeration of extensions corresponding to a coupling which has obstruction class zero. The first result is the basic construction principle.

Theorem 7.3.7 *Let A be a Lie algebroid on base M and let K be an LAB on M. Let $\nabla\colon A \to \mathfrak{D}_{\mathrm{Der}}\, K$ be a Lie derivation law such that $\natural \circ \nabla\colon A \to \mathrm{Out}\,\mathfrak{D}(K)$ has zero curvature, and let $R\colon A \oplus A \to K$ be an alternating 2–form on A with values in K. Then, if*

(i) *$\overline{R}_\nabla = \mathrm{ad} \circ R$, and*
(ii) *the Bianchi identity $\nabla(R) = 0$,*

hold, the formula

$$[X \oplus V, Y \oplus W] = [X, Y] \oplus \{\nabla_X(W) - \nabla_Y(V) + [V, W] - R(X, Y)\}$$

defines a bracket on $\Gamma(A \oplus K)$ which makes $A \oplus K$ a Lie algebroid on M with respect to the anchor $a' = a \circ \pi_1$, and an extension

$$K \overset{\iota_2}{>\!\!-\!\!\longrightarrow} A' \overset{\pi_1}{-\!\!\longrightarrow\!\!\gg} A$$

of A by K such that $\iota_1\colon A \to A'$ is a transversal with $\nabla^{\iota_1} = \nabla$ and $\overline{R}_{\iota_1} = R$.

Proof This is a straightforward calculation. We verify the Jacobi identity as an example.

Given $X_r \oplus V_r \in \Gamma A'$ for $r = 1, 2, 3$, we have for the K–component of $\mathfrak{S}\{[[X_1 \oplus V_1, X_2 \oplus V_2], X_3 \oplus V_3\}$ the expression

$$\mathfrak{S}\{\nabla_{[X_1, X_2]}(V_3) - \nabla_{X_3}(\nabla_{X_1}(V_2)) + \nabla_{X_3}(\nabla_{X_2}(V_1)) - \nabla_{X_3}([V_1, V_2])$$
$$+ \nabla_{X_3}(R(X_1, X_2)) + [\nabla_{X_1}(V_2), V_3] - [\nabla_{X_2}(V_1), V_3]$$
$$+ [[V_1, V_2], V_3] - [R(X_1, X_2), V_3] - R([X_1, X_2], X_3)\}.$$

The term $\mathfrak{S}[[V_1, V_2], V_3]$ vanishes by the Jacobi identity in K. The fifth term can be rewritten as $\mathfrak{S}\nabla_{X_1}(R(X_2, X_3))$, by cyclic permutation, and then cancels with the last, by the Bianchi identity (ii). Rewriting the first three terms via cyclic permutations, we have $\mathfrak{S}R_\nabla(X_1, X_2)(V_3)$ and this is, by (i), equal to $\mathfrak{S}[R(X_1, X_2), V_3]$, and so cancels with the second–last term. The remaining terms cancel by the equation $\nabla_X([V, W]) = [\nabla_X(V), W] + [V, \nabla_X(W)]$ for $X \in \Gamma A$, $V, W \in \Gamma K$, which characterizes the elements of $\mathfrak{D}_{\mathrm{Der}}\, K$.   □

We will treat the uniqueness of this construction in 7.3.10 below.

Theorem 7.3.7 gives, in particular, the following construction principle for transitive Lie algebroids with a prescribed curvature form.

Corollary 7.3.8 *Let L be an LAB on M and let $R \in \Gamma C^2(TM, L)$ be an alternating 2–form on M with values in L.*

Then there is a transitive Lie algebroid $L \rightarrowtail A \twoheadrightarrow TM$ and a connection γ in A with $\overline{R}_\gamma = R$ if and only if there is a Lie connection ∇ in L such that (i) $\overline{R}_\nabla = \mathrm{ad} \circ R$, and (ii) $\nabla(R) = 0$.

Consider in particular the case where L is abelian and trivial. One may take ∇ to be the standard flat connection and (i) is then satisfied while condition (ii) reduces to $\delta R = 0$. This case is thus part of the classical Weil Lemma (see 8.1.3). If, moreover, M is simply–connected, then any flat connection is equivalent to the standard flat connection, but for general M, this case of 7.3.8 is slightly more general than the classical result.

In the case where L is non–abelian it is neceessary that the 2–form take values in an LAB, rather than in a single Lie algebra, and this necessitates the introduction of ∇. This auxiliary connection ∇ and condition (i) are essential links between the algebraic properties and the curvature properties of an LAB which is the adjoint bundle of a transitive Lie algebroid. For example, it follows from 5.2.18(ii) that if L is abelian in 7.3.8, then it must be flat as a vector bundle, in order for a transitive Lie algebroid $L \rightarrowtail A \twoheadrightarrow TM$ to exist.

The classical Weil lemma is a criterion for the existence of a principal bundle. The question of when the Lie algebroid in 7.3.8 is integrable is dealt with in the next chapter.

Corollary 7.3.9 *Let A be a Lie algebroid on M, let K be an LAB on M, and let Ξ be a coupling of A with K. Then, if $\mathrm{Obs}(\Xi) = 0 \in \mathscr{H}^3(A, ZK)$, there is a Lie algebroid extension*

$$K \rightarrowtail A' \twoheadrightarrow A$$

of A by K, inducing the coupling Ξ, namely that constructed in 7.3.7, using any Lie derivation law ∇ which covers Ξ, and any lift $\Lambda = R$ of R_∇.

Proof This follows from 7.3.7 by applying 7.2.13 to the cocycle 0 in $\mathrm{Obs}(\Xi)$. $\qquad\square$

Corollary 7.3.9 and 7.3.5 together show that a coupling $\Xi \colon A \to \mathrm{Out}\mathfrak{D}(K)$ arises from an extension of A by K if and only if $\mathrm{Obs}(\Xi) = 0 \in \mathscr{H}^3(A, \rho^\Xi, ZK)$.

There are usually many distinct such extensions. We come now to the issue of their description.

Proposition 7.3.10 *Consider an extension* (9) *of Lie algebroids, with K any LAB. Then for any transversal* $\chi\colon A \to A'$, *the map*

$$\mathscr{E}_\chi\colon A \oplus K \to A', \quad X \oplus V \mapsto \chi(X) + \iota(V)$$

is an equivalence of (9) *with the extension* $K \overset{\iota_2}{>\!\!-\!\!-\!\!>} A \oplus K \overset{\pi_1}{-\!\!-\!\!\twoheadrightarrow} A$
constructed via 7.3.7 *from* ∇^χ *and* $\overline{R}_\chi$.

Proof For example, that $\mathscr{E}_\chi$ preserves the brackets merely asserts that

$$[\chi(X) + \iota(V), \chi(Y) + \iota(W)] =$$
$$\chi[X, Y] + \iota\left(\nabla^\chi_X(W) - \nabla^\chi_Y(V) + [V, W] - \overline{R}_\chi(X, Y)\right)$$

for all $X, Y \in \Gamma A$, $V, W \in \Gamma K$, and this is easily verified.   □

Proposition 7.3.10 is the correct uniqueness result for the problem of constructing transitive Lie algebroids with a preassigned curvature form and adjoint connection. However we will also need in what follows an enumeration of the extensions of A by K in terms of $\mathscr{H}^2(A, ZK)$.

From now until 7.3.18, let A be a fixed Lie algebroid on M, let K be an LAB on M, and let $\Xi\colon A \to \mathrm{Out}\mathfrak{D}(K)$ be a coupling of A with K such that $\mathrm{Obs}(\Xi) = 0 \in \mathscr{H}^3(A, \rho^\Xi, ZK)$.

Definition 7.3.11 An *operator extension* of A by K is an extension $K >\!\!-\!\!-\!\!> A' -\!\!-\!\!\twoheadrightarrow A$ which induces, via 7.3.4, the coupling Ξ.   □

The set of equivalence classes of operator extensions of A by K is denoted by $\mathscr{O}\mathrm{pext}(A, \Xi, K)$, or by $\mathscr{O}\mathrm{pext}(A, K)$ if Ξ is understood. We will show that $\mathscr{H}^2(A, \rho^\Xi, ZK)$ acts simply and transitively on the set $\mathscr{O}\mathrm{pext}(A, \Xi, K)$. It will then follow that $\mathscr{O}\mathrm{pext}(A, \Xi, K)$ can be put in bijective correspondence with $\mathscr{H}^2(A, \rho^\Xi, ZK)$, by the choice of any extension as reference point.

The first step is to define an action of $\mathscr{Z}^2(A, ZK)$ on the class of all operator extensions.

Definition 7.3.12 Let $K \overset{\iota}{>\!\!-\!\!-\!\!>} A' \overset{\pi}{-\!\!-\!\!\twoheadrightarrow} A$ be an operator extension, and let g be in $\mathscr{Z}^2(A, ZK)$. Then the *action* of g on the extension yields the extension

$$K \overset{\iota}{>\!\!-\!\!-\!\!>} A'_g \overset{\pi}{-\!\!-\!\!\twoheadrightarrow} A \tag{10}$$

where $A'_g = A'$ as vector bundles, the maps ι and π are the same in both extensions, the anchors $a'\colon A' \to TM$ and $a'_g\colon A'_g \to TM$ are equal, and the bracket $[\,,\,]_g$ on $\Gamma(A'_g)$ is given by

$$[X, Y]_g = [X, Y] + \iota i g(\pi X, \pi Y).$$

$\square$

The cocycle condition for g ensures that $[\,,\,]_g$ obeys the Jacobi identity. Since the values of g are in $ZK \subseteq K$, the maps ι and π remain morphisms with respect to the new structure, and the coupling is unchanged.

Proposition 7.3.13 *Continue the notation of 7.3.12. If* $\chi\colon A \to A'$ *is a transversal for* A', *then it is also a transversal for* A'_g, *and the two Lie derivation laws* $A \to \mathfrak{D}_{\mathrm{Der}}\, K$ *are equal. The curvatures of* χ *are related by*

$$\overline{R}^g_\chi = \overline{R}_\chi - \iota \circ g.$$

The proof is an easy calculation. It follows in particular that (10) is an operator extension. Clearly $(A'_g)_h = A'_{g+h}$.

Proposition 7.3.14 *Let* χ *be a transversal in an operator extension* $K \overset{\iota}{>\!\!\longrightarrow} A' \overset{\pi}{\longrightarrow\!\!\!\gg} A$, *and let* $\ell\colon A \to K$ *be a map. Then*

$$\overline{R}_{\chi + \iota \circ \ell} = \overline{R}_\chi - \left(\nabla^\chi(\ell) + [\ell, \ell] \right).$$

Proof This is formally identical to the proof of 5.2.20. $\square$

In particular, if ℓ takes values in ZK then

$$\overline{R}_\chi + \iota \circ i \circ \ell = \overline{R}_\chi - i \circ d\ell.$$

From this it follows that the action of $\mathscr{Z}^2(A, ZK)$ factors to an action of $\mathscr{H}^2(A, ZK)$.

Proposition 7.3.15 *Let* $K \overset{\iota}{>\!\!\longrightarrow} A' \overset{\pi}{\longrightarrow\!\!\!\gg} A$ *be an operator extension, and let* $h \in \Gamma C^1(A, ZK)$. *Then* $K >\!\!\longrightarrow A'_{dh} \longrightarrow\!\!\!\gg A$ *is equivalent to* $K >\!\!\longrightarrow A' \longrightarrow\!\!\!\gg A$.

Proof Let χ be a transversal of A'. Regarded as a transversal of A'_{dh}, it has curvature $\overline{R}_\chi - i \circ dh$ by 7.3.13. So, by 7.3.14, the transversal $\chi + \iota \circ i \circ h$ in A'_{dh} has curvature $\overline{R}_\chi$. Since h takes values in ZK, the Lie derivation law determined by $\chi + \iota \circ i \circ h$ in A'_{dh} is the same as that determined by χ in A'. So we have two extensions, A' and A'_{dh}, with two transversals, χ and $\chi + \iota \circ i \circ dh$, respectively, which have the same curvature and Lie derivation law. It now follows from 7.3.10 that the extensions are equivalent.   □

The equivalence $A' \to A'_{dh}$ is $\varepsilon_h \colon X \to X - (\iota \circ i \circ h)(\pi X)$. Briefly, $\varepsilon_h = \mathrm{id} - h \circ \pi$.

We therefore have an action of $\mathscr{H}^2(A, ZK)$ on the class of all operator extensions. It is easily checked that the action sends equivalent extensions to equivalent extensions, so we in fact have an action of $\mathscr{H}^2(A, ZK)$ on $\mathscr{O}pext(A, K)$.

We now prove that this action is free and transitive.

Theorem 7.3.16 *Let* $K \overset{\iota}{\rightarrowtail} A' \overset{\pi}{\twoheadrightarrow} A$ *be an operator extension, and let* $g \in \mathscr{Z}^2(A, ZK)$ *be a cocycle, and suppose that there is an equivalence* $\varphi \colon A' \to A'_g$. *Then* g *is cohomologous to zero,* $g = dh$, *and* $\varphi = \varepsilon_h$, *where* $h \in \Gamma C^1(A, ZK)$ *is the cochain determined by* $h = -\lambda \circ \varphi \circ \chi$ *for any transversal* χ *and associated retroversal* λ.

Proof From 7.3.12 it follows that

$$i \circ g = \overline{R}_\chi - \overline{R}^g_\chi$$

for any transversal χ. Now by the same calculation as in 5.2.21, we obtain

$$\overline{R}_\chi(\pi X, \pi Y) = \nabla^\chi_{\pi X}(\lambda Y) - \nabla^\chi_{\pi Y}(\lambda X) - \lambda[X, Y] + [\lambda X, \lambda Y]$$

for $X, Y \in \Gamma A'$.

Similarly, working in A'_g with the same transversal χ, and recalling from 7.3.12 that the two Lie derivation laws for χ, with respect to A' and A'_g, are equal, we get

$$\overline{R}^g_\chi(\pi X, \pi Y) = \overline{R}^g_\chi(\pi\varphi X, \pi\varphi Y)$$
$$= \nabla^\chi_{\pi X}(\lambda\varphi Y) - \nabla^\chi_{\pi Y}(\lambda\varphi X) - \lambda[\varphi X, \varphi Y] + [\lambda\varphi X, \lambda\varphi Y].$$

Substituting these in the equation for $i \circ g$, we get

$$(i \circ g)(\pi X, \pi Y) = \nabla^\chi_{\pi X}(\lambda Y - \lambda\varphi Y) - \nabla^\chi_{\pi Y}(\lambda X - \lambda\varphi X)$$
$$- (\lambda - \lambda\varphi)[X, Y] + \{[\lambda X, \lambda Y] - [\lambda\varphi X, \lambda\varphi Y]\}.$$

If $\lambda - \lambda\varphi$ is equal to $\theta\pi$ for a map $\theta \colon A \to K$ then $(\lambda - \lambda\varphi)\chi = \theta\pi\chi = \theta$

and since $\lambda\chi = 0$, it follows that $\theta = -\lambda\varphi\chi$ and $\lambda - \lambda\varphi = -\lambda\varphi\chi\pi$. Writing now $h = -\lambda\varphi\chi$ we have

$$
\begin{aligned}
(i \circ g)(\pi X, \pi Y) &= \nabla^{\chi}_{\pi X}(h\pi Y) - \nabla^{\chi}_{\pi Y}(h\pi X) - h\pi[X, Y] \\
&\quad + \{[\lambda\varphi X + h\pi X, \lambda\varphi Y + h\pi Y] - [\lambda\varphi X, h\varphi Y]\} \\
&= \{\nabla^{\chi}_{\pi X}(h\pi X) - \nabla^{\chi}_{\pi Y}(h\pi X) - h\pi[X, Y] + [h\pi X, h\pi Y]\} \\
&\quad + \{[\lambda\varphi X, h\pi Y] + [h\pi X, \lambda\varphi Y]\}.
\end{aligned}
$$

Set $\pi Y = 0$. Then the LHS and all terms on the RHS except the last, vanish. So $[h\pi X, \lambda\varphi Y] = 0$ for all $X, Y \in \Gamma A'$, and since π is onto A and $\lambda \circ \varphi$ is onto K (for λ is a surjective submersion), this proves that h takes values in ZK.

The equation for $i \circ g$ now reduces to $(i \circ g)(\pi X, \pi Y) = dh(\pi X, \pi Y)$, which proves that $g = dh$, since π is onto A.

After 7.3.15 it was remarked that $\varepsilon_h = \mathrm{id} - \iota \circ i \circ h \circ \pi$. Neglecting the i, we have

$$
\begin{aligned}
\varepsilon_h &= \mathrm{id} + \iota \circ \lambda \circ \varphi \circ \chi \circ \pi \\
&= \mathrm{id} + \iota \circ \lambda \circ \varphi \circ (\mathrm{id} - \iota \circ \lambda) \\
&= \mathrm{id} + \iota \circ \lambda \circ \varphi - \iota \circ \lambda) \qquad (\text{since } \varphi \circ \iota = \iota \text{ and } \lambda \circ \iota = \mathrm{id}) \\
&= \chi \circ \pi + \iota \circ \lambda \circ \varphi \\
&= \varphi \qquad (\text{since } \pi = \pi \circ \varphi \text{ and } \chi \circ \pi + \iota \circ \lambda = \mathrm{id}).
\end{aligned}
$$

$\square$

Theorem 7.3.17 *Let* $K \overset{\iota_r}{\rightarrowtail} A^r \overset{\pi_r}{\twoheadrightarrow} A$, $r = 1, 2$, *be two operator extensions. Let* χ_1 *and* χ_2 *be transversals of* A^1 *and* A^2, *respectively, which induce the same Lie derivation law (see 7.3.6), and define* $g \in \Gamma C^2(A, ZK)$ *by* $i \circ g = \overline{R}_{\chi_1} - \overline{R}_{\chi_2}$. *Then* g *is a cocycle,* $g \in \mathscr{Z}^2(A, ZK)$, *and* $\varphi = \iota_2 \circ \lambda_1 + \chi_2 \circ \pi_1$ *is an equivalence* $(A^1)_g \to A^2$.

Proof That $\overline{R}_{\chi_1} - \overline{R}_{\chi_2}$ takes values in ZK follows from the assumption that $\nabla^{\chi_1} = \nabla^{\chi_2}$, since $\mathrm{ad} \circ \overline{R}_{\chi_1} = \overline{R}_{\nabla \chi_1}$ and $\mathrm{ad} \circ \overline{R}_{\chi_2} = \overline{R}_{\nabla \chi_2}$ (by 7.3.3). Now

$$
\begin{aligned}
(i \circ dg)(X, Y, Z) &= i(\mathfrak{S}\{\rho(X)(g(Y, Z)) - g([X, Y], Z)\}) \\
&= \mathfrak{S}\{\nabla^{\chi_1}_X(\overline{R}_{\chi_1}(Y, Z)) - \overline{R}_{\chi_1}([X, Y], Z)\} \\
&\quad - \mathfrak{S}\{\nabla^{\chi_2}_X(\overline{R}_{\chi_2}(Y, Z)) - \overline{R}_{\chi_2}([X, Y], Z)\}
\end{aligned}
$$

which is zero by the Bianchi identities for χ_1 and χ_2 (compare 7.3.5).

To prove that φ preserves the brackets, requires manipulations of a type that will now be familiar.   □

Putting together 7.3.12 to 7.3.17, we have:

Theorem 7.3.18 *Let A be a Lie algebroid on M, let K be an LAB on M, and let Ξ be a coupling of A with K such that $\mathrm{Obs}(\Xi) = 0 \in \mathscr{H}^3(A, ZK)$. Then the additive group of $\mathscr{H}^2(A, \rho^\Xi, ZK)$ acts freely and transitively on $\mathscr{O}\mathrm{pext}(A, \Xi, K)$.*

Equivalently, $\mathscr{O}\mathrm{pext}(A, \Xi, K)$ is an affine space overlying the vector space $\mathscr{H}^2(A, \rho^\Xi, ZK)$.

Theorem 7.3.18 yields in particular a classification of transitive Lie algebroids up to equivalence. This should be contrasted with the Chern–Weil theory. Classically one tries to distinguish principal bundles by studying their connections. Infinitesimal connections actually belong in the Lie algebroids and one shows that two Lie algebroids are non-isomorphic by exhibiting a connection in one that cannot exist in the other. This is usually done by means of cohomological invariants derived from the curvature of all connections via the Weil morphism and in this context Lie algebroids cannot be distinguished from their reductions: such studies are topological rather than geometric.

In view of the application of 7.3.18 in the following chapter, some comments about semi–direct extensions are needed.

Definition 7.3.19 Let A be a Lie algebroid on M, let K be an LAB on M, and let $\nabla\colon A \to \mathfrak{D}_{\mathrm{Der}} K$ be a flat Lie derivation law; that is, let ∇ be a representation of A on K. Then the ∇–*semidirect extension* of A by K is the extension

$$K \overset{\iota_2}{\rightarrowtail} A \ltimes_\nabla K \overset{\pi_1}{\twoheadrightarrow} A$$

where $A \ltimes_\nabla K$ is the vector bundle $A \oplus K$ with anchor $a'(X \oplus V) = a^A(X)$ and bracket

$$[X \oplus V, Y \oplus W] = [X, Y] \oplus \{\nabla_X(W) - \nabla_Y(V) + [V, W]\}.$$

□

It follows immediately from 7.3.10 that every flat extension is equivalent to a semidirect extension.

Not every coupling with zero obstruction class has a flat Lie derivation law covering it. A typical example follows.

Example 7.3.20 Let Ω be the locally trivial Lie groupoid associated to the Hopf bundle $\mathbf{S}^7(\mathbf{S}^4, SU(2))$. Let Ξ denote the coupling of $T\mathbf{S}^4$ with $L\Omega = \frac{\mathbf{S}^7 \times \mathfrak{su}(2)}{SU(2)}$ induced by $L\Omega \rightarrowtail A\Omega \twoheadrightarrow T\mathbf{S}^4$. Then there is no flat Lie derivation law covering Ξ; in fact, $L\Omega$ admits no flat Lie connection.

To see this, note first that ad: $A\Omega \to \mathfrak{D}_{\mathrm{Der}} L\Omega$ is an isomorphism, since ad: $L\Omega \to \mathrm{Der}(L\Omega)$ is fibrewise the adjoint representation of $\mathfrak{su}(2)$, and $\mathfrak{su}(2)$ is semisimple. Now it is sufficent to observe that $\mathbf{S}^7(\mathbf{S}^4, SU(2))$ itself admits no flat connection, and this is elementary (since $\mathbf{S}^4$ is simply–connected, a flat connection would trivialize the bundle). ⊠

On the other hand, it can happen that a single coupling has nonequivalent semidirect extensions.

Example 7.3.21 Let M be a manifold with $H^2_{\mathrm{deRh}}(M) \neq 0$, and let $\mathfrak{g}$ be a nonabelian Lie algebra with centre $\mathfrak{z} \neq 0$. Define $K = M \times \mathfrak{g}$. Let ∇^0 be the standard flat Lie connection in K and let $\ell \colon TM \to K$ be a 1–form such that $\Lambda = (\nabla^0(\ell) + [\ell, \ell]) = (\delta\ell + [\ell, \ell])$ is closed, but not exact.

Let A^0 denote $TM \ltimes_{\nabla^0} K$ and let A' denote $(A^0)_{-\Lambda}$, in the notation of 7.3.12. Then A' is flat, for it is easily seen that $\chi(X) = X \oplus \ell(X)$ defines a flat transversal. Thus, by 7.3.9, A' is equivalent to the semidirect extension corresponding to $\nabla^\chi = \nabla^0 + j \circ \mathrm{ad} \circ \ell$. But, by 7.3.16, A' is not equivalent to A^0, for $-\Lambda$ is not exact. ⊠

A complete interpretation of $\mathscr{H}^3$ in Lie algebroid cohomology would require a notion of similarity for couplings, according to which couplings would define the same obstruction class if and only if they are similar, and a notion of effaceability for elements of $\mathscr{H}^3$, which would characterize those realizable as obstructions. These matters may have a geometric significance of their own, but we have not considered them here since they are not necessary for the results of the next chapter.

We close this section with a concept of produced Lie algebroid. This construction should be compared with the remarks concerning morphisms of locally trivial Lie groupoids at the end of §1.3.

Theorem 7.3.22 *Let* $L \rightarrowtail A \twoheadrightarrow TM$ *be a transitive Lie algebroid on* M *and let* $\varphi \colon L \to L'$ *be a morphism of LABs over* M. *Suppose that there exists a representation* $\rho' \colon A \to \mathfrak{D}_{\mathrm{Der}} L'$ *such that:*

(i) $\rho^+ = \text{ad}' \circ \varphi \colon L \to \text{Der}(L')$, *where* ad' *denotes the LAB adjoint* $L' \to \text{Der}(L')$; *and*

(ii) φ *is* A–*equivariant with respect to the actions* ad *and* ρ *of* A *on* L *and* L'.

Then there is a transitive Lie algebroid $L' \!\!>\!\!-\!\!-\!\!> A' \!-\!\!-\!\!\gg TM$ *and a morphism of Lie algebroids* $\widetilde{\varphi} \colon A \to A'$ *such that* $(\widetilde{\varphi})^+ \colon L \to L'$ *is equal to* φ, *and such that* $\rho = \text{ad}' \circ \widetilde{\varphi}$, *where* ad' *is now the adjoint representation* $A' \to \mathfrak{D}_{\text{Der}} L'$. *Furthermore,* A' *is uniquely determined up to equivalence by these conditions.*

Proof Let γ be a connection in A and define on the vector bundle $TM \oplus L'$ a bracket structure by

$$[X \oplus V', Y \oplus W'] = [X, Y] \oplus$$
$$\{\rho(\gamma X)(W') - \rho(\gamma Y)(V') + [V', W'] - \varphi \overline{R}_\gamma(X, Y)\}$$

where $X, Y \in \mathfrak{X}(M)$, $V', W' \in \Gamma L'$. It is easily checked that this makes $TM \oplus L'$ a transitive Lie algebroid on M; denote it by A^γ. Furthermore, $X \mapsto aX \oplus \varphi \omega X$ is a Lie algebroid morphism $A \to A^\gamma$, where ω is the connection reform in A corresponding to γ. The required properties are easily verified.

Now suppose that $L' \!\!>\!\!-\!\!-\!\!> A' \!-\!\!-\!\!\gg TM$ is another transitive Lie algebroid and $\widetilde{\varphi} \colon A \longrightarrow A'$ a morphism with the required properties. Then $\mathscr{E} \colon A^\gamma \longrightarrow A'$, $X \oplus V' \mapsto \widetilde{\varphi}\gamma X + V'$ may be checked to be an equivalence. (Note that $\rho = \text{ad}' \circ \widetilde{\varphi}$ ensures that $\rho(\gamma X)(W') = [\widetilde{\varphi}\gamma X, j'W']$.)
$\square$

The remark following 6.5.11 shows that every base–preserving morphism of transitive Lie algebroids is of the form constructed in 7.3.22.

Note that (ii) is, in part, a requirement that $\varphi \colon L \to L'$, regarded as a map $M \to \text{Hom}(L, L')$, be constant in the same sense in which one may say that the morphism $\varphi \colon \mathscr{I}\Omega \to \mathscr{I}\Omega'$ arising from a morphism of Lie groupoids is constant because φ commutes with inner automorphisms.

7.4 The spectral sequence of a transitive Lie algebroid

Given an extension of a transitive Lie algebroid by a vector bundle, there is an associated extension of the adjoint bundle, which forms a commutative diagram:

$$
\begin{array}{ccccc}
E & \overset{\iota^+}{\rightarrowtail} & L' & \overset{\pi^+}{\twoheadrightarrow} & L \\
\| & & {\scriptstyle j'}\downarrow & & {\scriptstyle j}\downarrow \\
E & \overset{\iota}{\rightarrowtail} & A' & \overset{\pi}{\twoheadrightarrow} & A \\
& & {\scriptstyle a'}\downarrow & & {\scriptstyle a}\downarrow \\
& & TM & =\!=\!= & TM.
\end{array}
\tag{11}
$$

The spectral sequence with which this section is concerned gives information as to when a given extension of the adjoint bundle can be lifted to an extension of the transitive Lie algebroid, and as to the construction of Lie algebroid extensions for which the extension of the adjoint bundles is trivial. These questions may be regarded as infinitesimal forms of the problems associated with the lifting of structure groups in a principal bundle.

Definition 7.4.1 Let A and A' be Lie algebroids on the same base M and let $\rho\colon A \to \mathfrak{D}(E)$ and $\rho'\colon A' \to \mathfrak{D}(E')$ be representations of A and A'. Then a *change of Lie algebroids* from (A, ρ, E) to (A', ρ', E') is a pair (φ, ψ) where $\varphi\colon A \to A'$ is a morphism of Lie algebroids over M and $\psi\colon E' \to E$ is a morphism of vector bundles over M, such that

$$
\psi(\rho'(\varphi(X))(\mu')) = \rho(X)(\psi(\mu'))
$$

for $\mu' \in E', X \in A$. □

A change of Lie algebroids induces a morphism of cochain complexes

$$
(\varphi, \psi)^{\#}\colon\ C^n(A', E') \to C^n(A, E), \qquad f' \mapsto \psi \circ f' \circ \varphi^n
$$

and hence morphisms $(\varphi, \psi)^*\colon \mathscr{H}^n(A', E') \to \mathscr{H}^n(A, E)$. If A and A' are transitive, then (φ^+, ψ) is also a change of Lie algebroids and induces morphisms $(\varphi^+, \psi)^*\colon H^n(L', E') \to H^n(L, E)$. If $E = E'$ and $\psi = \mathrm{id}$ we write $\varphi^{\#}, \varphi^*$ for $(\varphi, \mathrm{id})^{\#}, (\varphi, \mathrm{id})^*$.

Associated with a given transitive Lie algebroid A and representation $\rho\colon A \to \mathfrak{D}(E)$ there are two natural changes, $(j, \mathrm{id}_E)\colon (L, \rho^+, E) \to (A, \rho, E)$ and $(a, \subseteq)\colon (A, \rho, E) \to (TM, \overline{\rho}, E^L)$. Here $\overline{\rho}$ is the representation of TM on E^L induced by ρ (see 6.5.14). These induce maps

$$
j^*\colon \mathscr{H}^n(A, \rho, E) \to \Gamma H^n(L, \rho^+, E) \quad \text{and}
$$
$$
a^* = (a, \subseteq)^*\colon \mathscr{H}^n(TM, \overline{\rho}, E^l) \to \mathscr{H}^n(A, \rho, E)
$$

which are the *restriction* and *inflation* maps of (A, ρ, E). Note that $j^* \circ a^* = 0$. Our chief concern is with the kernels and cokernels of these maps.

For $n = 2$ the restriction and inflation maps have natural definitions in terms of extensions. Given a ρ-operator extension

$$E \rightarrowtail A' \overset{\pi}{\twoheadrightarrow} A, \qquad (12)$$

the extension $E \overset{\iota^+}{\rightarrowtail} L' \overset{\pi^+}{\twoheadrightarrow} L$ is a ρ^+-operator extension of L by E and for any transversal $\chi \colon A \to A'$ with cocycle $\overline{R}_\chi$, the map $\chi^+ \colon L \to L'$ is a transversal of π^+ and has cocycle $j^\#(\overline{R}_\chi)$. (Note that χ^+ is defined by virtue of $a' \circ \chi = a$.) It is easy to see that this defines a map $\mathscr{O}\mathrm{pext}(A, \rho, E) \to \Gamma\,\mathscr{O}\mathrm{pext}(L, \rho^+, E)$ which represents j^*. We call $E \overset{\iota^+}{\rightarrowtail} L' \overset{\pi^+}{\twoheadrightarrow} L$ the *restriction* of (12).

The inflation map a^* can be realized in terms of extensions in a similar way. Given an extension $E^L \rightarrowtail J \twoheadrightarrow TM$, construct the pullback extension as in 7.1.16:

$$
\begin{array}{ccccc}
E^L & \rightarrowtail & J & \twoheadrightarrow & TM \\
\| & & \uparrow & & {\scriptstyle a}\uparrow \\
E^L & \rightarrowtail & \overline{J} & \twoheadrightarrow & A
\end{array}
$$

and then the pushout as in 7.1.14:

$$
\begin{array}{ccccc}
E^L & \rightarrowtail & \overline{J} & \twoheadrightarrow & A \\
\downarrow & & \downarrow & & \| \\
E & \rightarrowtail & \widetilde{J} & \twoheadrightarrow & A
\end{array}
$$

Alternatively the steps may be reversed. We call $E \rightarrowtail \widetilde{J} \twoheadrightarrow A$ the *inflation* of the transitive Lie algebroid $E^L \rightarrowtail J \twoheadrightarrow TM$. It is easy to see directly that the restriction of the inflation $E \rightarrowtail \widetilde{J} \twoheadrightarrow A$ is semi–direct; in this sense the extension $E \rightarrowtail \widetilde{J} \twoheadrightarrow A$ has no algebraic component to its curvature and we will therefore also call it a *geometric extension* of A by E. An extension $E \rightarrowtail A' \twoheadrightarrow A$ whose restriction is semi–direct will be called a *restriction semi–direct*, or *RSD*, extension. Not all RSD extensions are geometric: see 7.4.16 below. The quotient space $\frac{\ker j^*}{\mathrm{im}\,a^*}$ is given by the term $E_3^{1,1}$ of the spectral sequence.

For the image of j^* there is first of all the following result.

Proposition 7.4.2 *The map* $j^* \colon \mathscr{H}^n(A, \rho, E) \to \Gamma H^n(L, \rho^+, E)$ *takes values in* $(\Gamma H^n(L, \rho^+, E))^A$, *where* A *acts on* $H^n(L, E)$ *via the Lie derivative* $\mathfrak{L}$ *of* 7.1.5.

Proof Observe that $j^\# \circ \mathfrak{L}_X = \mathfrak{L}_X \circ j^\# \colon \Gamma C^n(A, E) \to \Gamma C^n(L, E)$ for $X \in \Gamma A$. Here the $\mathfrak{L}_X$ on the left is defined in 7.1.2 and that on the right is the action of A on $C^n(L, E)$ defined in 7.1.4. It follows that for $f \in \mathscr{Z}^n(A, E)$,

$$\mathfrak{L}_X(j^\#(f)) = j^\#(\mathfrak{L}_X(f)) = j^\#(d \circ \iota_X(f)) + 0 = d(j^\# \circ \iota_X(f))$$

and so $\mathfrak{L}_X(j^*[f]) = [0]$. Here we used 7.1.3(v). □

For $n = 2$ the action $\mathfrak{L}$ can be transported to an action of A on the vector bundle $\mathscr{O}\mathrm{pext}(L, \rho^+, E)$ via 7.1.13. Since $\mathfrak{L}^+ = 0$ by 7.1.6 the equation $\mathfrak{L}_X(E \rightarrowtail L' \twoheadrightarrow L) = 0$ may be interpreted as signifying that the Lie algebra extension $E_x \rightarrowtail L'_x \twoheadrightarrow L_x$, as a function of $x \in M$, has zero derivatives in all directions $a(X) \in \Gamma TM$. Thus 7.4.2 implies that a necessary condition for an extension $E \rightarrowtail L' \twoheadrightarrow L$ of the adjoint bundle L to be the restriction of an extension of A is that the Lie algebra extensions $E_x \rightarrowtail L'_x \twoheadrightarrow L_x$ be constant with respect to x, where constancy is taken to refer to $\mathfrak{L}$. As in 6.5.10, this constancy is an abstraction of equivariance with respect to actions of adjoint type. See also 6.4.23.

There are two further conditions which must be imposed on an element of $(\Gamma H^2(L, E))^A$ before it can be guaranteed to lie in the image of j^*. These are most naturally formulated in terms of the spectral sequence, to which we now turn.

$$* \quad * \quad * \quad * \quad *$$

We deal with the spectral sequence of a canonically bounded descending filtration; thus $F^k C^n \supseteq F^{k+1} C^n, F^0 C^n = C^n$ and $F^{n+1} C^n = (0)$. We use an explicit approach; thus $E_r^{s,t} = Z_r^{s,t}/B_r^{s,t}$ where

$$Z_r^{s,t} = \{ f \in C^{s+t} \mid df \in F^{s+r} C^{s+t+1} \}, \quad B_r^{s,t} = dZ_{r-1}^{s-r+1, t+r-2} + Z_{r-1}^{s+1, t-1}$$

for $r \geqslant 1$, and $B_0^{s,t} = F^{s+1} C^{s+t}$. The isomorphism $E_{r+1}^{s,t} \to H^{s,t}(E_r^{*,*})$ induced by the inclusion $Z_{r+1}^{s,t} \subseteq Z_r^{s,t}$ is denoted $\sigma_r^{s,t}$. Such a spectral sequence is strongly convergent; the filtration

$$F^s H^n(C^\bullet) = \mathrm{im}(H^n(F^s C^\bullet) \to H^n(C^\bullet))$$

is also canonically bounded and the isomorphism

$$E_\infty^{s,t} \cong \frac{F^s H^{s+t}(C^\bullet)}{F^{s+1} H^{s+t}(C^\bullet)}$$

is, on the cochain level, the identity map. Here $E_\infty^{s,t} = Z_\infty^{s,t}/B_\infty^{s,t}$ where $Z_\infty^{s,t} = \{f \in F^s C^{s+t} \mid df = 0\}$ and

$$B_\infty^{s,t} = (F^s C^{s+t} \cap dC^{s+t-1}) + Z_\infty^{s+1,t-1}.$$

Note that $E_r^{s,t} = E_{r+1}^{s,t} = E_\infty^{s,t}$ for $r > \max\{s, t+1\}$. The edge morphisms are denoted by

$$e_B \colon E_2^{s,0} \longrightarrow\!\!\!\!\!\gg E_\infty^{s,0} \subseteq H^s(C^\bullet)$$

and

$$e_F \colon H^t(C^\bullet) \longrightarrow\!\!\!\!\!\gg E_\infty^{0,t} \succ\!\!\!\longrightarrow E_2^{0,t}.$$

The transgression relation $E_2^{0,n} \rightsquigarrow E_2^{n+1,0}$ is denoted tg^n. For $n = 1$ it is a well– and fully–defined map, namely $d_2^{0,1}$. If $E_{n+2}^{0,n} = (0) = E_{n+2}^{n+1,0}$ then $d_{n+1}^{0,n}$ is an isomorphism and the composite

$$E_2^{n+1,0} \longrightarrow\!\!\!\!\!\gg E_{n+1}^{n+1,0} \longrightarrow E_{n+1}^{0,n} \succ\!\!\!\longrightarrow E_2^{0,n}$$

may be considered a Cartan map κ^n with tg^n as inverse relation.

Let $L \succ\!\!\xrightarrow{j} A \xrightarrow{a}\!\!\!\gg TM$ be a transitive Lie algebroid and let $\rho \colon A \to \mathfrak{D}(E)$ be a representation of A.

We will filter the cochain complex $\Gamma C^\bullet(A, E)$: to cut down on notation, denote $\Gamma C^n(A, E)$ by $\Gamma^n(A, E)$. Define

$$F^s \Gamma^n(A, E) = \{f \in \Gamma^n(A, E) \mid f(X_1, \ldots, X_n) = 0 \text{ whenever}$$
$$(n - s + 1) \text{ or more of the } X_i \text{ are in } \ker a\}. \quad (13)$$

Then $F^s \Gamma^n(A, E) \supseteq F^{s+1}\Gamma^n(A, E)$. Clearly $F^0 \Gamma^n(A, E) = \Gamma^n(A, E)$ and $F^{n+1}\Gamma^n(A, E) = (0)$. By convention, $F^s \Gamma^n(A, E) = (0)$ for all $s > n + 1$.

Define $a^{s,t} \colon F^s \Gamma^{s+t}(A, E) \to \Gamma^s(TM, C^t(L, E))$ by

$$a^{s,t}(f)(X_1, \ldots, X_s)(V_1, \ldots, V_t) = f(jV_1, \ldots, jV_t, \gamma X_1, \ldots, \gamma X_s)$$

for any connection $\gamma \colon TM \to A$. It is easy to see that $a^{s,t}$ is independent of the choice of γ and has kernel $F^{s+1}\Gamma^{s+t}(A, E)$. It is also surjective: to see this, define

$$e^{s,t} \colon \Gamma^s(TM, C^t(L, E)) \to F^s \Gamma^{s+t}(A, E)$$

by

$$e^{s,t}(f)(X_1, \ldots, X_{s+t}) =$$
$$\frac{(-1)^{st}}{s!t!} \sum_\sigma \varepsilon_\sigma f(aX_{\sigma(t+1)}, \ldots, aX_{\sigma(t+s)})(\omega X_{\sigma(1)}, \ldots, \omega X_{\sigma(t)}),$$

where $\omega\colon A \to L$ is any connection reform and the summation is over all permutations of $\{1,\ldots,s+t\}$. It is straightforward to verify that $a^{s,t}(e^{s,t}(f)) = f$.

Therefore, $a^{s,t}$ quotients to an isomorphism $\alpha_0^{s,t}$ as in the diagram in 7.4.3.

Proposition 7.4.3 *The diagram*

$$
\begin{array}{ccc}
E_0^{s,t} & \xrightarrow{\ d_0^{s,t}\ } & E_0^{s,t+1} \\
{\scriptstyle \alpha_0^{s,t}}\Big\downarrow & & \Big\downarrow{\scriptstyle \alpha_0^{s,t+1}} \\
\Gamma^s(TM, C^t(L,E)) & \xrightarrow{\ \Gamma^s(TM,d)\ } & \Gamma^s(TM, C^{t+1}(L,E))
\end{array}
$$

commutes.

Proof Take $f \in Z_0^{s,t} = F^s\Gamma^{s+t}(A,E)$. We must show that, for all $X_1,\ldots,X_s \in \mathfrak{X}(M)$,

$$a^{s,t+1}(df)(X_1,\ldots,X_s) = d(a^{s,t}(f)(X_1,\ldots,X_s)).$$

Take $V_1,\ldots,V_{t+1} \in \Gamma L$. From $a^{s,t+1}(df)(X_1,\ldots,X_s)(V_1,\ldots,V_{t+1})$, we get, by expanding out,

$$
df(V_1,\ldots,V_{t+1},\gamma X_1,\ldots,\gamma X_s) =
$$
$$
\sum_{i=1}^{t+1}(-1)^{i+1}\rho^+(V_i)f(V_1,\ldots\hat{}\ldots,V_{t+1},\gamma X_1,\ldots,\gamma X_s)
$$
$$
+ \sum_{1\leqslant i<j\leqslant t+1}(-1)^{i+j}f([V_i,V_j],V_1,\ldots\hat{}\ldots,V_{t+1},\gamma X_1,\ldots,\gamma X_s)
$$
$$
+ \sum_{i=1}^{s}(-1)^{i+j}f([V_i,V_j],V_1,\ldots\hat{}\ldots,V_{t+1},\gamma X_1,\ldots,\gamma X_s)
$$
$$
+ \sum_{1\leqslant i<j\leqslant s}(-1)^{i+j}f([\gamma X_i,\gamma X_j],V_1,\ldots,V_{t+1},\gamma X_1,\ldots\hat{}\ldots,\gamma X_s)
$$
$$
+ \sum_{i=1}^{t+1}\sum_{j=1}^{s}(-1)^{i+j+t+1}f([V_i,\gamma X_i],V_1,\ldots\hat{}\ldots,V_{t+1},\gamma X_1,\ldots\hat{}\ldots,\gamma X_s).
$$

(Here the circumflexes refer to the arguments corresponding to the dummy indices.) In each of the last three summations each term vanishes, for each has $(t+1)$ arguments in ΓL and $f \in F^s\Gamma^{s+t}(A,E)$. The first two terms can be rewritten as

$$\sum_{i=1}^{t+1}(-1)^{i+1}\rho^+(V_i)(a^{s,t}(f)(X_1,\ldots,X_s)(V_1,\ldots,\widehat{\ldots},V_{t+1}))$$

$$+\sum_{1\leqslant i<j\leqslant t+1}(-1)^{i+j}a^{s,t}(f)(X_1,\ldots,X_s))(V_1,\ldots,V_{t+1}),$$

and are therefore equal to $d(a^{s,t}(f)(X_1,\ldots,X_s))(V_1,\ldots,V_{t+1})$, as required. $\qquad\square$

Identifying $E_1^{s,t}$ with $H^{s,t}(E_0^{*,*})$, we now have isomorphisms

$$\alpha_1^{s,t}\colon E_1^{s,t}\to\Gamma^s(TM,H^t(L,E)).$$

Following through the identifications, if $f\in Z_1^{s,t}$ represents a chosen element of $E_1^{s,t}$, then f is in $F^s\Gamma^{s+t}(A,E)$ and consequently $a^{s,t}(f)\in\Gamma^s(TM,C^t(L,E))$ takes values in $\Gamma Z^t(L,E)$; the resulting cohomology class $[a^{s,t}(f)(X_1,\ldots,X_s)]\in\Gamma H^t(L,E)$ is $\alpha_1^{s,t}([f])(X_1,\ldots,X_s)$. The proof of the next result is very similar to that of 7.4.3.

Proposition 7.4.4 *The diagram*

$$
\begin{array}{ccc}
E_1^{s,t} & \xrightarrow{\;d_1^{s,t}\;} & E_1^{s+1,t}\\[4pt]
{\scriptstyle\alpha_1^{s,t}}\Big\downarrow & & \Big\downarrow{\scriptstyle\alpha_1^{s+1,t}}\\[4pt]
\Gamma^s(TM,H^t(L,E)) & \xrightarrow{\;(-1)^t d\;} & \Gamma^s(TM,H^t(L,E))
\end{array}
$$

commutes.

Thus $\alpha_1^{*,t}$ induces isomorphisms $E_2^{s,t}\to\mathscr{H}^s(TM,H^t(L,E))$. Given $f\in Z_2^{s,t}\subseteq F^s\Gamma^{s+t}(A,E)$ representing $[f]\in E_2^{s,t}$, the class $\alpha_2^{s,t}([f])$ is represented by the cocycle which to $(X_1,\ldots,X_s)$ assigns the cohomology class in $\Gamma H^t(L,E)$ represented by $a^{s,t}(f)(X_1,\ldots,X_s)$.

Theorem 7.4.5 *For a transitive Lie algebroid A and representation $\rho\colon A\to\mathfrak{D}(E)$, there is a natural convergent spectral sequence*

$$\mathscr{H}^s(TM,H^t(L,E))\Longrightarrow\mathscr{H}^n(A,E).$$

Proof Only the naturality remains to be explained. If

$$(\varphi,\psi)\colon(A,\rho,E)\to(A',\rho',E')$$

is a change of transitive Lie algebroids, then $(\varphi,\psi)^{\#}\colon\Gamma^\bullet(A',E')\to\Gamma^\bullet(A,E)$ preserves the filtrations and so induces a morphism of spectral sequences $E_*^{*,*}(A',E')\to E_*^{*,*}(A,E)$. Also, there is a change of

Lie algebroids $(\mathrm{id}_{TM}, (\varphi^+, \psi)^*)$ from $TM \to \mathfrak{D}(H^\bullet(L, E))$ to $TM \to \mathfrak{D}(H^\bullet(L', E'))$. It is now straightforward to show that the induced morphisms $\mathscr{H}^s(TM, H^t(L', E')) \to \mathscr{H}^s(TM, H^t(L, E))$ commute with the morphisms of the spectral sequences, and similarly on the E_∞ level.

$\square$

In an exactly similar way, one may prove the next result.

Theorem 7.4.6 *For an exact sequence of Lie algebroids*

$$K \rightarrowtail A' \twoheadrightarrow A$$

and a representation of A' on a vector bundle E, there is a natural convergent spectral sequence

$$\mathscr{H}^s(A, H^t(K, E)) \implies \mathscr{H}^n(A', E).$$

In particular, for a regular Lie algebroid, $L \rightarrowtail A \twoheadrightarrow \Delta$, and representation of A on a vector bundle E, there is a natural convergent spectral sequence

$$\mathscr{H}^s(\Delta, H^t(L, E)) \implies \mathscr{H}^n(A, E).$$

For an involutive distribution Δ, a representation on a vector bundle E' is a partial flat connection along the leaves of Δ.

Proposition 7.4.7 *Figure 7.2(a) and Figure 7.2(b) commute.*

Proof Take $[f] \in E_2^{n,0}$ with $f \in Z_2^{n,0}$. Then $f \in F^n \Gamma^n(A, E)$ and

$$df \in F^{n+2} \Gamma^{n+1}(A, E) = (0).$$

So $f \in \mathscr{Z}^n(A, E)$ and $[f] \in \mathscr{H}^n(A, E)$ is $e_M([f])$. On the other hand, $\alpha_2^{n,0}([f])$ is represented by $a^{n,0}(f)$, which lies in $\mathscr{Z}^n(TM, E^L)$. Now $a^\#(a^{n,0}(f))(X_1, \ldots, X_n) = f(\gamma a X_1, \ldots, \gamma a X_n)$ for $X_i \in \Gamma A$, and since $\gamma a X = X - j\omega X$ and $f \in F^n \Gamma^n(A, E)$ vanishes whenever $n - n + 1$ or more arguments are in $j(L)$, it follows that $a^\#(a^{n,0}(f)) = f$.

The second half is proved similarly. Note that $\mathscr{H}^0(TM, H^n(L, E)) = (\Gamma H^n(L, E))^{TM}$ and this is equal to $(\Gamma H^n(L, E))^A$ by 7.1.6. $\square$

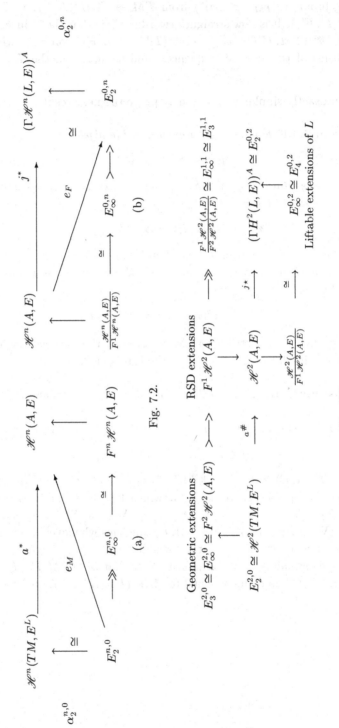

Fig. 7.2.

Fig. 7.3.

We can now express the images and kernels of a^* and j^* in terms of the spectral sequence. The image of a^* is $E_\infty^{n,0} \cong F^n \mathscr{H}^n(A, E)$ and this, for $n = 2$, characterizes the geometric extensions of A by E. Similarly, the kernel of j^* is $F^1 \mathscr{H}^n(A, E)$, so the RSD extensions of A by E are precisely those in $F^1 \mathscr{H}^2(A, E)$. Of course $F^1 \mathscr{H}^2(A, E) \supseteq F^2 \mathscr{H}^2(A, E)$ and the quotient, which represents the failure of RSD extensions to be geometric, is isomorphic to $E_\infty^{1,1}$.

Lastly, the image of j^* is isomorphic to $E_\infty^{0,n}$ and so $E_\infty^{0,2}$ represents those extensions of L which can be lifted to extensions of A.

For $n = 2$ these E_∞ spaces reduce: $E_\infty^{2,0} \cong E_3^{2,0}$, $E_\infty^{1,1} \cong E_3^{1,1}$, and $E_\infty^{0,2} \cong E_4^{0,2}$. In turn, $E_3^{2,0}$ is the cokernel of $d_2^{0,1} : E_2^{0,1} \to E_2^{2,0}$, while $E_3^{1,1}$ is the kernel of $d_2^{1,1} : E_2^{1,1} \to E_2^{3,0}$ and $E_4^{0,2}$ is the kernel of $d_3^{0,2} : E_3^{0,2} \to E_3^{3,0}$. Thus these spaces are all accessible in terms of d_2 and d_3.

The situation is summarized in Figure 7.3.

We now calculate $d_2^{r,1}$ for $r \geqslant 0$, when the action of L on E is trivial. For this we need the concept of pairing of spectral sequences. We summarize briefly the details.

Let A be an arbitrary Lie algebroid and let ρ^M, ρ^N, ρ^P be representations of A on vector bundles M, N, P. Then M and N are *paired* to P if there is a bilinear vector bundle map $M \oplus N \to P$, denoted $\wedge$, such that

$$\rho^P(X)(\mu_\wedge \nu) = \rho^M(X)(\mu)_\wedge \nu + \mu_\wedge \rho^N(X)(\nu) \tag{14}$$

for all $X \in \Gamma A$, $\mu \in \Gamma M$, $\nu \in \Gamma N$.

If M and N are paired to P and K is an ideal of A, then there is an induced pairing

$$C^m(K, M) \oplus C^n(K, N) \to C^{m+n}(K, P)$$

defined by

$$(f \wedge g)(X_1, \ldots, X_{m+n}) =$$

$$\frac{1}{m! n!} \sum \varepsilon_\sigma f(X_{\sigma(1)}, \ldots, X_{\sigma(m)}) \wedge g(X_{\sigma(m+1)}, \ldots, X_{\sigma(m+n)}), \tag{15}$$

where the sum is over all permutations on $\{1, \ldots, m + n\}$. Here the representation of A on $C^m(K, M)$ is by the Lie derivative 7.1.2 — if A is transitive then K, being an ideal, must be a subbundle of the adjoint bundle L of A and so, by 7.1.3(iii), $\mathfrak{L} : A \to \mathfrak{D}(C^m(K, M))$ is indeed a representation of A.

As for differential forms, we have

$$d(f \wedge g) = df \wedge g + (-1)^m f \wedge dg, \tag{16}$$

where $f \in \Gamma C^m(K, M)$, $g \in \Gamma C^n(K, N)$. There is therefore an induced pairing

$$H^m(K, M) \oplus H^n(K, N) \to H^{m+n}(K, P),$$

still denoted by $\wedge$.

Now consider a transitive Lie algebroid $L \rightarrowtail A \twoheadrightarrow TM$. Applying the above, there are pairings $C^m(L, M) \oplus C^n(L, N) \to C^{m+n}(L, P)$ and $H^m(L, M) \oplus H^n(L, N) \to H^{m+n}(L, P)$. There is also a map

$$C^m(A, M) \oplus C^n(A, N) \to C^{m+n}(A, P),$$

defined as in (15), which is bilinear and satisfies (16). Here the vector bundles do not (generally) admit representations of A and (14) has no meaning. We will call this map the *formal pairing* induced by the pairing $M \oplus N \to P$.

In particular there is a formal pairing

$$C^s(TM, H^m(L, M)) \oplus C^{s'}(TM, H^n(L, N)) \to C^{s+s'}(TM, H^{m+n}(L, E))$$

and, applying a form of (16), there is a bilinear map, the domain of which is

$$\mathscr{H}^s(TM, H^m(L, M)) \times \mathscr{H}^{s'}(TM, H^n(L, E))$$

and which has target $\mathscr{H}^{s+s'}(TM, H^{m+n}(L, E))$.

Proposition 7.4.8 *Let* $L \rightarrowtail A \twoheadrightarrow TM$ *be a transitive Lie algebroid, and let* ρ^M, ρ^N, ρ^P *be representations of* A *on vector bundles* M, N, P. *Let* $E(M), E(N), E(P)$ *denote the corresponding spectral sequences. Then*

$$F^s \Gamma^m(A, M) \wedge F^{s'} \Gamma^n(A, N) \subseteq F^{s+s'} \Gamma^{m+n}(A, P)$$

and

$$
\begin{array}{ccc}
E_2^{s,t}(M) \times E_2^{s',t'}(N) & \longrightarrow & E_2^{s+s',t+t'}(P) \\
\downarrow & & \downarrow \\
\mathscr{H}^s(TM, H^t(L,M)) \times \mathscr{H}^{s'}(TM, H^{t'}(L,N)) & \xrightarrow{(-1)^{st'}} & \mathscr{H}^{s+s'}(TM, H^{t+t'}(L,P))
\end{array}
$$

commutes, where the bottom row is the map described above multiplied by $(-1)^{st'}$.

Proof This is routine; one proves the corresponding result at the E_0 level and then follows through the formation of the homologies. □

We also need two elementary observations. Fix a transitive Lie algebroid $L \rightarrowtail A \twoheadrightarrow TM$ and a representation $\rho \colon A \to \mathfrak{D}(E)$. The next result is immediate from 7.1.13.

Lemma 7.4.9 *Let $L_{ab} \rightarrowtail A_{ab} \twoheadrightarrow TM$ be the quotient Lie algebroid $A_{ab} = A/[L, L]$, and let $\natural \colon A \to A_{ab}$ denote the natural projection. Then each connection $\gamma \colon TM \to A$ maps to a connection $\natural \circ \gamma$ in A_{ab} whose curvature $\overline{R}_{\natural \circ \gamma} = \natural^+ \circ \overline{R}_\gamma$ belongs to $\mathscr{Z}^2(TM, L_{ab})$ and which represents the cohomology class of $L_{ab} \rightarrowtail A_{ab} \twoheadrightarrow TM$.*

Denote the class of $L_{ab} \rightarrowtail A_{ab} \twoheadrightarrow TM$ in $\mathscr{H}^2(TM, L_{ab})$ by R_{ab}.

Lemma 7.4.10 *Let L be any totally intransitive Lie algebroid and let E be a vector bundle on the same base. Then, with respect to the zero representation of L on E,*

$$H^1(L, E) \cong C^1(L_{ab}, E).$$

Proof The coboundary $d \colon C^0(L, E) \to C^1(L, E)$ is zero and the next coboundary, $d \colon C^1(L, E) \to C^2(L, E)$ is $df(X, Y) = -f([X, Y])$. From the first formula it follows that $H^1(L, E) = Z^1(L, E)$ and from the second we see that $\Gamma Z^1(L, E)$ consists of those maps $\Gamma L \to \Gamma E$ that vanish on $\Gamma[L, L] = [\Gamma L, \Gamma L]$. Hence the result. □

Theorem 7.4.11 *Let $L \rightarrowtail A \twoheadrightarrow TM$ be a transitive Lie algebroid and ρ a representation of A on a vector bundle E for which $E^L = E$. Then the map*

$$\mathscr{H}^n(TM, H^1(L, E)) \to \mathscr{H}^{n+2}(TM, E)$$

corresponding to $d_2^{n,1}$ is given by

$$F \mapsto (-1)^n F_\wedge R_{ab}$$

where the pairing is that induced by $H^1(L, E) \oplus L/[L, L] \to E$ via 7.4.10.

Proof Let $f \in \mathscr{L}^n(TM, Z^1(L, E))$ represent F. Then $(\alpha_2^{n,1})^{-1}(F)$ is represented by $e^{n,1}(f) \in \Gamma^{n+1}(A, E)$, where

$$e^{n,1}(f)(X_1, \ldots, X_{n+1})$$
$$= (-1)^n \sum_{i=1}^{n+1} (-1)^i f(aX_1, \ldots, \widehat{aX_i}, \ldots, aX_{n+1})(\omega X_i).$$

Then $(\alpha_2^{n+2,0} \circ d_2^{n,1} \circ (\alpha_2^{n,1})^{-1})(F)$ is represented by $a^{n+2,0}(d(e^{n,1}(f)))$, and this reduces to

$$\sum_{i<j} (-1)^{i+j}((-1)^n f(X_1, \ldots, \widehat{X_i}, \ldots, \widehat{X_j}, \ldots, X_{n+2})(\omega[\gamma X_i, \gamma X_j]))$$

since all other terms in $d(e^{n,1}(f))(\gamma X_1, \ldots, \gamma X_{n+2})$ involve $\omega \circ \gamma$, and $\omega \circ \gamma = 0$. Now $\omega[\gamma X_i, \gamma X_j] = -\overline{R}_\gamma(X_i, X_j)$ so, recalling the definition of the pairing $H^1(L, E) \oplus L_{ab} \to E$, the sum

$$(-1)^n \sum_{i<j} (-1)^{i+j+1} f(X_1, \ldots, \widehat{X_i}, \ldots, \widehat{X_j}, \ldots, X_{n+2})(\overline{R}_\gamma(X_i, X_j))$$

is seen to be the value at $(X_1, \ldots, X_{n+2})$ of the cocycle representing $(-1)^n F_\wedge R_{ab}$. $\square$

Remark 7.4.12 The condition $E^L = E$ forces E to be flat.

This result decomposes $d_2^{n,1}$ into the pairing — which concerns only the adjoint bundle L and the coefficient bundle E and may be regarded as a purely algebraic matter — and the class R_{ab}, which is a topological invariant. Thus we have the following corollaries.

Corollary 7.4.13 (i) *If $L = [L, L]$, for example if L is semisimple, then $d_2^{n,1} = 0$ for all $n \geqslant 0$.*
(ii) *If the abelianized Lie algebroid $L_{ab} \rightarrowtail A_{ab} \twoheadrightarrow TM$ is flat, then $d_2^{n,1} = 0$ for all $n \geqslant 0$.*

The sequence of terms of low degree is (without any assumption on the coefficients)

$$\mathscr{H}^1(TM, E^L) \overset{a^*}{\rightarrowtail} \mathscr{H}^1(A, E) \overset{j^*}{\longrightarrow} (\Gamma H^1(L, E))^A \longrightarrow \cdots$$
$$\cdots \overset{tg^1}{\longrightarrow} \mathscr{H}^2(TM, E^L) \overset{a^*}{\longrightarrow} \mathscr{H}^2(A, E).$$

The transgression here is simply $d_2^{0,1}$ and so if either of the conditions of 7.4.13 is satisfied, it follows that $\mathscr{H}^2(TM, E)$ is injected into $\mathscr{H}^2(A, E)$

and the space of geometric extensions of A by E may be identified with $\mathscr{H}^2(TM, E)$.

In the general case, the map $(\Gamma H^1(L, E))^A \to \mathscr{O}\text{pext}(TM, E)$ can be interpreted as assigning to suitable $f \in \Gamma Z^1(L, E)$ the pushout of $L \rightarrowtail A \twoheadrightarrow TM$ over $f \colon L \to E^L$. These pushouts are those extensions $E^L \rightarrowtail J \twoheadrightarrow TM$ which, when inflated, give the semidirect extension of A by E.

Corollary 7.4.14 *If $E^L = E$ and either condition of 7.4.13 is satisfied, then the space of geometric extensions of A by E is isomorphic to $\mathscr{H}^2(TM, E)$.*

Corollary 7.4.15 *If $E^L = E$ and either condition of 7.4.13 is satisfied, then the space of RSD extensions, modulo the space of geometric extensions, is isomorphic to $\mathscr{H}^1(TM, H^1(L, E)) \cong \mathscr{H}^1(TM, C^1(L_{ab}, E))$.*

In general $d_2^{1,1}$ maps $\mathscr{H}^1(TM, H^1(L, E))$ into $\mathscr{H}^3(TM, E^L)$; if the element of $\mathscr{H}^3(TM, E^L)$ is zero, then there is an extension of TM by E whose inflation is the given RSD extension.

The following example illustrates circumstances in which $E_3^{1,1} \neq 0$.

Example 7.4.16 Let $\mathfrak{g}$ be a reductive Lie algebra with a one–dimensional centre, for example $\mathfrak{gl}(n, \mathbb{R})$. Let M be any manifold for which $H^3_{\text{deRh}}(M) = 0$ and $H^1_{\text{deRh}}(M) \neq 0$. Let A be a transitive Lie algebroid on M with adjoint bundle $L = M \times \mathfrak{g}$ and let ρ be the trivial representation of A. Then $E_2^{1,1} \cong \mathscr{H}^1(TM, H^1(L, M \times \mathbb{R})) = \mathscr{H}^1(TM, M \times \mathbb{R})$, since $\mathfrak{g} \cong [\mathfrak{g}, \mathfrak{g}] \oplus \mathfrak{z}$, and so this is isomorphic to $H^1_{\text{deRh}}(M)$ and hence non–zero. On the other hand, $E_2^{3,0} \cong H^3_{\text{deRh}}(M) = 0$. So $E_3^{1,1} \cong \ker d_2^{1,1} = E_2^{1,1} \neq 0$. ⊠

If we seek general criteria for an extension

$$E \rightarrowtail L' \twoheadrightarrow L \tag{17}$$

of the adjoint bundle L to lift to an extension of A, on the basis of the results so far we can only say that in addition to the condition that (17) lies in $(\Gamma H^2(L, E))^A$, there are the two consecutive conditions that $d_2^{0,2}$ and $d_3^{0,2}$ map the extension to the zero extension.

Proposition 7.4.17 *Let $L \rightarrowtail A \twoheadrightarrow TM$ be an arbitrary transitive Lie algebroid and let ρ be any representation of A on E. Then if $\mathscr{H}^2(TM, H^1(L, E)) = 0$ and $\mathscr{H}^3(TM, E^L) = 0$, every extension of L by E which lies in $(\Gamma H^2(L, E))^A$ lifts to an extension of A by E.*

Proof The space of such extensions has been identified with $E_4^{0,2}$ and this is isomorphic to the kernel of $d_3^{0,2} \colon E_3^{0,2} \to E_3^{3,0}$. In turn, $E_3^{3,0}$ is a quotient of $E_2^{3,0}$ and so is zero, and $E_3^{0,2}$ is the kernel of $d_2^{0,2} \colon E_2^{0,2} \to E_2^{2,1}$. Now $E_2^{2,1} = 0$ also, by hypothesis, so we finally have $E_4^{0,2} \cong E_2^{0,2} \cong (\Gamma H^2(L,E))^A$. $\qquad\square$

The conditions of 7.4.17 are fulfilled if M is simply–connected and $H^2_{\mathrm{deRh}}(M) = H^3_{\mathrm{deRh}}(M) = 0$, or if M is simply–connected, L is semi-simple and $H^3_{\mathrm{deRh}}(M) = 0$. In particular:

Corollary 7.4.18 *Let $L \rightarrowtail A \twoheadrightarrow TM$ be a transitive Lie algebroid on a simply–connected base M for which $H^3_{\mathrm{deRh}}(M) = 0$ and for which either $H^2_{\mathrm{deRh}}(M) = 0$ or L is semisimple. Let ρ be a representation of A on a vector bundle E. Then if $V \rightarrowtail \mathfrak{g}' \twoheadrightarrow \mathfrak{g}$ is an operator extension of the fibre type of L by the fibre type of E, there is an operator extension $E \rightarrowtail A' \twoheadrightarrow A$ whose restriction $E \rightarrowtail L' \twoheadrightarrow L$ has fibre type $V \rightarrowtail \mathfrak{g}' \twoheadrightarrow \mathfrak{g}$.*

Proof Notice that, because M is simply–connected, we have

$$(\Gamma H^2(L,E))^A \cong H^2(\mathfrak{g},V)^{\mathfrak{g}} = H^2(\mathfrak{g},V)$$

by 6.5.16 and 7.1.6. $\qquad\square$

For the case where $\pi_1 M \neq 0$, the comments in 6.4.24 apply.

Concerning the multiplicity of lifts possible for a given (17), this is measured by $\ker j^*$, which can in principle be constructed from $E_3^{2,0}$ and $E_3^{1,1}$; when $E^L = E$ and either one of the conditions of 7.4.13 is satisfied, we have seen that these spaces reduce to $E_2^{2,0}$ and $E_2^{1,1}$.

*   *   *   *   *

For the Lie algebroid of a locally trivial Lie groupoid, this spectral sequence is closely related to the Leray–Serre spectral sequence in de Rham cohomology of the associated principal bundle.

Let Ω be a locally trivial Lie groupoid on M, and let $\rho \colon \Omega \to \Phi(E)$ be a representation of Ω on a vector bundle E. Choose $m \in M$ and write $P = \Omega_m$, $G = \Omega_m^m$, $V = E_m$.

There is a natural action of Ω on $H^\bullet(L\Omega, E)$. Each $\xi \in \Omega$ induces a change of Lie algebras (as in 7.4.1) $(\mathrm{Ad}\xi^{-1}, \rho(\xi))$ from $(\rho_{\beta\xi}^{\beta\xi})_* \colon L\Omega|_{\beta\xi} \to \mathrm{End}(E_{\beta\xi})$ to $(\rho_{\alpha\xi}^{\alpha\xi})_* \colon L\Omega|_{\alpha\xi} \to \mathrm{End}(E_{\alpha\xi})$. Hence ξ induces an isomorphism $(\mathrm{Ad}\xi^{-1}, \rho(\xi))^* \colon H^\bullet(L\Omega, E)_{\alpha\xi} \to H^\bullet(L\Omega, E)_{\beta\xi}$, and it is routine to verify that this defines a smooth action. In particular, G acts on $H^\bullet(\mathfrak{g}, V)$ and it follows from 1.6.5 that $H^\bullet(L\Omega, E)$ is equivariantly isomorphic to $\frac{P \times H^\bullet(\mathfrak{g}, V)}{G}$. Note that this bundle is flat.

In 5.3.11 it is shown that the cochain complex $\Gamma C^\bullet(A\Omega, E)$ is naturally isomorphic to the G–equivariant de Rham complex $\Omega^\bullet(P, V)^G$ and it follows from 7.1.7 that $\mathscr{H}^\bullet(A\Omega, \rho_*, E) \cong H^n_{\mathrm{deRh}}(P, V)^G$. The next result follows immediately.

Theorem 7.4.19 *Let $P(M, G)$ be a principal bundle and let G act on a vector space V. Then there is a natural convergent spectral sequence*

$$\mathscr{H}^s \left(TM, \frac{P \times H^t(\mathfrak{g}, V)}{G} \right) \Longrightarrow H^n_{\mathrm{deRh}}(P, V)^G.$$

If M is simply-connected then the E_2 term simplifies to

$$H^s_{\mathrm{deRh}}(M, H^t(\mathfrak{g}, V)) = H^s_{\mathrm{deRh}}(M) \otimes H^t(\mathfrak{g}, V).$$

If, in addition, G is compact then $H^\bullet_{\mathrm{deRh}}(P, V)^G = H^\bullet_{\mathrm{deRh}}(P, V)$ and $H^t(\mathfrak{g}, V) \cong H^t_{\mathrm{deRh}}(G) \otimes V$ so we obtain the standard Leray–Serre spectral sequence in de Rham cohomology.

We close this section with some brief comments on cases in which the Lie algebroid spectral sequence collapses to a Gysin sequence. Let $L \!>\!\!-\!\!\!-\!\!> A \longrightarrow\!\!\!\!\!\twoheadrightarrow TM$ be an arbitrary transitive Lie algebroid and let ρ be any representation of A on a vector bundle E.

Assume firstly that $H^n(L, E) = 0$ for $n \geqslant 2$. Then the sequence of terms of low degree can be continued:

$$\cdots \longrightarrow \mathscr{H}^1(TM, H^1(L, E)) \xrightarrow{d_2^{1,1}} \mathscr{H}^3(TM, E^L) \xrightarrow{a^*} \mathscr{H}^3(A, E) \longrightarrow \cdots$$

$$\cdots \longrightarrow \mathscr{H}^n(TM, E^L) \xrightarrow{a^*} \mathscr{H}^n(A, E) \longrightarrow \cdots$$

$$\longrightarrow \mathscr{H}^{n-1}(TM, H^1(L, E)) \xrightarrow{d_2^{n-1,1}} \mathscr{H}^{n+1}(TM, E^L) \longrightarrow \cdots \quad (18)$$

When $E^L = E$, 7.4.11 applies and the identification of $d_2^{n,1}$ with $(-1)^n$ times the map $- \wedge R_{ab}$ shows that R_{ab} may, in this case, be regarded as a generalization of the Euler class of a circle bundle: Suppose that $E = E^L = M \times \mathbb{R}$ with ρ the trivial representation, and also assume that $H^1(L, M \times \mathbb{R}) \cong M \times \mathbb{R}$. By 7.4.10, this last assumption is equivalent to $L/[L, L] \cong M \times \mathbb{R}$. Now $R_{ab} \in \mathscr{H}^2(TM, M \times \mathbb{R}) = H^2_{\mathrm{deRh}}(M)$ and the pairing $H^1(L, M \times \mathbb{R}) \oplus L_{ab} \to M \times \mathbb{R}$ is reduced to the multiplication $\mathbb{R} \times \mathbb{R} \to \mathbb{R}$. Also, $tg^1 \colon (\Gamma H^1(L, M \times \mathbb{R}))^A \to \mathscr{H}^2(TM, M \times \mathbb{R})$ becomes $\mathbb{R} \to H^2_{\mathrm{deRh}}(M)$, $t \mapsto tR_{ab}$ and so R_{ab} is the image under tg^1 of a generator of $(\Gamma H^1(L, M \times \mathbb{R}))^A$. It may thus be considered to be the Euler class of $L \succ\!\!\!\longrightarrow A \longrightarrow\!\!\!\!\!\gg TM$.

Secondly, assume that $H^n(L, E) = 0$ for $n = 1, 2$. By the Whitehead lemmas for Lie algebras, this is the case for semisimple L. We then have $\mathscr{H}^2(TM, E^L) \overset{a^*}{\cong} \mathscr{H}^2(A, E)$, so that all extensions are geometric, and there is an exact sequence

$$\mathscr{H}^3(TM, E^L) \overset{a^*}{\succ\!\!\!\longrightarrow} \mathscr{H}^3(A, E) \overset{j^*}{\longrightarrow} (\Gamma H^3(L, E))^A \overset{tg^3}{\longrightarrow} \cdots$$

$$\cdots \longrightarrow \mathscr{H}^4(TM, E^L) \overset{a^*}{\longrightarrow\!\!\!\!\!\gg} \mathscr{H}^4(A, E). \quad (19)$$

If, in addition, $H^n(L, E) = 0$ for $n \geqslant 4$, then this sequence can be continued, and it then includes the Gysin sequence for $SU(2)$–bundles.

7.5 Schouten Calculus

Consider a Lie algebroid A on base M, not necessarily transitive, and denote the kth exterior power by $\Lambda^k(A)$. We refer to a section of $\Lambda^k(A)$ as a *k–multisection of* A. The anchor and bracket on A extend to give $\Gamma\Lambda(A)$ the structure of a Schouten algebra in the following sense.

Definition 7.5.1 Let R be a ring and let C be an R–algebra. A *Schouten algebra* over C is a $\mathbb{Z}$–graded module $\mathscr{A} = \bigoplus_{i \in \mathbb{Z}} \mathscr{A}^i$ over C together with an associative and graded commutative product

$$a \wedge b \in \mathscr{A}^{i+j}, \quad b \wedge a = (-1)^{ij} a \wedge b, \quad \text{for } a \in \mathscr{A}^i, \ b \in \mathscr{A}^j,$$

and a bracket $[a, b] \in \mathscr{A}^{i+j-1}$ for $a \in \mathscr{A}^i$, $b \in \mathscr{A}^j$, such that

$$[a, b] = -(-1)^{(i-1)(j-1)}[b, a], \quad \text{for} \quad a \in \mathscr{A}^i, \ b \in \mathscr{A}^j, \quad (20)$$

and, for $a \in \mathscr{A}^i$, $b \in \mathscr{A}^j$, $c \in \mathscr{A}^k$,

$$(-1)^{(i-1)(k-1)}[[a, b], c] + (-1)^{(j-1)(i-1)}[[b, c], a]$$
$$+ (-1)^{(k-1)(j-1)}[[c, a], b] = 0, \quad (21)$$

subject to the condition that

$$[a, b \wedge c] = [a, b] \wedge c + (-1)^{(i-1)j} b \wedge [a, c], \quad (22)$$

holds for all $a \in \mathscr{A}^i$, $b \in \mathscr{A}^j$, $c \in \mathscr{A}^k$. $\qquad\square$

Schouten algebras are also known as *Gerstenhaber algebras*.

Condition (21) is the graded Jacobi identity for the relabelling of $\mathscr{A}$ given by $\mathscr{A}^{(i)} = \mathscr{A}^{i+1}$. We refer to condition (22) as the *graded Leibniz identity*.

Now let $\mathscr{A} = \Gamma\wedge^\bullet(A)$ be the module of multisections of the Lie algebroid A, with the standard wedge product. We extend the Lie algebroid bracket to $\mathscr{A}$ by writing $[X, f] = a(X)(f)$ for $X \in \mathscr{A}^1 = \Gamma A$ and $f \in \mathscr{A}^0 = C^\infty(M)$, and then using (22) and (20) inductively. It is then a straightforward exercise to prove the first half of the next result.

Theorem 7.5.2 *Let A be a vector bundle on M and let $\mathscr{A}$ denote the algebra of sections of $\wedge^\bullet(A)$ with the standard wedge product. Then the construction immediately above gives a bijective correspondence between Lie algebroid structures on A and brackets on $\mathscr{A}$ which, together with the exterior algebra structure, make $\mathscr{A}$ a Schouten algebra.*

Proof Suppose $\mathscr{A}$ has a Schouten algebra structure. Since $[\mathscr{A}^1, \mathscr{A}^1] \subseteq \mathscr{A}^1$, we have a bracket on ΓA, and by (20) and (21) it is skew–symmetric and satisfies Jacobi. Take $X \in \Gamma A$ and $f, g \in C^\infty(M) = \mathscr{A}^0$. Then from (22) it follows that

$$[X, fg] = [X, f]g + f[X, g]$$

so $[X, -]$ is a vector field on M; denote the resulting map $\Gamma A \to \mathfrak{X}(M)$ by a. Now, by (22),

$$a(gX)(f) = [gX, f] = g[X, f] + [f, g]X$$

and $[f, g] = 0$ since $\mathscr{A}^{-1} = \{0\}$. So a induces a vector bundle map $a\colon A \to TM$. The Lie algebroid Leibniz rule follows directly from (22) and that a preserves the brackets follows from (21). $\qquad\square$

The bracket on $\Gamma\wedge^\bullet(A)$ arising from a Lie algebroid structure on A is the *Schouten bracket of multisections* of A. In the case of the tangent bundle Lie algebroid, this is the classical calculus of Schouten.

We will use the Schouten calculus on Lie algebroids in Part III. At this point, we set out various formulas which will be needed there.

Given a vector bundle A, denote the pairing between a $\varphi \in \Gamma\wedge^k(A^*)$ and a k–multisection ξ of A by $\langle\varphi, \xi\rangle$ or $\langle\xi, \varphi\rangle$. When ξ is decomposable and given explicitly as $X_1 \wedge \cdots \wedge X_k$, we may write $\varphi(X_1, \ldots, X_k)$ as usual. In terms of decomposable elements, the pairing is

$$\langle\varphi_1 \wedge \cdots \wedge \varphi_k, \; X_1 \wedge \cdots \wedge X_k\rangle = \det(\langle\varphi_i, X_j\rangle).$$

From here until the end of the section, we consider a single Lie algebroid A. For A, the exterior algebra on $\Gamma(A^*)$ is the algebra of real–valued cochains $\bigoplus_{i \geqslant 0} \Gamma C^i(A, M \times \mathbb{R})$ as described in §7.1. Here and in Part III, we will usually write $\Gamma \Lambda^i(A^*)$ in place of $\Gamma C^i(A, M \times \mathbb{R})$.

For the coboundary $d \colon \Gamma \Lambda^\bullet(A^*) \to \Gamma \Lambda^{\bullet+1}(A^*)$ defined in 7.1.1 we now have, as in (16),

$$d(\varphi \wedge \psi) = (d\varphi) \wedge \psi + (-1)^i \varphi \wedge d\psi \tag{23}$$

for $\varphi, \psi \in \Gamma \Lambda^\bullet(A^*)$ with φ of degree i.

There are two *insert* or *contraction* operators. First, extend ι_X, for $X \in \Gamma A$, by defining

$$\iota_{X_1 \wedge \cdots \wedge X_i} = \iota_{X_1} \circ \cdots \circ \iota_{X_i}. \tag{24}$$

This extends to give a well–defined map $\iota_\xi \colon \Gamma \Lambda^k(A^*) \to \Gamma \Lambda^{k-i}(A^*)$ for $\xi \in \Gamma \Lambda^i(A)$. For $\xi \in \Gamma \Lambda^i(A)$, $\eta \in \Gamma \Lambda^j(A)$, $\varphi \in \Gamma \Lambda^{i+j}(A^*)$. we have

$$\langle \iota_\xi(\varphi), \eta \rangle = (-1)^{[\frac{i}{2}]} \langle \varphi, \xi \wedge \eta \rangle. \tag{25}$$

Next, given $\varphi \in \Gamma(A^*)$, define $\iota_\varphi \colon \Gamma \Lambda^k(A) \to \Gamma \Lambda^{k-1}(A)$, for $\xi \in \Gamma \Lambda^k(A)$ and $\psi \in \Gamma \Lambda^{k-1}(A^*)$ by $\langle \psi, \iota_\varphi(\xi) \rangle = \langle \varphi \wedge \psi, \xi \rangle$, and define

$$\iota_{\varphi_1 \wedge \cdots \wedge \varphi_i} = \iota_{\varphi_1} \circ \cdots \circ \iota_{\varphi_i}. \tag{26}$$

This likewise extends to a well–defined map $\iota_\varphi \colon \Gamma \Lambda^k(A) \to \Gamma \Lambda^{k-i}(A)$ for $\varphi \in \Gamma \Lambda^i(A^*)$. For $\varphi \in \Gamma \Lambda^i(A^*)$, $\psi \in \Gamma \Lambda^j(A^*)$, $\xi \in \Gamma \Lambda^{i+j}(A)$, we have

$$\langle \psi, \iota_\varphi(\xi) \rangle = (-1)^{[\frac{i}{2}]} \langle \varphi \wedge \psi, \xi \rangle. \tag{27}$$

Note that, for $\xi \in \Gamma \Lambda^i(A)$, $\varphi \in \Gamma \Lambda^i(A^*)$,

$$\iota_\xi(\varphi) = (-1)^{[\frac{i}{2}]} \langle \varphi, \xi \rangle = \iota_\varphi(\xi). \tag{28}$$

Of the two following equations, the first is proved in the same way as for differential forms, and the second is similar:

$$\iota_X(\varphi \wedge \psi) = \iota_X(\varphi) \wedge \psi + (-1)^i \varphi \wedge \iota_X(\psi), \tag{29}$$

$$\iota_\theta(\xi \wedge \eta) = \iota_\theta(\xi) \wedge \eta + (-1)^i \xi \wedge \iota_\theta(\eta). \tag{30}$$

Here $X \in \Gamma A$, $\theta \in \Gamma A^*$, $\varphi, \psi \in \Gamma \Lambda(A^*)$ and $\xi, \eta \in \Gamma \Lambda(A)$ with φ and ξ of degree i.

Next, by an induction on the degree of η,

$$[\mathfrak{L}_X, \iota_\eta] = \iota_{[X, \eta]} \tag{31}$$

for $X \in \Gamma A$, $\eta \in \Gamma \Lambda^\bullet(A)$. This extends (vii) of 7.1.3.

Define the *Lie derivative of multisections* by

$$\mathcal{L}_X(\xi) = [X, \xi]$$

for $X \in \Gamma A$ and $\xi \in \Gamma \Lambda^\bullet(A)$. It follows from (20)—(22) above that the degree zero map $\mathcal{L}_X \colon \Gamma \Lambda^\bullet(A) \to \Gamma \Lambda^\bullet(A)$ has the following properties:

$$\mathcal{L}_X(\xi \wedge \eta) = \mathcal{L}_X(\xi) \wedge \eta + \xi \wedge \mathcal{L}_X(\eta) \qquad (32)$$

$$[\mathcal{L}_X, \mathcal{L}_Y] = \mathcal{L}_{[X,Y]} \qquad (33)$$

$$\mathcal{L}_X(f\xi) = f\mathcal{L}_X(\xi) + a(X)(f)\xi \qquad (34)$$

$$\mathcal{L}_{fX}(\xi) = f\mathcal{L}_X(\xi) - X \wedge \iota_{df}(\xi) \qquad (35)$$

where $X, Y \in \Gamma A$, $\xi, \eta \in \Gamma \Lambda^\bullet(A)$, $f \in C^\infty(M)$.

Note that (35) is not a form of (1) on p. 262, which we restate:

$$\mathcal{L}_{fX}(\varphi) = f\mathcal{L}_X(\varphi) + df \wedge \iota_X(\varphi).$$

Here $\mathcal{L}$ is the Lie derivative of cochains $\varphi \in \Gamma \Lambda^\bullet(A^*)$.

The Lie derivative of multisections and the Lie derivative of cochains are related by

$$\mathcal{L}_X(\langle \varphi, \xi \rangle) = \langle \mathcal{L}_X(\varphi), \xi \rangle + \langle \varphi, \mathcal{L}_X(\xi) \rangle. \qquad (36)$$

for $\varphi \in \Gamma \Lambda^\bullet(A^*)$, $\xi \in \Gamma \Lambda^\bullet(A)$, $X \in \Gamma A$. This follows from 7.1.2 and

$$\mathcal{L}_X(Y_1 \wedge \ldots \wedge Y_k) = \sum_{i=1}^{k} Y_1 \wedge \ldots \wedge [X, Y_i] \wedge \ldots \wedge Y_k,$$

which may be proved inductively.

We now need to define a Lie derivative by multisections. This is assisted by defining the following bracket for graded endomorphisms. An $\mathbb{R}$-linear map $D \colon \Gamma \Lambda(A^*) \to \Gamma \Lambda(A^*)$ is a *graded endomorphism of degree* p if D maps each $\Gamma \Lambda^i(A^*)$ to $\Gamma \Lambda^{i+p}(A^*)$. For graded endomorphisms D_1, D_2 of degrees p, q, define

$$[D_1, D_2] = D_1 \circ D_2 - (-1)^{pq} D_2 \circ D_1. \qquad (37)$$

Then $[D_1, D_2]$ is a graded endomorphism of degree $p + q$. As usual, we extend to sums of graded endomorphisms of differing degrees.

Now for $\xi \in \Gamma \Lambda^i(A)$, the *Lie derivative* $\Gamma \Lambda^\bullet(A^*) \to \Gamma \Lambda^{\bullet - i + 1}(A^*)$ is defined to be:

$$\mathcal{L}_\xi = [\iota_\xi, \delta] = \iota_\xi \circ \delta - (-1)^i \delta \circ \iota_\xi. \qquad (38)$$

Theorem 7.5.3 *For any Lie algebroid A, and for any $\xi, \eta \in \Gamma\Lambda^\bullet(A)$,*

$$[\mathfrak{L}_\xi, \iota_\eta] = \iota_{[\xi, \eta]}. \tag{39}$$

Proof First assume that $\eta = Y \in \Gamma A$, and let ξ be of degree i. Expanding out the LHS of (39), we obtain

$$\iota_\xi \circ d \circ \iota_Y + (-1)^i \iota_Y \circ \iota_\xi \circ d - (-1)^i d \circ \iota_\xi \circ \iota_Y - \iota_Y \circ d \circ \iota_\xi$$
$$= \iota_\xi \circ d \circ \iota_Y + \iota_\xi \circ \iota_Y \circ d - (-1)^i d \circ \iota_\xi \circ \iota_Y - \iota_Y \circ d \circ \iota_\xi.$$

Using $\mathfrak{L}_Y = \iota_Y \circ d + d \circ \iota_Y$ (see 7.1.3) in the first and last terms, and (31), the result follows.

Now assume that (39) holds for η of degree j and take $Y \in \Gamma A$. The inductive hypothesis is, when written out,

$$\iota_\xi \circ d \circ \iota_\eta - (-1)^i d \circ \iota_\xi \circ \iota_\eta - (-1)^{(i-1)j} \iota_\eta \circ \iota_\xi \circ d + (-1)^{(i-1)j+1} \iota_\eta \circ d \circ \iota_\xi = \iota_{[\xi, \eta]}.$$

For $[[\iota_\xi, d], \iota_{\eta \wedge Y}]$ we have, since $\iota_{\eta \wedge Y}$ has degree $-(j+1)$,

$$\iota_\xi \circ d \circ \iota_\eta \circ \iota_Y - (-1)^i d \circ \iota_\xi \circ \iota_\eta \circ \iota_Y$$
$$- (-1)^{(i-1)(j+1)} \iota_\eta \circ \iota_Y \circ \iota_\xi \circ d + (-1)^{(i-1)(j+1)+i} \iota_\eta \circ \iota_Y \circ d \circ \iota_\xi.$$

In the last two terms, use $\mathfrak{L}_Y = \iota_Y \circ d + d \circ \iota_Y$ to move ι_Y to the extreme right. Then, by the inductive hypothesis, and (31) again, this becomes the insertion operator for $[\xi, \eta] \wedge Y - (-1)^{(i-1)j} \eta \wedge [Y, \xi] = [\xi, \eta \wedge Y]$, as required. $\qquad\square$

7.6 Notes

§7.1 — §7.4. The first four sections of this chapter have been taken from §2, §3 and §5, of Chapter IV of [Mackenzie, 1987a], in each case with only minor revision.

The observation that de Rham cohomology and Chevalley–Eilenberg cohomology are defined by formally identical complexes is what led many authors to introduce the concept of Lie pseudoalgebra. Treatments of the cohomology of Lie pseudoalgebras in terms of the standard cochain complex used here may thus be found in [Palais, 1961], [Hermann, 1967], [Nelson, 1967], and [Teleman, 1972].

At about the same time, Rinehart [1963] gave a vastly superior treatment of Lie pseudoalgebra cohomology, proving a Poincaré–Birkhoff–Witt theorem — the first major result for Lie pseudoalgebras — and obtaining the cohomology as derived functors. Rinehart's results were put into the context of sheaves by Kamber and Tondeur [1971], and have since been further developed by Huebschmann [1999b] (using the terminology Lie–Rinehart algebra).

However the account given in [Mackenzie, 1987a, Chap. IV] was the first to work with Lie algebroids — that is, with Lie pseudoalgebras that arise from smooth, finite–rank vector bundles — and to show that all the necessary constructions remain within that category. In particular, the results on the cohomology of transitive Lie algebroids, with coefficients in Lie algebroid representations, such as 7.1.5, 7.1.6, 7.1.7, and related results in §7.3, were new when they appeared in [Mackenzie, 1987a].

For Chevalley–Eilenberg cohomology of Lie algebras, see, for example, [Cartan and Eilenberg, 1956, XIII§8].

The cohomology of principal bundles with compact structure groups was the subject of intense work through the 1950s and 1960s; see [Greub et al., 1976] and [Guillemin and Sternberg, 1999] for modern accounts. It is by no means clear to what extent this work can be reformulated in terms of (transitive) Lie algebroids, or to what extent that might be of value.

What I have here called a retroversal was called in [Mackenzie, 1987a] a *back-transversal.*

The flat connection in 7.1.6 is nowadays sometimes called a Gauss–Manin connection.

For the notion of Baer sum of extensions see [Mac Lane, 1995, pp. 113–114].

For the homology of Lie algebroids see [Evens, Lu, and Weinstein, 1999] and [Xu, 1999].

The terminology *coupling*, already used in [Mackenzie, 1987a], was taken from [Robinson, 1982]. The terminology of Mac Lane [1995] was *abstract kernel*. A coupling of Lie algebroids is a particular case of a crossed module of Lie algebroids, and the construction of the obstruction class given here may be extended to any crossed module; see [Androulidakis, 2001].

The cohomological formalism developed in §7.3 is of a standard type, and closely follows the theory of nonabelian Lie algebra extensions. The most useful sources are often the oldest: I particularly used [Hochschild, 1954b,a], [Mori, 1953], and [Shukla, 1966]. Some aspects of this formalism have been noted before, by [Palais, 1961], [Hermann, 1967], and [Teleman, 1972], but, as noted above, these authors worked with Lie pseudoalgebras and were not concerned with whether the constructions remained within the category of vector bundles (or finitely generated projective modules).

The spectral sequence of an extension of Lie algebras is due to Hochschild and Serre [1953] and I have largely followed their original treatment. For the Leray–Serre spectral sequence of a principal bundle in de Rham cohomology see, for example, [Greub, Halperin, and Vanstone, 1976].

For the general question of lifting the structure group in a principal bundle see the discussion preceding 8.3.10 and the Notes to Chapter 8.

For spectral sequences see [Cartan and Eilenberg, 1956], [Mac Lane, 1995], and [Greub et al., 1976].

The filtration (13) is the standard filtration on an exact sequence of Lie algebras as in [Hochschild and Serre, 1953] and is also of the same type as the filtration associated with a $\mathfrak{g}$–DGA [Greub et al., 1976, 9.1]

In the case of Lie algebras, [Hochschild and Serre, 1953] have an elegant device by which to simplify the proof of 7.4.3, but it cannot be properly formulated in the case of transitive Lie algebroids.

Regarding Example 7.4.16, it is interesting that the Lie algebroids of the examples of [Milnor, 1958] with positive genus are instances.

The Gysin sequence (18) is a straightforward analogue of Theorem 7 in [Hochschild and Serre, 1953].

A fundamentally new cohomology theory which controls deformations of Lie algebroids has been introduced in [Crainic and Moerdijk, arXiv:0403434].

The main results of these four sections were new when they appeared in [Mackenzie, 1987a].

For the Chern–Weil construction for Lie groupoids and Lie algebroids, see [Mackenzie, 1988b], [Huebschmann, 1999b].

§7.5. The Schouten bracket of multivector fields on a manifold goes back to Schouten [1954] and Nijenhuis [1955]. The same algebraic structures on the cohomology of associative algebras were introduced by Gerstenhaber [1963]. The formalism is now widely used; see [Koszul, 1984], [Bhaskara and Viswanath, 1988a,b], [Vaisman, 1994], [Cartier, 1994], [Kosmann–Schwarzbach, 1996a], [Evens, Lu, and Weinstein, 1999]. for example.

Some of the material of this section has been taken from [Mackenzie and Xu, 1994, §2], but different conventions are used here.

For Theorem 7.5.2, see [Kosmann–Schwarzbach and Magri, 1990] and [Xu, 1999] and the references given there. Theorem 7.5.3 is proved in [Koszul, 1984]; in some accounts it is taken as the definition of the Schouten bracket.

8

The cohomological obstruction
to integrability

This chapter is concerned with the integrability problem for transitive Lie algebroids, and the single cohomological obstruction which gives a complete solution in that case. For general Lie algebroids, the problem of integrability has recently been completely solved; see the Notes to this chapter and the Appendix for references on the general problem.

We begin in §8.1 with a rapid summary of the most simple interesting case; most readers will know this material from accounts of geometric prequantization. In §8.2 we prove that a transitive Lie algebroid on a contractible base admits a flat connection; in terms of §5.4 this shows that every transitive Lie algebroid is locally trivial. At the same time 8.2.1 is a strong form of local integrability for transitive Lie algebroids. This result leads to a classification of transitive Lie algebroids in terms of what we call systems of transition data; these consist of a family of local Maurer–Cartan forms subject to a (nonabelian) cocycle condition twisted by a cocycle for the adjoint bundle. This classification is an exact infinitesimal analogue of the classification of principal bundles by transition functions.

For the Lie algebroid of a locally trivial Lie groupoid, the system of transition data is obtained from a groupoid cocycle by differentiation. For an abstract transitive Lie algebroid, attempting to reverse this process leads to the cohomological integrability obstruction class. This is the subject of §8.3.

8.1 The classical case

The first Chern class of a principal $U(1)$–bundle has several different aspects which will be significant in what follows. We recall them here.

We denote Čech cohomology of a manifold M with coefficients in an abelian group R by $\check{H}^q(M,\underline{R})$; that is, the coefficients are local smooth functions from M to R. Čech cohomology with coefficients in the constant functions $M \to R$ is denoted $\check{H}^q(M,R)$.

Aspect 1

Consider a principal $U(1)$–bundle $P(M,U(1))$. Choose a simple open cover $\{U_i\}$ of M and denote by $\{s_{ij} \colon U_{ij} \to U(1)\}$ the transition functions. Let $\hat{s}_{ij} \colon U_{ij} \to \mathbb{R}$ be any (smooth) maps covering the s_{ij}; that is, $\exp(2\pi i\,\hat{s}_{ij}) = s_{ij}$, or $\hat{s}_{ij} = \log s_{ij}$ with respect to some branch of the logarithm. Then on any $U_{ijk} \neq \emptyset$ the function $e_{ijk} \colon U_{ijk} \to \mathbb{R}$

$$e_{ijk} = \hat{s}_{jk} - \hat{s}_{ik} + \hat{s}_{ij}$$

is integer–valued and hence constant. It therefore defines a Čech 2–cocycle with respect to $\{U_i\}$, and thereby an element $e \in \check{H}^2(M,\mathbb{Z})$. This is the *first Chern class* of $P(M,U(1))$, or the *Euler class* of the corresponding $SO(2)$–bundle.

The construction of e from $P(M,U(1))$ is often identified with the map $\check{H}^1(M,\underline{U(1)}) \to \check{H}^2(M,\mathbb{Z})$ induced by the exact sequence

$$\mathbb{Z} \rightarrowtail \mathbb{R} \twoheadrightarrow U(1).$$

This map is an isomorphism, since $\check{H}^q(M,\underline{\mathbb{R}}) = 0$ for $q \geqslant 1$. Since $\check{H}^1(M,\underline{U(1)})$ is the set of equivalence classes of circle bundles on M, the first Chern class is a complete invariant of $U(1)$–bundles.

Aspect 2

As a particular case of the foregoing, the bundle $P(M,U(1))$ is trivializable if and only if $e = 0 \in \check{H}^2(M,\mathbb{Z})$. If P is trivializable, then it is a quotient of the trivial principal bundle $M \times \mathbb{R}(M,\mathbb{R})$ and, conversely, any quotient over $\mathbb{Z}$ of a principal bundle $Q(M,\mathbb{R})$ will inherit the triviality of Q. Thus we also have:

Proposition 8.1.1 *A principal bundle $P(M,U(1))$ with first Chern class e is a quotient of an $\mathbb{R}$–principal bundle if and only if $e = 0$.*

That is, e also measures the obstruction to lifting the structure group of P to its universal cover.

Aspect 3

Now consider the Atiyah sequence of $P(M, U(1))$,

$$M \times \mathbb{R} \rightarrowtail \frac{TP}{U(1)} \twoheadrightarrow TM. \tag{1}$$

Take any connection γ in (1). Since the adjoint bundle is abelian, the adjoint connection ∇^γ is flat. Assume that $\pi_1 M = 0$, so that ∇^γ may be transformed to the standard flat connection ∇^0 by an automorphism of $M \times \mathbb{R}$. Then the curvature R_γ is a real–valued 2–form on M, and $[R_\gamma] \in H^2(TM, M \times \mathbb{R}, \nabla^0) = H^2_{\mathrm{deRh}}(M, \mathbb{R})$ is independent of the choice of γ. With the notation from §5.4, we have

$$\gamma|_{U_i} = \omega_i + \Theta_i$$

where the $\omega_i \in \Omega^1(U_i)$ are the local connection 1–forms and Θ_i may be regarded as the pullback of the Maurer–Cartan form of $\mathfrak{g}$ over the decomposing section $\sigma_i \colon U_i \to P$. Furthermore, the restriction of R_γ to U_i is given by

$$R_\gamma|_{U_i} = -\delta\omega_i.$$

Now, also from §5.4,

$$\omega_i - \omega_j = \chi_{ij} = \delta s_{ij}.$$

So the Čech class e corresponds precisely to the de Rham class $[R_\gamma]$ under the standard isomorphism.

Aspect 4

Lastly, recall the construction of a $U(1)$–bundle from a suitable closed 2–form $R \in \Omega^2(M)$.

As above, we take a simple open cover $\{U_i\}$. For any closed R, its restriction to any U_i is exact, and we can write $R|_{U_i} = -\delta\omega_i$ for some $\omega_i \in \Omega^1(U_i)$. Again, whenever $U_{ij} \neq \emptyset$, the difference form $\omega_i - \omega_j$ is closed and so we can write $\omega_i - \omega_j = \delta s_{ij}$ for some function $s_{ij} \colon U_{ij} \to \mathbb{R}$. Whenever $U_{ijk} \neq \emptyset$, we have

$$\delta(s_{jk} - s_{ik} + s_{ij}) = 0$$

and so $e_{ijk} = s_{jk} - s_{ik} + s_{ij}$ is a constant real number. We thus have a Čech class $e \in \check{H}^2(M, \mathbb{R})$ which of course is the Čech class corresponding to $[R]$ under the standard isomorphism.

Now suppose that there exists $\lambda \in \mathbb{R}\backslash\{0\}$ and a simple open cover

$\{U_i\}$ such that $e_{ijk} \in \lambda \mathbb{Z}$ for all $U_{ijk} \neq \emptyset$. Define $p \colon \mathbb{R} \to U(1)$ by $p(t) = \exp(2\pi i t/\lambda)$. Then the functions $p \circ s_{ij} \colon U_{ij} \to U(1)$ are a cocycle and define a principal bundle $P(M, U(1))$. Furthermore, this bundle has a connection γ defined locally by the ω_i, and the curvature of this γ is the given R.

Definition 8.1.2 A cohomology class $e \in \check{H}^2(M, \mathbb{R})$ is *discrete* if there exists $\lambda \in \mathbb{R}\backslash\{0\}$ and a simple open cover $\{U_i\}$ such that $e_{ijk} \in \lambda \mathbb{Z}$ for all $U_{ijk} \neq \emptyset$. □

This last aspect of the first Chern class can therefore be stated:

Theorem 8.1.3 *A closed 2-form $R \in \Omega^2(M)$ is the curvature of a connection in a $U(1)$-bundle if and only if the corresponding Čech class in $\check{H}^2(M, \mathbb{R})$ is discrete.*

Theorem 8.1.3 is the classical Weil Lemma.

Remark 8.1.4 Let $P(M, U(1))$ be a circle bundle. Then the vector bundle E associated to the standard action of $U(1)$ on $\mathbb{C}$ has the complex structure $J(\lfloor u, w \rfloor) = \lfloor u, iw \rfloor$, and the Hermitian metric defined by

$$\langle \lfloor u, w_1 \rfloor, \lfloor u, w_2 \rfloor \rangle = w_1 \overline{w_2},$$

where $u \in P$, $w_1, w_2 \in \mathbb{C}$. A vector bundle with a complex structure and a Hermitian metric, and which has complex rank 1, is a *Hermitian line bundle*.

If E is a Hermitian line bundle on M, then the vertex bundles of the Lie groupoid of Hermitian frames $\Phi_U(E)$ are circle bundles. (Recall 1.7.10.) Because $U(1)$ is abelian, the inner automorphisms of $\Phi_U(E)$ give canonical isomorphisms between any two vertex groups.

8.2 Transition forms and transition data

The first theorem of this section leads to a classification of transitive Lie algebroids by systems of transition data.

Theorem 8.2.1 *Let $L \rightarrowtail A \twoheadrightarrow TM$ be a transitive Lie algebroid on a contractible manifold M. Then A admits a flat connection.*

This result is an infinitesimal analogue of the well–known result that a principal bundle — or locally trivial Lie groupoid — on a contractible base admits a global section. The principal bundle result is achieved by contracting the base, and using the homotopy classification of bundles; the proof of 8.2.1 achieves a similar end by using the cohomology theory of Chapter 7.

We begin with some observations which apply to any transitive Lie algebroid on any manifold.

Definition 8.2.2 Let $L \rightarrowtail A \twoheadrightarrow TM$ be a transitive Lie algebroid. Then the α–connected Lie subgroupoid of $\Phi_{\text{Aut}}(L)$ which corresponds to $\text{ad}(A) \leqslant \mathfrak{D}[L]$ is denoted $\text{Int}(A)$, and called the *Lie groupoid of inner automorphisms* of A, or the *interior Lie groupoid* of A. $\qquad\square$

For any connection γ in A, the adjoint connection ∇^γ takes values in $\text{ad}(A)$, and so, by 6.3.15, may be said to reduce to $\text{Int}(A)$.

Now assume that M is contractible. The idea of the proof is roughly as follows: Since M is contractible, we presume that $\mathscr{H}^2(TM, ZL)$ is zero. Granted this, there will be precisely one equivalence class of operator extensions of TM by L. So if we can show that there is an operator extension of TM by L that is flat, then the given Lie algebroid, being equivalent to it, must also be flat.

To carry out this idea, two matters must be arranged rather carefully. Firstly, in order to obtain $\mathscr{H}^2(TM, ZL) = (0)$, it is necessary to realize $\mathscr{H}^2(TM, ZL)$ as the de Rham cohomology of M. Since M is contractible, it is certainly true that ZL is isomorphic to the trivial bundle $M \times \mathfrak{z}$; it must also be shown that the representation of TM on ZL transports to the trivial representation. Since M is simply–connected, this can be achieved by 5.4.7.

Secondly, in order to obtain a flat Lie algebroid equivalent to the given one, it must be shown that the coupling $TM \to \text{Out}\mathfrak{D}(L)$ of the given Lie algebroid is covered by a flat Lie connection $TM \to \mathfrak{D}_{\text{Der}}(L)$ (see 7.3.19 to 7.3.21). Although L is isomorphic as an LAB to a trivial bundle $M \times \mathfrak{g}$, one cannot apply 5.4.7, for it is not known that the given Lie algebroid has an adjoint connection which is flat. Neither can one apply 5.4.7 to the coupling itself, for there is no known general construction of a Lie groupoid whose Lie algebroid is $\text{Out}\mathfrak{D}(L)$. (This is so even in the case of Lie algebras.) However, all these difficulties can be circumvented simultaneously.

Consider, therefore, a transitive Lie algebroid $L \overset{j}{\rightarrowtail} A \overset{a}{\twoheadrightarrow} TM$ on a contractible base M. We may as well assume from the outset that L is a trivial LAB $M \times \mathfrak{g}$. Since $\text{Int}(A)$ is a locally trivial Lie groupoid on the contractible base M, it is trivializable; equivalently, it admits a global decomposing section $\sigma \colon M \to \text{Int}(A)_m$, for some $m \in M$. From σ we obtain, as in §5.4, a global morphism $\theta \colon M \times M \to \text{Int}(A)$ and a flat connection θ_* in $\text{ad}(A)$.

Since $\text{ad}(A) \leqslant \mathfrak{D}_{\text{Der}}(L)$, we may also consider θ_* to be a flat Lie connection in $L = M \times \mathfrak{g}$. Therefore, by 5.4.7, there is an automorphism $\psi \colon M \times \mathfrak{g} \to M \times \mathfrak{g}$ which maps θ_* to the standard flat connection ∇^0. That is,

$$\psi(\theta_*(X)(V)) = X(\psi(V)) \tag{2}$$

for $X \in \mathfrak{X}(M)$, $V \colon M \to \mathfrak{g}$.

Let $\gamma \colon TM \to A$ be a connection in A. Since ∇^γ and θ_* are both connections in $\text{ad}(A)$, there is a map $\ell \colon TM \to M \times \mathfrak{g}$ such that

$$\nabla^\gamma = \theta_* + \text{ad} \circ \ell, \tag{3}$$

Our intention is to transform (3) into $\nabla^\gamma = \nabla^0 + \text{ad} \circ \ell'$ by using (2).

Define a new embedding of $L = M \times \mathfrak{g}$ into A by $j' = j \circ \psi^{-1}$.

Lemma 8.2.3 *Let* $L \overset{j}{\rightarrowtail} A \overset{a}{\twoheadrightarrow} TM$ *be a transitive Lie algebroid on an arbitrary manifold* M. *Let* $\psi \colon L \to L$ *be an LAB automorphism of* L, *and let* $j' = j \circ \psi^{-1}$.

Then for any connection $\gamma \colon TM \to A$, *the adjoint connections* ∇^γ *and* $\nabla'^{,\gamma}$ *induced in* L *by the two Lie algebroids* $L \overset{j}{\rightarrowtail} A \overset{a}{\twoheadrightarrow} TM$ *and* $L \overset{j'}{\rightarrowtail} A \overset{a}{\twoheadrightarrow} TM$, *are related by*

$$\nabla'^{\gamma}_X(\psi(V)) = \psi(\nabla^\gamma_X(V)).$$

Proof For $X \in \mathfrak{X}(M)$ and $V \colon M \to \mathfrak{g}$,

$$j'(\nabla'^{\gamma}_X(\psi(V)) = [\gamma(X), j'(\psi(v))] = [\gamma(X), j(V)]$$
$$= j(\nabla^\gamma_X(V)) = j'(\psi(\nabla^\gamma_X(V))).$$

$\square$

Returning to the proof of 8.2.1, we have from (3) that

$$\nabla^\gamma_X(V) = \theta_*(X)(V) + [\ell(X), V].$$

Therefore

$$\psi(\nabla^\gamma_X(V)) = (\psi \circ \theta_*)(X)(V) + \psi([\ell(X), V])$$

and therefore by the lemma, and equation (2),

$$\nabla'^{\,\gamma}_X(\gamma(V)) = X(\psi(V)) + [(\psi \circ \ell)(X), \psi(V)].$$

Since ψ is surjective, this establishes that

$$\nabla'^{,\gamma} = \nabla^0 + \mathrm{ad}(\ell') \tag{4}$$

where $\ell' = \psi \circ \ell$.

From now on we work exclusively with $M \times \mathfrak{g} \xrightarrow{j'} A \xrightarrow{a} TM$. Equation (4) implies immediately that, firstly, the representation of TM on $M \times \mathfrak{z}$ induced by A is the trivial one, and, secondly, that the coupling Ξ of TM with $M \times \mathfrak{g}$ induced by A admits ∇^0 as a Lie connection covering it.

We therefore have $\mathscr{H}^2(TM, \nabla^0, M \times \mathfrak{z}) \cong H^2_{deRh}(M, \mathfrak{z}) = (0)$ and so, by 7.3.17, there is a unique equivalence class of Ξ–operator extensions of TM by $M \times \mathfrak{g}$. Since Ξ admits ∇^0 as a covering connection we can, by 7.3.18, construct the semidirect Ξ operator extension

$$M \times \mathfrak{g} \rightarrowtail TM \ltimes_{\nabla^0} (M \times \mathfrak{g}) \twoheadrightarrow TM,$$

and this must be equivalent to $M \times \mathfrak{g} \xrightarrow{j'} A \xrightarrow{a} TM$. So there is an isomorphism of Lie algebroids $TM \ltimes_{\nabla^0} (M \times \mathfrak{g}) \to A$, and A therefore admits a flat connection. This completes the proof of 8.2.1. $\qquad\square$

The proof has actually established that A is isomorphic to a trivial Lie algebroid $TM \oplus (M \times \mathfrak{g})$. This stronger formulation can in any case be deduced from 8.2.1 by an application of 5.4.7.

Consider now a transitive Lie algebroid $L \xrightarrow{j} A \xrightarrow{a} TM$ on an arbitrary base M. Given any cover $\{U_i\}$ of M by contractible open sets, there is an isomorphism of Lie algebroids over U_i

$$S_i \colon TU_i \oplus (U_i \times \mathfrak{g}) \to A_{U_i}$$

Thus 8.2.1 has established that transitive Lie algebroids are locally trivial as defined in 5.4.1.

The map $S_i^+ \colon U_i \times \mathfrak{g} \to L_{U_i}$ is an LAB chart; denote it by ψ_i. Denote by Θ^i the flat connection $X \mapsto S_i(X \oplus 0)$ induced in A_{U_i} by S_i. As in

§5.4 the Θ^i are *local flat connections* in A. Again as in §5.4, we define *transition forms* χ_{ij} by

$$\Theta^j = \Theta^i + j \circ \ell_{ij}, \qquad \chi_{ij} = \psi_i^{-1} \circ \ell_{ij}$$

for every nonvoid U_{ij}.

Theorem 8.2.4 *Let* $\{U_i, \psi_i, \Theta^i\}$ *be a Lie algebroid atlas for the transitive Lie algebroid* $L \!>\!\!\!-\!\!\!\to\! A \longrightarrow\!\!\!\!\!> TM$, *let* $\{a_{ij}\}$ *be the transition functions for* L *and let* $\{\chi_{ij}\}$ *be the transition forms just defined. Then:*

(TD1) *each* χ_{ij} *is a Maurer-Cartan form;*

(TD2) $\chi_{ik} = \chi_{ij} + a_{ij}(\chi_{jk})$ *whenever* $U_{ijk} \neq \emptyset$;

(TD3) $\Delta(a_{ij}) = \mathrm{ad} \circ \chi_{ij}$ *for all* $U_{ij} \neq \emptyset$, *where* Δ *is the Darboux derivative for the group* $\mathrm{Aut}(\mathfrak{g})$.

Proof The first was proved as 5.4.2, and the second is a simple calculation. We prove the third. Consider the Lie algebroid isomorphisms S_i and S_j arising from Θ^i, ψ_i and Θ^j, ψ_j. Then the overlap isomorphism $S_i^{-1} \circ S_j \colon TU_{ij} \oplus (U_{ij} \times \mathfrak{g}) \to TU_{ij} \oplus (U_{ij} \times \mathfrak{g})$ is

$$X \oplus V \mapsto X \oplus \{\chi_{ij}(X) + a_{ij}(V)\}.$$

This is an automorphism of a trivial Lie algebroid so must obey the three equations of 3.3.3. (In particular, we recover the Maurer–Cartan equation for χ_{ij}.) The third equation, the compatibility condition, is now

$$X(a_{ij}(V)) - a_{ij}(X(V)) + [\chi_{ij}(X), a_{ij}(V)] = 0$$

where $X \in \mathfrak{X}(U_{ij})$, $V \colon U_{ij} \to \mathfrak{g}$. It is shown in §5.1 that this is precisely the equation in (TD3). $\qquad\qquad\square$

Definition 8.2.5 Given a manifold M and a Lie algebra $\mathfrak{g}$, a *system of transition data for a transitive Lie algebroid* consists of an open cover $\{U_i\}$ of M, a cocycle $\{a_{ij} \colon U_{ij} \to \mathrm{Aut}(\mathfrak{g})\}$, and a set of local $\mathfrak{g}$–valued 1–forms $\{\chi_{ij} \in \Omega^1(U_{ij}, \mathfrak{g})\}$ which together satisfy conditions (TD1), (TD2), (TD3) above. $\qquad\qquad\square$

Usually we simply write *a system of transition data*. Note that a system of transition data may exist for an open cover by sets which are not contractible. In the case of 8.2.1, the contractibility ensured that the adjoint bundle was trivializable. For a general open cover $\{U_i\}$, one must not only look for local flat connections but also ensure that the $\{U_i\}$ are domains for an LAB chart; furthermore, if the U_i are not simply–connected, the compatibility equation (28) on p. 206 may not automatically be satisfied. Even when the open cover is by contractible sets, it is better to think of a system of transition data as arising from a Lie algebroid atlas, rather than from a family of local flat connections.

Theorem 8.2.6 *Let M be a manifold and let $\mathfrak{g}$ be a Lie algebra. Let $\{U_i\}$, $\{a_{ij} : U_{ij} \to \mathrm{Aut}(\mathfrak{g})\}$, and $\{\chi_{ij} \in \Omega^1(U_{ij}, \mathfrak{g})\}$ constitute a system of transition data.*

Then there is a transitive Lie algebroid $L \rightarrowtail A \twoheadrightarrow TM$ on M whose adjoint bundle L is the LAB corresponding to $\{a_{ij}\}$ and which admits local flat connections $\gamma_i : TM|_{U_i} \to A|_{U_i}$ such that

$$\gamma_j = \gamma_i + \psi_i \circ \chi_{ij}$$

where $\{\psi_i : U_i \times \mathfrak{g} \to L_{U_i}\}$ is an LAB atlas for L with $\{a_{ij}\}$ as cocycle.

Proof For each i, let A^i be the *set* $TU_i \oplus (U_i \times \mathfrak{g})$ and on the disjoint sum $\coprod_i A^i$ define an equivalence relation $\sim$ by

$$(i, X \oplus V) \sim (j, Y \oplus W) \iff X = Y \text{ and } W = \chi_{ji}(X) + a_{ji}(V).$$

Denote the quotient set by A and equivalence classes by $\langle i, X \oplus V \rangle$.

Define a map $q : A \to M$ by $q(\langle i, X \oplus V \rangle) = x$ where $X \in T_x(U_i)$. Then it is easy to see that

$$\overline{\psi}_i : TU_i \oplus (U_i \times \mathfrak{g}) \to q^{-1}(U_i), \qquad X \oplus V \mapsto \langle i, X \oplus V \rangle$$

is a bijection. Give A the smooth structure induced from the manifolds $TU_i \oplus (U_i \times \mathfrak{g})$ via the $\overline{\psi}_i$.

Now (A, q, M) is a vector bundle and the map $a : A \to TM$ defined by $\langle i, X \oplus V \rangle \mapsto X$, is well defined and a surjective vector bundle morphism over M. Denote the kernel of a by L. The $\overline{\psi}_i$ restrict to charts

$$\psi_i : U_i \times \mathfrak{g} \to L_{U_i}, \qquad V \mapsto \langle i, 0 \oplus V \rangle$$

for L and the atlas $\{\psi_i\}$ has $\{a_{ij}\}$ as cocycle.

Now we define a bracket in ΓA. For $\mu, \nu \in \Gamma A$ and $x \in M$, choose U_i containing x and write $\mu = \langle i, X \oplus V \rangle$, $\nu = \langle i, Y \oplus W \rangle$ where X and Y are vector fields on U_i, and V and W are maps $U_i \to \mathfrak{g}$. Define

$$[\langle i, X \oplus V \rangle, \langle i, Y \oplus W \rangle] = \langle i, [X, Y] \oplus \{X(W) - Y(V) + [V, W]\} \rangle.$$

It is an instructive exercise to verify that this is well defined and makes A a transitive Lie algebroid on M with adjoint bundle L.

The local flat connections γ_i are defined by $X \mapsto \langle i, X \oplus 0 \rangle$. The remainder of the proof is straightforward. □

Consider again a transitive Lie algebroid $L \rightarrowtail A \twoheadrightarrow TM$, an open cover $\{U_i\}$, and suppose that $\{\Theta_i, \psi_i\}$ as above and another $\{\Theta_i' : TU_i \to A_{U_i}, \ \psi_i' : U_i \times \mathfrak{g} \to L_{U_i}\}$ are both Lie algebroid charts for A with respect to $\{U_i\}$. Then, writing

$$\Theta_i' = \Theta_i + \psi_i \circ m_i,$$

where $m_i \in \Omega^1(U_i, \mathfrak{g})$, and

$$\psi_i' = \psi_i \circ n_i,$$

where $n_i : U_i \to \mathrm{Aut}(\mathfrak{g})$, we obtain

(ETD1) $$\delta m_i + [m_i, m_i] = 0,$$

(ETD2) $$\Delta(n_i) = \mathrm{ad} \circ m_i,$$

(ETD3) $$\chi_{ij}' = n_i^{-1}\{-m_i + \chi_{ij} + a_{ij}m_j\},$$

(ETD4) $$a_{ij}' = n_i^{-1}a_{ij}n_j.$$

Definition 8.2.7 Let $\mathfrak{g}$ be a Lie algebra and let M be a manifold. Let $\{\chi_{ij}, a_{ij}\}$ and $\{\chi_{ij}', a_{ij}'\}$ be two systems of transition data on M with values in $\mathfrak{g}$ and with respect to the one open cover $\{U_i\}$ of M. Then $\{\chi_{ij}, a_{ij}\}$ and $\{\chi_{ij}', a_{ij}'\}$ are *equivalent* if there exists a system of Maurer–Cartan forms $m_i \in \Omega^1(U_i, \mathfrak{g})$ and a system of functions $n_i : U_i \to \mathrm{Aut}(\mathfrak{g})$ such that (ETD2), (ETD3), (ETD4) are satisfied. □

Proposition 8.2.8 *Let $\{\chi_{ij}, a_{ij}\}$ and $\{\chi_{ij}', a_{ij}'\}$ be two equivalent systems of transition data on a manifold M with values in a Lie algebra $\mathfrak{g}$ and with respect to the same open cover. Then the Lie algebroids constructed from $\{\chi_{ij}, a_{ij}\}$ and $\{\chi_{ij}', a_{ij}'\}$ are equivalent.*

Proof Let $m_i \in \Omega^1(U_i, \mathfrak{g})$, $n_i : U_i \to \mathrm{Aut}(\mathfrak{g})$ be the data establishing the equivalence. Let A and A' denote the Lie algebroids constructed via 8.2.6 from $\{\chi_{ij}, a_{ij}\}$ and $\{\chi_{ij}', a_{ij}'\}$.

Define $\varphi : A' \to A$ locally by

$$(i, X \oplus V)' \mapsto (i, X \oplus (m_i(X) + n_i(V)));$$

it is easily verified that this is well defined, and gives the desired equivalence. □

Clearly one may modify 8.2.7 and 8.2.8 to take account of systems of transition data defined with respect to different open covers, and one may take an inductive limit; we leave it to the reader to work out the details.

Remark 8.2.9 Theorem 8.2.6 should be compared with 7.3.8. Both are construction principles for transitive Lie algebroids: the one by local, Čech–type data, the other by global data of de Rham type. The cocycle condition (TD2) corresponds to the Bianchi identity (condition (ii) of 7.3.8) and the compatibility condition (TD3) to condition (i) of 7.3.8, with the Aut($\mathfrak{g}$)–cocycle of 8.2.6 corresponding to the Lie connection in 7.3.8. Comments similar to those which follow 7.3.8 apply to the local construction also.

There is a similar correspondence between the local and global notions of equivalence for transitive Lie algebroids (8.2.8 and §7.3).

For Lie algebroids with abelian adjoint bundle, the equivalence between the local and global descriptions of transitive Lie algebroids is (most of) the equivalence between de Rham and Čech cohomology in low degrees. ⊠

Examples of transition forms may be obtained easily, by taking the right–derivative of transition functions of known examples of locally trivial Lie groupoids. For example, with Ω the Lie groupoid of the Hopf bundle $SU(2)(S^2, U(1))$ and charts defined by stereographic projection in the usual way, the transition form for $S^2 \times \mathbb{R} \to \frac{TSU(2)}{U(1)} \to TS^2$ is essentially the Maurer–Cartan form for $U(1)$. Transition forms for $\mathfrak{D}(E), \mathfrak{D}_{\mathrm{Der}}(L), \mathfrak{D}_{\mathfrak{so}}(E)$, and so on, can be constructed directly from transition functions for the bundles E, L, and so on.

We conclude the section with two direct applications of systems of transition data. First, a stronger version of 6.5.3 which is of independent interest.

Theorem 8.2.10 *Let $\varphi \colon A \to A'$ be a morphism of transitive Lie algebroids over M. Then there is an open cover $\{U_i\}$ of M and isomorphisms*

$$S_i \colon TU_i \oplus (U_i \times \mathfrak{g}) \to A_{U_i}, \qquad S_i' \colon TU_i \oplus (U_i \times \mathfrak{g}') \to A_{U_i}',$$

such that $(S_i')^{-1} \circ \varphi \circ S_i$ is characterized by the zero Maurer–Cartan form and a map $U_i \to \mathrm{Hom}(\mathfrak{g}, \mathfrak{g}')$ that is constant.

Proof Let $\{U_i\}$ by any cover by contractible open sets, and let S_i be given by 8.2.1. Denote by θ^i and ψ_i the flat connections and LAB charts associated with S_i. Define $\theta'^i = \varphi \circ \theta^i$; then θ'^i is a flat connection in A'_{U_i}. Now the adjoint connection $\nabla^{\theta'^i}$ is a flat Lie connection in a trivializable LAB L'_{U_i} on a simply-connected base U_i; by an obvious modification of 5.4.7 there is an LAB chart $\psi'_i \colon U_i \times \mathfrak{g}' \to L'_{U_i}$ which maps the standard flat connection ∇^0 to $\nabla^{\theta'^i}$. Denote by S'_i the isomorphism $TU_i \oplus (U_i \times \mathfrak{g}') \to A'_{U_i}$ defined by ψ'_i and θ'^i.

We now have a morphism of trivial Lie algebroids

$$(S'_i)^{-1} \circ \varphi \circ S_i \colon TU_i \oplus (U_i \times \mathfrak{g}) \to TU_i \oplus (U_i \times \mathfrak{g}');$$

denote by $f_i \colon U_i \times \mathfrak{g} \to U_i \times \mathfrak{g}'$ its restriction $(\psi'_i)^{-1} \circ \varphi^+ \circ \psi_i$. To determine the Maurer–Cartan form for $(S'_i)^{-1} \circ \varphi \circ S_i$, calculate

$$((S'_i)^{-1} \circ \varphi \circ S_i)(X \oplus V) = (S'_i)^{-1}(\varphi(\theta^i(X) + j\psi_i(V))$$
$$= (S'_i)^{-1}(\theta'^i(X) + j'\varphi^+\psi_i(V))) = X \oplus f_i(V).$$

Thus the Maurer–Cartan form is zero.

Now the compatibility equation for $\omega_i = 0$ and f_i is

$$X(f_i(W)) - f_i(X(W)) = 0$$

and, by (7) on p. 183, this implies that the Lie derivatives $X(f_i)$ of f_i as $\mathrm{Hom}(\mathfrak{g}, \mathfrak{g}')$–valued maps on U_i, are zero. Since U_i is connected, it follows that f_i is constant. $\square$

A similar proof can be given for 6.5.12.

We conclude with an infinitesimal reduction theorem.

Theorem 8.2.11 *Let $L \rightarrowtail A \relbar\joinrel\twoheadrightarrow TM$ be a transitive Lie algebroid. Then if A has a connection γ whose curvature $\overline{R}_\gamma$ takes values in a sub–LAB K which is stable under ∇^γ, then A has a system of transition forms taking values in the fibre type of K.*

Proof Granted the local triviality proved in 6.4.18, this follows by the same method as in 8.2.1. $\square$

The local description of connections is as given in §5.4.

8.3 The obstruction class

Throughout this section, all base manifolds are connected. We call a Lie algebroid A *integrable* if there is a Lie groupoid Ω such that $A\Omega$ is isomorphic to A over M. It was proved in 3.5.18 that a Lie groupoid on a connected base for which the Lie algebroid is transitive must be locally trivial.

In §5.4 we showed that the Darboux derivatives $\Delta(s_{ij})$ of a cocycle $\{s_{ij}\}$ for a locally trivial Lie groupoid Ω are transition forms for the Lie algebroid $A\Omega$. In the previous section we showed that an abstract transitive Lie algebroid A on an arbitrary base M admits a system of transition forms χ_{ij}. Our problem now is as follows: Given an abstract transitive Lie algebroid A and a system of transition forms χ_{ij}, is it possible to integrate the χ_{ij} to functions s_{ij} which obey the cocycle condition? If this can be accomplished, then the resulting Lie groupoid will have A as its Lie algebroid, by the classification theorem 8.2.6.

Consider, therefore, a transitive Lie algebroid $L \rightarrowtail A \twoheadrightarrow TM$ on an arbitrary (connected) base M. Let $\mathfrak{g}$ denote the fibre type of L, and let $\{U_i\}$ be a simple open cover of M.

By §8.2, there are local flat connections $\Theta_i \colon TU_i \to A_{U_i}$ and LAB charts $\psi_i \colon U_i \times \mathfrak{g} \to L_{U_i}$ such that $\nabla_X^{\Theta_i}(\psi_i(V)) = \psi_i(X(V))$ identically. Let $\chi_{ij} \in \Omega^1(U_{ij}, \mathfrak{g})$ and $a_{ij} \colon U_{ij} \to \mathrm{Aut}(\mathfrak{g})$ be the resulting system of transition data. From §8.2 we have the three conditions (TD1), (TD2) and (TD3). Let $\widetilde{G}$ be the connected and simply connected Lie group with Lie algebra $\mathfrak{g}$. From (TD1) and the simple–connectivity of U_{ij}, it follows that there are functions $s_{ij} \colon U_{ij} \to \widetilde{G}$ such that $\Delta(s_{ij}) = \chi_{ij}$; such functions are unique, up to right–translation by constants. From (10) on p. 183, it follows that

$$\Delta(\mathrm{Ad} \circ s_{ij}) = \mathrm{ad} \circ \Delta(s_{ij}) = \mathrm{ad} \circ \chi_{ij} = \Delta(a_{ij}),$$

where the first and last Δ refer to the group $\mathrm{Aut}(\mathfrak{g})$, and so, by the uniqueness result just referred to, there are elements $\varphi_{ij} \in \mathrm{Aut}(\mathfrak{g})$ such that

$$a_{ij} = (\mathrm{Ad} \circ s_{ij}) \circ \varphi_{ij} \tag{5}$$

Suppose that the a_{ij} take values in $\mathrm{Ad}(\widetilde{G}) = \mathrm{Int}(\mathfrak{g}) \leqslant \mathrm{Aut}(\mathfrak{g})$. Then the equation $\Delta(\mathrm{Ad} \circ s_{ij}) = \Delta(a_{ij})$ may be solved with respect to $\mathrm{Ad}(\widetilde{G})$, and it will be possible to take $\varphi_{ij} = \mathrm{id}_{\mathfrak{g}}$ in (5). We now show that this can be done whenever M is simply connected. In fact we have the following general result, which is a refinement of 8.2.1.

Theorem 8.3.1 *Let $L \rightarrowtail A \twoheadrightarrow TM$ be a transitive Lie algebroid on an arbitrary base M. Let $\psi\colon U \to \mathrm{Int}(A)_m$ be a decomposing section of the Lie groupoid $\mathrm{Int}(A)$ over a contractible open set U. Let ψ also denote the chart $U \times \mathfrak{g} \to L_U$ obtained by regarding $\mathrm{Int}(A)$ as a reduction of $\Phi_{\mathrm{Aut}}(L)$. (Here $\mathfrak{g} = L_m$.) Then there is a local flat connection $\Theta\colon TU \to A_U$ such that $\nabla^\Theta = \psi_*(\nabla^0)$.*

Proof The decomposing section $\psi\colon U \to \mathrm{Int}(A)_m$ induces, as in §5.4, a local flat connection $\kappa\colon TU \to \mathrm{ad}(A)_U$. By (28) on p. 206, applied to $\Omega = \mathrm{Int}(A)$, the induced chart

$$\mathrm{Ad}(\psi)\colon U \times \mathrm{ad}(\mathfrak{g}) \to \mathrm{ad}(L)_U$$

maps ∇^0 to ∇^κ. Here Ad refers to the Lie groupoid $\mathrm{Int}(A)$; note that $\mathrm{Ad}(\psi)_x\colon \mathrm{ad}(\mathfrak{g}) \to \mathrm{ad}(L_x)$, for $x \in U$, is $T_{\mathrm{id}}(I_{\psi(x)})\colon T_{\mathrm{id}}(\mathrm{Int}(\mathfrak{g})) \to T_{\mathrm{id}}(\mathrm{Int}(L_x))$ and since $I_{\psi(x)}$ is linear, it is its own tangent and so

$$\mathrm{Ad}(\psi)_x(\varphi) = \psi(x) \circ \varphi \circ \psi(x)^{-1}$$

for $\varphi \in \mathrm{ad}(\mathfrak{g})$.

Choose any connection γ in A; since the adjoint connection ∇^γ and $\psi_*(\nabla^0)$ are both in $\mathrm{ad}(A)$ we can write

$$\nabla^\gamma = \psi_*(\nabla^\circ) + \mathrm{ad}(\psi \circ \ell) \tag{6}$$

for some $\ell \in \Omega^1(U, \mathfrak{g})$. This equation shows that $\psi_*(\nabla^0)$ and ∇^γ cover the same coupling and so, by 7.3.6, there is a connection $\gamma'\colon TU \to A_U$ such that $\nabla^{\gamma'} = \psi_*(\nabla^0)$. It remains to show that there is such a connection which is flat.

Let A' be the semidirect extension $TU \ltimes_\nabla L_U$, where $\nabla = \nabla^{\gamma'} = \psi_*(\nabla^0)$. Now A' and A_U both define elements of $\mathscr{O}\mathrm{pext}(TU, L_U)$ and since $\mathscr{H}^2(TU, ZL_U) \cong H^2_{\mathrm{deRh}}(U, \mathfrak{z}) = (0)$, by the contractibility of U, it follows that A' and A_U are equivalent. Thus there is an isomorphism $\varphi\colon A' \to A_U$ such that

$$
\begin{array}{ccccc}
L_U & \overset{j}{\rightarrowtail} & A_U & \overset{a}{\twoheadrightarrow} & TU \\
\| & & \varphi \uparrow & & \| \\
L_U & \overset{j'}{\rightarrowtail} & A' & \overset{a'}{\twoheadrightarrow} & TU
\end{array}
$$

commutes. Define $\Theta\colon TU \to A_U$ by $\Theta(X) = \varphi(X \oplus 0)$. Then Θ is a flat connection and

$$j(\nabla_X^\theta(V)) = [\theta(X), j(V)] = [\varphi(X \oplus 0), \varphi j'(V)]$$
$$= \varphi(0 \oplus \nabla_X(V)) = \varphi j'(\nabla_X(V)) = j(\nabla_X(V)).$$

So $\nabla^\theta = \nabla = \psi_*(\nabla^0)$, as required. $\qquad\square$

It is interesting to note that the full force of the classification of extensions by $\mathscr{H}^2$ and $\mathscr{H}^3$ is used in this proof.

We now make a fresh start. Assume that M is simply connected, and take a simple open cover U_i of M. There are decomposing sections $\psi_i \colon U_i \to \mathrm{Int}(A)_m$ and, from 8.3.1, local flat connections Θ_i with $(\psi_i)_*(\nabla^0) = \nabla^{\Theta_i}$. Proceeding as before, we now have $a_{ij} \colon U_{ij} \to \mathrm{Int}(A)_m^m$. Since $\mathrm{Int}(A)$ is α–connected and M is simply connected, it follows that $\mathrm{Int}(A)_m^m$ is connected. It is therefore equal to $\mathrm{Int}(\mathfrak{g})$, and so to $\mathrm{Ad}(\widetilde{G})$. Now the equation $\Delta(\mathrm{Ad} \circ s_{ij}) = \Delta(a_{ij})$ may be solved with respect to the group $\mathrm{Ad}(\widetilde{G})$ and we have $\varphi_{ij} = \mathrm{Ad}(g_{ij})$ for $g_{ij} \in \widetilde{G}$. So, redefining s_{ij} as $s_{ij}g_{ij}$, we have

$$a_{ij} = \mathrm{Ad} \circ s_{ij}. \tag{7}$$

Consider the map $s_{jk}s_{ik}^{-1}s_{ij} \colon U_{ijk} \to \widetilde{G}$ for a non–void U_{ijk}. Using formulas (2) and (3) on p. 182 and the cocycle equation (TD2), we have

$$\begin{aligned}
\Delta(s_{jk}s_{ik}^{-1}s_{ij}) &= \Delta(s_{jk}) + \mathrm{Ad}(s_{jk})\Delta(s_{ik}^{-1}s_{ij}) \\
&= \chi_{jk} + a_{jk}\{\Delta(s_{ik}^{-1}) + a_{ik}^{-1}\Delta(s_{ij})\} \\
&= \chi_{jk} + a_{jk}\{-a_{ik}^{-1}\chi_{ik} + a_{ik}^{-1}\chi_{ij}\} \\
&= \chi_{jk} + a_{ji}(\chi_{ij} - \chi_{ik}) \\
&= \chi_{jk} - a_{ji}a_{ij}\chi_{jk} = 0.
\end{aligned}$$

Since U_{ijk} is connected, it follows that $s_{jk}s_{ik}^{-1}s_{ij}$ is constant; write

$$e_{ijk} = s_{jk}s_{ik}^{-1}s_{ij} \tag{8}$$

for its value. Clearly $e_{ijk} \in Z\widetilde{G}$, for $\mathrm{Ad}(e_{ijk}) = a_{jk}a_{ik}^{-1}a_{ij} = \mathrm{id}_{\mathfrak{g}}$. In fact, $\{e_{ijk}\}$ is a Čech 2–cocycle. For, if $U_{ijk\ell} \neq \emptyset$, then

$$e_{jk\ell}e_{ik\ell}^{-1}e_{ij\ell}e_{ijk}^{-1} = (s_{k\ell}s_{j\ell}^{-1}s_{jk})(s_{ik}^{-1}s_{i\ell}s_{k\ell}^{-1})(s_{j\ell}s_{i\ell}^{-1}s_{ik}s_{jk}^{-1}).$$

Interchanging the first two bracketed terms, this becomes

$$(s_{ik}^{-1}s_{i\ell}s_{j\ell}^{-1}s_{jk})(s_{j\ell}s_{i\ell}^{-1}s_{ik}s_{jk}^{-1}).$$

Now the second bracketed term is $e_{ij\ell}e_{ijk}^{-1}$ and is central, so we can interchange s_{jk} with it, and the expression then collapses to the identity.

Thus we have $e = \{e_{ijk}\} \in \check{H}^2(M, Z\widetilde{G})$. It remains to prove that e is well defined. This requires a little care. Let $\{\psi_i' \colon U_i \to \mathrm{Int}(A)_m\}$ be a second section–atlas for $\mathrm{Int}(A)$ with respect to the same open cover. Write $\psi_i' = \psi_i n_i$ where $n_i \colon U_i \to \mathrm{Int}(A)_m^m = \mathrm{Int}(\mathfrak{g}) = \mathrm{Ad}(\widetilde{G})$. Let $\{\Theta_i'\}$ be a second family of local flat connections, compatible with $\{\psi_i'\}$. Write $\Theta_i' = \Theta_i + \psi_i \circ m_i$. We then have equations (EDT1)—(EDT4) of §8.2. Since m_i is a Maurer–Cartan form, we can integrate and get $r_i \colon U_i \to \widetilde{G}$ with $\Delta(r_i) = m_i$. Then $\Delta(\mathrm{Ad} \circ r_i) = \mathrm{ad} \circ \Delta(r_i) = \mathrm{ad} \circ m_i = \Delta(n_i)$ so there exists $\varphi_i \in \mathrm{Ad}\widetilde{G}$ such that $n_i = (\mathrm{Ad} \circ r_i)\varphi_i$. Writing $\varphi_i = \mathrm{Ad}g_i$ and redefining r_i to be $r_i g_i$, we now have

$$n_i = \mathrm{Ad} \circ r_i. \tag{9}$$

Now, by using equations (2) and (3) on p. 182, one easily sees that (EDT3) is equivalent to

$$\Delta(s_{ij}') = \Delta(r_i^{-1} s_{ij} r_j)$$

so there are elements $c_{ij} \in \widetilde{G}$ such that

$$s_{ij}' = r_i^{-1} s_{ij} r_j c_{ij}.$$

Applying Ad to this equation, we find that $c_{ij} \in Z\widetilde{G}$. It is now straight-forward to verify that

$$e_{ijk}' = e_{ijk}(c_{jk} c_{ik}^{-1} c_{ij})$$

and so $\{e_{ijk}'\}$ and $\{e_{ijk}\}$ represent the same element of $\check{H}^2(M, Z\widetilde{G})$. It is also straightforward to show that this element is well defined with respect to the inductive limit. There is thus a well defined element $e \in \check{H}^2(M, Z\widetilde{G})$, independent of the choice of section atlas for $\mathrm{Int}(A)$. We call e the *cohomological integrability obstruction* of A on account of the following theorem.

Theorem 8.3.2 *Let* $L \rightarrowtail A \twoheadrightarrow TM$ *be a transitive Lie algebroid on a simply connected base* M. *Then there is a Lie groupoid* Ω *such that* $A\Omega \cong A$ *if and only if* e *lies in* $\check{H}^2(M, D)$ *for some discrete subgroup* D *of* $Z\widetilde{G}$.

Proof $(\Longrightarrow)$ This requires some work. First note that Ω may be assumed to be α–connected, and it then follows that $\mathrm{Ad}(\Omega) = \mathrm{Int}(A\Omega)$. Choose $m \in M$ and denote Ω_m^m by G; since M is simply connected and Ω is α–connected, G is connected. Let $\widetilde{G}$ denote the simply connected covering group.

Choose an atlas $\{\sigma_i \colon U_i \to \Omega_m\}$ for Ω. Then $\{\psi_i = \mathrm{Ad}\sigma_i\}$ is an atlas for $\mathrm{Ad}(\Omega) = \mathrm{Int}(A\Omega)$. Let θ_i denote the local morphism $U_i \times U_i \to \Omega_{U_i}^{U_i}$ as in §5.4. Then, by (28) on p. 206, $(\theta_i)_*$ is compatible with ψ_i, and so we can use $\{\psi_i, (\theta_i)_*\}$ to define e. Write $\chi_{ij} = \Delta(\widetilde{s}_{ij})$ where $\widetilde{s}_{ij} \colon U_{ij} \to \widetilde{G}$. Note that the $\widetilde{s}_{ij}$ are the would–be cocycles found from $A\Omega$, whereas s_{ij} are the actual cocycles for Ω. Now $\Delta(p \circ \widetilde{s}_{ij}) = \chi_{ij} = \Delta(s_{ij})$, where $p \colon \widetilde{G} \to G$ is the covering projection. So $p \circ \widetilde{s}_{ij} = s_{ij} w_{ij}$ for some element $w_{ij} \in G$. Now, applying Ad to this and noting that $\mathrm{Ad} \circ s_{ij} = a_{ij}$ and $\mathrm{Ad} \circ p \circ \widetilde{s}_{ij} = \widetilde{\mathrm{Ad}} \circ \widetilde{s}_{ij} = a_{ij}$ (by (7)), we find that $w_{ij} \in ZG$. Now the covering projection p maps $Z\widetilde{G}$ onto ZG, by general Lie group considerations, and so we can write $w_{ij} = p(\widetilde{w}_{ij})$ where $\widetilde{w}_{ij} \in Z\widetilde{G}$. Now redefine $\widetilde{s}_{ij}$ to be $\widetilde{s}_{ij}\widetilde{w}_{ij}^{-}$ and we have

$$p \circ \widetilde{s}_{ij} = s_{ij}.$$

(Note that this redefinition does not affect condition (7), since $\widetilde{w}_{ij}$ is central.) Now

$$p(e_{ijk}) = p(\widetilde{s}_{jk}\widetilde{s}_{ik}^{-1}\widetilde{s}_{ij}) = s_{jk}s_{ik}^{-1}s_{ij} = 1 \in G$$

so e takes its values in the discrete subgroup $\ker(p)$ of $Z\widetilde{G}$.

($\Longleftarrow$) Assume that $e \in \check{H}^2(M, D)$ for a discrete subgroup D of $Z\widetilde{G}$. Define $G = \widetilde{G}/D$ and let p be the covering projection. Let $s_{ij} \colon U_{ij} \to \widetilde{G}$ be the system of maps which define a representative $\{e_{ijk}\}$ for which $e_{ijk} \in D$. Define $\overline{s}_{ij} = p \circ s_{ij}$. Then $\{\overline{s}_{ij} \colon U_{ij} \to G\}$ satisfies the cocycle condition $\overline{s}_{jk}\overline{s}_{ik}^{-1}\overline{s}_{ij} = 1 \in G$; let Ω be the resulting Lie groupoid. The Lie algebroid $A\Omega$ has transition forms $\Delta(\overline{s}_{ij}) = p_* \circ \Delta(s_{ij}) = \Delta(s_{ij})$ and so, by 8.2.6, is isomorphic to A. $\qquad\square$

The case of a multiply–connected base space can now be handled by lifting back to the universal cover.

Theorem 8.3.3 *Let Υ be an α–simply connected locally trivial Lie groupoid with base the total space of a principal bundle $P(M, G)$.*

Suppose that for all $g \in G$ there is given a Lie algebroid automorphism $\widetilde{R}_g \colon A\Upsilon \to A\Upsilon$ which defines the structure of a PBG–Lie algebroid on $A\Upsilon$. Then there is a natural structure of PBG–groupoid on Υ which induces on $A\Upsilon$ the given PBG–Lie algebroid structure.

Proof The PBG–Lie algebroid structure defines for each $g \in G$ a Lie algebroid automorphism $\widetilde{R}_g \colon A\Upsilon \to A\Upsilon$ over $R_g \colon P \to P$. This can be considered as a base–preserving isomorphism from $A\Upsilon$ to the pullback

$R_g^{!!}A\Upsilon$, which is canonically isomorphic to $A(R_g^{\Downarrow}\Upsilon)$ by 4.3.11. Since both Υ and $R_g^{\Downarrow}\Upsilon$ are α–simply connected, we obtain by integration (see 6.2.4) a unique isomorphism $\Upsilon \to R_g^{\Downarrow}\Upsilon$ which yields an automorphism $\check{R}_g \colon \Upsilon \to \Upsilon$ over R_g. Using the uniqueness of these automorphisms, it is straightforward to show that this defines an action of G on Υ. That the action is smooth follows from observing that, by construction,

$$\check{R}_g(\mathrm{Exp}\, tX(u)) = (\mathrm{Exp}\, t\widetilde{R}_g(X))(ug),$$

for all $g \in G$, $X \in \Gamma A\Upsilon$ and possible $t \in \mathbb{R}$ and $u \in P$, and remembering that every element of Υ is of the form $\mathrm{Exp}\, tX(u)$ by 3.6.3. Thus Υ has the structure of a PBG–groupoid, and it certainly induces the given PBG–Lie algebroid structure. $\square$

Theorem 8.3.4 *Let M be a connected manifold with universal cover $c \colon \widetilde{M} \to M$. Let A be a transitive Lie algebroid on M. Then A is integrable if and only if the pullback Lie algebroid $c^{!!}A$ on $\widetilde{M}$ is integrable.*

Proof ($\Longrightarrow$) If $A = A\Omega$ for Ω a locally trivial Lie groupoid on M, then $c^{!!}A \cong A(c^{\Downarrow}\Omega)$ by 4.3.11.

($\Longleftarrow$) Observe firstly that the pullback Lie algebroid $c^{!!}A$ is canonically isomorphic to the action Lie algebroid $A \vartriangleleft c$, the underlying vector bundle of which is the pullback vector bundle $c^!A$; see 4.1.3.

Suppose now that $c^{!!}A = A\Omega'$, where Ω' is an α–simply connected and locally trivial Lie groupoid on $\widetilde{M}$. The pullback Lie algebroid $c^{!!}A$ is a PBG–Lie algebroid for the principal bundle $\widetilde{M}(M, \pi_1 M, c)$, under the canonical action of $\pi_1 M$ on the pullback vector bundle $c^!A$. By 8.3.3, it now follows that Ω' has a PBG–groupoid structure for $\widetilde{M}(M, \pi_1 M, c)$, and so from 2.5.5 the quotient manifold $\Omega = \Omega'/\pi_1 M$ exists and has a locally trivial Lie groupoid structure for which, by 4.5.8, $A\Omega \cong c^{!!}A/\pi_1 M \cong A$ as required. $\square$

For ease of reference, we state the combined form of these results.

Definition 8.3.5 Let $L \rightarrowtail A \longrightarrow\!\!\!\gg TM$ be a transitive Lie algebroid on a connected base M. Denote the universal covering for M by $c \colon \widetilde{M} \to M$ and the fibre type of L by $\mathfrak{g}$.

The *cohomological obstruction class* of A is the cohomological obstruction class $e \in \check{H}^2(\widetilde{M}, Z\widetilde{G})$ of the pullback Lie algebroid $c^{!!}A$. $\square$

Theorem 8.3.6 *Let $L \rightarrowtail A \longrightarrow\!\!\!\gg TM$ be a transitive Lie algebroid* on a connected base M. Write $\mathfrak{g}$ for the fibre type of L and write $\widetilde{G}$ for the universal Lie group corresponding to $\mathfrak{g}$. Then there is a Lie groupoid Ω such that $A\Omega \cong A$ if and only if e lies in $\check{H}^2(\widetilde{M}, D)$ for some discrete subgroup D of $Z\widetilde{G}$.

We may abbreviate 'e lies in $\check{H}^2(\widetilde{M}, D)$ for some discrete subgroup D of $Z\widetilde{G}$' to 'e is discrete in $\check{H}^2(\widetilde{M}, Z\widetilde{G})$.'

There are a number of remarks to be made about 8.3.6. First note that if the centre of $\mathfrak{g}$ is trivial, then $Z\widetilde{G}$ itself is discrete:

Corollary 8.3.7 *Let* $L \rightarrowtail A \twoheadrightarrow TM$ *be a transitive Lie algebroid on base* M, *with semisimple adjoint bundle* L. *Then* A *is integrable.*

At the other extreme, consider a transitive Lie algebroid

$$E \overset{j}{\rightarrowtail} A \overset{a}{\twoheadrightarrow} TM \tag{10}$$

in which the adjoint bundle E is abelian. Denote the fibre type of E by V, and the cohomological obstruction class by $e \in \check{H}^2(\widetilde{M}, V)$.

For any $\lambda \in \mathbb{R}$, $\lambda \neq 0$, define $\mu_\lambda \colon E \to E$ by $\mu_\lambda(e) = \lambda e$. Then (10) is equivalent to the transitive Lie algebroid

$$E \overset{j\circ\mu_\lambda}{\rightarrowtail} A \overset{a}{\twoheadrightarrow} TM. \tag{11}$$

Following through the various constructions of this section, it is easy to see that:

Proposition 8.3.8 *The cohomological obstruction class of* (11) *is* $\frac{1}{\lambda}e$.

Thus in the abelian case — and in particular in the classical case of the Weil Lemma — there is a genuine scaling factor in the cohomological obstruction class which cannot be removed. It is for this reason that we prefer the term 'discrete' rather than 'integral' for the condition which ensures integrability. In the case of nonabelian LABs, automorphisms of this kind may exist on an abelian sub–LAB.

Note that once a transitive Lie algebroid is known to be integrable, it has a natural adjoint bundle, arising from its presentation in either of the forms

$$L\Omega \rightarrowtail A\Omega \twoheadrightarrow TM \qquad \text{or} \qquad \frac{P \times \mathfrak{g}}{G} \rightarrowtail \frac{TP}{G} \twoheadrightarrow TM$$

and the curvature of any connection (for example) is then well defined. However, for an abstract transitive Lie algebroid $L \rightarrowtail A \twoheadrightarrow TM$, the curvature of a connection $\gamma \colon TM \to A$ depends not only on A but also on the choice of L. One should, strictly speaking, distinguish between a transitive Lie algebroid $A \xrightarrow{\;a\;} TM$ and what one might call a *Lie algebroid sequence* $L \xmapsto{\;j\;} A \xrightarrow{\;a\;} TM$.

Consider now the classical Weil Lemma itself, 8.1.3. This is usually stated in terms of the existence of a connection — either in an Hermitian line bundle, or in a principal $U(1)$–bundle — with prescribed curvature form. However, Theorem 8.3.6 above makes no mention of connections or curvature forms. Combining it now with 7.3.8, we obtain:

Theorem 8.3.9 *Let M be a connected manifold, and let L be an LAB on M. Let R be an L–valued 2–form on M. Then R is the curvature of a connection in a principal bundle $P(M, G)$ with $\frac{P \times \mathfrak{g}}{G} \cong L$ if and only if:*

(i) there exists a Lie connection ∇ in L such that $\overline{R}_\nabla = \mathrm{ad} \circ R$ and $\nabla(R) = 0$; and

(ii) the cohomological obstruction $e \in \check{H}^2(\widetilde{M}, Z\widetilde{G})$, defined by the transitive Lie algebroid corresponding to ∇ and R, lies in $\check{H}^2(\widetilde{M}, D)$ for some discrete subgroup D of $Z\widetilde{G}$.

One can legitimately regard 8.3.9, rather than 8.3.6, as the non–abelian version of the Weil Lemma. This result is of value only when L has a non–trivial centre; we have the somewhat uncommon situation of a result which is of most interest when a Lie algebra is abelian, and of least when it is semisimple.

The fact that 8.3.6 implies the classical Weil Lemma is due to the facts that, when the adjoint bundle is abelian, there is only one curvature form up to equivalence and that, given a curvature form in this case, there is only one connection — in the sense described below — which could induce it.

Assume that, in 8.3.9, L is abelian and M is simply connected. Then ∇ must be flat, L must be trivializable, and there is a trivialization $M \times V \cong L$ which maps the standard flat connection in $M \times V$ to ∇ (see 5.4.7). Thus in this case, 7.1.13 shows that the Lie algebroid is determined uniquely by the closed 2–form R and 7.1.17 shows that the connections in that Lie algebroid with curvature R are determined by $\mathscr{H}^1(TM, M \times V) \cong H^1_{\mathrm{deRh}}(M, V) = 0$. There is therefore a uniquely determined Lie algebroid with a connection having curvature R and, when the Lie algebroid is integrable, there is a unique principal bundle with simply connected total space.

If L is not abelian, or if M is not simply connected, then uniqueness in this strong sense fails. The appropriate results can be obtained by following back through the results of §7.3.

There is a difference between the classical Weil Lemma and the specialization of Theorem 8.3.9 to $L = M \times \mathbb{R}$: in 8.3.9 R is allowed to be closed with respect to any flat connection in $M \times \mathbb{R}$; if M is not simply connected, there may be such connections which are not gauge equivalent to the standard flat connection.

Aspect 2 of §8.1 also extends to general transitive Lie algebroids. Consider firstly any principal bundle $P(M, G)$ and a Lie group extension

$$K \rightarrowtail H \overset{p}{\twoheadrightarrow} G$$

in which K is discrete and (hence) central in H. Take any transition functions $s_{ij} \colon U_{ij} \to G$ for P and consider any smooth functions $h_{ij} \colon U_{ij} \to H$ with $p \circ h_{ij} = s_{ij}$ for all i, j. Then, using the same method by which integrability obstructions were defined,

$$\lambda_{ijk} = h_{jk} h_{ik}^{-1} h_{ij}$$

gives a well defined Čech class $\lambda \in \check{H}^2(M, K)$ and this class is null if and only if there is a principal bundle $Q(M, H)$ and a map $\overline{p} \colon Q \to P$ such that $\overline{p}(\mathrm{id}_M, p)$ is a morphism of principal bundles. This λ is the *obstruction to lifting* $P(M, G)$ *to* H.

Proposition 8.3.10 *Consider a principal bundle $P(M, G, \pi)$ in which M is simply connected and P and (therefore) G are connected. Let A denote the Atiyah–Lie algebroid of $P(M, G, \pi)$. Then the integrability obstruction of A and the obstruction to lifting $P(M, G)$ to $\widetilde{G}$ coincide.*

Proof Take transition functions $s_{ij} \colon U_{ij} \to G$ for P and consider any smooth functions $\widetilde{s}_{ij} \colon U_{ij} \to \widetilde{G}$ with $p \circ \widetilde{s}_{ij} = s_{ij}$ for all i, j. Because p is étale,

$$\Delta(\widetilde{s}_{ij}) = \Delta(s_{ij}) \qquad \text{and} \qquad \mathrm{Ad} \circ \widetilde{s}_{ij} = \mathrm{Ad} \circ s_{ij}$$

and so $\Delta(\widetilde{s}_{ij})$ and $\mathrm{Ad} \circ \widetilde{s}_{ij}$ form a system of transition data for A. Now the functions $\widetilde{s}_{ij}$ may be taken as integrating this system and so

$$e_{ijk} = \widetilde{s}_{jk} \widetilde{s}_{ik}^{-1} \widetilde{s}_{ij} = \lambda_{ijk}$$

is both the integrability obstruction and the obstruction to the lifting of $P(M, G)$ to $\widetilde{G}$. $\qquad \square$

This point of view may be inverted. Suppose given a transitive Lie algebroid $L \rightarrowtail A \longrightarrow\!\!\!\gg TM$ and for convenience assume that M is simply connected. Denote the fibre type of L by $\mathfrak{g}$, and let G be any connected Lie group with Lie algebra $\mathfrak{g}$. Then one may define, in a straightforward manner, an element of $\check{H}^2(M, ZG)$ which is the obstruction class to the existence of a principal bundle $P(M, G)$ with $\frac{TP}{G} \cong A$. This class is the image of the cohomological obstruction class of A under the map on cohomology induced by $Z\widetilde{G} \to ZG$.

8.4 Notes

§8.1. The relationship between Čech and de Rham cohomology and the parallel correspondence between $U(1)$–bundles and real valued 2–forms goes back at least to Weil [1958]. His account and that of Kostant [1970] are still most definitely worth reading. More recent accounts of this material are given in many places, such as [Bott and Tu, 1982] and [Woodhouse, 1992].

Cohomological obstructions to the lifting of structure groups go back to Haefliger [1956]. This material is less widely available; one account is [Greub and Petry, 1978].

Each of the four aspects of this section can be usefully extended to the case of arbitrary abelian groups; this was done by Androulidakis [2000].

§8.2. The material of this section comes from III§5, IV§4 and Chapter V of [Mackenzie, 1987a]; it was new when it appeared there. In particular, 8.2.1 was the first local integrability proof for a class of Lie algebroids. In [Mackenzie, 1987a] the interior Lie groupoid of a transitive Lie algebroid was called the *adjoint groupoid*.

§8.3. This section is an expansion of V§1 of [Mackenzie, 1987a]; again, that material was new when it appeared there. The treatment of the multiply–connected case is from [Mackenzie, 1987b].

Other possible definitions of integrability exist, at least for specific classes of Lie algebroids; some discussion of this is given in [Mackenzie, 1995].

The question of realizing given curvature quantities as the curvatures of actual geometries is a long–established and large field. Theorem 8.3.9 gives a complete answer to the question *When is an LAB–valued 2–form the curvature form of a connection in a principal bundle?*, but it should be noted that this gives no information whatever in the Riemannian case, where the structure groups are semisimple.

The lifting obstruction is an example of the Čech class associated to a crossed module of principal bundles; the general case is treated in [Mackenzie, 1988b]. This paper also shows that the concept of crossed module clarifies the construction of the integrability obstruction. Suppose given a transitive Lie algebroid $L \rightarrowtail A \longrightarrow\!\!\!\gg TM$. Quotienting over ZL gives the associated coupling, as in §7.3, and equivalently a crossed module. The linearity of A/ZL enables it to be integrated by 6.2.1 and a little extra work makes this a crossed module of Lie groupoids, which is an integration of the crossed module of the given Lie algebroid. This crossed module of Lie groupoids has its obstruction class in Čech cohomology with variable coefficients, and is the obstruction to the existence of a Lie groupoid with the given crossed module; we want, however, a Lie groupoid with not only the given crossed module but also with the given Lie algebroid. This additional differential condition imposes the constancy condition on the Čech cohomology.

The integrability question for transitive Lie algebroids.

For many years the major outstanding problem in the theory of Lie groupoids and Lie algebroids was to provide a full proof of a result announced by Pradines [1968], that every Lie algebroid is (isomorphic to) the Lie algebroid of a Lie groupoid (possibly non–Hausdorff). From the perspective of the 21st century, it may be hard to credit that the general consensus from the 1960s to the mid 1980s was that this announcement was valid; nonetheless most workers in the field during that period believed that a detailed proof of the integrability of all Lie algebroids would emerge in due course. No distinction between the transitive and general cases seems to have been considered.

Though Pradines' announcement was made in 1968, forms of the problem had been the subject of earlier work: see, for example, [Rodrigues, 1962]. Ehresmann's concept of Lie groupoid and Pradines' concept of Lie algebroid, however, enabled the abstract problem to be separated from its incarnation in terms of transformations, and this in itself was an important step. It ought also to be remembered that Pradines' programme [Pradines, 1966, 1967a,b, 1968] has otherwise been vindicated, and shown to be of very wide–ranging importance.

That not all Lie algebroids are integrable was demonstrated conclusively by Almeida and Molino in a note [Almeida and Molino, 1985] from a seminar of 1985. Their examples arose from work of Molino on transversally complete foliations, in which transitive Lie algebroids are constructed as infinitesimal invariants of transversally complete foliations, which need not necessarily be associated to a Lie groupoid; see [Molino, 1988] and [Moerdijk and Mrčun, 2003].

The construction of the cohomological obstruction to integrability goes back to my work as a graduate student in the late 1970s; my approach was, from the outset, concerned with utilizing the special features of the transitive case. The conference announcement [Mackenzie, 1980] constructed the elements e_{ijk} of §8.3, and observed that if they lie in a discrete subgroup of the centre of the universal covering group, then the Lie algebroid is integrable. I was able to speak on this construction on several occasions during the early 1980s, in Australia and in Brazil.

The distinction between the construction of counterexamples to integrability and the construction of the cohomological obstruction needs to be emphasized because some accounts have conflated the two.

In 2001 the subject was transformed again by the appearance of the preprint [Crainic and Fernandes, 2003] which provides two criteria, readily applicable, for the integrability of an arbitrary Lie algebroid. Some further comment on [Crainic and Fernandes, 2003] is given in the Appendix.

The evolution of the cohomological obstruction

The construction of the invariant e, and the cohomology theory of Chapter 7 on which it depends, grew out of the strategy, referred to above, which I developed in the mid–late 1970s for proving the integrability of transitive Lie algebroids by generalizing the cohomological proof of the integrability of Lie algebras due to van Est [1953, 1955b]. I believe it will be of interest to describe this process and, by so doing, to set this integrability result in the context of the ongoing evolution of the group concept.

The result which asserts the integrability of Lie algebras — Lie's third theorem — has a long and continuing history in mathematics, corresponding to the evolution of the group concept from its first rigorous formulation to the point where it is again capable of being applied to the study of partial differential equations. Concerning Lie groups in the now standard meaning of the term, prior to [van Est, 1953, 1955b] there were essentially two different methods of proving the converse of Lie's third theorem. One method — let us call it the *structural proof* — first uses the Levi–Mal'cev decomposition to reduce the problem to the two separate cases of solvable and semisimple Lie algebras. Lie's third theorem for these two cases is straightforward: the solvable case is reduced, by virtue of the chain condition, to the case of 1–dimensional Lie algebras, where the result is trivial; in the semisimple case, the adjoint representation is faithful and the result follows from the subgroup/subalgebra correspondence for a general linear group. (For details, see, for example, [Varadarajan, 1974, 3.15].) This proof is essentially a rigorous reformulation of Cartan's 1930 proof, in which Cartan was chiefly concerned to complete a proof of Lie's, valid only when the adjoint representation is faithful. The other method, for which I know no classical reference, integrates the given general Lie algebra directly to a Lie group germ, and must then show that this Lie group germ can be globalized. For the integration step, see, for example [Malliavin, 1972, pp.232–4] or [Greub et al., 1973, pp.368–9]; the globalization may be accomplished by the method of P. A. Smith [1952] — note that this depends on the fact that $\pi_2(G) = 0$ for a (semisimple) Lie group G. Call this second method the *geometric proof* since it depends on $\pi_2(G) = 0$, rather than on the structure theory of Lie algebras.

Whichever of these proofs one follows, it is clear that Lie's third theorem is an integration result; so it was to Lie (for example, [Cohn, 1957, Chapter V]), and so it remains. The geometric method divides the proof into an 'integration' step which yields a local group, and a 'globalization' step which depends on a deep topological result; the structural proof uses deep results of Lie algebra structure theory to reduce the integration to that of the subgroup/subalgebra correspondence — that is, to the Frobenius theorem. (Similarly Ado's theorem, an even deeper result from the structure theory of Lie algebras and itself depending on the Levi–Mal'cev decomposition, can be used for the same purpose.) Having noted the element of integration present in both proofs, note also that the other main steps in the two proofs are formally analogous: The existence of Lie subalgebras depends on the semi–directness of the extension resulting from quotienting the Lie algebra over its radical; that the extension is semi–direct follows from the second Whitehead lemma (and the first, via the theorem of Weyl; see, for example, [Varadarajan, 1974, 3.14.1]), and the second Whitehead lemma, that is, $H^2(\mathfrak{g}, V) = 0$ for semisimple $\mathfrak{g}$, may be regarded as analogous to $\pi_2(G) = 0$ for semisimple G, the condition which is crucial for the globalization step of the geometric proof.

To describe van Est's proof, it is necessary to first summarize the results from which he deduces the converse of Lie's third theorem. In a note [van Est, 1955b] which reformulates earlier results [van Est, 1953, 1955a], van Est constructed two convergent spectral sequences:

$$H^s(G, V) \otimes H^t_{\mathrm{deRh}}(G) \Longrightarrow H^{s+t}(\mathfrak{g}, V) \tag{12}$$

$$H^s(G, V) \otimes H^t_{\mathrm{deRh}}(G/K) \Longrightarrow H^{s+t}(\mathfrak{g}, \mathfrak{k}, V) \tag{13}$$

for a Lie group G and a representation of G on a vector space V. Here $H^*_{\mathrm{deRh}}(M)$ denotes the de Rham cohomology of the manifold M, and $H^*(G, V)$ denotes the smooth Eilenberg–MacLane cohomology; K is a compact subgroup of G with Lie algebra $\mathfrak{k} \subseteq \mathfrak{g}$. Both spectral sequences arise from double complexes in the standard manner; the double complex for (13) is a K–variant subcomplex of that for (12).

When G is connected and K is a maximal compact subgroup, G/K is diffeomorphic to a Euclidean space by the Iwasawa decomposition and the second spectral sequence therefore collapses to isomorphisms $H^*(G, V) \simeq H^*(\mathfrak{g}, \mathfrak{k}, V)$. This result was reproved in a more modern context by Hochschild and Mostow [1962] and was generalized extensively in the 1970s to model pseudogroups on $\mathbb{R}^n$ (see, for example, the survey articles [Lawson, 1974, §6] and [Stasheff, 1978, §7]).

Consider the first spectral sequence (12). In the first paper of the series, van Est [1953] noted that for a Lie group G with $H^1_{\mathrm{deRh}}(G) = H^2_{\mathrm{deRh}}(G) = 0$ a similar collapse leads to $H^2(G, V) \simeq H^2(\mathfrak{g}, V)$. Now in general $\pi_2(G) = 0$, and so if G is connected and simply connected, the Hurewicz theorem gives $H^1_{\mathrm{deRh}}(G) = H^2_{\mathrm{deRh}}(G) = 0$ and one therefore has $H^2(G, V) \overset{e_B}{\simeq} H^2(\mathfrak{g}, V)$ or, equivalently,

$$\mathcal{O}\mathrm{pext}(G, V) \simeq \mathcal{O}\mathrm{pext}(\mathfrak{g}, V)$$

under the map which assigns to $V \rightarrowtail^{\iota} H \xrightarrow{\pi} G$ the differentiated extension $V \rightarrowtail^{\iota_*} \mathfrak{h} \xrightarrow{\pi_*} \mathfrak{g}$. Hochschild [1951] had proved that $\mathcal{O}\mathrm{pext}(G, V) \simeq \mathcal{O}\mathrm{pext}(\mathfrak{g}, V)$ for connected and simply connected G, by use of Lie's third theorem; van Est now shows that the process can be reversed: any Lie algebra $\mathfrak{h}$ is an extension

$$\mathfrak{z} \longrightarrow \mathfrak{h} \longrightarrow \mathrm{ad}\mathfrak{h}$$

and thus defines an element of $H^2(\mathfrak{g}, \mathfrak{z})$ where the representation of $\mathfrak{g} = \mathrm{ad}\mathfrak{h}$ on $\mathfrak{z}$ is the trivial one. With $\widetilde{G}$ the universal covering group of $\mathrm{Int}(\mathfrak{h})$, represented trivially on the vector space $\mathfrak{z}$, he thus obtains an extension $\mathfrak{z} \rightarrowtail H \longrightarrow\!\!\!\gg \widetilde{G}$ with $\mathfrak{h}$ the Lie algebra of H. Since H is an extension of a connected and simply connected Lie group by a vector space, it is itself connected and simply connected. Thus Lie's third theorem is proved.

Once again the integration step has been reduced to the subgroup/subalgebra correspondence for a general linear group. van Est's procedure thus uses the (deep) topological fact that $\pi_2(G) = 0$ for any Lie group G but avoids the direct consideration of local groups needed for the geometric proof, and uses no deep result of Lie algebra theory.

A clear analysis of the importance of $\pi_2(G) = 0$, and other relevant points was given some time later by van Est [1962]. For other, more general, forms of the group concept, theorems of 'Lie third theorem' type have since been proved: for example, [Goldschmidt, 1978], [Quê and Rodrigues, 1975] (Lie equations and transitive Lie algebras), [Pommaret, 1977]. These results also represent a combination of cohomology and integration.

My original strategy to prove the integrability of transitive Lie algebroids was to construct a cohomology theory for Lie groupoids by means of which a straightforward generalization of van Est's spectral sequence (12) and of the ensuing argument, could be given. Much of this argument does hold for Lie groupoids and Lie algebroids. For example, given a transitive Lie algebroid, the exact sequence $ZL \rightarrowtail A \longrightarrow\!\!\!\gg A/ZL \cong \mathrm{ad}A$ exists, and, by 6.2.1, $\mathrm{ad}A$ can be integrated to the Lie subgroupoid $\mathrm{Int}(A)$ of $\Phi_{\mathrm{Aut}}(L)$. The induced representation $\mathrm{ad}(A) \to \mathfrak{D}(ZL)$ cannot be said to be trivial, since there is no concept of trivial representation unless ZL is trivializable as a vector bundle, but the representation does integrate to $\mathcal{M}\mathrm{Int}(A) \to \Phi(ZL)$.

The crucial problem therefore is to construct a satisfactory cohomology theory for Lie groupoids. A continuous cohomology for locally trivial topological groupoids with coefficients in vector bundles was presented in [Mackenzie, 1978], and the corresponding constructions for Lie groupoids can be given and follow the same pattern. (The cohomology of [Mackenzie, 1978] was called *rigid* since the topology of the extensions which it classifies is determined 'rigidly' by that of the quotient and the kernel.) It is proved in [Mackenzie, 1978, §7] that this cohomology classifies all extensions (satisfying some natural weak conditions) of locally trivial groupoids by vector bundles.

Nonetheless this theory was not adequate for the application to the Lie third theorem. If $L \rightarrowtail A \longrightarrow\!\!\!\gg TM$ is an abelian Lie algebroid on a simply connected base then $ZL = L$ and $\mathcal{M}\mathrm{Int}(A) = M \times M$, and the problem is to find a Lie groupoid Ω on M with $A\Omega \cong A$. The rigid cohomology can only cope with groupoids which are extensions $L \rightarrowtail \Omega \longrightarrow\!\!\!\gg M \times M$ and all such groupoids are trivializable.

The explanation is, of course, that the coefficient bundle ZL must itself be integrated, and a cohomology theory which will classify all extensions of Lie groupoids by Lie group bundles is needed. It may now (2004) finally be clear how to do this, but for the integrability question, a full cohomology theory is not needed.

Out of this strategy the existence of transition forms and the construction of the elements e_{ijk} emerged, and it is interesting to observe that these results themselves divide into the same two steps. Namely, for adjoint Lie algebroids the problem is (comparatively) easily solved — for every transitive Lie algebroid A the adjoint $\mathrm{ad}(A)$ is integrable by 6.2.1 — and the problem is to lift this across $ZL \rightarrowtail A \longrightarrow\!\!\!\gg \mathrm{ad}A$.

Furthermore, in 8.2.1, the existence of the required flat connections is easily established on the adjoint level; the difficulty is in lifting these connections to the given Lie algebroid. It is this lifting process which the cohomological apparatus describes.

PART THREE:
THE POISSON AND
SYMPLECTIC THEORIES

9

Double vector bundles
and their duality

The most fundamental example of a double vector bundle, the tangent manifold TE of a vector bundle, has already been met in §3.4. This object, it will be remembered, provides a systematic way of dealing with connections, derivative endomorphisms, and linear vector fields, and does so because of the two vector bundle structures on TE and the interaction between them. In this chapter we study in detail the general concept of double vector bundle, which will be used repeatedly in the rest of the book.

A certain amount of the general theory of double vector bundles consists of a straightforward (but necessary) upgrading of constructions well–known for ordinary vector bundles. What is decisively different however, is the theory of duality. The duality of double vector bundles behaves in entirely new and unexpected ways and is the principal reason for the importance of double vector bundles whenever the relations between Lie algebroids and Poisson structures are considered.

In §9.1 we give the basic definitions, including the fundamental concept of core. The general theory of duality for double vector bundles is in §9.2. In §9.3, §9.4 and §9.5, we treat the two duals of TA, and the canonical isomorphisms which relate them.

Considered as a double vector bundle, the double tangent bundle is unusual in two respects: because the side bundles and the core bundle are equal to each other, it is easy to confuse the different roles which they play; and whereas the general concept of double vector bundle does not include any notion of bracket structure, it is precisely the bracket structures on the double tangent bundle which are the main focus of its interest. §9.6 gives the main results relating the bracket structures on a double tangent bundle and its duals to the canonical isomorphisms of §9.3 and §9.5.

In §9.7 we show that a structure of Lie algebroid on a bundle $A \to M$ induces a Lie algebroid on the prolongation bundle $TA \to TM$; if $A = AG$ is the Lie algebroid of a Lie groupoid G, then this structure on TAG is canonically isomorphic to the Lie algebroid of the tangent groupoid $TG \implies TM$. Thus $TAG \cong ATG$ may be thought of as a generalization of the double tangent bundle and in §9.8 we show that the calculus of complete and vertical lifts has a natural extension to this setting. This extended calculus will be used in Chapter 11.

Throughout the chapter we eschew local coordinate formulations of the traditional type. Already in the context of this chapter, and even more so in the more general contexts to which this provides an introduction, the use of local coordinates obscures essential differences by reducing all objects to strings of parameter values. The formalism used here is both global and intrinsic. It is often necessary to separate arguments into two or more cases (for example, separate consideration of linear and vertical vector fields) but the distinguishing features of these cases are an essential part of the geometric structure.

9.1 General double vector bundles

Definition 9.1.1 A *double vector bundle* $(D; A, B; M)$ is a system of four vector bundle structures

$$
\begin{array}{ccc}
D & \xrightarrow{\ q_B^D\ } & B \\
{\scriptstyle q_A^D}\big\downarrow & & \big\downarrow{\scriptstyle q_B} \\
A & \xrightarrow[\ q_A\ ]{} & M
\end{array}
\tag{1}
$$

in which D has two vector bundle structures, on bases A and B, which are themselves vector bundles on M, such that each of the four structure maps of each vector bundle structure on D (namely the bundle projection, addition, scalar multiplication and the zero section) is a morphism of vector bundles with respect to the other structure. □

Several comments about notation need to be made. First, we commit the usual abuses, and in particular denote (1) by D when no confusion seems likely. We refer to A and B as the *side bundles* of D, and to M as the *double base*. In the two side bundles we denote addition, scalar multiplication and subtraction by the usual symbols $+$, juxtaposition, and $-$; we usually denote elements of A by letters $a, a_1, a_2, \ldots$, and elements of B by letters $b, b_1, b_2, \ldots$, which should make clear which bundle is relevant. We distinguish the two zero–sections, writing $0^A \colon M \to A$, $m \mapsto 0^A_m$, and $0^B \colon M \to B$, $m \mapsto 0^B_m$. We may denote an element $d \in D$ by $(d; a, b; m)$ to indicate that $a = q^D_A(d)$, $b = q^D_B(d)$, $m = q_B(q^D_B(d)) = q_A(q^D_A(d))$; the various projections make up the *outline* of d.

The notation q^D_A is clear; when the base of the bundle is the double base we write q_A, for example, rather than q^A_M.

In the *vertical bundle structure* on D with base A the vector bundle operations are denoted $\underset{A}{+}$, $\underset{A}{\cdot}$, $\underset{A}{-}$, with $\widetilde{0}^A \colon A \to D$, $a \mapsto \widetilde{0}^A_a$, for the zero–section. Similarly, in the *horizontal bundle structure* on D with base B we write $\underset{B}{+}$, $\underset{B}{\cdot}$, $\underset{B}{-}$ and $\widetilde{0}^B \colon B \to D$, $b \mapsto \widetilde{0}^B_b$. For $m \in M$ the *double zero* $\widetilde{0}^A_{0^A_m} = \widetilde{0}^B_{0^B_m}$ is denoted $\odot_m$ or 0^2_m. The two structures on D, namely (D, q^D_B, B) and (D, q^D_A, A), will occasionally be denoted $\widetilde{D}_B$ and $\widetilde{D}_A$, respectively.

In dealing with general double vector bundles such as (1), we thus usually label objects and operations in the two structures on D by the symbol for the base over which they take place. The words 'horizontal' and 'vertical' may be used as an alternative, but need to be referred to the arrangement in the diagram (1) or the sequence in $(D; A, B; M)$. When considering examples in which $A = B$, the words 'horizontal' and 'vertical' become necessary, and we use H and V as labels to distinguish the two structures on D.

Although the concept of double vector bundle is symmetric, most actual examples are not; in some processes it is important to distinguish between (1) and its *flip*

$$
\begin{array}{ccc}
D^F & \xrightarrow{\;\;q^D_A\;\;} & A \\[2pt]
{\scriptstyle q^D_B}\Big\downarrow & & \Big\downarrow{\scriptstyle q_A} \\[2pt]
B & \xrightarrow[\;\;q_B\;\;]{} & M
\end{array}
\qquad (2)
$$

in which the arrangement of the two structures is reversed. In such processes it is not the starting arrangement which is significant, but the distinction between whichever arrangement is taken at the start, and its flip.

Another point to note about the diagram (1) is that it should generally be read as a single object, consisting of four related vector bundles. Each arrow should be read as denoting the whole bundle structure, not merely the projection, just as the notation $G \Longrightarrow M$ denotes a groupoid structure, not merely the source and target maps. Only occasionally will a diagram such as (1) have its conventional meaning as representing a morphism of ordinary vector bundles (either horizontally or vertically). In cases where confusion might arise, we place a small square in the centre to indicate that the diagram should be read as a double structure, or a double arrow to show that it should be read as a morphism (see (15) below and Figure 11.5).

The compatibility conditions on the four structures can now be written explicitly as follows. The bundle projection $q_B^D \colon \widetilde{D}_A \to B$ is required to be a vector bundle morphism over $q_A \colon A \to M$, and the zero–section $\widetilde{0}^B \colon B \to \widetilde{D}_A$ must be a vector bundle morphism over $0^A \colon M \to A$. Taking the pullback in the category of vector bundles shown in Figure 9.1(a), we require that addition $\underset{B}{+} \colon \widetilde{D}_A \oplus_B \widetilde{D}_A \to \widetilde{D}_A$ be a morphism of vector bundles over addition $+ \colon A \oplus_M A \to A$.

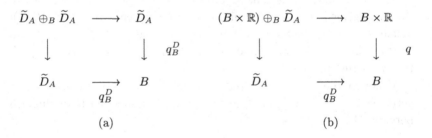

(a) (b)

Fig. 9.1.

Lastly, taking the pullback shown in Figure 9.1(b) where $B \times \mathbb{R}$ is regarded as a vector bundle over $M \times \mathbb{R}$ in the obvious way, and q is the projection, regarded as a morphism of vector bundles over $M \times \mathbb{R} \to M$, the scalar multiplication $\underset{B}{\cdot} \colon (B \times \mathbb{R}) \oplus_B \widetilde{D}_A \to \widetilde{D}_A$ must be a morphism of vector bundles over the scalar multiplication $(M \times \mathbb{R}) \oplus_M A \to A$. Once these four conditions are satisfied, the corresponding conditions in the vertical structure follow.

It is often useful to display elements of D with their associated projections to the side bundles indicated; thus Figure 9.2(i) shows that $d \in D$ has $q_B^D(d) = b$, $q_A^D(d) = a$, and that $q_A(a) = q_B(b) = m$. If d' in (ii) is another element of D and $a = a'$, then we have $d \underset{A}{+} d'$ as in (iii) and if instead $b = b'$, then $d \underset{B}{+} d'$ is as in (iv). Similarly, for $t \in \mathbb{R}$ the effects

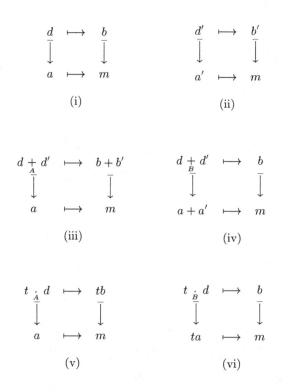

Fig. 9.2.

of the two scalar multiplications are shown in (v) and (vi). Finally, the zero elements are shown in Figure 9.3.

We now write out in detail the conditions interlinking the two structures on D. The condition that each addition in D is a morphism with respect to the other is:

$$(d_1 \underset{B}{+} d_2) \underset{A}{+} (d_3 \underset{B}{+} d_4) = (d_1 \underset{A}{+} d_3) \underset{B}{+} (d_2 \underset{A}{+} d_4) \qquad (3)$$

for quadruples $d_1, \ldots, d_4 \in D$ such that $q_B^D(d_1) = q_B^D(d_2)$, $q_B^D(d_3) = q_B^D(d_4)$, $q_A^D(d_1) = q_A^D(d_3)$, and $q_A^D(d_2) = q_A^D(d_4)$. Next,

$$t \underset{A}{\cdot} (d_1 \underset{B}{+} d_2) = t \underset{A}{\cdot} d_1 \underset{B}{+} t \underset{A}{\cdot} d_2, \qquad (4)$$

for $t \in \mathbb{R}$ and $d_1, d_2 \in D$ with $q_B^D(d_1) = q_B^D(d_2)$; similarly

$$t \underset{B}{\cdot} (d_1 \underset{A}{+} d_2) = t \underset{B}{\cdot} d_1 \underset{A}{+} t \underset{B}{\cdot} d_2, \tag{5}$$

for $t \in \mathbb{R}$ and $d_1, d_2 \in D$ with $q_A^D(d_1) = q_A^D(d_2)$. The two scalar multiplications are related by

$$t \underset{A}{\cdot} (u \underset{B}{\cdot} d) = u \underset{B}{\cdot} (t \underset{A}{\cdot} d), \tag{6}$$

where $t, u \in \mathbb{R}$ and $d \in D$.

Lastly, for compatible $a, a' \in A$ and compatible $b, b' \in B$, and $t \in \mathbb{R}$,

$$\widetilde{0}_{a+a'}^A = \widetilde{0}_a^A \underset{B}{+} \widetilde{0}_{a'}^A, \qquad \widetilde{0}_{ta}^A = t \underset{B}{\cdot} \widetilde{0}_a^A, \tag{7}$$

and

$$\widetilde{0}_{b+b'}^B = \widetilde{0}_b^B \underset{A}{+} \widetilde{0}_{b'}^B, \qquad \widetilde{0}_{tb}^B = t \underset{A}{\cdot} \widetilde{0}_b^B. \tag{8}$$

Equations (3)–(8) are known as the *interchange laws*.

$$
\begin{array}{ccc}
\widetilde{0}_b^B \longmapsto b & \widetilde{0}_a^A \longmapsto 0_m^B & \odot_m \longmapsto 0_m^B \\
\downarrow \qquad \downarrow & \downarrow \qquad \downarrow & \downarrow \qquad \downarrow \\
0_m^A \longmapsto m & a \longmapsto m & 0_a^A \longmapsto m
\end{array}
$$

Fig. 9.3.

Definition 9.1.2 A *morphism of double vector bundles*

$$(\varphi; \varphi_A, \varphi_B; f) \colon (D; A, B; M) \to (D'; A', B'; M')$$

consists of maps $\varphi \colon D \to D'$, $\varphi_A \colon A \to A'$, $\varphi_B \colon B \to B'$, $f \colon M \to M'$, such that each of (φ, φ_B), (φ, φ_A), (φ_A, f) and (φ_B, f) is a morphism of the relevant vector bundles.

If $M = M'$ and $f = \mathrm{id}_M$, we say that φ is *over M*; if, further, $A = A'$ and $\varphi_A = \mathrm{id}_A$, we say that φ is *over A* or *preserves A*. If $A = A'$ and $B = B'$ and both φ_A and φ_B are identities, we say that φ *preserves the side bundles*. □

Example 9.1.3 For an ordinary vector bundle (E, q, M) the tangent double vector bundle structure on TE, as defined in §3.4, is a double vector bundle. We continue to denote the operations over TM by $+\!\!\!+$, $\cdot$ and $-$. There is no preferred arrangement for the side bundles of TE. ⊠

Example 9.1.4 Let A, B and C be any three vector bundles on the one base M, and write D for the pullback manifold $A \times_M B \times_M C$ over M. Then D may be regarded as the direct sum $q_A^! B \oplus q_A^! C$ over A, and as the direct sum $q_B^! A \oplus q_B^! C$ over B, and with respect to these two structures, D is a double vector bundle with side bundles A and B. We call this the *trivial double vector bundle over A and B with core C.* It is tempting, but incorrect, to denote it by $A \oplus B \oplus C$. ☒

Example 9.1.5 A double vector bundle $(D; A, B; M)$ may be pulled back over both of its side structures simultaneously. Suppose given vector bundles $(A', q_{A'}, M')$ and $(B', q_{B'}, M')$ and morphisms $\varphi \colon A' \to A$ and $\psi \colon B' \to B$, both over a map $f \colon M' \to M$. Let D' denote the set of all (a', d, b') such that $\varphi(a') = q_A^D(d)$, $\psi(b') = q_B^D(d)$ and $q_{A'}(a') = q_{B'}(b')$. Then, with the evident structures, $(D'; A', B'; M')$ is a double vector bundle and the projection $D' \to D$ is a morphism over φ, ψ and f. ☒

Further examples follow later in the chapter.

The core and the core sequences

Until 9.1.7, we regard as fixed the double vector bundle $(D; A, B; M)$. Each of the bundle projections is a morphism with respect to the other structure and so has a kernel (in the ordinary sense); denote by C the intersection of the two kernels:

$$C = \{c \in D \mid \exists m \in M \text{ such that } q_B^D(c) = 0_m^B,\ q_A^D(c) = 0_m^A\}.$$

This is an embedded submanifold of D. We will show that it has a well–defined vector bundle structure with base M, projection q_C which is the restriction of $q_B \circ q_B^D = q_A \circ q_A^D$ and addition and scalar multiplication which are the restrictions of either of the operations on D.

Note first that the two additions coincide on C since

$$c \underset{B}{+} c' = (c \underset{A}{+} \bigcirc m) \underset{B}{+} (\bigcirc m \underset{A}{+} c') = (c \underset{B}{+} \bigcirc m) \underset{A}{+} (\bigcirc m \underset{B}{+} c') = c \underset{A}{+} c',$$

for $c, c' \in C$ with $q_C(c) = q_C(c')$, using (3). From this it follows that $t \underset{B}{.} c = t \underset{A}{.} c$ for integers t, and consequently for rational t, and thence for all real t by continuity.

It will often be helpful to distinguish between $c \in C$, regarding C as a distinct vector bundle, and the image of c in D, which we will denote by $\bar{c}$. Given $c, c' \in C$ with $q_C(c) = q_C(c')$ there is a unique $c + c' \in C$ with

$$\overline{c + c'} = \bar{c} \underset{B}{+} \bar{c'} = \bar{c} \underset{A}{+} \bar{c'},$$

and given $t \in \mathbb{R}$ there is a unique $tc \in C$ such that

$$\overline{tc} = t \underset{B}{\,\cdot\,} \bar{c} = t \underset{A}{\,\cdot\,} \bar{c}.$$

It is now easy to prove that (C, q_C, M) is a (smooth) vector bundle, which we call the *core* of $(D; A, B; M)$.

Theorem 9.1.6 *There is an exact sequence*

$$q_A^! C \overset{\tau_A}{\rightarrowtail} \widetilde{D}_A \overset{(q_B^D)^!}{\twoheadrightarrow} q_A^! B \qquad (9)$$

of vector bundles over A, and an exact sequence

$$q_B^! C \overset{\tau_B}{\rightarrowtail} \widetilde{D}_B \overset{(q_A^D)^!}{\twoheadrightarrow} q_B^! A \qquad (10)$$

of vector bundles over B, where the injections are $\tau_A \colon (a, c) \mapsto \widetilde{0}_a^A \underset{B}{+} \bar{c}$ and $\tau_B \colon (b, c) \mapsto \widetilde{0}_b^B \underset{A}{+} \bar{c}$, respectively, and $(q_B^D)^!$ and $(q_A^D)^!$ denote the maps induced by q_B^D and q_A^D into the pullback bundles.

Proof Take $a \in A_m$, $c \in C_m$ where $m \in M$. Then both $\widetilde{0}_a^A$ and $\bar{c}$ project under q_B^D to 0_m^B. So $\widetilde{0}_a^A \underset{B}{+} \bar{c}$ is defined and also projects under q_B^D to 0_m^B. That τ_A is linear over A follows from the interchange laws.

Suppose that $d \in D$ has $q_B^D(d) = 0_m^B$ for some $m \in M$. Write $a = q_A^D(d)$. Then $d \underset{B}{-} \widetilde{0}_a^A$ is defined and $q_B^D(d \underset{B}{-} \widetilde{0}_a^A) = 0_m^B$. On the other hand, $q_A^D(d \underset{B}{-} \widetilde{0}_a^A) = a - a = 0_m^A$. So $d \underset{B}{-} \widetilde{0}_a^A \in C_m$. This establishes the exactness of (9). The proof of (10) is similar. $\qquad \square$

We refer to (9) as the *core sequence of D over A*, and to (10) as the *core sequence of D over B*. The two injections τ_A and τ_B are the *translation maps* over A and B, respectively.

As well as being linear as stated in Theorem 9.1.6, the translation maps each possess two further properties, which are easily proved:

$$\tau_A(a + a', c + c') = \tau_A(a, c) \underset{B}{+} \tau_A(a', c'), \qquad (11)$$

$$\tau_A(ta, tc) = t \underset{B}{\,\cdot\,} \tau_A(a, c), \qquad (12)$$

$$\tau_B(b + b', c + c') = \tau_B(b, c) \underset{A}{+} \tau_B(b', c'), \qquad (13)$$

$$\tau_B(tb, tc) = t \underset{A}{\,\cdot\,} \tau_B(b, c), \qquad (14)$$

for $c, c' \in C$, $a, a' \in A$ and $b, b' \in B$ suitably compatible, and $t \in \mathbb{R}$.

If $(\varphi; \varphi_A, \varphi_B; f) \colon (D; A, B; M) \to (D'; A', B'; M')$ is a morphism of double vector bundles, then $\varphi \colon D \to D'$ maps C into C', the core of D'. It is clear that the restriction, $\varphi_C \colon C \to C'$, is a morphism of the vector bundle structures on the cores, over $f \colon M \to M'$.

It will be useful to reformulate the exactness of the sequences in Theorem 9.1.6 in terms of sections. Denote the $C^\infty(A)$–module of sections of $\widetilde{D}_A$ by $\Gamma_A D$, and the $C^\infty(B)$–module of sections of $\widetilde{D}_B$ by $\Gamma_B D$.

Given $c \in \Gamma(C)$, define $c^A \in \Gamma_A D$ by

$$c^A(a) = \tau_A(a, c(q_A(a))), \qquad a \in A.$$

Then $c \mapsto c^A$, $\Gamma(C) \to \Gamma_A D$ is additive, and $(fc)^A = (f \circ q_A) \underset{A}{\cdot} c^A$ for $f \in C^\infty(M)$. Indeed the map $C^\infty(A) \otimes_{C^\infty(M)} \Gamma(C) \to \Gamma_A D$ induced by τ_A is $\sum F^i \otimes c_i \mapsto \sum F^i \underset{A}{\cdot} c_i^A$.

Note that c^A projects under q_B^D to the zero section of B. Conversely, if an element of $\Gamma_A D$ projects to the zero section of B, then it is of the form $\sum F^i \underset{A}{\cdot} c_i^A$ for $F^i \in C^\infty(A)$, $c_i \in \Gamma(C)$. We call c^A the *core section over A corresponding to c.*

Similarly, given $c \in \Gamma(C)$, define $c^B \in \Gamma_B D$ by

$$c^B(b) = \tau_B(b, c(q_B(b))), \qquad b \in B.$$

Then $c \mapsto c^B$, $\Gamma(C) \to \Gamma_B D$ is additive, and $(fc)^B = (f \circ q_B) \underset{B}{\cdot} c^B$ for $f \in C^\infty(M)$. Again, the map $C^\infty(B) \otimes_{C^\infty(M)} \Gamma(C) \to \Gamma_B D$ induced by τ_B is $\sum G^i \otimes c_i \mapsto \sum G^i \underset{B}{\cdot} c_i^H$.

Likewise, c^B projects under q_A^D to the zero section of A, and if an element of $\Gamma_B D$ projects to the zero section of A, then it is of the form $\sum G^i \underset{B}{\cdot} c_i^B$ for $G^i \in C^\infty(B)$, $c_i \in \Gamma(C)$.

Examples 9.1.7 For E an ordinary vector bundle, consider the tangent double vector bundle $(TE; E, TM; M)$. The kernel of $T(q)$ consists of the vertical tangent vectors and the kernel of p_E consists of the vectors tangent along the zero section; their intersection can be naturally identified with E itself as in §3.4. Given $\mu \in \Gamma E$, the core section over E is the vertical lift $\mu^\uparrow \in \mathfrak{X}(E)$ and the core section over TM is $\widehat{\mu} \in \Gamma_{TM}(TE)$.

For $\varphi \colon E \to E'$ a morphism of vector bundles over $f \colon M \to M'$, the morphism $T(\varphi)$ of the tangent double vector bundles induces $\varphi \colon E \to E'$ on the cores. In the case where φ and f are surjective submersions, the vertical subbundles form a double vector subbundle (in an obvious sense) $(T^\varphi E; E, T^f M; M)$ of TE, the core of which is the kernel (in the ordinary sense) of φ.

The trivial double vector bundle $A \times_M B \times_M C$ of 9.1.4 has core C.

$\boxtimes$

9.2 Duals of double vector bundles

Throughout this section we consider a double vector bundle as in (1), with core bundle C. We will show that dualizing either structure on D leads again to a double vector bundle; in the case of the dual of the structure over A we denote this by

$$
\begin{array}{ccc}
D \,\bar{*}\,A & \xrightarrow{\quad q_{C^*}^{\bar{*}A} \quad} & C^* \\[2mm]
q_A^{\bar{*}A} \downarrow & \square & \downarrow q_{C^*} \\[2mm]
A & \xrightarrow{\quad q_A \quad} & M,
\end{array}
\tag{15}
$$

Here C^* is the ordinary dual of C as a vector bundle over M. We denote the dual of D as a vector bundle over A by $D \,\bar{*}\,A$; this modification of the usual notation for duals is better adapted to iteration.

The vertical structure in (15) is the usual dual of the bundle structure on D with base A, and $q_{C^*} \colon C^* \to M$ is the usual dual of $q_C \colon C \to M$. The additions and scalar multiplications in the side bundles of (15) will be denoted by the usual plain symbols. The zero of $D \,\bar{*}\,A$ above $a \in A$ is denoted $\tilde{0}_a^{\bar{*}A}$.

The unfamiliar projection $q_{C^*}^{\bar{*}A} \colon (D \,\bar{*}\,A) \to C^*$ is defined by

$$
\langle q_{C^*}^{\bar{*}A}(\Phi), c \rangle = \langle \Phi, \tilde{0}_a^A \underset{B}{+} \bar{c} \rangle
\tag{16}
$$

where $c \in C_m$, $\Phi \colon (q_A^D)^{-1}(a) \to \mathbb{R}$ and $a \in A_m$. The addition $\underset{C^*}{+}$ in $D \,\bar{*}\,A \to C^*$ is defined by

$$
\langle \Phi \underset{C^*}{+} \Phi', d \underset{B}{+} d' \rangle = \langle \Phi, d \rangle + \langle \Phi', d' \rangle
\tag{17}
$$

Here $\Phi, \Phi' \in D \,\bar{*}\,A$ are given, say with the forms $(\Phi; a, \kappa; m)$ and $(\Phi'; a', \kappa; m)$. Any element with outline $(d''; a + a', b''; m)$ can be written as $d'' = d \underset{B}{+} d'$ where $(d; a, b''; m)$ is arbitrary of this form, and $d' = d'' \underset{B}{-} d$. It needs to be shown that (17) is independent of the choice of d.

Suppose that $d \underset{B}{+} d' = d_1 \underset{B}{+} d'_1$ for elements $(d_1; a, b''; m)$ and $(d'_1; a', b''; m)$. Then $d_1 = d \underset{A}{+} (\widetilde{0}^A_a \underset{B}{+} \overline{c})$ and, since Φ is linear with respect to $\underset{A}{+}$, this implies that

$$\langle \Phi, d_1 \rangle = \langle \Phi, d \rangle + \langle \Phi, \widetilde{0}^A_a \underset{B}{+} \overline{c} \rangle = \langle \Phi, d \rangle + \langle \kappa, c \rangle.$$

Since $d \underset{B}{+} d' = d_1 \underset{B}{+} d'_1$, we also have $d' = d'_1 \underset{A}{+} (\widetilde{0}^A_{a'} \underset{B}{+} \overline{c})$ for the same c, and it follows from $q^{\bar{*}A}_{C*}(\Phi) = q^{\bar{*}A}_{C*}(\Phi') = \kappa$ that the RHS of (17) is well–defined.

Similarly, define

$$\langle t \underset{C*}{\cdot} \Phi, t \underset{H}{\cdot} d \rangle = t \langle \Phi, d \rangle,$$

for $t \in \mathbb{R}$ and $d \in D$ with $q^D_A(d) = \widetilde{q}^{\bar{*}A}_A(\Phi)$.

The zero above $\kappa \in C^*_m$ is denoted $\widetilde{0}^{\bar{*}A}_\kappa$ and is defined by

$$\langle \widetilde{0}^{\bar{*}A}_\kappa, \widetilde{0}^B_b \underset{A}{+} \overline{c} \rangle = \langle \kappa, c \rangle \tag{18}$$

where $b \in B_m$, $c \in C_m$. The core element $\overline{\psi}$ corresponding to $\psi \in B^*_m$ is

$$\langle \overline{\psi}, \widetilde{0}^B_b \underset{A}{+} \overline{c} \rangle = \langle \psi, b \rangle.$$

It is straightforward to verify that (15) is a double vector bundle, and that its core is B^*. We call (15) the *vertical dual* or *dual over A* of (1).

As for any double vector bundle, there are exact sequences

$$q^!_A B^* \overset{\sigma_A}{\rightarrowtail} D \bar{*} A \overset{(q^{\bar{*}A}_{C*})^!}{\longrightarrow\!\!\!\!\!\rightarrow} q^!_A C^*, \tag{19}$$

of vector bundles over A and

$$q^!_{C*} B^* \overset{\sigma_{C*}}{\rightarrowtail} D \bar{*} A \overset{(q^{\bar{*}A}_A)^!}{\longrightarrow\!\!\!\!\!\rightarrow} q^!_{C*} A, \tag{20}$$

of vector bundles over C^*. Here the translation maps are given by

$$\sigma_A(a, \psi) = \widetilde{0}^{\bar{*}A}_a \underset{C*}{+} \overline{\psi}, \qquad \sigma_{C*}(\kappa, \psi) = \widetilde{0}^{\bar{*}A}_\kappa \underset{A}{+} \overline{\psi},$$

where $a \in A, \psi \in B^*, \kappa \in C^*$. It is easily seen that

$$\langle \sigma_A(a, \psi), d \rangle = \langle \psi, q^D_B(d) \rangle$$

for $d \in D$ and so σ_A is precisely the dual of $(q^D_B)^!$. It is clear from the definition of $q^{\bar{*}A}_{C*}$ that $(q^{\bar{*}A}_{C*})^! = \tau^*_A$. Thus (19) is precisely the dual of the core exact sequence (9).

For the sequence over C^* we have

$$\langle \sigma_{C^*}(\kappa, \psi), \widetilde{0}_b^B \underset{A}{+} \overline{c}\rangle = \langle \kappa, c\rangle + \langle \psi, b\rangle$$

for $\kappa \in C_m^*$, $\psi \in B_m^*$, $x \in B_m$, $c \in C_m$.

The proof of the following result is straightforward. In Figure 9.4 and in similar figures in future, we omit arrows that are the identity.

Proposition 9.2.1 *Consider a morphism of double vector bundles, as in Figure 9.4(a), which preserves the horizontal side bundles, and which*

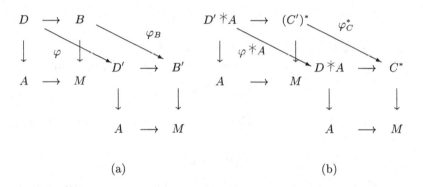

(a)   (b)

Fig. 9.4.

has core morphism $\varphi_C\colon C \to C'$, where C' is the core of D'. Dualizing φ as a morphism of vector bundles over A, we obtain a morphism $\varphi{*}A$ of double vector bundles over A and φ_C^*, as in Figure 9.4(b), with core morphism φ_B^*.

This completes the description of the vertical dual of (1). There is of course also a *horizontal dual*

$$
\begin{array}{ccc}
D\!\ast\!B & \xrightarrow{\;q_B^{\ast B}\;} & B \\[2pt]
{\scriptstyle q_{C^*}^{\ast B}}\Big\downarrow & & \Big\downarrow{\scriptstyle q_B} \\[6pt]
C^* & \xrightarrow[\;q_{C^*}\;]{} & M,
\end{array} \qquad (21)
$$

with core $A^* \to M$, defined in an analogous way. For reference, we provide the following formulas. The unfamiliar projection is given by

$$\langle q_{C^*}^{\bar{*}B}(\Psi), c \rangle = \langle \Psi, \tilde{0}_b^B \underset{B}{+} \bar{c} \rangle \tag{22}$$

where $c \in C_m$, $\Psi \colon (q_B^D)^{-1}(b) \to \mathbb{R}$ and $b \in B_m$. The addition $\underset{C^*}{+}$ and scalar multiplication in $D \bar{*} B \to C^*$ are defined by

$$\langle \Psi \underset{C^*}{+} \Psi', d \underset{A}{+} d' \rangle = \langle \Psi, d \rangle + \langle \Psi', d' \rangle \tag{23}$$

$$\langle t \underset{C^*}{.} \Psi, t \underset{A}{.} d \rangle = t \langle \Psi, d \rangle, \tag{24}$$

for suitable elements.

The zero above $\kappa \in C_m^*$ is denoted $\tilde{0}_\kappa^{\bar{*}B}$ and is defined by

$$\langle \tilde{0}_\kappa^{\bar{*}B}, \tilde{0}_a^A \underset{B}{+} \bar{c} \rangle = \langle \kappa, c \rangle \tag{25}$$

where $a \in A_m$, $c \in C_m$. The core element $\bar{\varphi}$ corresponding to $\varphi \in A_m^*$ is

$$\langle \bar{\varphi}, \tilde{0}_a^A \underset{B}{+} \bar{c} \rangle = \langle \varphi, a \rangle.$$

The following result is a phenomenon with no analogue in ordinary duality.

Theorem 9.2.2 *There is a natural (up to sign) pairing $\lfloor\, , \,\rfloor$ between the bundles $D \bar{*} A$ and $D \bar{*} B$ over C^* given by*

$$\lfloor \Phi, \Psi \rfloor = \langle \Phi, d \rangle_A - \langle \Psi, d \rangle_B \tag{26}$$

where $\Phi \in D \bar{*} A$, $\Psi \in D \bar{*} B$ have $q_{C^*}^{\bar{*}A}(\Phi) = q_{C^*}^{\bar{*}B}(\Psi)$ and d is any element of D with $q_A^D(d) = q_A^{\bar{*}A}(\Phi)$ and $q_B^D(d) = q_B^{\bar{*}B}(\Psi)$.

Each of the pairings on the RHS of (26) is a canonical pairing of an ordinary vector bundle with its dual, the subscripts there indicating the base over which the pairing takes place.

Proof Let Φ and Ψ have the forms $(\Phi; a, \kappa; m)$ and $(\Psi; \kappa, b; m)$. Then d must have the outline $(d; a, b; m)$. If d' also has outline $(d'; a, b; m)$ then there is a $c \in C_m$ such that $d = d' \underset{A}{+} (\tilde{0}_a^A \underset{B}{+} \bar{c})$, and so

$$\langle \Phi, d \rangle_A = \langle \Phi, d' \rangle_A + \langle \kappa, c \rangle$$

by (16). By the interchange law (3) we also have $d = d' \underset{B}{+} (\tilde{0}_b^B \underset{A}{+} \bar{c})$ and so

$$\langle \Psi, d \rangle_B = \langle \Psi, d' \rangle_B + \langle \kappa, c \rangle.$$

Thus (26) is well defined. To check that it is bilinear is routine. It remains to prove that it is non–degenerate.

Suppose Φ, given as above, is such that $\lceil \Phi, \Psi \rfloor = 0$ for all $\Psi \in (q_{C^*}^{*B})^{-1}(\kappa)$. Take any $\varphi \in A_m^*$ and consider $\Psi = \widetilde{0}_\kappa^{*B} \underset{B}{+} \overline{\varphi}$. Then, taking $d = \widetilde{0}_a^A$ we find $\langle \Phi, d \rangle_A = 0$ and $\langle \Psi, d \rangle_B = \langle \varphi, a \rangle$. Thus $\langle \varphi, a \rangle = 0$ for all $\varphi \in A_m^*$ and so $a = 0_m^A$. It therefore follows from the horizontal exact sequence for $D \,\widetilde{*}\, A$ that

$$\Phi = \widetilde{0}_\kappa^{*A} \underset{A}{+} \overline{\psi}$$

for some $\psi \in B_m^*$. Now taking any $c \in C_m$ and defining $d = \widetilde{0}_b^B \underset{A}{+} \overline{c}$, we find that

$$\langle \Phi, d \rangle_A = \langle \kappa, c \rangle + \langle \psi, b \rangle \qquad \text{and} \qquad \langle \Psi, d \rangle_B = \langle \kappa, c \rangle.$$

So $\langle \psi, b \rangle = 0$ for all $b \in B_m$, since a suitable Ψ exists for any given b. It follows that $\psi = 0 \in B_m^*$ and so Φ is indeed the zero element over κ. Thus the pairing (26) is non–degenerate. $\qquad\square$

Note several special cases:

$$\lceil \widetilde{0}_\kappa^{*A}, \widetilde{0}_\kappa^{*B} \rfloor = 0, \qquad \lceil \widetilde{0}_a^{*A}, \widetilde{0}_b^{*B} \rfloor = 0, \qquad (27)$$

$$\lceil \widetilde{0}_a^{*A}, \overline{\varphi} \rfloor = -\langle \varphi, a \rangle, \qquad \lceil \overline{\psi}, \widetilde{0}_b^{*B} \rfloor = \langle \psi, b \rangle, \qquad \lceil \overline{\psi}, \overline{\varphi} \rfloor = 0, \qquad (28)$$

$$\lceil \widetilde{0}_\kappa^{*A} \underset{A}{+} \overline{\psi}, \Psi \rfloor = \langle \psi, b \rangle, \qquad \lceil \Phi, \widetilde{0}_\kappa^{*B} \underset{B}{+} \overline{\varphi} \rfloor = -\langle \varphi, a \rangle, \qquad (29)$$

where $b \in B$, $a \in A$, $\varphi \in A^*$, $\psi \in B^*$ and we have $(\Psi; \kappa, b; m) \in D \,\widetilde{*}\, B$ and $(\Phi; a, \kappa; m) \in D \,\widetilde{*}\, A$.

Although we have proved that $D \,\widetilde{*}\, A$ and $D \,\widetilde{*}\, B$ are dual as vector bundles over C^*, we have not yet considered the relationships between the other structures present. This is taken care of by the following result, the proof of which is straightforward.

Proposition 9.2.3 *Let $(D; A, B; M)$ and $(E; A, W; M)$ be double vector bundles with a side bundle A in common, and with cores C and L respectively. Suppose given a non–degenerate pairing $\lceil \;,\; \rfloor$ of D over A with E over A, and two further non–degenerate pairings, both denoted $\langle \,,\, \rangle$, of B with L and of C with W, such that:*

(i) *for all $b \in B$, $\ell \in L$, $\lceil \widetilde{0}_b^B, \overline{\ell} \rfloor = \langle b, \ell \rangle$;*

(ii) *for all $c \in C$, $w \in W$, $\lceil \overline{c}, \widetilde{0}_w^W \rfloor = \langle c, w \rangle$;*

(iii) *for all $c \in C$, $\ell \in L$, $\lceil \overline{c}, \overline{\ell} \rfloor = 0$;*

(iv) *for all $d_1, d_2 \in D$, $e_1, e_2 \in E$ such that $q_B^D(d_1) = q_B^D(d_2)$,*
$q_W^E(e_1) = q_W^E(e_2)$, $q_A^D(d_1) = q_A^E(e_1)$, $q_A^D(d_2) = q_A^E(e_2)$, we have

$$\mathbf{I}\, d_1 \underset{B}{+} d_2, e_1 \underset{W}{+} e_2 \,\mathbf{I} = \mathbf{I}\, d_1, e_1 \,\mathbf{I} + \mathbf{I}\, d_2, e_2 \,\mathbf{I};$$

(v) *for all $d \in D$, $e \in E$ such that $q_A^D(d) = q_A^E(e)$ and all $t \in \mathbb{R}$, we have*

$$\mathbf{I}\, t \underset{B}{\,.\,} d, t \underset{W}{\,.\,} e \,\mathbf{I} = t \mathbf{I}\, d, e \,\mathbf{I}.$$

(In all the above conditions we assume the various elements of the side bundles lie in compatible fibres over M.)

Then the map $Z \colon D \to E\,\overline{*}\,A$ defined by $\langle Z(d), e \rangle_A = \mathbf{I}\, d, e\,\mathbf{I}$ *is an isomorphism of double vector bundles, with respect to* $\mathrm{id} \colon A \to A$ *and the isomorphisms $B \to L^*$ and $C \to W^*$ induced by the pairings in* (i) *and* (ii).

A pairing $\mathbf{I}\,,\,\mathbf{I}$ satisfying the conditions of 9.2.3 is called a *pairing of the double vector bundles.*

Applying this result to the pairing (26) of $D\,\overline{*}\,A$ and $D\,\overline{*}\,B$, we find that the induced pairing of B with B^* is the standard one, but that of A^* with A is the negative of the standard pairing. Hence the signs in the following result are unavoidable.

Corollary 9.2.4 *The pairing* (26) *induces isomorphisms of double vector bundles*

$$Z_A \colon D\,\overline{*}\,A \to D\,\overline{*}\,B\,\overline{*}\,C^*, \qquad \langle Z_A(\Phi), \Psi \rangle_{C^*} = \mathbf{I}\, \Phi, \Psi\,\mathbf{I}$$

$$Z_B \colon D\,\overline{*}\,B \to D\,\overline{*}\,A\,\overline{*}\,C^*, \qquad \langle Z_B(\Psi), \Phi \rangle_{C^*} = \mathbf{I}\, \Phi, \Psi\,\mathbf{I}$$

with $(Z_A)\,\overline{*}\,C^* = Z_B$. Both isomorphisms induce the identity on the sides $C^* \to C^*$.

Z_A is the identity on the cores $B^* \to B^*$, and induces $-\mathrm{id}$ on the side bundles $A \to A$.

Z_B is the identity on the side bundles $B \to B$, and induces $-\mathrm{id}$ on the cores $A^* \to A^*$.

Example 9.2.5 Consider a trivial double vector bundle

$$D = A \times_M B \times_M C.$$

Let $\Phi = (a, \psi, \kappa)$ be an element of $D\,\overline{*}\,A = A \times_M B^* \times_M C^*$ and let

$\Psi = (\varphi, b, \kappa)$ be an element of $D\,{}^{*}\!B$. Then taking any $d = (a, b, c) \in D$, we find that

$$\mathbf{I}\, \Phi, \Psi \,\mathbf{I} = \langle \psi, b \rangle - \langle \varphi, a \rangle.$$

The associated maps are given by

$$Z_A \colon A \times_M B^* \times_M C^* \to A \times_M B^* \times_M C^*, \quad (a, \psi, \kappa) \mapsto (-a, \psi, \kappa);$$
$$Z_B \colon A^* \times_M B \times_M C^* \to A^* \times_M B \times_M C^*, \quad (\varphi, b, \kappa) \mapsto (-\varphi, b, \kappa).$$

The following result is essentially equivalent to Theorem 9.2.2, but deserves independent statement.

Theorem 9.2.6 *For any double vector bundle* $(D; A, B; M)$ *there is a canonical isomorphism* Q *from* D *to the flip of* $(D\,{}^{*}\!A\,{}^{*}\!C^*\,{}^{*}\!B)$ *which preserves the side bundles* A *and* B *and is* $-\mathrm{id}$ *on the cores* C.

Proof Let $P = Z_A\,{}^{*}\!A$ be the dualization of Z_A over A. Denote by $F \colon D \to D$ the map $d \mapsto \underset{B}{-}\, d$, and define $Q = (F \circ P)^{-1}$. $\qquad\square$

Thus the process of taking duals of double vector bundles does not return to a structure naturally isomorphic to the starting structure after two steps, but requires three steps (or six if the identification of the cores is required to be consistent).

9.3 The prolongation dual

Throughout this section we consider a fixed vector bundle (A, q, M). We saw in §3.4 that the canonical pairing of A with A^* induces a canonical pairing $\langle\!\langle\ ,\ \rangle\!\rangle$ of TA with $T(A^*)$ over TM. We now need to establish that this is a pairing of the double vector bundles.

Proposition 9.3.1 *The tangent pairing* $\langle\!\langle\ ,\ \rangle\!\rangle$ *of* $T(A^*)$ *with* TA *over* TM *satisfies the conditions of Proposition 9.2.3. In particular, for* $m \in M$ *and* $\varphi_1, \varphi_2 \in A_m^*$, $a_1, a_2 \in A_m$,

$$\langle\!\langle \overline{\varphi}, \overline{a} \rangle\!\rangle = 0, \qquad \langle\!\langle \widetilde{0}_\varphi, \widetilde{0}_a \rangle\!\rangle = 0,$$
$$\langle\!\langle \widetilde{0}_\varphi, \overline{a} \rangle\!\rangle = \langle \varphi, a \rangle, \qquad \langle\!\langle \overline{\varphi}, \widetilde{0}_a \rangle\!\rangle = \langle \varphi, a \rangle.$$

and

$$\langle\!\langle \tau_*(\varphi_1, \varphi_2), \tau(a_1, a_2) \rangle\!\rangle = \langle \varphi_1, a_2 \rangle + \langle \varphi_2, a_1 \rangle,$$

where τ_* *and* τ *are the translation maps of* $T(A^*)$ *and* $T(A)$.

Proof These are easily verified from the definition. For example, $\overline{\varphi} = \frac{d}{dt}(t\varphi)\big|_0$ and $\overline{a} = \frac{d}{dt}(ta)\big|_0$ so

$$\langle\!\langle \overline{\varphi}, \overline{a} \rangle\!\rangle = \frac{d}{dt} t^2 \langle \varphi, a \rangle \bigg|_0 = 0$$

whereas $\widetilde{0}_\varphi = \frac{d}{dt}\varphi\big|_0$ so

$$\langle\!\langle \widetilde{0}_\varphi, \overline{a} \rangle\!\rangle = \frac{d}{dt} t \langle \varphi, a \rangle \bigg|_0 = \langle \varphi, a \rangle.$$

The bilinearity conditions are easily verified and the final equation follows. $\qquad\square$

Thus the pairing of the cores of $T(A^*)$ and $T(A)$ is the zero pairing, and so too is the pairing of the zero sections above A^* and A. However the core of $T(A^*)$ and the zero section of $T(A)$ are paired under the standard pairing, and the same is true of the zero section of $T(A^*)$ and the core of $T(A)$.

It now follows that there is an isomorphism of double vector bundles from $T(A^*)$ to the dual $TA \,\overline{\star}\, TM$ of TA over TM. For convenience we denote this simply by $T^\bullet A$ and call it the *prolongation dual of* TA. The next result follows from the general theory of §9.2.

Proposition 9.3.2 *The map* $I\colon T(A^*) \to T^\bullet(A)$ *defined by*

$$\langle I(\mathscr{X}), \eta \rangle_{TM} = \langle\!\langle \mathscr{X}, \eta \rangle\!\rangle$$

where $\mathscr{X} \in T(A^*)$, $\eta \in TA$, *is an isomorphism of double vector bundles preserving the side bndles* A^* *and* TM *and the core bundles* A^*.

When a name is needed we call I the *internalization map*. In future we will almost always work with $T(A^*)$ and the tangent pairing, rather than with $T^\bullet A$ and I. There are, however, a few occasions on which it is useful to work directly with $T^\bullet A$ and so we summarize its properties and notation here. In such cases we use the notation of Figure 9.5.

The fibre of $T(q)_\bullet$ over $x \in TM$ is denoted $T^\bullet_x(A)$. An element $\mathfrak{f} \in T^\bullet_x(A)$ is now a linear map $T(q)^{-1}(x) \to \mathbb{R}$, and addition and scalar multiplication in this bundle is the standard addition and multiplication of linear maps. The zero element in $T^\bullet_x(A)$ is denoted $\widetilde{0}_x$.

The map $r_\bullet\colon T^\bullet A \to A^*$ is defined as follows. Take $\mathfrak{f}\colon T(q)^{-1}(x) \to \mathbb{R}$, where $x \in T_m(M)$, and define

$$r_\bullet(\mathfrak{f})(X) = \mathfrak{f}(\upsilon(x, X)) = \mathfrak{f}(T(0)(x) + \overline{X}), \qquad X \in A_m.$$

$$
\begin{array}{ccc}
T^{\bullet}A & \xrightarrow{\;T(q)_{\bullet}\;} & TM \\[4pt]
{\scriptstyle r_{\bullet}}\big\downarrow & & \big\downarrow{\scriptstyle p} \\[6pt]
A^{*} & \xrightarrow[\;q_{*}\;]{} & M,
\end{array}
$$

Fig. 9.5.

Given $\mathfrak{f} \in T^{\bullet}_{x}(A)$, $\mathfrak{s} \in T^{\bullet}_{y}(A)$ with $r_{\bullet}(\mathfrak{f}) = r_{\bullet}(\mathfrak{s}) = \varphi \in A^{*}_{m}$, define $\mathfrak{f} \underset{A^{*}}{+} \mathfrak{s} \in T^{\bullet}_{x+y}(A)$ by

$$(\mathfrak{f} \underset{A^{*}}{+} \mathfrak{s})(\xi + \eta) = \mathfrak{f}(\xi) + \mathfrak{s}(\eta),$$

where $\xi, \eta \in T(A)$ have $T(q)(\xi) = x$, $T(q)(\eta) = y$ and $p_{A}(d) = p_{A}(\eta)$.

Similarly, scalar multiplication of $\mathfrak{f}$ as above by $t \in \mathbb{R}$ is given by

$$(t \underset{A^{*}}{\cdot} \mathfrak{f})(t\xi) = t\mathfrak{f}(\xi), \qquad \xi \in T(q)^{-1}(x).$$

The zero element of $r_{\bullet}^{-1}(\varphi)$, where $\varphi \in A^{*}_{m}$, is $\widetilde{0}^{\bullet}_{\varphi} \colon T(q)^{-1}(0^{T}_{m}) \to \mathbb{R}$, defined by

$$\widetilde{0}^{\bullet}_{\varphi}(\widetilde{0}_{X} \underset{A^{*}}{+} \overline{Y}) = \varphi(Y), \qquad X, Y \in A_{m}.$$

The core of $T^{\bullet}A$ can be canonically identified with A^{*}, with $\varphi \in A^{*}_{m}$ corresponding to $\overline{\varphi} \in T^{\bullet}A$ where

$$\overline{\varphi}(\widetilde{0}_{X} \underset{A^{*}}{+} \overline{Y}) = \varphi(X), \qquad X, Y \in A_{m}.$$

The core sequence for $T^{\bullet}A$ over TM is now

$$
p^{!}A^{*} \xrightarrow{\;(p^{!}_{A})^{*}\;} T^{\bullet}A \xrightarrow{\;r^{!}_{\bullet}=v^{*}\;} p^{!}A^{*}, \tag{30}
$$

and this is the dual of the core sequence for TA over TM (see (20) on p. 112). The core sequence over A^{*} is

$$
q^{!}_{*}A^{*} \xrightarrow{\;\chi\;} T^{\bullet}A \xrightarrow{\;T(q)^{!}_{\bullet}\;} q^{!}_{*}TM, \tag{31}
$$

where $\chi(\varphi, \psi) = \widetilde{0}^{\bullet}_{\varphi} \underset{A^{*}}{+} \overline{\psi}$ and $(\widetilde{0}^{\bullet}_{\varphi} \underset{A^{*}}{+} \overline{\psi})(\widetilde{0}_{X} \underset{A^{*}}{+} \overline{Y}) = \varphi(Y) + \psi(X)$ for $\varphi, \psi \in A^{*}_{m}$, $X, Y \in A_{m}$.

If A is the trivial bundle $M \times V$, we can write $T^{\bullet}A \to TM$ as $TM \times V^{*} \times V^{*} \to TM$ where $\mathfrak{f} = (x, \varphi, \varphi')$ acts on $d = (x, X, X') \in$

$TM \times V \times V$ by

$$\langle (x, \varphi, \varphi'), (x, X, X') \rangle = \varphi(X) + \varphi'(X').$$

The operations in this bundle are accordingly similar to those given in equations (22) on p. 113 for $TM \times V \times V \to TM$, namely

$$T(q)_\bullet : (x, \varphi, \varphi') \mapsto x,$$
$$(x, \varphi, \varphi') + (x, \psi, \psi') = (x, \varphi + \varphi', \psi + \psi'),$$
$$t(x, \varphi, \varphi') = (x, t\varphi, t\varphi'),$$
$$\widetilde{0}_x = (x, 0, 0),$$
$$-(x, \varphi, \varphi') = (x, -\varphi, -\varphi'),$$

where $x \in TM$, $\varphi, \varphi', \psi, \psi' \in V$, $t \in \mathbb{R}$.

Likewise a straightforward calculation shows that the operations in $(TM \times V^* \times V^*, r_\bullet, M \times V^*)$ are given by

$$r_\bullet : (x, \varphi, \varphi') \mapsto (px, \varphi'),$$
$$(x, \varphi, \varphi') \underset{A^*}{+} (y, \psi, \varphi') = (x + y, \varphi + \psi, \varphi'),$$
$$t \underset{A^*}{\cdot} (x, \varphi, \varphi') = (tx, t\varphi, \varphi'),$$
$$\widetilde{0}^\bullet_{(m,\varphi)} = (0^T_m, 0, \varphi),$$
$$\underset{A^*}{-} (x, \varphi, \varphi') = (-x, -\varphi, \varphi'),$$

where $m \in M$, $\varphi, \varphi', \psi \in V^*$ and $x, y \in TM$ have $p(x) = p(y)$.

Given $(m, \varphi) \in M \times V^*$, the corresponding core element is

$$\overline{(m, \varphi)} = (0^T_m, \varphi, 0).$$

9.4 The cotangent double vector bundle

Throughout this section we consider a single vector bundle (A, q, M). Since the core of the double vector bundle TA is A, dualizing the structure over A leads to a double vector bundle of the form

$$
\begin{array}{ccc}
& r_A & \\
T^*A & \longrightarrow & A^* \\
c_A \downarrow & & \downarrow q_{A^*} \\
A & \longrightarrow & M; \\
& q_A &
\end{array}
\tag{32}
$$

We refer to this as the *cotangent dual* of TA. We will give a detailed description of the structures involved. Although this is a special case of the general construction, this example is so basic to the remaining chapters of the book that it merits a specific treatment.

In (32) the vertical bundle (T^*A, c_A, A) is the standard cotangent bundle of A, and the notation $T_X^*(A)$ will always refer to the fibre with respect to c_A. In this bundle we use standard notation, and denote the zero element of $T_X^*(A)$ by $\widetilde{0}_X^*$. We drop the subscripts A from the maps when no confusion is likely.

The map $r \colon T^*A \to A^*$ sends $\Phi \in T_X^*(A)$, where $X \in A_m$, to $r(\Phi) \in A_m^*$ where

$$\langle r(\Phi), Y \rangle = \langle \Phi, \widetilde{0}_X \mathbin{+\!\!+} \overline{Y} \rangle = \langle \Phi, Y^{\uparrow}(X) \rangle \tag{33}$$

for $Y \in A_m$. Given $\Phi \in T_X^*(A)$, $\Psi \in T_Y^*(A)$ with $r(\Phi) = r(\Psi) \in A_m^*$, the sum $\Phi \mathbin{\underset{A^*}{+}} \Psi \in T_{X+Y}^*(A)$ has

$$\langle \Phi \mathbin{\underset{A^*}{+}} \Psi, \xi \mathbin{+\!\!+} \eta \rangle = \langle \Phi, \xi \rangle + \langle \Psi, \eta \rangle,$$

where $\xi \in T_X(A)$, $\eta \in T_Y(A)$, and $T(q)(\xi) = T(q)(\eta)$. Similarly,

$$\langle t \mathbin{\underset{A^*}{\cdot}} \Phi, t \cdot \xi \rangle = t \langle \Phi, \xi \rangle,$$

for $t \in \mathbb{R}$ and $\xi \in T_X(A)$. The zero element of $r^{-1}(\varphi)$, where $\varphi \in A_m^*$, is $\widetilde{0}_\varphi^r \in T_{0_m}^*(A)$ where

$$\langle \widetilde{0}_\varphi^r, T(0)(x) + \overline{X} \rangle = \langle \varphi, X \rangle$$

for $x \in T_m(M), X \in A_m$.

We know from §9.2 that (32) is a double vector bundle, and that its core is T^*M. Given $\omega \in T_m^*(M)$, the corresponding core element $\overline{\omega}$ has

$$\langle \overline{\omega}, T(0)(x) + \overline{X} \rangle = \langle \omega, x \rangle,$$

for $x \in T_m(M)$, $X \in A_m$. The translation map over A,

$$q^! T^*M \to T^*A, \quad (X, \omega) \mapsto \widetilde{0}_X^* \mathbin{\underset{A^*}{+}} \overline{\omega},$$

is precisely the dual of $T(q)^!$; that is to say, it is the map corresponding to the lifting of 1–forms from M to A. Thus $\widetilde{0}_X^* \mathbin{\underset{A^*}{+}} \overline{\omega}$ is the pullback of $\omega \in T_m^*(M)$ to A at the point $X \in A_m$, and for $\omega \in \Omega^1(M)$, the corresponding core section of $T^*A \to A$ is $q^*\omega \in \Omega^1(A)$.

The core exact sequence for c is

$$q^! T^*M \rightarrowtail T^*A \xrightarrow{\; r^! = \tau^* \;} q^! A^*, \tag{34}$$

and this is the dual of the core exact sequence (see (19) on p. 112) for TA and p_A. The other core exact sequence is

$$q_*^! T^* M \rightarrowtail T^* A \xrightarrow{c^!} q_*^! A, \qquad (35)$$

where each bundle here is over A^*. The translation map $q_*^! T^* M \to T^* A$ is $(\varphi, \omega) \mapsto \widetilde{0}_\varphi^r + \overline{\omega}$ and $\langle \widetilde{0}_\varphi^r + \overline{\omega}, T(0)(x) + \overline{X} \rangle = \langle \varphi, X \rangle + \langle \omega, x \rangle$. For $\omega \in \Omega^1(M)$, we denote the corresponding core section of $T^* A \to A^*$ by $\check{\omega}$.

For A a trivial bundle $M \times V$, the cotangent bundle $T^* A \to A$ is $T^* M \times V \times V^* \to M \times V$ where $\Phi = (\omega, X, \varphi)$ acts on $d = (x, X, X') \in TM \times V \times V$ by

$$\langle (\omega, X, \varphi), (x, X, X') \rangle = \omega(x) + \varphi(X').$$

The operations in this bundle are now given by

$$c_{M \times V} \colon (\omega, X, \varphi) \mapsto (c\omega, X),$$
$$(\omega, X, \varphi) + (\chi, X, \psi) = (\omega + \chi, X, \varphi + \psi),$$
$$t(\omega, X, \varphi) = (t\omega, X, t\varphi),$$
$$\widetilde{0}_{(m,X)}^* = (0_m^T, X, 0),$$
$$-(\omega, X, \varphi) = (-\omega, X, -\varphi),$$

where $m \in M$, $X \in V$, $\varphi, \psi \in V^*$, $t \in \mathbb{R}$ and $\omega, \chi \in T^* M$ have $c(\omega) = c(\chi)$.

Likewise a straightforward calculation shows that the operations in $(T^* M \times V \times V^*, r, M \times V^*)$ are given by

$$r \colon (\omega, X, \varphi) \mapsto (c\omega, \varphi),$$
$$(\omega, X, \varphi) \underset{A^*}{+} (\chi, Y, \varphi) = (\omega + \chi, X + Y, \varphi),$$
$$t \underset{A^*}{\cdot} (\omega, X, \varphi) = (t\omega, tX, \varphi),$$
$$\widetilde{0}_{(m,\varphi)}^r = (0_m^T, 0, \varphi),$$
$$\underset{A^*}{-} (\omega, X, \varphi) = (-\omega, -X, \varphi),$$

where $m \in M$, $X, Y \in V$, $\varphi \in V^*$, $t \in \mathbb{R}$ and $\omega, \chi \in T^* M$ have $c(\omega) = c(\chi)$.

Given $\omega \in T^* M$, the corresponding core element is

$$\overline{\omega} = (\omega, 0, 0).$$

Elements of $T^*(A)$ may also be described in terms of sections of A^*. Given $\varphi \in \Gamma(A^*)$, define $\ell_\varphi \colon A \to \mathbb{R}$ to be the fibrewise–linear function $X \mapsto \langle \varphi(qX), X \rangle$. Then for $X \in A$, we have $\delta \ell_\varphi(X) \in T_X^*(A)$. It is

$$\begin{array}{ccccccc} D\ast A = & T^*A & \longrightarrow & A^* & \qquad D\ast TM = & T^\bullet A & \longrightarrow & TM \\ & \downarrow & & \downarrow & & \downarrow & & \downarrow \\ & A & \longrightarrow & M & & A^* & \longrightarrow & M \end{array}$$

Fig. 9.6.

easily verified that $r(\delta\ell_\varphi(X)) = \varphi_m$, for $X \in A_m$, and so we in fact have

$$\begin{array}{ccc} \delta\ell_\varphi(X) & \longmapsto & \varphi_m \\ \bar{\downarrow} & & \bar{\downarrow} \\ X & \longmapsto & m. \end{array}$$

Note also that $\delta\ell_\varphi(0_m) = \widetilde{0}^r_{\varphi_m}$. This makes clear, in particular, that not all elements of $T^*(A)$ are of the form $\delta\ell_\varphi(X)$. The proof of the following is immediate.

Proposition 9.4.1 *For $(\Phi; X, \varphi(m); m) \in T^*(A)$ and any $\varphi \in \Gamma(A^*)$ which takes the value $\varphi(m)$, there exists $\omega \in \Omega^1(M)$ such that*

$$\Phi = \delta\ell_\varphi(X) + (q^*\omega)(X) = \delta\ell_\varphi(X) \underset{A^*}{+} \breve{\omega}(\varphi(m)).$$

9.5 The reversal isomorphism

Again we consider an ordinary vector bundle (A, q, M). Applying Theorem 9.2.2 to $D = TA$ we get the two dual double vector bundles shown in Figure 9.6 and the pairing

$$[\![\Phi, \mathfrak{f}]\!] = \langle \Phi, \xi \rangle_A - \langle \mathfrak{f}, \xi \rangle_{TM} \tag{36}$$

for suitable $\xi \in TA$. Composing the isomorphism Z_A from 9.2.4 with the dual over A^* of the isomorphism I from §9.3, we get an isomorphism of double vector bundles

$$(I \ast A^*) \circ Z_A \colon T^*A \to T^*(A^*);$$

denote this temporarily by S^{-1}. For $\Phi \in T^*A$ we have

$$\langle S^{-1}(\Phi), \mathscr{X} \rangle_{A^*} = \langle (I \stackrel{*}{} A^*) \circ Z_A(\Phi), \mathscr{X} \rangle_{A^*} = \langle Z_A(\Phi), I(\mathscr{X}) \rangle_{A^*}$$
$$= \mathbf{|}\, \Phi, I(\mathscr{X}) \,\mathbf{|} = \langle \Phi, \xi \rangle_A - \langle I(\mathscr{X}), \xi \rangle_{TM} = \langle \Phi, \xi \rangle_A - \langle\!\langle \mathscr{X}, \xi \rangle\!\rangle.$$

Here we used the definition of Z_A, the definition (36), and the definition of I. It follows that for $\mathfrak{F} \in T^*(A^*)$, writing $\Phi = S(\mathfrak{F})$, we have

$$\langle \mathfrak{F}, \mathscr{X} \rangle_{A^*} = \langle S(\mathfrak{F}), \xi \rangle_A - \langle\!\langle \mathscr{X}, \xi \rangle\!\rangle.$$

Recall that I, and hence its dual, preserves both sides and the core, whereas Z_A induces $-\mathrm{id}$ on the sides A. We therefore define

$$R \colon T^*(A^*) \to T^*(A), \qquad R(\mathfrak{F}) = S(\underset{A^*}{-}\,\mathfrak{F}).$$

To summarize:

Theorem 9.5.1 *The map R just defined is an isomorphism of double vector bundles, preserving the side bundles A and A^*, and inducing $-\mathrm{id} \colon T^*M \to T^*M$ on the cores. Further, for all $\xi \in TA$, $\mathscr{X} \in T(A^*)$, $\mathfrak{F} \in T^*(A^*)$ such that ξ and $\mathscr{X}$ have the same projection into TM, $\mathscr{X}$ and $\mathfrak{F}$ have the same projection into A^*, and $\mathfrak{F}$ and ξ have the same projection into A,*

$$\langle\!\langle \mathscr{X}, \xi \rangle\!\rangle = \langle R(\mathfrak{F}), \xi \rangle_A + \langle \mathfrak{F}, \mathscr{X} \rangle_{A^*} \tag{37}$$

We call R the *reversal isomorphism*.

We now wish to prove that R is an antisymplectomorphism with respect to the canonical symplectic structures on T^*A and T^*A^*. This is an immediate consequence of the following.

Theorem 9.5.2 *Let ν_A and ν_{A^*} be the canonical 1–forms on T^*A and T^*A^* respectively. Then*

$$R^*(\nu_A) + \nu_{A^*} = \delta P$$

where $P \colon T^*A^* \to \mathbb{R}$ is the map $\mathfrak{F} \mapsto \langle \varphi, X \rangle$ for $(\mathfrak{F}; \varphi, X; m) \in T^*A^*$.

Proof Consider any element $\mathfrak{X} \in T(T^*A^*)$. Write $\mathfrak{X} = \frac{d}{dt}\mathfrak{F}_t\big|_0$, where $\mathfrak{F}_t$ has the form $(\mathfrak{F}_t; \varphi_t, X_t; m_t)$. Thus

$$p_{T^*A^*}(\mathfrak{X}) = \mathfrak{F}_0, \qquad T(c_{A^*})(\mathfrak{X}) = \frac{d}{dt}\varphi_t\bigg|_0, \qquad T(r_{A^*})(\mathfrak{X}) = \frac{d}{dt}X_t\bigg|_0,$$

where c_{A^*} is the projection of the cotangent bundle of A^*. Evaluating ν_{A^*} on $\mathfrak{X}$, we get

$$\langle \nu_{A^*}, \mathfrak{X} \rangle = \langle \mathfrak{F}_0, \mathscr{X} \rangle \qquad \text{where} \qquad \mathscr{X} = \left. \frac{d}{dt} \varphi_t \right|_0 .$$

On the other hand, considering $T(R)(\mathfrak{X})$, we have

$$p_{T^*A}(T(R)(\mathfrak{X})) = R(\mathfrak{F}_0), \qquad T(r_A)(T(R)(\mathfrak{X})) = \left. \frac{d}{dt} \varphi_t \right|_0 ,$$

$$T(c_A)(T(R)(\mathfrak{X})) = \left. \frac{d}{dt} X_t \right|_0 ,$$

and hence

$$\langle \nu_A, T(R)(\mathfrak{X}) \rangle = \langle \mathfrak{F}_0, \xi \rangle \qquad \text{where} \qquad \xi = \left. \frac{d}{dt} X_t \right|_0 .$$

Lastly,

$$\langle \delta P, \mathfrak{X} \rangle = \left. \frac{d}{dt} P(\mathfrak{F}_t) \right|_0 = \left. \frac{d}{dt} \langle \varphi_t, X_t \rangle \right|_0 = \langle\!\langle \mathscr{X}, \xi \rangle\!\rangle ,$$

and so, applying (37), the result follows. $\square$

The final result of this section, which we will need in a later chapter, is not so immediate a corollary of (37) as one might expect.

Proposition 9.5.3 *For $X \in \Gamma A$ and $\varphi \in \Gamma A^*$,*

$$R(\delta \ell_X(\varphi(m))) = \delta \ell_\varphi(X(m)) - q^*(\delta\langle \varphi, X \rangle)(X(m)).$$

Proof Apply (37) with $\mathfrak{F} = \delta \ell_X(\varphi(m)) \in T^*A^*$ and any compatible $\xi \in TA$. This gives

$$\langle R(\delta \ell_X(\varphi(m))), \xi \rangle_A = \langle\!\langle \mathscr{X}, \xi \rangle\!\rangle - \langle \mathfrak{F}, \mathscr{X} \rangle_{A^*},$$

for any compatible $\mathscr{X} \in T(A^*)$. Now the compatibility conditions which ξ and $\mathscr{X}$ are obliged to satisfy enable us to apply 3.4.6, giving

$$\langle\!\langle \mathscr{X}, \xi \rangle\!\rangle = \mathscr{X}(\ell_X) + \xi(\ell_\varphi) - x(\langle \varphi, X \rangle)$$

where $x = T(q)(\xi) = T(q_*)(\mathscr{X})$. Now $\mathscr{X}(\ell_X) = \langle \mathfrak{F}, \mathscr{X} \rangle_{A^*}$, so these terms cancel and we have

$$\langle R(\delta \ell_X(\varphi(m))), \xi \rangle_A = \langle \delta \ell_\varphi(X(m)), \xi \rangle - \langle \delta s, T(q)(\xi) \rangle,$$

where $s = \langle \varphi, X \rangle \colon M \to \mathbb{R}$. Since this is true for all ξ, the result follows.
                                                                        $\square$

9.6 The double tangent bundle and its duals

The double tangent bundle of a manifold M is the double vector bundle obtained by taking the tangent double vector bundle of $A = TM$. We consider elements $(\xi; X, x; m) \in T^2M$ as second derivatives

$$\xi = \frac{\partial^2 m}{\partial t \, \partial u}(0,0)$$

where $m \colon \mathbb{R}^2 \to M$ is a smooth square of elements of M, and the notation means that m is first differentiated with respect to u, yielding a curve $X_t = \frac{\partial m}{\partial u}(t,0)$ in TM with $\left. \frac{d}{dt} X_t \right|_0 = \xi$. (In this context the notation $m \colon \mathbb{R}^2 \to M$ will be taken to mean a smooth map defined in a connected neighbourhood of the origin.) Thus

$$\frac{\partial m}{\partial u}(0,0) = p_{TM}(\xi), \qquad \frac{\partial m}{\partial t}(0,0) = T(p_M)(\xi).$$

Now define a map $J \colon T^2M \to T^2M$, called the *canonical involution on* T^2M, for ξ as above, by

$$J(\xi) = \frac{\partial^2 m}{\partial u \, \partial t}(0,0).$$

The proof of the following result is left to the reader.

Theorem 9.6.1 *The canonical involution* $J \colon T^2M \to T^2M$ *is an isomorphism of double vector bundles, preserving the sides and the core, from* T^2M *to its flip, as shown in Figure 9.7.*

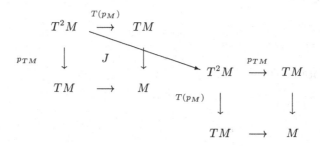

Fig. 9.7.

If we dualize this vertically (say) in terms of the arrangement shown in Figure 9.7, we obtain a morphism of vector bundles over TM,

$$J^*: T^\bullet(TM) \to T^*(TM).$$

Here there is no clarity to be gained by writing J^*TM or J^*V. Dualizing in the other direction produces the inverse.

Definition 9.6.2 The *Tulczyjew isomorphism* is the vector bundle morphism

$$\Theta: T(T^*M) \to T^*(TM), \qquad \Theta = J^* \circ I.$$

$\square$

The properties of I and J immediately yield the following:

Proposition 9.6.3 *The morphism Θ is an isomorphism of double vector bundles*

$$\Theta: (T(T^*M); TM, T^*M; M) \to (T^*(TM); TM, T^*M, M),$$

where the domain is the tangent double vector bundle of T^*M and the target is the cotangent dual of T^2M. The map induced on the cores is the identity.

Proposition 9.6.4 *For $\mathfrak{X} \in T(T^*M)$ and $\xi \in T^2M$ with $T(c)(\mathfrak{X}) = T(p)(\xi)$, the tangent pairing is given by*

$$\langle\!\langle \mathfrak{X}, \xi \rangle\!\rangle = \langle \Theta(\mathfrak{X}), J(\xi) \rangle,$$

where $\langle\, , \rangle$ is the standard pairing between the cotangent and tangent bundles of TM.

Proof Let $\langle\, , \rangle_{TM}$ temporarily denote the standard pairing over TM between $T^\bullet(TM)$ and T^2M. Then

$$\langle \Theta(\mathfrak{X}), J(\xi) \rangle = \langle I(\mathfrak{X}), J^2(\xi) \rangle_{TM} = \langle I(\mathfrak{X}), \xi \rangle_{TM} = \langle\!\langle \mathfrak{X}, \xi \rangle\!\rangle,$$

by the definition of I. $\square$

The next lemma will be needed later in the section.

Lemma 9.6.5 *Take $\xi \in T^2M$ with $p_{TM}(\xi) = X$ and $T(p)(\xi) = x$. Then for any 1–form φ on M,*

$$J(\xi)(\ell_\varphi) = \xi(\ell_\varphi) + \delta\varphi(X, x). \tag{38}$$

Proof Suppose that $\varphi = \delta f$ for $f \in C^\infty(M)$, so that ℓ_φ is the function $TM \to \mathbb{R}$, $X \mapsto X(f)$. In this case (38) is the identity of second partials. Now it suffices to verify that if (38) holds for φ then it holds for $g\varphi$ for any $g \in C^\infty(M)$, and this is straightforward. □

Consider a vector field X on M. Applying the tangent functor yields a section $T(X)$ of $T(p)$, and composing this with the canonical involution gives a vector field $J \circ T(X)$ on TM.

Proposition 9.6.6 *This vector field $J \circ T(X)$ is $\tilde{X}$, the complete lift vector field of 3.4.8. If $\{\varphi_t\}$ is a (local) flow for X, then $\{T(\varphi_t)\}$ is a (local) flow for $\tilde{X}$.*

Proof It is immediate that $\{T(\varphi_t)\}$ is a flow for some vector field on TM. Take any $x \in TM$ and let $c(u)$ be a curve with $\dfrac{d}{dt}c(u)\Big|_0 = x$. Consider the map $\mathbb{R}^2 \to M$ by $(t, u) \mapsto \varphi_t(c(u))$. Then

$$\frac{d}{dt}T(\varphi_t)(x)\Big|_0 = \frac{\partial^2}{\partial t\, \partial u}\varphi_t(c(u))\Big|_{(0,0)}$$

whereas

$$\frac{\partial^2}{\partial u\, \partial t}\varphi_t(c(u))\Big|_{(0,0)} = \frac{d}{du}X(c(u))\Big|_0 = T(X)(x).$$

Thus the vector field for which $\{T(\varphi_t)\}$ is the flow is $J \circ T(X)$.

Now a direct calculation with this flow shows that $J \circ T(X)$ acts on functions of the form ℓ_φ and $f \circ p$ in the same way (see 3.4.8) as does $\tilde{X}$, and so $J \circ T(X) = \tilde{X}$. □

In addition to the formulas for $\tilde{X}$ given in 3.4.8, it is worth noting that

$$\widetilde{X + Y} = \tilde{X} + \tilde{Y}, \qquad \widetilde{fX} = (f \circ p)\tilde{X} + \tilde{f}X^{\uparrow}, \qquad \tilde{X}(\tilde{f}) = \widetilde{X(f)}, \quad (39)$$

where $X, Y \in \mathfrak{X}(M)$, $f \in C^\infty(M)$ and $\tilde{f}\colon TM \to \mathbb{R}$ is the function $X \mapsto X(f)$.

The theorem which follows gives an important alternative expression for Θ. Recall (or see (8) on p. 386) that for a symplectic manifold (S, ω), there is a canonical isomorphism of vector bundles $\omega^\flat\colon TS \to T^*S$ which we take with a sign as follows:

$$\langle \omega^\flat(X), Y \rangle = -\omega(X, Y),$$

for $X, Y \in \mathfrak{X}(S)$. In the theorem, $\delta\nu$ is the symplectic structure defined by the canonical 1–form ν on T^*M (see 10.1.8).

Theorem 9.6.7 *The Tulczyjew isomorphism* $\Theta\colon T(T^*M) \to T^*(TM)$ *is equal to* $R \circ (\delta\nu)^\flat$.

Proof Recall that $\Theta = J^{\divideontimes} \circ I$. Consider the composition

$$T(T^*M) \xrightarrow{I} T^\bullet(TM) \xrightarrow{J^{\divideontimes}} T^*(TM) \xrightarrow{R^{-1}} T^*(T^*M);$$

denote this temporarily by W. We will show that

$$\langle W(\mathscr{X}), \mathscr{Y} \rangle = \langle (\delta\nu)^\flat(\mathscr{X}), \mathscr{Y} \rangle = -\delta\nu(\mathscr{X}, \mathscr{Y})$$

for all $\mathscr{X}, \mathscr{Y} \in T(T^*M)$.

Consider the LHS. Write $\Phi = (J^{\divideontimes})(I(\mathscr{X}))$. From the definition of R we have that

$$R^{-1}(\Phi) = \underset{T^*M}{\overline{}}\, ((I \divideontimes A^*)(Z_A(\Phi)))$$

where $A = TM$ but we retain the A for clarity. Thus, expanding out the definitions,

$$\langle R^{-1}(\Phi), \mathscr{Y} \rangle = -\langle (I \divideontimes A^*)(Z_A(\Phi)), \mathscr{Y} \rangle = -\lfloor \Phi, I(\mathscr{Y}) \rfloor,$$

where the $\underset{T^*M}{\overline{}}$ may be taken outside because the relevant pairing is over $A^* = T^*M$. Next, replacing Φ by its definition,

$$-\lfloor (J^{\divideontimes})(I(\mathscr{X})), I(\mathscr{Y}) \rfloor = -\langle (J^{\divideontimes})(I(\mathscr{X})), \xi \rangle + \langle I(\mathscr{Y}), \xi \rangle_{\mathrm{prol}} =$$
$$- \langle I(\mathscr{X}), J(\xi) \rangle_{\mathrm{prol}} + \langle\!\langle \mathscr{Y}, \xi \rangle\!\rangle = -\langle\!\langle \mathscr{X}, J(\xi) \rangle\!\rangle + \langle\!\langle \mathscr{Y}, \xi \rangle\!\rangle.$$

Here $\langle\ ,\ \rangle_{\mathrm{prol}}$ denotes the pairing of $T^\bullet(TM)$ with T^2M. The element $\xi \in T^2M$ must have $p_{TM}(\xi) = x$ and $T(p_M)(\xi) = y$ where $x = T(c)(\mathscr{X})$ and $y = T(c)(\mathscr{Y})$, this c being the cotangent bundle projection.

Now we use (32) on p. 117 to get

$$\langle W(\mathscr{X}), \mathscr{Y} \rangle = \mathscr{Y}(\ell_x) - \mathscr{X}(\ell_y) + \xi(\ell_\varphi) - J(\xi)(\ell_\varphi) + x\langle \varphi, y \rangle - y\langle \varphi, x \rangle. \quad (40)$$

Here $\mathscr{X}$ and $\mathscr{Y}$ are single tangent vectors at $\varphi \in T_m^*M$, while x and y denote vector fields on M taking the given values at m and φ is a 1–form taking the given value at m. Using (38) this reduces to

$$\langle W(\mathscr{X}), \mathscr{Y} \rangle = \mathscr{Y}(\ell_x) - \mathscr{X}(\ell_y) + \langle \varphi, [x, y] \rangle. \quad (41)$$

Now consider the RHS. To use the usual formula for $\delta\nu$, both $\mathscr{X}$ and

$\mathscr{Y}$ must be vector fields. The vector fields on T^*M are generated by the complete lifts H_x for $x \in \mathfrak{X}(M)$ and the vertical lifts $\psi^\uparrow$ for $\psi \in \Omega^1(M)$: see §3.4. Assume firstly that $\mathscr{X} = H_x$ and $\mathscr{Y} = H_y$; then x, y have the same meaning as above. Clearly,

$$\langle \delta\nu, \mathscr{Y} \rangle(\varphi) = \langle \varphi, y(m) \rangle = \ell_y(\varphi)$$

for all $\varphi \in T^*M$, so $\langle \delta\nu, \mathscr{Y} \rangle = \ell_y$. Similarly $\langle \delta\nu, \mathscr{X} \rangle = \ell_x$. Finally $\langle \delta\nu, [\mathscr{X}, \mathscr{Y}] \rangle = \ell_{[x,y]}$ since $[H_x, H_y] = H_{[x,y]}$, and $H_x(\ell_y) = \ell_{[x,y]}$, so

$$\delta\nu(\mathscr{X}, \mathscr{Y})(\varphi) = \ell_{[x,y]}(\varphi) = \langle \varphi, [x, y] \rangle.$$

Now (41) becomes

$$\langle W(H_x(\varphi)), H_y(\varphi) \rangle = H_y(\ell_x)(\varphi) - H_x(\ell_y)(\varphi) + \langle \varphi, [x, y] \rangle = -\langle \varphi, [x, y] \rangle$$

and this completes the proof for such $\mathscr{X}, \mathscr{Y}$. The other cases are simple variations of this argument. $\qquad\square$

9.7 The tangent prolongation Lie algebroid

Consider a Lie algebroid A on M, with anchor $a\colon A \to TM$. We intend to put a Lie algebroid structure on the prolongation bundle $TA \to TM$ (see §3.4). For the anchor, define $a_T = J \circ T(a)\colon TA \to T^2M$; then a_T is a morphism of vector bundles over TM. We define the bracket in terms of sections of $TA \to TM$ of the two canonical types $T(X)$ and $\widehat{X}$. To see that this is sufficient, consider the vector bundle morphism $p_A\colon TA \to A$ over $p_M\colon TM \to M$. Given any $\xi \in \Gamma_{TM}(TA)$ we can write

$$p_A^!(\xi) = \sum F_i \otimes X_i$$

where $X_i \in \Gamma A$, $F_i \in C^\infty(TM)$. Now $\xi - \sum F_i T(X_i)$ is a vertical section and is therefore a sum $\sum G_j \widehat{Y_j}$ for suitable $G_j \in C^\infty(TM)$, $Y_j \in \Gamma A$. It is thus sufficient to define a bracket by

$$[T(X), T(Y)] = T([X, Y]), \quad [T(X), \widehat{Y}] = [X, Y]^{\widehat{}}, \quad [\widehat{X}, \widehat{Y}] = 0, \quad (42)$$

for $X, Y \in \Gamma(A)$, and to extend over $C^\infty(TM)$ by the Leibniz property

$$[\xi, F \cdot \eta] = F \cdot [\xi, \eta] + a_T(\xi)(F) \cdot \eta, \quad (43)$$

for all $\xi, \eta \in \Gamma_{TM}(TA)$, $F \in C^\infty(TM)$.

Theorem 9.7.1 *Let A be a Lie algebroid on M. Then the anchor $a_T = J \circ T(a)\colon TA \to T^2M$ and the bracket on $\Gamma_{TM}TA$ defined by (42) make*

TA *a Lie algebroid on* TM. *The bundle projection* $p_A \colon TA \to A$ *is a morphism of Lie algebroids over* $p \colon TM \to M$.

Proof Notice first that $a_T(T(X)) = \widetilde{a(X)}$ and $a_T(\widehat{X}) = a(X)^\uparrow$ for $X \in \Gamma(A)$; the latter uses the fact that the map induced on the cores by $T(a) \colon TA \to T^2M$ is $a \colon A \to TM$ itself. From these and 3.4.8 it follows that $a_T([\xi, \eta]) = [a_T(\xi), a_T(\eta)]$ holds for ξ and η of the form $T(X)$ or $\widehat{X}$. Similarly the Jacobi identity is easily verified on sections of these forms.

We must prove that the definitions (42) are consistent with respect to different representations of sections. First consider the case where $\xi = T(X)$, $F = f \circ p$, $\eta = \widehat{Y}$. We have

$$
\begin{aligned}
[T(X), (f \circ p) \cdot \widehat{Y}] &= [T(X), (fY)\,\widehat{\;}\,] \\
&= [X, fY]\,\widehat{\;} \\
&= (f \circ p) \cdot [X, Y]\,\widehat{\;} + (a(X)(f) \circ p) \cdot \widehat{Y} \\
&= (f \circ p) \cdot [T(X), \widehat{Y}] + \widetilde{a(X)}(f \circ p) \cdot \widehat{Y} \\
&= (f \circ p) \cdot [T(X), \widehat{Y}] + a_T(T(X))(f \circ p) \cdot \widehat{Y},
\end{aligned}
$$

so (42) is consistent with (43). A similar proof applies in the case where $\xi = \widehat{X}$, $F = f \circ p$, $\eta = \widehat{Y}$.

Likewise consider the identity

$$
T(fY) = (f \circ p) \cdot T(Y) + \widetilde{f} \cdot \widehat{Y},
$$

which is equivalent to the second equation in (39). Applying (42), we get

$$
[T(X), T(fY)] = T([X, fY]) = T(f[X, Y] + a(X)(f)Y)
$$

whereas applying (43) we obtain

$$
\begin{aligned}
[T(X), (f \circ p) \cdot T(Y) &+ \widetilde{f} \cdot \widehat{Y}] \\
&= (f \circ p) \cdot [T(X), T(Y)] + a_T(T(X))(f \circ p) \cdot T(Y) \\
&\qquad + \widetilde{f} \cdot [T(X), \widehat{Y}] + a_T(T(X))(\widetilde{f}) \cdot \widehat{Y}
\end{aligned}
$$

and the two expressions are equal.

To prove that $p_A \colon TA \to A$ is a morphism of Lie algebroids, first observe that p_A commutes with the anchors. Next, in this case the module of sections of the domain is generated by the projectable sections $T(X)$ and $\widehat{X}$, and we have $T(X) \sim X$ and $\widehat{X} \sim 0$. Thus it is clear from the definition (42) that for sections of these two types, if $\xi \sim X$ and $\eta \sim Y$, then $[\xi, \eta] \sim [X, Y]$. This completes the proof. □

This structure is the *tangent prolongation Lie algebroid*; we usually omit the word prolongation.

Example 9.7.2 In the case of $A = TM$, this construction equips the bundle $T(p) \colon T^2M \to TM$ with a Lie algebroid structure with anchor $J \colon T^2M \to T^2M$ and bracket

$$[\xi, \eta] = J[J\xi, J\eta].$$

⊠

Now suppose that A is the Lie algebroid of a Lie groupoid $G \rightrightarrows M$. Consider the tangent groupoid $TG \rightrightarrows TM$ of 1.1.16. The bundle projection $p_G \colon TG \to G$ is a groupoid morphism over p_M and applying the Lie functor gives $A(p_G) \colon ATG \to AG$. We define a vector bundle structure on ATG, for which this map is the projection, by applying the Lie functor to the vector bundle operations in $TG \to G$. For example, the addition $TG \times_G TG \to TG$ is a groupoid morphism, regarding the domain as the pullback groupoid of the diagram

$$TG$$

$$\downarrow p_G$$

$$TG \xrightarrow{\quad\quad} G.$$
$$p_G$$

Applying the Lie functor gives $ATG \times_{AG} ATG = A(TG \times_G TG) \to ATG$; this is the addition in $ATG \to AG$ and is automatically a morphism of Lie algebroids. Similarly the zero over $X \in AG$ is $A(0)(X)$. We use the notation $+\!\!\!+ \, , \, -\!\!\!- \, , \, \cdot$ for the operations in $ATG \to AG$.

That $ATG \to AG$ is a vector bundle follows easily from the fact that A preserves diagrams and pullbacks, as does the rest of the next proposition.

Proposition 9.7.3 *With respect to the structure just defined, and the standard structure over TM of the Lie algebroid of $TG \rightrightarrows TM$, the* *diagram*

$$
\begin{array}{ccc}
 & q_{TG} & \\
ATG & \longrightarrow & TM \\
A(p_G) \Big\downarrow & & \Big\downarrow p \\
AG & \longrightarrow & M. \\
 & q &
\end{array}
\tag{44}
$$

is a double vector bundle.

The core may be calculated to be AG using the following easy lemma.

Lemma 9.7.4 *Let* $f\colon M \to N$ *be a surjective submersion. Then the map* $J\colon T^2M \to T^2M$ *restricts to an isomorphism from the double vector bundle* $(T^{T(f)}TM; TM, T^fM; M)$ *of 9.1.7 to the tangent double vector bundle of* T^fM.

Theorem 9.7.5 *Let* G *be a Lie groupoid on base* M. *Then there is a canonical isomorphism of double vector bundles* $j_G\colon TAG \to ATG$, *where* ATG *is as above and* TAG *is the tangent double vector bundle of* $AG \to M$, *which induces the identities on the side bundles* AG *and* TM *and on the cores* AG. *Further,* j_G *is an isomorphism of Lie algebroids over* TM, *where* $TAG \to TM$ *has the tangent Lie algebroid structure of Theorem 9.7.1 and* $ATG \to TM$ *is the Lie algebroid of* $TG \rightrightarrows TM$.

Proof The Lie algebroid ATG is defined by the pullback diagram in Figure 9.8(a), where ι_{TG} is usually regarded as an inclusion, and fits

$$
\begin{array}{ccc}
 & \iota_{TG} & \\
ATG & \longrightarrow & T^{T(\alpha)}TG \\
q_{TG} \Big\downarrow & & \Big\downarrow p_{TG} \\
TM & \longrightarrow & TG, \\
 & T(1) &
\end{array}
\qquad\qquad
\begin{array}{ccc}
 & \iota_G & \\
AG & \longrightarrow & T^\alpha G \\
q_G \Big\downarrow & & \Big\downarrow p_G \\
M & \longrightarrow & G, \\
 & 1 &
\end{array}
$$

$$
\hspace{3cm}(a)\hspace{5cm}(b)
$$

Fig. 9.8.

into the morphism of double vector bundles in Figure 9.9 as the top face, the bottom face being the pullback diagram defining AG itself. Here we are denoting the restrictions of maps by the same symbols.

On the other hand, we can apply the tangent functor to the pullback diagram in Figure 9.8(b) to obtain a morphism of double vector bundles as in Figure 9.10. From 9.7.4 we know that the two front faces of Figure 9.9 and Figure 9.10 are isomorphic under a restriction of the canonical involution $J\colon T^2G \to T^2G$. Since in both diagrams the top and bottom faces are pullbacks, it follows that J restricts to the required isomorphism $j_G\colon TAG \to ATG$. That j_G preserves the side bundles and the core is now evident.

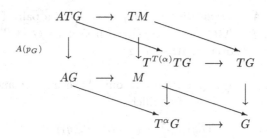

Fig. 9.9.

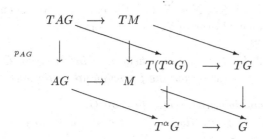

Fig. 9.10.

That $a_{TG} \circ j_G = J_M \circ T(a_G)$ follows easily by considering both a_{TG} and $T(a_G)$ as restrictions of $T^2(\beta)$ and using $J_M \circ T^2(\beta) = T^2(\beta) \circ J_G$. To show that j_G maps the bracket (42) on $TAG \to TM$ to the bracket on $ATG \to TM$, we need only consider sections of TAG of the form $T(X)$ and $\widehat{X}$ for $X \in \Gamma AG$. Using the arrow notation for both the right–invariant vector fields on G, and the right–invariant vector fields on TG, it is only necessary to check that

$$\overrightarrow{j_G \circ T(X)} = \widetilde{\overrightarrow{X}}, \qquad \overrightarrow{j_G \circ \overleftarrow{X}} = (\overrightarrow{X})^\uparrow, \tag{45}$$

for $X \in \Gamma AG$. Now the result follows from comparing (42) with 3.4.8.

$\square$

9.8 Vector fields on Lie groupoids

Consider a Lie groupoid G on base M.

Definition 9.8.1 A *multiplicative vector field* on G is a pair of vector fields (ξ, x) where $\xi \in \mathfrak{X}(G)$, $x \in \mathfrak{X}(M)$, such that $\xi \colon G \to TG$ is a morphism of groupoids over $x \colon M \to TM$.

$\square$

Example 9.8.2 For a group G, we can only have $x = 0$ and so a multiplicative vector field is one for which

$$X(gh) = T(L_g)(X(h)) + T(R_h)(X(g))$$

for all $g, h \in G$.

For a pair groupoid $M \times M$, the multiplicative vector fields are those of the form $x \times x$, where $x \in \mathfrak{X}(M)$.

Given $X \in \Gamma AG$, let $\xi = \overrightarrow{X} + \overleftarrow{X}$ (see the end of §3.5). Then ξ is a multiplicative vector field over $a(X)$, the anchor of X.

$\boxtimes$

Proposition 9.8.3 *Let ξ be a vector field on a Lie groupoid G and x a vector field on its base M. Then the following are equivalent:*

(i) *(ξ, x) is a multiplicative vector field on G;*
(ii) *the flows φ_t of ξ are (local) Lie groupoid automorphisms over the flows f_t of x.*

Proof ($\Longrightarrow$) Assume for simplicity that φ_t and f_t are global flows for ξ and x. Since ξ projects to x under the groupoid source projection α, it follows that $\alpha \circ \varphi_t = f_t \circ \alpha$. Similarly $\beta \circ \varphi_t = f_t \circ \beta$.

Denote $\{(h, g) \in G \times G \mid \alpha(h) = \beta(g)\}$ by $G * G$. Define a vector field $\xi * \xi$ on $G * G$ by $\xi * \xi(h, g) = (\xi(h), \xi(g))$; since $T(\alpha)(\xi(h)) = x(\alpha(h)) = x(\beta(g)) = T(\beta)(\xi(g))$, $\xi * \xi$ is tangent to $G * G$. Evidently $\xi * \xi$ has flow $\psi_t(h, g) = (\varphi_t(h), \varphi_t(g))$. Denoting the groupoid composition by $\kappa \colon G * G \to G$, we know that $\xi * \xi$ projects to ξ under κ. It follows that

$\varphi_t(h)\varphi_t(g) = \varphi_t(hg)$ for all $(h,g) \in G * G$, and so φ_t is a Lie groupoid automorphism.

The converse is established by retracing these steps in the reverse order. □

Call $F\colon G \to \mathbb{R}$ a *multiplicative function* if it is a groupoid morphism from G into the abelian group $\mathbb{R}$.

Corollary 9.8.4 *Let* (ξ, x) *be a multiplicative vector field on* G, *and let* $F\colon G \to \mathbb{R}$ *be a multiplicative function. Then* $\xi(F)$ *is also a multiplicative function.*

It follows from Proposition 9.8.3 that if (ξ, x) and (η, y) are multiplicative vector fields, then $([\xi, \eta], [x, y])$ is also.

Proposition 9.8.5 *Let* (ξ, x) *be a multiplicative vector field and take* $X \in \Gamma AG$ *with corresponding right– and left–invariant vector fields* $\overrightarrow{X}$ *and* $\overleftarrow{X}$. *Then* $[\xi, \overrightarrow{X}]$ *is right–invariant and* $[\xi, \overleftarrow{X}]$ *is left–invariant.*

Proof First consider any α–vertical vector field $Z \in \mathfrak{X}(G)$. Define a vector field $(Z, 0)$ on $G * G$ by $(Z, 0)(h, g) = (Z(h), 0_g)$. In terms of the tangent groupoid, $Z(h) \bullet 0_g = T(R_g)(Z(h))$, so Z is right–invariant if and only if $Z * 0 \overset{\kappa}{\sim} Z$. Now let ξ be multiplicative. We have, as in 9.8.3, that $\xi * \xi \overset{\kappa}{\sim} \xi$. So $[\xi * \xi, \overrightarrow{X} * 0] \overset{\kappa}{\sim} [\xi, \overrightarrow{X}]$, whence $[\xi, \overrightarrow{X}] * 0 \overset{\kappa}{\sim} [\xi, \overrightarrow{X}]$, and so $[\xi, \overrightarrow{X}]$ is right–invariant.

A similar proof applies in the left–invariant case. □

For a multiplicative vector field (ξ, x) there is now a map

$$D_\xi \colon \Gamma AG \to \Gamma AG, \qquad \overrightarrow{D_\xi(X)} = [\xi, \overrightarrow{X}].$$

The proof of the properties which follow is straightforward.

Proposition 9.8.6 (i) *The map* $D_\xi \colon \Gamma AG \to \Gamma AG$ *is a derivation of the bracket structure.*

(ii) *For* $X \in \Gamma AG$ *and any vector field* $\overline{X}$ *on* G *with* $\overline{X}|_M = X$, *we have* $D_\xi(X) = [\xi, \overline{X}]|_M$.

(iii) *If* (η, y) *is a second multiplicative vector field, then* $D_{[\xi, \eta]} = [D_\xi, D_\eta]$.

The operator D_ξ has been constructed directly from ξ. We now consider a similar construction for the vector field induced on AG by ξ.

For any multiplicative vector field (ξ, x), applying the Lie functor produces a section

$$
\begin{array}{ccc}
ATG & \longrightarrow & TM \\[2mm]
A(\xi) \uparrow & & \uparrow \; x \\[2mm]
AG & \longrightarrow & M.
\end{array}
\tag{46}
$$
$$\qquad\qquad q_G$$

Define $\widetilde{\xi} = (j_G)^{-1} \circ A(\xi)$; it is clear that $(\widetilde{\xi}, x)$ is a linear vector field on AG. In the case where $G = M \times M$ and $\xi = x \times x$ for $x \in \mathcal{X}(M)$, this $\widetilde{\xi}$ is precisely the complete lift as in §9.6. The linear vector field $(\widetilde{x}, x)$ corresponds to the Lie derivative $\mathfrak{L}_x \colon \mathcal{X}(M) \to \mathcal{X}(M)$, $y \mapsto [x, y]$. From Proposition 3.4.5 we therefore have

$$
\tau(x(m), [x, y](m)) = \widetilde{y}(x(m)) - T(x)(y(m)).
\tag{47}
$$

If φ_t is a (local) flow for ξ, then $A(\varphi_t)$ is a (local) flow for $\widetilde{\xi}$; this follows by a modification of 9.6.6. Since $a_{TG} \circ j_G = J_M \circ T(a_G)$, the vector field $\widetilde{\xi}$ on AG projects under a_G to $\widetilde{x}$ on TM.

Since $(\widetilde{\xi}, x)$ is linear, it induces $D_{\widetilde{\xi}} \colon \Gamma AG \to \Gamma AG$ as in (27) on p. 115.

Theorem 9.8.7 *For any multiplicative vector field (ξ, x) on G,*

$$
D_{\widetilde{\xi}} = D_\xi.
$$

Proof Applying (47) to ξ and $\overrightarrow{X}$ for $X \in \Gamma AG$, at $1_m \in G$, and recalling that $\xi(1_m) = T(1)(x(m))$, we have

$$
\tau(X(m), [\xi, \overrightarrow{X}](1_m)) = T(\overrightarrow{X})(T(1)(x(m)) - \widetilde{\xi}(X(m));
$$

note that τ and the tildes here refer to the double vector bundle $T^2 G$. On the other hand, in TAG we have, from Proposition 3.4.5,

$$
\tau(X(m), D_{\widetilde{\xi}}(X)(m)) = T(X)(x(m)) - \widetilde{\xi}(X(m)).
$$

Note that $\widetilde{\xi}$ as a vector field on AG is the restriction of $\widetilde{\xi}$ as a vector field on TG.

From (45) we have

$$
\overrightarrow{\widetilde{X}} = \overrightarrow{j_G \circ T(X)},
$$

where the tilde on the left refers to T^2G and the arrow on the right refers to the tangent groupoid TG. So $T(\overrightarrow{X}) = J_G(\overrightarrow{j_G \circ T(X)})$. Since $j_G \colon TAG \to ATG$ is a restriction of $J_G \colon T^2G \to T^2G$, it follows that $T(\overrightarrow{X})(T(1)(x(m))) = T(X)(x(m))$. Regarding $ATG \subseteq T^2G$ and noting that the τ for TAG is then the restriction of the τ for T^2G, we have $D_{\widetilde{\xi}}(X)(m) = [\xi, \overrightarrow{X}](1_m)$. □

For any multiplicative vector fields (ξ, x) and (η, y) on G and for any $f \in C^\infty(M)$, $\varphi \in \Gamma A^*G$, we now have

$$\widetilde{\xi + \eta} = \widetilde{\xi} + \widetilde{\eta}, \qquad \widetilde{\xi}(f \circ q) = x(f) \circ q, \qquad \widetilde{\xi}(\ell_\varphi) = \ell_{D_\xi^{(*)}(\varphi)}. \tag{48}$$

For $F \colon G \to \mathbb{R}$ any multiplicative function, denote $A(F) \colon AG \to \mathbb{R}$ by $\widetilde{F}$.

Proposition 9.8.8 *For* $(\xi, x), (\eta, y)$ *multiplicative vector fields on* G, $X, Y \in \Gamma AG$, *and* F *a multiplicative function on* G,

$$[\widetilde{\xi}, \widetilde{\eta}] = \widetilde{[\xi, \eta]}, \quad [\widetilde{\xi}, X^\uparrow] = D_\xi(X)^\uparrow, \quad [X^\uparrow, Y^\uparrow] = 0, \quad \widetilde{\xi}(\widetilde{F}) = \widetilde{\xi(F)}.$$

Proof The third equation is known from (21) on p. 112 and the last follows from Proposition 9.8.3.

For the first and second equations, it suffices to verify equality on functions of the forms ℓ_φ, $\varphi \in \Gamma A^*G$, and $f \circ q$, $f \in C^\infty(M)$. For the second equation and $\varphi \in \Gamma A^*G$ we have

$$\begin{aligned} [\widetilde{\xi}, X^\uparrow](\ell_\varphi) &= \widetilde{\xi}(\langle \varphi, X \rangle \circ q) - X^\uparrow(\ell_{D_\xi^{(*)}(\varphi)}) \\ &= x(\langle \varphi, X \rangle) \circ q - \langle D_\xi^{(*)}(\varphi), X \rangle \circ q \\ &= \langle \varphi, D_\xi(X) \rangle \circ q \\ &= D_\xi(X)^\uparrow(\ell_\varphi), \end{aligned}$$

whilst for $f \in C^\infty(M)$,

$$[\widetilde{\xi}, X^\uparrow](f \circ q) = 0 = D_\xi(X)^\uparrow(f \circ q).$$

The first equation is proved in a similar way. □

Multiplicative vector fields are rather special and we now briefly consider two more general types of vector field on Lie groupoids. In what follows we omit most proofs, which are similar to those given above.

Definition 9.8.9 A *star vector field* on G is a pair of vector fields (ξ, x) where $\xi \in \mathfrak{X}(G)$, $x \in \mathfrak{X}(M)$, such that $T(\alpha) \circ \xi = x \circ \alpha$ and $\xi \circ 1 = T(1) \circ x$. □

Lemma 9.8.10 *Given any vector field x on M, there is a star vector field (ξ, x) on G.*

Proof Define a vector field η on G by setting $\eta(1_m) = T(1)(x(m))$ for $m \in M$, and extending over G. Then $\mu\colon G \to TM$ defined by $\mu(g) = T(\alpha)(\eta(g)) - x(\alpha g)$ is a section of the pullback bundle $\alpha^! TM$. Since α is a surjective submersion, there is a vector field ζ on G with $T(\alpha) \circ \zeta = \mu$; we can also require that ζ vanish on all $1_m \in G$. Now $\xi = \eta - \zeta$ is a star vector field over x. $\qquad\square$

Proposition 9.8.11 *Let ξ be a vector field on a Lie groupoid G and x a vector field on its base M. Then the following are equivalent:*

  (i) *(ξ, x) is a star vector field on G;*
  (ii) *the flows φ_t of ξ are (local) star maps over the flows f_t of x.*

It follows as before that if (ξ, η) and (η, y) are star vector fields, then $([\xi, \eta], [x, y])$ is also. If (ξ, x) is a star vector field and $X \in \Gamma AG$, then it is clear that $[\xi, \vec{X}]$ is α–vertical. We now define $D_\xi\colon \Gamma AG \to \Gamma AG$ by

$$D_\xi(X) = [\xi, \vec{X}] \circ 1.$$

Again, $D_\xi(X) = [\xi, \overline{X}] \circ 1$ for any vector field $\overline{X}$ on G such that $\overline{X} \circ 1 = X$.

Given a star vector field (ξ, x), we can still apply the Lie functor and obtain a linear vector field $\widetilde{\xi} = (j_G)^{-1} \circ A(\xi)$ on AG. If φ_t is a (local) flow for ξ, then $A(\varphi_t)$ is still a (local) flow for $\widetilde{\xi}$. The proof of the following result follows the method of 9.8.7.

Theorem 9.8.12 *For any star vector field (ξ, x) on G, $D_{\widetilde{\xi}} = D_\xi$.*

Equations (48) continue to hold for star vector fields (ξ, x) and (η, y) and $f \in C^\infty(M)$, $\varphi \in \Gamma A^*G$.

Proposition 9.8.13 *Given star vector fields $(\xi, x), (\eta, y)$ on G, and $X, Y \in \Gamma AG$,*

$$[\widetilde{\xi}, \widetilde{\eta}] = \widetilde{[\xi, \eta]}, \qquad [\widetilde{\xi}, X^\uparrow] = D_\xi(X)^\uparrow, \qquad [X^\uparrow, Y^\uparrow] = 0.$$

In view of the following result, these equations determine the bracket structure for all vector fields on AG.

Proposition 9.8.14 *The vector fields of the form $\widetilde{\xi}$, where (ξ, x) is a star vector field, together with those of the form $X^\uparrow$, where $X \in \Gamma AG$, generate $\mathfrak{X}(AG)$.*

Proof Take $\Xi \in TAG$ with $T(q)(\Xi) = x(m)$ and $p_{AG}(\Xi) = X$. Extend $x(m) \in TM$ to a vector field x on M. By Lemma 9.8.10, there is a star vector field (ξ, x) on G. We now have $T(q)(\widetilde{\xi}(X)) = x(m)$ and so, by the sequence (19) on p. 112 we have

$$\Xi = \widetilde{\xi}(X) + Y^\uparrow(X)$$

for some $Y \in \Gamma AG$. $\qquad \square$

Lastly in this section, we briefly consider the notion of affine vector field on a Lie groupoid.

Definition 9.8.15 A vector field ξ on a Lie groupoid $G \rightrightarrows M$ is *affine* if

$$T(\alpha)(\xi(g)) = T(\alpha)(\xi(1_{\alpha g})) \text{ and } T(\beta)(\xi(g)) = T(\beta)(\xi(1_{\beta g})) \qquad (49)$$

for all $g \in G$ and if, for all $g, h \in G$ with $\alpha(g) = \beta(h) = m$, we have

$$\xi(gh) = \xi(g) \bullet \xi(1_m)^{-1} \bullet \xi(h).$$

$\qquad \square$

Proposition 9.8.16 *A vector field ξ on G is affine if and only if* (49) *holds and for all $g, h, k \in G$ with $\alpha(g) = \alpha(k) = m$ and $\beta(k) = \beta(h) = n$ we have*

$$\xi(gk^{-1}h) = \xi(g) \bullet \xi(k)^{-1} \bullet \xi(h).$$

Proof Assume that ξ is affine. Then

$$\begin{aligned}
\xi(gk^{-1}h) &= \xi(g) \bullet \xi(1_m)^{-1} \bullet \xi(k^{-1}h) \\
&= \xi(g) \bullet \xi(1_m)^{-1} \bullet \xi(k^{-1}) \bullet \xi(1_n)^{-1} \bullet \xi(h)
\end{aligned}$$

and so we need only prove that $\xi(1_m)^{-1} \bullet \xi(k^{-1}) \bullet \xi(1_n)^{-1} = \xi(k)^{-1}$. But $\xi(k) \bullet \xi(1_m)^{-1} \bullet \xi(k^{-1}) = \xi(1_n)$ by the affine condition, and the result follows. $\qquad \square$

Proposition 9.8.17 *A vector field ξ on G is affine if and only if for all $g, h \in G$ with $\alpha(g) = \beta(h) = m$ we have*

$$\xi(gh) = T(L_\sigma)(\xi(h)) + T(R_\tau)(\xi(g)) - T(L_\sigma)T(R_\tau)(\xi(1_m)),$$

where σ, τ *are (local) bisections with* $\sigma(\alpha g) = g$, $\tau(\alpha h) = h$.

Proof Suppose that the condition holds. Then because $\beta \circ L_\sigma = \beta \circ \sigma \circ \beta$ and $\beta \circ R_\tau = \beta$, we have $T(\beta)(\xi(gh)) = T(\beta)(\xi(g))$ and setting $h = g^{-1}$ yields the second of (49). Likewise with the first.

The remainder of the proof is an application of 1.4.14.   □

A multiplicative vector field is affine (see 1.4.14) and for any $X \in \Gamma AG$, both $\overrightarrow{X}$ and $\overleftarrow{X}$ are affine vector fields. It is clear that the sum and scalar multiples of affine vector fields are affine. Further, affine vector fields are closed under the bracket of vector fields.

Note that conditions (49) are very much weaker than requiring ξ to be projectable under α and β. Likewise, there is no requirement that $\xi(1_m)$ ever be tangent to 1_M.

On a Lie group, an affine vector field ξ is multiplicative if and only if $\xi(1) = 0$. On a Lie group any affine vector field is a sum of a multiplicative vector field and a right–invariant vector field.

Since multiplicative and right–invariant vector fields on a Lie groupoid are always both α– and β–projectable, the following result is the best that can be expected in the general case. It is easy to construct affine vector fields on a pair groupoid $M \times M$ that are not projectable.

Proposition 9.8.18 *Let* ξ *be an affine vector field on the Lie groupoid* G, *and suppose that* ξ *is both* α– *and* β–*projectable. Then* ξ *is the sum of a multiplicative and a right–invariant (or left–invariant) vector field.*

Proof Define $X \in \Gamma AG$ by $X(m) = \xi(1_m) - T(1)T(\alpha)(\xi(1_m))$, where $m \in M$, and write $\eta = \xi - \overrightarrow{X}$. We prove that η is multiplicative. Let x, y be the vector fields on M for which $\xi \overset{\alpha}{\sim} x$, $\xi \overset{\beta}{\sim} y$. Then $\overrightarrow{X} \overset{\beta}{\sim} y - x$ and so $\eta \overset{\alpha}{\sim} x$ and $\eta \overset{\beta}{\sim} x$. Clearly $\eta(1_m) = T(1)(x(m))$ for $m \in M$.

Since ξ and $\overrightarrow{X}$ are affine, it follows that η is affine. In this case, the affine condition implies that $\eta(gh) = \eta(g) \bullet \eta(h)$. Thus η is multiplicative.   □

9.9 Notes

§9.1. The concept of a double vector bundle is just below the surface of many accounts of the connection theory of vector bundles from the 1960s onward; see, for example, the account in [Dieudonné, 1972]. It is explicit, though not fully developed, in [Besse, 1978]. A thorough development of many basics of the theory, though not of duality, was given by Pradines in detail in [Pradines, 1974a], and in short summary [Pradines, 1974b,c]. In particular, the crucial concept of core (*cœur*) of a double vector bundle was introduced in [Pradines, 1974a, C§2].

§9.2. The construction of the dual of a double vector bundle is a special case (not explicitly noted) of the construction of the dual of a $\mathscr{VB}$–groupoid due to Pradines [1988]. The account given here (which puts a greater emphasis on the role of the core) follows that in [Mackenzie, 1999, §3].

The duality between the two duals of a double vector bundle was found by the author [Mackenzie, 1999] and by Konieczna and Urbański [1999].

§9.3—§9.5. The material of these three sections is a reworking of that in the paper [Mackenzie and Xu, 1994]. Here I have derived the canonical isomorphisms I and R from the general theory of §9.2, rather than giving direct constructions. Theorem 9.5.1 is Theorem 5.5 of [Mackenzie and Xu, 1994], and Proposition 9.5.3 is Proposition 6.4 of [Mackenzie and Xu, 1994].

A considerable body of work by Tulczyjew, summarized in [Tulczyjew, 1989], makes extensive use of TA and T^*A for A a vector bundle, concentrating on the cases where A is a tangent or cotangent bundle.

§9.6. The material in this section is more or less well–known, but we present it here in a way consistent with the general apparatus of double structures. Theorem 9.6.7 is Theorem 7.4 of [Mackenzie and Xu, 1994].

The Tulczyjew isomorphism Θ is the map denoted α in [Abraham and Marsden, 1985, p. 423], where it provides a canonical symplectomorphism from the tangent lift to $T(T^*M)$ of the canonical symplectic structure on T^*M to the canonical cotangent structure on $T^*(TM)$. The definition of Θ given here in terms of J and I is the same as that in [Tulczyjew, 1989, Ch. 11].

Lemma 9.6.5 is from [Abraham and Marsden, 1985, p. 122].

There is a considerable literature on 'the geometry of the tangent bundle'; see, for example, [Yano and Ishihara, 1973], [Morandi, Ferrario, Lo Vecchio, Marmo, and Rubano, 1990]. Much of this work proceeds entirely in coordinates.

§9.7. The contents of this section are from §5 and §7 of [Mackenzie and Xu, 1994].

§9.8. The material on multiplicative and star vector fields comes from §3 of the paper [Mackenzie and Xu, 1998]. The terminology *star vector field* comes from [Higgins and Mackenzie, 1990a], where a *star map* of groupoids is a map which preserves the α–fibration (sometimes called the star projection) and the identities.

For affine vector fields, see [Weinstein, 1990], [Xu, 1995] and [Mackenzie and Xu, 1998]. The treatment here includes some new features.

Another approach to multiplicativity conditions on (multi)vector fields and forms has been given very recently in [Bursztyn, Crainic, Weinstein, and Zhu, arXiv:0303180].

10

Relations between
Poisson structures and
Lie algebroids

Poisson geometry has developed from three principal sources: it is a more natural and convenient framework for many of the basic constructions and ideas of symplectic geometry; it embodies a theory dual to Lie algebra theory; and it is a semi–classical limit of modern quantum geometry.

Associated with any symplectic manifold (M, ω) is the Poisson bracket $\{u, v\} = \omega(X_u, X_v) = X_u(v)$ of functions on M, where X_u is the Hamiltonian vector field for u; this bracket makes $C^\infty(M)$ a real Lie algebra and in each variable it is of course a derivation of the algebra structure on $C^\infty(M)$. Furthermore, the symplectic structure can easily be recovered from the Poisson bracket. An abstract *Poisson manifold* is a manifold M equipped with a bracket of smooth functions which satisfies these two properties (see 10.1.1). Certain crucial constructions in symplectic geometry — most notably reduction — proceed more naturally if one works with the Poisson bracket rather than with the symplectic form, and have Poisson formulations from which the classical symplectic results can be retrieved easily.

The other fundamental example of a Poisson manifold is the vector space dual of a Lie algebra. The bracket on $\mathfrak{g}$ induces a Poisson bracket on the smooth functions on $M = \mathfrak{g}^*$ (see 10.3.7) which is *linear* in the sense that the Poisson bracket of two linear functions is linear. These linear Poisson manifolds are in an exact duality with Lie algebras, and one may work entirely in terms of these duals rather than in terms of the Lie algebras. The symplectic structure of the coadjoint orbits of a Lie group emerges quickly from this linear Poisson structure.

Finally, Poisson geometry represents a first step in modern theories of quantization; or, viewed from the opposite perspective, Poisson geometry is the penultimate stage in the limit as quantum approaches classical geometry. This is true in the precise sense that the first–order term in a deformation quantization is a Poisson bracket, but there is also a more vague, but useful, analogy between Poisson geometry and quantum constructions which means that each can throw light on the other, though it seldom allows an exact correspondence.

The single word which best describes the relationship between Lie algebroid theory and Poisson geometry is duality. The duality between Lie algebras and linear Poisson manifolds extends to a duality between abstract Lie algebroids and vector bundles with a suitably linear Poisson structure, but there is a further sense in which the concepts may be said to be dual.

Any Poisson structure on a manifold P gives rise to a Lie algebra structure on $C^\infty(P)$. However $C^\infty(P)$ is infinite–dimensional and the properties of this Lie algebra are of no serious value in understanding the Poisson manifold. The concept of Lie algebroid captures much more closely the inherent structure of a Poisson manifold: a Poisson structure on P induces a Lie algebroid structure on the cotangent bundle T^*P. This *cotangent Lie algebroid* of P is both a substantial tool in working with the Poisson structure and an important source of examples of Lie algebroids.

There are therefore two processes that link Lie algebroids and Poisson manifolds. Composition leads to two functors: one from a Poisson manifold P to its tangent bundle with the prolongation of the Poisson structure from P, and the other from a Lie algebroid A on base M to the cotangent bundle $T^*(A^*)$ of its dual; this is a Lie algebroid over A^*. Though these two compositions result in larger structures, the processes which convert Lie algebroids to Poisson manifolds and Poisson manifolds to Lie algebroids are remarkably useful.

It is usual in work with manifolds to regard the tangent functor as a well–behaved and useful process, and the cotangent as a more awkward and secondary device. In Poisson geometry the use of Lie algebroids has crystallized what was already noticeable: it is the cotangent bundle which is the more basic and useful structure. Poisson geometry is essentially a contravariant theory and the use of the cotangent Lie algebroid enables this contravariance to be largely undone; it is often surprising to see a statement for Poisson structures translate covariantly into a statement for Lie algebroids (see, for example, §11.1). This process gains its full force by the use of symplectic groupoids, which will be described in the next chapter.

This leads to what may be called the *cotangent philosophy*: results and problems in Poisson geometry should be reformulated in terms of the cotangent Lie algebroids, where the methods of the Lie theory of Lie groupoids and Lie algebroids may be applied. It is also true that some problems in Lie algebroid theory are simplified by working with the dual Poisson structures; an immediate example is the characterization 10.4.9 of morphisms of Lie algebroids, but see also the Appendix.

The value of the two processes converting Poisson phenomena into terms of Lie algebroids, and vice versa, stems from the nontrivial nature of these processes: neither can be expressed solely in terms of the two categories, of Poisson structures and of Lie algebroids.

In this book we will be almost entirely concerned with Poisson geometry as it relates to Lie algebroid theory (and thus, of course, to Lie groupoid theory). In §10.1 we summarize the basic notions concerning Poisson structures and the notation which we use. This section makes the book technically self–contained as regards Poisson geometry, but a reader who is completely new to Poisson structures should also consult one of the sources listed in the Notes at the end of the Chapter.

Work in Poisson geometry is often conducted wholly in terms of the bracket of functions or the Poisson 2–tensor. The cotangent Lie algebroid is defined in terms of the bracket of 1–forms and the map which we call the Poisson anchor, and we work primarily in these terms.

In §10.2 we briefly consider Poisson cohomology in terms of Lie algebroid cohomology. The two sections which follow give the main details of the various correspondences between Lie algebroids and Poisson structures.

10.1 Poisson structures

Definition 10.1.1 A *Poisson structure on a manifold* P is a bracket of smooth functions $\{\ ,\ \}\colon C^\infty(P) \times C^\infty(P) \to C^\infty(P)$ with respect to which $C^\infty(P)$ is an $\mathbb{R}$–Lie algebra, and such that for all $u, v, w \in C^\infty(P)$,

$$\{u, vw\} = v\{u, w\} + w\{u, v\}. \tag{1}$$

$\square$

Thus the Lie bracket is a derivation in each variable for the associative multiplication on $C^\infty(P)$. Each function $u \in C^\infty(P)$ accordingly defines a vector field X_u on P by

$$X_u(v) = \{u, v\}. \tag{2}$$

This X_u is the *Hamiltonian vector field with energy* u. The Jacobi identity for $\{\ ,\ \}$ implies that $u \mapsto X_u$ maps the Poisson bracket to the bracket of vector fields. Equation (1) implies that the bracket of any smooth function with the function constant at 1 is zero and hence, by $\mathbb{R}$–linearity, the bracket of any constant function with any smooth function is zero. It follows that $u \mapsto X_u$ can be lifted to a map of 1–forms

$$\pi^{\#} \colon \Omega^1(P) \to \mathfrak{X}(P), \qquad u\,\delta v \mapsto uX_v,$$

which defines a linear map $T^*P \to TP$, also denoted $\pi^{\#}$. In practice we usually abbreviate $\pi^{\#}(\theta)$, for $\theta \in \Omega^1(P)$, to $\theta^{\#}$. The proof of the following result should come from the reader.

Theorem 10.1.2 *The map* $\pi^{\#} \colon T^*P \to TP$ *is the anchor for a Lie algebroid structure on* T^*P *the bracket of which is given by (the additive extension of)*

$$[u\,\delta v, u'\,\delta v'] = uu'\,\delta\{v,v'\} + uX_v(u')\,\delta v' - u'X_{v'}(u)\,\delta v. \qquad (3)$$

This is the *cotangent Lie algebroid of the Poisson manifold* P. We refer to $\pi^{\#}$ as the *Poisson anchor*. To give the bracket (3) of 1–forms in intrinsic form, first define a 2–vector field π by

$$\pi(u\,\delta v, u'\,\delta v') = uu'\{v,v'\}; \qquad (4)$$

the corresponding $\pi \colon T^*P \oplus T^*P \to P \times \mathbb{R}$ or $P \to TP \wedge TP$ is the *Poisson 2–tensor*. The bracket (3) of 1–forms may now be given in intrinsic form by

$$[\theta_1, \theta_2] = \mathcal{L}_{\theta_1^{\#}}(\theta_2) - \mathcal{L}_{\theta_2^{\#}}(\theta_1) - \delta(\pi(\theta_1, \theta_2))$$
$$= \iota_{\theta_1^{\#}}(\delta\theta_2) - \iota_{\theta_2^{\#}}(\delta\theta_1) + \delta(\pi(\theta_1, \theta_2)). \qquad (5)$$

The first of these can be proved by verifying that the two sides are equal on exact differentials and that both sides satisfy the Lie algebroid conditions; the second then follows by standard exterior calculus. The Poisson tensor in terms of the anchor is given by

$$\pi(\theta_1, \theta_2) = \langle \theta_2, \pi^{\#}(\theta_1) \rangle. \qquad (6)$$

Proposition 10.1.3 *In terms of the Schouten bracket of multivector fields (see §7.5),*

$$[\pi, \pi] = 0.$$

Proof We apply 7.5.3. Expanding out $[[\iota_\pi, \delta], \iota_\pi] = \iota_{[\pi,\pi]}$ we have

$$2\iota_\pi \circ \delta \circ \iota_\pi - \iota_\pi \circ \iota_\pi \circ \delta - \delta \circ \iota_\pi \circ \iota_\pi = \iota_{[\pi,\pi]}.$$

Take $\varphi \in \Omega^3(P)$. Evaluating at φ, the final term of the LHS vanishes. Take $f, g, h \in C^\infty(P)$ and set $\varphi = \delta f \wedge \delta g \wedge \delta h$. Then the second term on the LHS also vanishes and we have

$$2(\iota_\pi \circ \delta \circ \iota_\pi)(\varphi) = -\langle \varphi, [\pi, \pi] \rangle.$$

For any $X \in \mathfrak{X}(P)$, we have $\langle \iota_\pi(\varphi), X \rangle = -\langle \varphi, \pi \wedge X \rangle = -\langle \iota_X \varphi, \pi \rangle$ and

$$\iota_X \varphi = X(f) \delta g \wedge \delta h + X(g) \delta h \wedge \delta f + X(h) \delta f \wedge \delta g$$

so

$$\langle \iota_X \varphi, \pi \rangle = X(f)\{g, h\} + X(g)\{h, f\} + X(h)\{f, y\},$$

using (4). Thus we have

$$-\iota_\pi(\varphi) = \{g, h\} \delta f + \{h, f\} \delta g + \{f, g\} \delta h$$

and so

$$-\delta(\iota_\pi(\varphi)) = \delta\{g, h\} \wedge \delta f + \delta\{h, f\} \wedge \delta g + \delta\{f, g\} \wedge \delta h.$$

Lastly,

$$-\iota_\pi(\delta(\iota_\pi(\varphi))) = \{\{g, h\}, f\} + \{\{h, f\}, g\} + \{\{f, g\}, h\} = 0.$$

Thus $\iota_{[\pi,\pi]}(\varphi) = 0$ for all $\varphi = \delta f \wedge \delta g \wedge \delta h$ and hence, by linearity, for all $\varphi \in \Omega^3(P)$. □

It is clear from the proof that $[\pi, \pi] = 0$ is also a sufficient condition for a skew–symmetric 2–vector field to define a Poisson structure by (4).

From the first line of (5) it follows that for a regular Poisson manifold, the adjoint bundle $\ker(\pi^\#)$ is always abelian. For any Poisson manifold P and any $m \in P$, the vertex Lie algebra $\ker(\pi_m^\#)$ of T^*P may be an arbitrary Lie algebra; see 10.3.7 below. In particular, if $\pi^\#$ is of zero rank at some $m \in P$, then T_m^*P is a Lie algebra.

A Lie algebroid structure on a cotangent bundle need not correspond to a Poisson structure: that $a^* = -a$ is clearly a necessary condition. The following gives a usable criterion.

Proposition 10.1.4 *Let M be a manifold. Suppose that T^*M has a Lie algebroid structure such that $a^* = -a$ and such that*

$$[\delta u, \delta v] = \delta(a(\delta u)(v))$$

for all $u, v \in C^\infty(M)$. Then $\{u, v\} = a(\delta u)(v)$ defines a Poisson struc-ture on M for which the Lie algebroid structure is the given one.

Proof That the bracket of functions is skew–symmetric follows from $a^* = -a$; that the bracket is a derivation in each variable then follows. To verify Jacobi, first note that

$$\{\{u, v\}, w\} = -a(\delta w)(a(\delta u)(v)).$$

Lastly, the second hypothesis can be written $[\delta u, \delta v] = \delta(\{u, v\})$ and this shows that the Lie algebroid structure from $\{\ ,\ \}$ is the given one.

$\square$

Definition 10.1.5 A smooth map $f\colon P' \to P$ of Poisson manifolds is a *Poisson map* if $\{u \circ f, v \circ f\} = \{u, v\} \circ f$ for all $u, v \in C^\infty(P)$. It is an *antiPoisson map* if $\{u \circ f, v \circ f\} = -\{u, v\} \circ f$ for all $u, v \in C^\infty(P)$. $\square$

Equivalently, f is Poisson if and only if

$$\pi^{\#}_{P'}(f^*\omega) \stackrel{f}{\sim} \pi^{\#}_P(\omega)$$

for all $\omega \in \Omega^1(P)$, or if and only if

$$\pi_{P'}(f^*\omega_1, f^*\omega_2) = \pi_P(\omega_1, \omega_2) \circ f$$

for all $\omega_1, \omega_2 \in \Omega^1(P)$.

Given a Poisson map $f\colon P' \to P$, define a Lie algebroid action of T^*P on f by

$$\omega \mapsto \omega^\dagger = (f^*\omega)^\#, \quad \Omega^1(P) \to \mathfrak{X}(P'). \tag{7}$$

The dual of $T(f)^!\colon TP' \to f^! TP$ is a vector bundle map

$$C(f)\colon f^! T^*P \to T^*P'$$

with induced map $\Gamma T^*P \to \Gamma T^*P'$ precisely $\omega \mapsto f^*\omega$. It is immediate that $C(f)$ is a comorphism from T^*P' to T^*P, as defined in 4.3.16, and that the induced action of T^*P on f is precisely (7). Thus $C(f)$ is a morphism of Lie algebroids $T^*P \lessdot f \to T^*P'$. In particular,

$$[f^*\omega_1, f^*\omega_2] = f^*[\omega_1, \omega_2]$$

for all $\omega_1, \omega_2 \in \Omega^1(P)$.

A submanifold S of a Poisson manifold P is a *Poisson submanifold* if it has itself a Poisson structure and the inclusion map is Poisson.

We now need the basic terminology for symplectic manifolds.

Definition 10.1.6 A *symplectic structure on a manifold* M is a 2–form $\omega \in \Omega^2(M)$ which is non–degenerate and closed; that is, $\delta\omega = 0$. □

That ω is non–degenerate means that the map $\omega^\flat \colon TM \to T^*M$ defined by $\omega^\flat(X) = -\iota_X\omega$, or equivalently by

$$\langle \omega^\flat(X), Y \rangle = -\omega(X, Y), \tag{8}$$

is an isomorphism. We claim that the 2–vector field π defined by $\pi^\# = (\omega^\flat)^{-1}$ is a Poisson structure. First note that, if this is so, then the bracket of functions will be given by

$$\{f, g\} = \langle \delta g, \pi^\#(\delta f) \rangle = \langle \delta g, (\omega^\flat)^{-1}(\delta f) \rangle.$$

In accordance with the conventions above, write $X_f = (\omega^\flat)^{-1}(\delta f)$ for $f \in C^\infty(M)$. We thereby have $\{f, g\} = X_f(g)$ for all $f, g \in C^\infty(M)$.

Equivalently, $\omega^\flat(X_f) = \delta f$, and this can be rewritten

$$\omega(X_f, Y) = -Y(f) \tag{9}$$

for all $Y \in \mathfrak{X}(M)$. It follows immediately that

$$\omega(X_f, X_g) = -X_g(f) = -\{g, f\}$$

and this gives the skew–symmetry of $\{\, , \,\}$.

We now expand $\delta\omega(X_f, X_g, X_h) = 0$. For the first group of terms we have, typically,

$$X_f(\omega(X_g, X_h)) = \{f, \{g, h\}\}.$$

In the second group we have, using (9),

$$\omega([X_f, X_g], X_h) = +[X_f, X_g](h) = \{f, \{g, h\}\} - \{g, \{f, h\}\}$$

and likewise for the other terms. Altogether we obtain the Jacobi identity for $\{\, , \,\}$, and this completes the proof of the following:

Proposition 10.1.7 *Given a symplectic structure ω on a manifold M, the 2–vector field π such that $\pi^\# = (\omega^\flat)^{-1}$ is a Poisson structure on M.*

Example 10.1.8 For any manifold M, the cotangent bundle T^*M has a canonical 1–form ν, the *Liouville form*, defined as follows. Given $\xi \in T(T^*M)$, write $X = T(c)(\xi) \in TM$, where $c \colon T^*M \to M$ is the projection, and $\varphi = p_{T^*M}(\xi) \in T^*M$; then $\nu(\xi) = \langle \varphi, X \rangle$.

To show that $\omega = \delta\nu$ is a symplectic structure on T^*M, it is only necessary to show that ω is non–degenerate, and this follows from 9.6.7.

⊠

Now return to a general Poisson manifold P. Since T^*P and TP have the same rank, the Lie algebroid T^*P is transitive only if $\pi^\#$ is an isomorphism (of vector bundles, and of Lie algebroids). In this case the Poisson structure is symplectic, with symplectic form

$$\omega(X, Y) = -\langle (\pi^\#)^{-1}(X), Y \rangle.$$

That ω is closed follows from an argument similar to that above.

Next consider a transitivity orbit $\mathcal{O}$ of T^*P, for a general Poisson manifold P. For any $m \in \mathcal{O}$, we claim that $\pi_m^\#: T_m^*(P) \to T_m(\mathcal{O})$ descends to $T_m^*(\mathcal{O})$. First, the kernel of $T_m^*P \to T_m^*\mathcal{O}$ is $(T_m\mathcal{O})^\circ$ and, second, given any $\varphi \in (T_m\mathcal{O})^\circ$ and any $\psi \in T_m^*P$ we have

$$\langle \psi, \pi^\#(\varphi) \rangle = -\langle \varphi, \pi^\#(\psi) \rangle = 0.$$

Now the descended map $T_m^*(\mathcal{O}) \to T_m(\mathcal{O})$ is surjective and hence an isomorphism. It is easy to check that it defines a Poisson structure on $\mathcal{O}$, making $\mathcal{O}$ a Poisson submanifold.

Thus we have:

Proposition 10.1.9 *Each transitivity orbit of the Lie algebroid T^*P has a canonical symplectic structure with respect to which it is a Poisson submanifold of P.*

With this structure, the transitivity orbits are usually called the *symplectic leaves* of P.

A Poisson structure is *regular* if T^*P is regular as a Lie algebroid.

10.2 Poisson cohomology

Consider a Poisson manifold P with Poisson tensor π. A number of standard constructions in Poisson geometry follow immediately from the Lie algebroid structure on T^*P.

Definition 10.2.1 The *Poisson cohomology* of (P, π) is the cohomology of the Lie algebroid T^*P, with coefficients in the trivial representation. It is denoted $H_\pi^\bullet(P)$. □

Remark 10.2.2 The trivial representation of T^*P is the anchor $\pi^\#$, considered as a map assigning to $\varphi \in \Gamma T^*P$ the Lie derivative $\mathcal{L}_{\varphi^\#}$ acting as a derivation on the trivial line bundle $P \times \mathbb{R}$.

One may also, of course, consider cohomology of T^*P with coefficients in any representation $\rho\colon T^*P \to \mathfrak{D}(E)$. Representations of T^*P are sometimes called *flat contravariant connections* in E.

The cochain complex for the Poisson cohomology is the Schouten algebra of multivector fields on P; for clarity we denote this here by $\mathscr{X}^\bullet(P)$. The coboundary of this complex, which we denote by d_π, may be expressed succinctly in terms of the Poisson tensor:

Theorem 10.2.3

$$d_\pi = [\pi, -].$$

Proof First consider $X \in \mathfrak{X}(P)$. For $\varphi, \psi \in \Omega^1(P)$ we have

$$\langle \varphi \wedge \psi, d_\pi(X) \rangle = \mathfrak{L}_{\varphi\#}(\langle \psi, X \rangle) - \mathfrak{L}_{\psi\#}(\langle \varphi, X \rangle) - \langle [\varphi, \psi], X \rangle.$$

Using the definition of $[\varphi, \psi]$ and the derivation property of the Lie derivative, this becomes

$$\langle \psi, \mathfrak{L}_{\varphi\#}(X) \rangle - \langle \varphi, \mathfrak{L}_{\psi\#}(X) \rangle + X(\pi(\varphi, \psi)). \tag{10}$$

On the other hand,

$$\begin{aligned}
\langle \varphi \wedge \psi, [\pi, X] \rangle &= -\langle \varphi \wedge \psi, \mathfrak{L}_X(\pi) \rangle \\
&= -\mathfrak{L}_X \langle \varphi \wedge \psi, \pi \rangle + \langle \mathfrak{L}_X(\varphi \wedge \psi), \pi \rangle \tag{11} \\
&= -X(\pi(\varphi, \psi)) + \pi(\mathfrak{L}_X(\varphi), \psi) + \pi(\varphi, \mathfrak{L}_X(\psi)).
\end{aligned}$$

Now $\pi(\mathfrak{L}_X(\varphi), \psi) = -\langle \mathfrak{L}_X(\varphi), \psi^\# \rangle$. Using this, the derivation property of $\mathfrak{L}_X$ again, and noting that $\mathfrak{L}_X(\varphi^\#) = -\mathfrak{L}_{\varphi\#}(X)$, it follows that the expressions in (11) and (10) are equal.

Now suppose the result has been proved for $\xi \in \Gamma\mathfrak{X}^k(P)$, and take $X \in \mathfrak{X}(P)$. From (22) on p. 304,

$$[\pi, \xi \wedge X] = [\pi, \xi] \wedge X + (-1)^k \xi \wedge [\pi, X].$$

And from (23) on p. 306,

$$d_\pi(\xi \wedge X) = (d_\pi \xi) \wedge X + (-1)^k \xi \wedge d_\pi X.$$

The result follows by induction. □

The map $\pi^\#\colon T^*P \to TP$ extends to a map $\pi^\#\colon \Omega^\bullet(P) \to \mathfrak{X}^\bullet(P)$ in the obvious way, namely

$$\pi^\#(\varphi_1 \wedge \cdots \wedge \varphi_k) = \pi^\#(\varphi_1) \wedge \cdots \wedge \pi^\#(\varphi_k).$$

This is, in effect, the map induced on the cochain complexes corresponding to the change of Lie algebroids $\pi^\#$ (see 7.4.1). From the definition
of d_π it is immediate that

$$d_\pi \circ \pi^\# = \pi^\# \circ \delta, \tag{12}$$

and so there is an induced map $H^\bullet_{\mathrm{deRh}}(P) \to H^\bullet_\pi(P)$.

10.3 Linear Poisson structures

We begin with a notion that concerns any double vector bundle.

Definition 10.3.1 Let $(D; A, B; M)$ be a double vector bundle. A
linear section of the horizontal structure of D is a pair (ξ, X) where
$X \in \Gamma A$ and $\xi \in \Gamma_B(D)$, such that (ξ, X) is a morphism of vector
bundles.   □

A linear section of the vertical structure is defined in the evident way.
For example, in a tangent double vector bundle $(TA; A, TM; M)$ any
linear vector field is a linear section of the vertical structure, while any
section $\mu \in \Gamma A$ induces a linear section $(T(\mu), \mu)$ of the horizontal
structure. In a cotangent double vector bundle $(T^*A; A, A^*; M)$, any
section $\varphi \in \Gamma A^*$ induces a linear section $(\delta \ell_\varphi, \varphi)$.

Definition 10.3.2 Let (E, q, M) be a vector bundle and π a Poisson
structure on E. The Poisson structure is *linear*, and E together with
π is a *Poisson vector bundle*, if there exists a map $a \colon E^* \to TM$, necessarily a vector bundle map over M, such that $\pi^\#$ is a morphism of
double vector bundles over E and a.   □

Consider such a linear Poisson structure. Each $X \in \Gamma E^*$ induces the
linear section $(\delta \ell_X, X)$ of T^*E. Composing with $\pi^\#$ we obtain a linear
vector field $(\pi^\#(\delta \ell_X), a(X))$ on E; write $H_X = \pi^\#(\delta \ell_X)$ for brevity.
Now ℓ_Y, for any other $Y \in \Gamma E^*$, is a linear function, so $H_X(\ell_Y)$ is also
a linear function. Define $[X, Y] \in \Gamma E^*$ by

$$H_X(\ell_Y) = \ell_{[X,Y]}. \tag{13}$$

It follows that

$$\{\ell_X, \ell_Y\} = \ell_{[X,Y]}, \qquad \{\ell_X, f \circ q\} = a(X)(f) \circ q \tag{13'}$$

for $f \in C^\infty(M)$. Next, from the skew–symmetry of π and 9.2.1, it
follows that the core morphism of $\pi^\#$ is $-a^*$. Each $\omega \in \Omega^1(M)$ defines

the core 1–form $q^*\omega$, the image of which under $\pi^\#$ must be the vertical vector field corresponding to $-a^*(\omega)$; thus

$$\pi^\#(q^*\omega) = -a^*(\omega)^\uparrow.$$

From the properties of vertical vector fields, it follows that

$$\{f \circ q, g \circ q\} = 0 \qquad (14)$$

for all $f, g \in C^\infty(M)$. This proves half the next result.

Proposition 10.3.3 *Let (E, q, M) be a vector bundle and π a Poisson structure on E. The Poisson structure is linear if and only if the following three conditions on the bracket of functions hold:*

- *the bracket of two linear functions is linear;*
- *the bracket of a linear function with a pullback function is a pullback function;*
- *the bracket of two pullback functions is zero.*

Proof Assume that the three conditions hold. Let ℓ be any linear function on E. Then for any other linear function ℓ',

$$\pi^\#(\delta\ell)(\ell') = \langle \delta\ell', \pi^\#(\delta\ell) \rangle = \{\ell, \ell'\}$$

is linear. And for any $f \in C^\infty(M)$,

$$\pi^\#(\delta\ell)(f \circ q) = \langle \delta(f \circ q), \pi^\#(\delta\ell) \rangle = \{\ell, f \circ q\}$$

is a pullback function. It follows from 3.4.2 that $\pi^\#(\delta\ell)$ is a linear vector field. For $\ell = \ell_X$, $X \in \Gamma E^*$, denote it by H_X and denote the vector field on M to which it projects by $a(X)$. Then H_X is linear in X, and a defines a linear map $E^* \to TM$.

Now take any $f \in C^\infty(M)$. As above, for $X \in \Gamma E^*$,

$$\pi^\#(q^*\delta f)(\ell_X) = \{f \circ q, \ell_X\}$$

is a pullback function, and $\pi^\#(q^*\delta f)(g \circ q) = 0$ for all $g \in C^\infty(M)$. Hence $\pi^\#(q^*\delta f)$ is a vertical vector field on E; denote it by $b(\delta f)^\uparrow$. Now we have both

$$\{\ell_X, f \circ q\} = H_X(f \circ q) = a(X)(f) \circ q$$

and

$$\{f \circ q, \ell_X\} = \langle \delta\ell_X, \pi^\#(q^*\delta f) \rangle = \langle \delta\ell_X, b(\delta f)^\uparrow \rangle$$
$$= \langle X, b(\delta f) \rangle \circ q = (b^*X)(f) \circ q.$$

Hence, by the skew–symmetry of the bracket, $a = -b^*$.

Now we can prove that $\pi^\#\colon T^*E \to TE$ is linear over $a\colon E^* \to TM$. That $T(q) \circ \pi^\# = a \circ r$, where r is the bundle projection $T^*E \to E^*$ follows from the facts that H_X projects to $a(X)$ and $-a^*(\delta f)^\uparrow$ projects to zero.

To see that $\pi^\#(\Phi \underset{E^*}{+} \Psi) = \pi^\#(\Phi) \mathbin{+\!\!\!+} \pi^\#(\Psi)$ when $r(\Phi) = r(\Psi)$, note first that for any $X \in \Gamma E^*$, the linear section $\delta \ell_X$ is mapped to the linear vector field H_X; and that core elements of T^*E are mapped by $\pi^\#$ to core elements of TE. Using 9.4.1, it follows that $\pi^\#$ is linear. $\qquad\square$

Theorem 10.3.4 *Let (E, q, M) be a vector bundle with a linear Poisson structure π inducing $a\colon E^* \to TM$. Then the bracket on ΓE^* defined by (13) defines a Lie algebroid structure on E^* with anchor a.*

Proof Firstly, $\ell_{[X,Y]} = \langle \delta \ell_Y, \pi^\#(\delta \ell_X) \rangle = \pi(\delta \ell_X, \delta \ell_Y)$, so the bracket on ΓE^* is skew–symmetric since the Poisson tensor is so. Secondly,

$$H_X(\ell_{fY}) = H_X((f \circ q)\ell_Y) = (f \circ q)\ell_{[X,Y]} + (a(X)(f) \circ q)\ell_Y$$

and from this the Leibniz condition follows.

Next, from the equations (13′) we have

$$a([X,Y])(f) \circ q = \{\ell_{[X,Y]}, f \circ q\} = \{\{\ell_X, \ell_Y\}, f \circ q\}$$

and using the Jacobi identity for $\{\ ,\ \}$ and (13′) again, we have that $a([X,Y]) = [aX, aY]$.

The Jacobi identity in ΓE^* is proved in a similar way by expanding $\ell_{[[X,Y],Z]}$ according to (13′), and using the Jacobi identity for $\{\ ,\ \}$. $\qquad\square$

Now consider any Lie algebroid (A, q, M). From 9.4.1, a map on $T^*(A^*)$ is uniquely determined by its values on linear forms and the pullbacks from M. Define $\pi^\#\colon T^*(A^*) \to T(A^*)$ by

$$\pi^\#(\delta \ell_X) = H_X, \qquad \pi^\#(q_*^*\omega) = -a^*(\omega)^\uparrow, \tag{15}$$

for $X \in \Gamma A$, $\omega \in \Omega^1(M)$, where H_X is the linear vector field on A^* of 3.4.8 corresponding to X, and $a^*(\omega)^\uparrow$ is the core vector field on A^* defined by $a^*(\omega)$.

In terms of the bracket of functions, these are equivalent, by (6), to

$$\{\ell_X, \ell_Y\} = \ell_{[X,Y]}, \quad \{\ell_X, f \circ q_*\} = a(X)(f) \circ q_*, \quad \{f \circ q_*, g \circ q_*\} = 0, \tag{16}$$

for $X, Y \in \Gamma A$, $f, g \in C^\infty(M)$.

Theorem 10.3.5 *The structure just defined is a linear Poisson structure on A^*.*

Proof For skew–symmetry, we have, for $X \in \Gamma A$, $\omega \in \Omega^1(M)$,

$$\langle q_*^*\omega, \pi^\#(\delta\ell_X)\rangle = \langle q_*^*\omega, H_X\rangle = \langle \omega, aX\rangle \circ q_*$$

since $H_X \sim a(X)$, and on the other hand,

$$\langle \delta\ell_X, \pi^\#(q_*^*(\omega))\rangle = \langle \delta\ell_X, -a^*(\omega)^\top\rangle = -a^*(\omega)^\top(\ell_X)$$
$$= -\langle a^*(\omega), X\rangle \circ q_* = -\langle \omega, aX\rangle \circ q_*,$$

using (21) on p. 112. Thus $\pi(\delta\ell_X, q_*^*\omega) = -\pi(q_*^*\omega, \delta\ell_X)$. The other two cases are immediate from (16).

The Jacobi identity for any three linear functions ℓ_X, ℓ_Y, ℓ_Z follows immediately from the Jacobi identity in ΓA. For any three pullback functions it is trivial. For the case $\ell_X, \ell_Y, f \circ q_*$, we have, for the first term,

$$\{\ell_X, \{\ell_Y, f \circ q_*\}\} = \{\ell_X, a(Y)(f) \circ q_*\} = a(X)(a(Y)(f)) \circ q_*$$

and Jacobi follows from $a[X, Y] = [aX, aY]$. Similarly with the other mixed case.

This proves that π is a Poisson structure; the linearity follows from 10.3.3. $\qquad\square$

These two constructions are mutually inverse and give a bijective correspondence between Lie algebroid structures on A and linear Poisson structures on A^*. In terms of the bracket of 1–forms, (16) becomes:

$$[\delta\ell_X, \delta\ell_Y] = \delta\ell_{[X,Y]}, \quad [\delta\ell_X, q_*^*\theta] = q_*^*(\mathcal{L}_{aX}(\theta)), \quad [q_*^*\omega, q_*^*\theta] = 0, \quad (17)$$

for $X, Y \in \Gamma A$, $\omega, \theta \in \Omega^1(M)$.

Proposition 10.3.6 *Let (E, q, M) be a vector bundle with a linear Poisson structure π inducing $a \colon E^* \to TM$. Then $r \colon T^*E \to E^*$ is a morphism (and fibration) of Lie algebroids over q.*

Proof The anchor condition $a \circ r = T(q) \circ \pi^\#$ is part of the condition that $\pi^\#$ be linear over a. Since q is a surjective submersion and r is fibrewise surjective, we can apply 4.3.8.

From (33) on p. 358, it follows that $\Phi \in \Omega^1(E)$ projects to $X \in \Gamma E^*$ if and only if

$$\langle \Phi, \mu^\top\rangle = \langle \mu, X\rangle \circ q$$

for all $\mu \in \Gamma E$, and from 9.4.1 it follows that if $\varphi \in \Omega^1(E)$ projects to $X \in \Gamma E^*$, then $\Phi = \delta \ell_X + F q^* \omega$ for some $F \in C^\infty(E)$ and some $\omega \in \Omega^1(M)$. Consider such a Φ and another $\Psi \in \Omega^1(E)$ with $\Psi = \delta \ell_Y + G q^* \theta$. Expanding out $\langle [\Phi, \Psi], \mu^\uparrow \rangle$, we have

$$\langle [\delta \ell_X, \delta \ell_Y], \mu^\uparrow \rangle + \langle [F q^* \omega, \delta \ell_Y], \mu^\uparrow \rangle + \langle [\delta \ell_X, G q^* \theta], \mu^\uparrow \rangle + \langle [F q^* \omega, G q^* \theta], \mu^\uparrow \rangle.$$

The first term is $\langle \delta \ell_{[X,Y]}, \mu^\uparrow \rangle = \langle [X, Y], \mu \rangle \circ q$ by (21) on p. 112. The third term is

$$\langle G[\delta \ell_X, q^* \theta], \mu^\uparrow \rangle + \langle H_X(G) q^* \theta, \mu^\uparrow \rangle = G \langle q^* (\mathcal{L}_{aX} \theta), \mu^\uparrow \rangle + H_X(G) \langle q^* \theta, \mu^\uparrow \rangle$$

and both pairing terms vanish since $\mu^\uparrow$ is vertical. Likewise the second and the fourth terms vanish. Altogether we have that

$$\langle [\Phi, \Psi], \mu^\uparrow \rangle = \langle \mu, [X, Y] \rangle \circ q$$

for all $\mu \in \Gamma E$ and so $[\Phi, \Psi]$ projects to $[X, Y]$ as required. $\qquad\square$

Examples 10.3.7 For $A = \mathfrak{g}$, a Lie algebra, the definition $\{\ell_X, \ell_Y\} = \ell_{[X,Y]}$ extends readily to

$$\{F, G\}(\theta) = \langle \theta, [D(F)(\theta), D(G)(\theta)] \rangle,$$

for any $F, G \in C^\infty(M)$ and $\theta \in \mathfrak{g}^*$, where $D(F)(\theta)$ denotes the elementary 'directional derivative' $\mathfrak{g}^* \to \mathbb{R}$, regarded as an element of $\mathfrak{g}$.

Now consider a cotangent bundle $T^* M$, with the Poisson structure defined by the symplectic structure $\delta \nu$, where ν is the canonical 1–form on $T^* M$. It was proved in 9.6.7 that $(\delta \nu)^\flat = \Theta^{-1} \circ R$. Both Θ and R are isomorphisms of double vector bundles over TM and $T^* M$ (and Θ preserves the core, while R is $-\mathrm{id}$ on the core). Hence $(\delta \nu)^\flat$ is a linear Poisson structure and therefore induces a Lie algebroid structure on TM. Since the map $TM \to TM$ induced by $(\delta \nu)^\flat$ is the identity, the anchor of this structure on TM is also. Hence it is the standard Lie algebroid structure on TM.

This gives the following explicit formulas for the brackets: For $X \in \mathcal{X}(M)$, denote by ℓ_X the corresponding function $T^* M \to \mathbb{R}$; then

$$\{\ell_X, \ell_Y\} = \ell_{[X,Y]}, \qquad \{\ell_X, f \circ c\} = X(f) \circ c, \qquad \{f \circ c, g \circ c\} = 0, \quad (18)$$

where $c \colon T^* M \to M$ is the projection and $f, g \in C^\infty(M)$. $\qquad\boxtimes$

Example 10.3.8 Let $\mathfrak{g} \to \mathcal{X}(M)$ be an action of a Lie algebra on a smooth manifold and let $A = \mathfrak{g} \ltimes M$ be the action Lie algebroid. Then $A^* = M \times \mathfrak{g}^*$ has a linear Poisson structure, the *semi–direct Poisson structure corresponding to the action* $\mathfrak{g} \to \mathcal{X}(M)$. $\qquad\boxtimes$

The following result will be needed in §12.2.

Proposition 10.3.9 *Let A be a Lie algebroid on M and write R for $R_{A^*} \colon T^*A \to T^*A^*$.*

For $\varphi \in \Gamma A^*$ and any $X, Y \in A_m$, $m \in M$,

$$\pi_{A^*}(R(\delta\ell_\varphi(X)),\ R(\delta\ell_\varphi(Y))) = -d\varphi(X, Y).$$

For $X, Y \in A_m$, $m \in M$, and $\omega \in T_m^*M$,

$$\pi_{A^*}(R((q^*\omega)(X)), R((q^*\omega)(Y))) = \langle \omega, aY - aX \rangle,$$

where $(q^*\omega)(X)$ is the pullback of ω to X.

Proof Let $X, Y \in \Gamma A$ be any sections with the specific values at m. Then, by 9.5.3,

$$\pi_{A^*}(R(\delta\ell_\varphi(X(m))),\ R(\delta\ell_\varphi(Y(m)))) =$$
$$\pi_{A^*}(\delta\ell_X(\varphi(m)) - q_*^*(\delta\langle\varphi, X\rangle)(\varphi(m)), \delta\ell_Y(\varphi(m)) - q_*^*(\delta\langle\psi, Y\rangle)(\varphi(m)).$$

Using the formulas (16), this becomes

$$\langle[X, Y], \varphi\rangle - a(X)(\langle\varphi, Y\rangle) + a(Y)(\langle\varphi, X\rangle),$$

all evaluated at $\varphi(m)$, and this is $-d\varphi(X, Y)(\varphi(m))$ and depends only on the given $X(m)$ and $Y(m)$.

For the second half, note first that $(q^*\omega)(X) = \widetilde{0}_X +_{A^*} \overline{\omega} \in T_X^*A$.

Since R is a double vector bundle morphism reversing the core, it follows that $R(\widetilde{0}_X +_{A^*} \overline{\omega}) = \widetilde{0}_X^r - \overline{\omega}$. Now $\widetilde{0}_X^r = \delta\ell_X(0_m)$, where $X \in \Gamma A$ is any section passing through the given X. So we have

$$\pi_{A^*}(\delta\ell_X(0_m) - \overline{\omega}, \delta\ell_Y(0_m) - \overline{\omega}) = \ell_{[X,Y]}(0_m) - \langle\omega, aX\rangle + \langle\omega, aY\rangle,$$

whence the result. $\square$

We now consider the tangent prolongation of Poisson structures. We will need the following lemma, now and later.

Lemma 10.3.10 *For any manifold M, smooth function $f \in C^\infty(M)$, and 1-form $\varphi \in \Omega^1(M)$, we have*

$$\Theta \circ T(\delta f) = \delta\widetilde{f}, \qquad \Theta \circ \widehat{\varphi} = p^*\varphi. \tag{19}$$

Proof For the first equation, calculate $\langle \Theta(T(\delta f)(X)), \xi \rangle$ and $\langle \delta \tilde{f}(X), \xi \rangle$ in the two cases where ξ is a complete lift $\tilde{x}(X)$ and a vertical lift $x^\uparrow(X)$. Use equation (21) on p. 112, equation (32) on p. 117, and Example 3.4.8, together with (38) from p. 364.

For the second statement note that $\hat{\varphi}$ is a core section, and Θ preserves the core; the pullback form $p^*\varphi$ is the corresponding core section of the cotangent double vector bundle. $\qquad\square$

It is important to note that $\Theta \circ T(\varphi) = \delta \ell_\varphi$ is not true for general 1-forms φ.

Now consider a Poisson manifold P. Since $T^*P \to P$ is a Lie algebroid, its dual $TP \to P$ has a linear Poisson structure, the bracket of which is given explicitly by

$$\{\ell_\varphi, \ell_\psi\} = \ell_{[\varphi,\psi]}, \quad \{\ell_\varphi, f \circ p\} = \pi^\#(\varphi)(f) \circ p, \quad \{f \circ p, g \circ p\} = 0, \quad (20)$$

for $\varphi, \psi \in \Omega^1(P)$, $f, g \in C^\infty(P)$. Here p is the bundle projection. Consider in particular $\varphi = \delta f$, $\psi = \delta g$. We then have

$$\{\tilde{f}, \tilde{g}\} = \{f, g\}^\sim, \quad \{\tilde{f}, g \circ p\} = X_f(g) \circ p, \quad \{f \circ p, g \circ p\} = 0, \quad (21)$$

where $\tilde{f} = \ell_{\delta f}$ is the linear function $TP \to \mathbb{R}$ corresponding to δf. (Notice that the first equation in (21) is essentially $[\delta f, \delta g] = \delta\{f, g\}$ from (3). It will clarify matters later to maintain the distinction between $\tilde{f}$ and δf.)

To paraphrase: the Hamiltonian of $\tilde{f}$ is the complete lift of the Hamiltonian of f, and the Hamiltonian of $p^*\omega$ is the vertical lift of $\omega^\#$.

Definition 10.3.11 The Poisson structure on TP just defined is the *tangent lift* or *tangent prolongation* of the Poisson structure on P. $\qquad\square$

Proposition 10.3.12 *Given a Poisson structure $\pi^\#$ on a manifold P, the anchor $\pi^\#_{TP}$ for the tangent lift Poisson structure makes the following diagram commute.*

$$
\begin{array}{ccc}
& T(\pi^\#) & \\
T(T^*P) & \longrightarrow & T(TP) \\
\\
\Theta \downarrow & & \downarrow J \qquad (22) \\
\\
T^*(TP) & \longrightarrow & T(TP) \\
& \pi^\#_{TP} &
\end{array}
$$

Proof This follows by verification on the generating elements $T(\delta f)$ and $\widehat{\varphi}$ in $T(T^*P)$, using (19). □

Proposition 10.3.13 *For any Poisson manifold P, the map*

$$\Theta \colon T(T^*P) \to T^*(TP)$$

is an isomorphism of Lie algebroids over TP from the tangent prolongation of $T^*P \to P$ to the cotangent Lie algebroid of TP.

Proof The anchor on the domain is $J \circ T(\pi^\#)$ and the anchor on the target is $\pi_{TP}^\#$; thus (22) is precisely anchor commutativity.

For the brackets, it suffices to consider the sections $T(\delta f)$ and $\widehat{\varphi}$ of the domain, for $f \in C^\infty(M)$, $\varphi \in \Omega^1(M)$. The verifications are straightforward. □

Theorem 10.3.14 (i) *If $E \to M$ is a Poisson vector bundle, then the tangent Poisson structure on TE makes $TE \to TM$ a Poisson vector bundle.*

(ii) *Let (A, q, M) be a Lie algebroid, and give $TA \to TM$ the tangent Lie algebroid structure of Theorem 9.7.1, and $A^* \to M$ the dual Poisson structure of Theorem 10.3.5. Then $I_A \colon T(A^*) \to T^\bullet(A)$ is a Poisson isomorphism with respect to the tangent Poisson structure on $T(A^*)$ induced from the Poisson structure on A^*, and the dual Poisson structure on $T^\bullet(A)$ induced from the tangent Lie algebroid structure on $TA \to TM$.*

Proof It is instructive to prove (i) directly from the definition 10.3.2. Applying the tangent functor to the morphism of double vector bundles $\pi^\#$ gives a morphism of triple vector bundles, and linearity with respect to a different 'double face' is what is required.

Alternatively, one may apply 10.3.3. First consider the functions $TE \to \mathbb{R}$ which are pullbacks across $T(q) \colon TE \to TM$. These are generated by those of the types $\widetilde{f} \circ T(q)$ and $f \circ p_M \circ T(q)$ for $f \in C^\infty(M)$. For the functions $TE \to \mathbb{R}$ which are linear with respect to $TE \to TM$ we need the sections of the dual bundle $T^\bullet(E) \to TM$. Using I, these are generated by those of the forms

$$I \circ T(X) \qquad \text{and} \qquad I \circ \widehat{X}$$

for $X \in \Gamma E^*$. For clarity we denote the corresponding functions by $\ell_{I \circ T(X)}^\bullet$ and $\ell_{I \circ \widehat{X}}^\bullet$.

Next notice that

$$\ell^{\bullet}_{I \circ T(X)} = \widetilde{\ell_X} \qquad \text{and} \qquad \ell^{\bullet}_{I \circ \widehat{X}} = \ell_X \circ p_E.$$

To prove the first, consider any $\xi = \frac{d}{dt}\varphi_t\big|_0 \in TE$ where $\varphi_t \in E$. Let $q(\varphi_t) = m_t$ so that $x = T(q)(\xi) = \frac{d}{dt}m_t\big|_0$. Now

$$\ell^{\bullet}_{I \circ T(X)}(\xi) = \langle I(T(X)(x)), \xi \rangle = \langle\!\langle T(X)(x), \xi \rangle\!\rangle$$

$$= \frac{d}{dt}\langle X(m_t), \varphi_t \rangle\bigg|_0 = \frac{d}{dt}\ell_X(\varphi_t)\bigg|_0 = \xi(\ell_X) = \widetilde{\ell_X}(\xi).$$

The proof of the second is similar.

It is important to note that the functions $\ell_X \circ p_E$ are linear for $TE \to TM$ although they are pullbacks for $TE \to E$. Likewise $\widetilde{f \circ T(q_E)} = \widetilde{f} \circ q_E$ is a complete lift for $TE \to E$ although it is a pullback for $TE \to TM$. Bearing this in mind, and using (21), we can calculate the various types of brackets and verify that the conditions of 10.3.3 hold. For example, both

$$\{\widetilde{\ell_X}, \widetilde{\ell_Y}\} = \widetilde{\{\ell_X, \ell_Y\}} = \widetilde{\ell_{[X,Y]}}$$

and $\{\ell_X \circ p_E, \ell_Y \circ p_E\} = 0$ are linear with respect to $TE \to TM$. Likewise,

$$\{\widetilde{\ell_X}, f \circ q_E \circ p_E\} = \{\ell_X, f \circ q_E\} \circ p_E$$

$$= a(X)(f) \circ q_E \circ p_E = a(X)(f) \circ p_M \circ T(q_E)$$

is a pullback function with respect to $TE \to TM$.

For (ii), proceed in a similar way, now using

$$\widetilde{\ell_X} = \ell^{\bullet A}_{T(X)} \circ I_A, \qquad \ell_X \circ p_* = \ell^{\bullet A}_{\widehat{X}} \circ I_A,$$

where $\ell^{\bullet A}$ refers to the duality between $TA \to TM$ and $T^*A \to TM$. $\qquad\square$

Morphisms are considered in the section which follows.

10.4 Coisotropic submanifolds, Lie subalgebroids, and morphisms

Consider a closed embedded submanifold C of a manifold P. We denote by $(TC)^{\circ}$ the *conormal bundle of C in P*: that subset of $T^*P|_C$ which consists of the elements which annihilate TC. As a vector bundle, $(TC)^{\circ}$ is isomorphic to the dual of the normal bundle $TP|_C/TC$.

Definition 10.4.1 Let P be a Poisson manifold and let $C \subseteq P$ be a closed embedded submanifold. Then C is a *coisotropic submanifold of P* if it satisfies one and hence all of the following equivalent conditions:

  • the subset of $C^\infty(P)$ consisting of functions which vanish on C is a Lie subalgebra of $C^\infty(P)$;

  • whenever $u \in C^\infty(P)$ vanishes on C, the vector field X_u, restricted to C, is tangent to C;

  • the Poisson anchor $\pi^\# \colon T^*P \to TP$, when restricted to $(TC)^\circ$, goes into TC.   □

Theorem 10.4.2 *Let C be a closed embedded coisotropic submanifold of a Poisson manifold P. Then $(TC)^\circ$ is a Lie subalgebroid of T^*P over C.*

Proof We must verify the three conditions of 4.3.14. The first is the definition that C is coisotropic. For the second, take any $X \in \mathfrak{X}(X)(C)$ and extend it to a vector field on P, also denoted X. Take $\varphi, \psi \in \Omega^1(P)$ such that $\varphi|_C, \psi|_C$ are sections of $(TC)^\circ$. From (5) we have

$$\langle [\varphi, \psi], X \rangle = \langle \mathfrak{L}_{\varphi^\#}(\psi), X \rangle - \langle \mathfrak{L}_{\psi^\#}(\varphi), X \rangle - \langle \delta(\pi(\varphi, \psi)), X \rangle. \tag{23}$$

The first term is

$$\langle \mathfrak{L}_{\varphi^\#}(\psi), X \rangle = \mathfrak{L}_{\varphi^\#}(\langle \psi, X \rangle) - \langle \varphi, [\psi^\#, X] \rangle.$$

On the RHS, the first term is zero on C because $\langle \psi, X \rangle$ is constant on C. In the second term, both $\psi^\#$ and X are tangent to C on C, and so their Lie bracket is also; hence the pairing with φ is zero on C.

The second term of (23) is likewise zero on C. The third term is $-X(\pi(\varphi, \psi))$ and $\pi(\varphi, \psi) = \langle \psi, \varphi^\# \rangle$ vanishes on C since $\varphi^\#$ is tangent to C on C.

The third condition is verified in a similar way.   □

With this structure, $(TC)^\circ$ is the *conormal Lie algebroid of C in P*. The anchor is the restriction to $(TC)^\circ \to TC$ of $\pi^\#$. The dual Poisson structure on the normal bundle is the *linearized Poisson structure*.

Examples 10.4.3 If m_0 is a point of P at which π has rank zero, then $\{m_0\}$ is a coisotropic submanifold of P. In this case the conormal Lie algebroid is a Lie algebra structure on $T^*_{m_0}P$ itself.

The Poisson structure on a Poisson Lie group G vanishes at the identity and the conormal Lie algebra structure on $\mathfrak{g}^*$ is the structure constructed by other means in §11.1.

More generally, anticipating §11.4, given a Poisson groupoid $G \rightrightarrows P$, the submanifold 1_P is coisotropic in G and the conormal Lie algebroid is the Lie algebroid dual A^*G.   ☒

In place of the Lie subalgebroid TC of TP defined by a submanifold, now consider the wide Lie subalgebroid $T^f P$ defined by a surjective submersion $f \colon P \relbar\joinrel\twoheadrightarrow Q$. This is the Lie algebroid of the Lie subgroupoid $R(f) \subseteq P \times P$ used in Chapter 2. The notation $\overline{P}$ below denotes P equipped with the Poisson structure $-\pi$; this is the *opposite Poisson structure*.

Theorem 10.4.4 *Let P be a Poisson manifold and let $f \colon P \relbar\joinrel\twoheadrightarrow Q$ be a surjective submersion. Then there is a (necessarily unique) Poisson structure on Q making f a Poisson map if and only if $R(f)$ is a coisotropic submanifold of $P \times \overline{P}$.*

Proof Recall that $T(R(f)) = R(T(f)) = TP \times_{TQ} TP$. Hence $T(R(f))^\circ$ is the set of all $(\varphi_1, \varphi_2) \in T^*_{m_1} P \times T^*_{m_2} P$ such that $(m_1, m_2) \in R(f)$ and such that (φ_1, φ_2) satisfies the condition:

for $X_1 \in T_{m_1} P$, $X_2 \in T_{m_2} P$,

$$T(f)(X_1) = T(f)(X_2) \implies \langle \varphi_1, X_1 \rangle = \langle \varphi_2, X_2 \rangle. \quad (24)$$

Assume that $R(f)$ is coisotropic. Take any $\psi \in T^*_n Q$. Take any $m_1 \in f^{-1}(n)$ and form $f^* \psi \in T^*_{m_1} P$. Define

$$\pi^{\#}_Q(\psi) = T(f)(\pi^{\#}_P(f^* \psi)).$$

This is well defined because given another choice $m_2 \in f^{-1}(n)$, we have $(m_1, m_2) \in R(f)$ and, writing φ_1 and φ_2 for the pullbacks $f^* \psi$ to m_1 and m_2 respectively, we have that (φ_1, φ_2) satisfies (24). This is immediate, for $\langle \varphi_1, X_1 \rangle = \langle \psi, T(f)(X_1) \rangle$ for $X_1 \in T_{m_1} P$, and similarly for φ_2. The verification that $\pi^{\#}_Q$ is indeed a Poisson structure on Q is straightforward, and the equation by which it was defined shows that f is now a Poisson map.

Conversely suppose that f is Poisson with respect to a Poisson structure on Q. Take $(m_1, m_2) \in R(f)$ and $\varphi_1 \in T^*_{m_1} P$, $\varphi_2 \in T^*_{m_2} P$ such that (24) holds. Define $\psi \in T^*_n Q$, where $n = f(m_1) = f(m_2)$ by

$$\langle \psi, T(f)(X_1) \rangle = \langle \varphi_1, X_1 \rangle.$$

This is well defined for $X_1 \in T_{m_1} P$ because given any other $X'_1 \in T_{m_1} P$ with $T(f)(X'_1) = T(f)(X_1)$, both X_1 and X'_1 can be substituted in the

LHS of (24). Now $\varphi_1 = f_{m_1}^* \psi$, the pullback of ψ to m_1. Similarly $\varphi_2 = f_{m_2}^* \psi$, the pullback of ψ to m_2. Therefore,

$$T(f)(\varphi_1^{\#}) = T(f)(\pi_P^{\#}(f_{m_1}^* \psi)) = \pi_Q^{\#}(\psi)$$

since f is a Poisson map. But, equally, $T(f)(\varphi_2^{\#}) = \pi_Q^{\#}(\psi)$. This proves that the Poisson structure on $P \times \overline{P}$ maps (φ_1, φ_2) into $T(R(f))$. So $R(f)$ is coisotropic. $\square$

Associated with any surjective submersion $f \colon P \longrightarrow\!\!\!\!\gg Q$ is the exact sequence $T^f P \gg\!\!\!\longrightarrow TP \longrightarrow\!\!\!\!\gg TQ$. This has dual

$$f^! T^* Q \gg\!\!\!\longrightarrow T^* P \longrightarrow\!\!\!\!\gg (T^f P)^*$$

and when f is a Poisson map of Poisson manifolds, the injection is a morphism of Lie algebroids over P from $T^* Q \lhd f$ to $T^* P$. In general $T^* Q \lhd f$ is not an ideal system of $T^* P$.

Notice that both in the case of a coisotropic submanifold $C \subseteq P$ and a coisotropic equivalence relation $R(f) \subseteq P \times \overline{P}$, the corresponding Lie algebroid construction is a Lie subalgebroid. One might expect a cotangent process to interchange subobjects with quotients, but the relationship between Poisson constructions and Lie algebroid constructions is essentially a direct one. There is one further result of this type.

For any vector bundle A on M, and wide vector subbundle B of A, denote by B° the subbundle of A^* which annihilates B.

Proposition 10.4.5 *Let A be a Lie algebroid on M and let B be a wide vector subbundle of A. Then B is a Lie subalgebroid of A if and only if B° is coisotropic in A^*.*

Proof Of the functions ℓ_X for $X \in \Gamma A$, those which vanish on B° are precisely those for which $X = Y \in \Gamma B$. If B is a Lie subalgebroid of A then $\{\ell_{Y_1}, \ell_{Y_2}\} = \ell_{[Y_1, Y_2]}$ and so the linear functions which vanish on B° are closed under the Poisson bracket. Of the pullback functions, only the zero function vanishes on B°. Thus B° is coisotropic in A^*. The converse is similar. $\square$

$$*\quad *\quad *\quad *\quad *$$

It was already noted at the end of §10.1 that if $f \colon P \to Q$ is a Poisson map then $C(f) \colon T^* Q \lhd f \to T^* P$ is a Lie algebroid morphism. We now turn to consider the relationships between a morphism of Lie algebroids and the induced map of the duals. There are a number of results. We start with the simplest.

Proposition 10.4.6 *Let A and B be Lie algebroids on the base M and consider a morphism of vector bundles $F\colon A \to B$ over M. Then F is a morphism of Lie algebroids if and only if $F^*\colon B^* \to A^*$ is a Poisson map.*

Proof Assume that F is a Lie algebroid morphism. Note that, for $X \in \Gamma A$, $\ell_X \circ F^* = \ell_{FX}$. So, in the Poisson structure on B^*,

$$\{\ell_{X_1} \circ F^*, \ell_{X_2} \circ F^*\} = \{\ell_{FX_1}, \ell_{FX_2}\}$$
$$= \ell_{[FX_1, FX_2]} = \ell_{[X_1, X_2]} \circ F^* = \{\ell_{X_1}, \ell_{X_2}\} \circ F^*.$$

Similarly the anchor preservation property implies that

$$\{\ell_X \circ F^*, f \circ q_{A^*} \circ F^*\} = \{\ell_X, f \circ q_{A^*}\} \circ F^*,$$

for $f \in C^\infty(M)$. Thus F^* is Poisson. The converse is similar. $\qquad\square$

The problem in extending this result to general morphisms of Lie algebroids is that the dual is no longer a map defined on the duals, but a relation, as we define now.

Definition 10.4.7 Consider manifolds P and Q. A *relation from P to Q* is a submanifold $R \subseteq Q \times P$. If P and Q are Poisson manifolds, the relation is *coisotropic* if R is a coisotropic submanifold of $Q \times \overline{P}$. $\qquad\square$

The concept of relation is in itself symmetric but we use the terminology 'from P to Q' in order to show the order in which we take the factors. The next result shows that the definition is consistent with the notion of Poisson map.

Proposition 10.4.8 *Let P and Q be Poisson manifolds and consider a smooth map $f\colon P \to Q$. Then f is Poisson if and only if its graph*

$$\mathrm{Gr}(f) = \{(f(m), m) \mid m \in P\}$$

is coisotropic in $Q \times \overline{P}$.

Proof Write $S = \mathrm{Gr}(f)$. Notice firstly that $TS = \mathrm{Gr}(T(f)) \subseteq TQ \times TP$, so $(TS)^\circ$ is the set of all $(\psi, \varphi) \in T^*_{f(m)}Q \times T^*_m P$ such that $m \in M$ and such that for all $X \in T_m P$,

$$\langle \psi, T(f)(X) \rangle = \langle \varphi, X \rangle.$$

That is,

$$(TS)^\circ = \{(\psi, f_m^* \psi) \in T_{f(m)}^* Q \times T_m^* P \mid m \in M, \ \psi \in T_{f(m)}^* Q\}.$$

The condition that $\pi_Q^\# \times \overline{\pi_P^\#}$ maps $(\psi, f_m^* \psi) \in (TS)^\circ$ into TS is that

$$T(f)(\pi_P^\#(f_m^* \psi)) = \pi_Q^\#(\psi)$$

and this, for all ψ, is precisely the condition that f is a Poisson map. This proves the equivalence. $\qquad\square$

Now consider a general morphism of vector bundles $F \colon A \to B$ over $f \colon M \to N$. Define a relation F^* from B^* to A^* as the set of all $(\varphi, \psi) \in A_m^* \times B_{f(m)}^*$ such that $m \in M$ and such that $\langle \varphi, X \rangle = \langle \psi, F(X) \rangle$ for all $X \in A_m$. That is,

$$F^* = \{(F_m^* \psi, \psi) \mid m \in M, \ \psi \in B_{f(m)}^*\}, \tag{25}$$

where $F_m^* \psi = \psi \circ F_m \in A_m^*$. Alternatively, F^* is the set of all $(\varphi, \psi) \in A^* \times B^*$ such that

$$f(q_{A^*}(\varphi)) = q_{B^*}(\psi) \quad \text{and} \quad \langle \varphi, X \rangle = \langle \psi, F(X) \rangle \text{ for all } X \in A_{q_{A^*}(\varphi)}. \tag{26}$$

Call F^* the *relation dual to F*. It is straightforward to extend this notion to define the dual of an arbitrary relation, and the relationship between a relation and its dual is then symmetric.

Theorem 10.4.9 *Consider Lie algebroids A on M and B on N, and let $F \colon A \to B$, $f \colon M \to N$ be a vector bundle morphism. Then (F, f) is a morphism of Lie algebroids if and only if the dual relation F^* is coisotropic in $A^* \times \overline{B}^*$.*

Proof We work with the first formulation in 10.4.1. First we need to describe F^* in terms of the functions which vanish on it.

For any $w \in C^\infty(M)$, define

$$Z_w \colon A^* \times B^* \to \mathbb{R}, \qquad Z_w(\varphi, \psi) = w(f(q_{A^*}(\varphi))) - w(q_{B^*}(\psi)).$$

Clearly Z_w vanishes on F^*. Furthermore, a point (φ, ψ) of $A^* \times B^*$ satisfies the first equation in (26) if and only if Z_w vanishes at (φ, ψ) for all $w \in C^\infty(M)$.

Next, given $X \in \Gamma A$, denote by $\widetilde{\ell}_X$ the pullback of $\ell_X \colon A^* \to \mathbb{R}$ to

$A^* \times B^*$. Given $\eta \in \Gamma(f^! B)$, say $\eta = \sum u_i \otimes Y_i$ where $u_i \in C^\infty(M)$ and $Y_i \in \Gamma B$, likewise define $\widetilde{\ell}_\eta \colon A^* \times B^* \to \mathbb{R}$ by

$$\widetilde{\ell}_\eta(\varphi, \psi) = \sum u_i(m)\langle \psi, Y_i(n)\rangle$$

for $\varphi \in A_m^*$, $\psi \in B_n^*$. Note that

$$\widetilde{\ell}_\eta = \sum \widetilde{u}_i \widetilde{\ell}_{Y_i}$$

where $\widetilde{u}_i$ is the pullback to $A^* \times B^*$ of $u_i \circ q_{A^*} \colon A^* \to \mathbb{R}$, and $\widetilde{\ell}_{Y_i}$ is the pullback to $A^* \times B^*$ of $\ell_{Y_i} \colon B^* \to \mathbb{R}$.

Now take any $X \in \Gamma A$ and any $\eta \in \Gamma(f^! B)$. On points of F^* we have

$$(\widetilde{\ell}_X - \widetilde{\ell}_\eta)(F_m^* \psi, \psi) = \langle \psi, F(X(m))\rangle - \sum u_i(m)\langle \psi, Y_i(f(m))\rangle$$

and so

$$\widetilde{\ell}_X - \widetilde{\ell}_\eta \text{ vanishes on } F^* \text{ if and only if } F^!(X) = \eta. \tag{27}$$

Further, a point (φ, ψ) which satisfies the first equation of (26) also satisfies the second if and only if $\widetilde{\ell}_X - \widetilde{\ell}_\eta$, where $\eta = F^!(X)$, vanishes on (φ, ψ) for all $X \in \Gamma A$.

Now assume that F^* is coisotropic in $A^* \times \overline{B^*}$. First we prove that $b \circ F = T(f) \circ a$. Take any $X \in \Gamma A$ and write $\eta = F^!(X)$. Since F^* is coisotropic, $\{\widetilde{\ell}_X - \widetilde{\ell}_\eta, Z_w\}$ must vanish on F^*. Expanding this out, we have

$$\{\widetilde{\ell}_X, w \circ f \circ \widetilde{q}_{A^*}\} - \{\widetilde{\ell}_X, w \circ \widetilde{q}_{B^*}\} - \{\widetilde{\ell}_\eta, w \circ f \circ \widetilde{q}_{A^*}\} + \{\widetilde{\ell}_\eta, w \circ \widetilde{q}_{B^*}\},$$

where $\widetilde{q}_{A^*}$ denotes the pullback of q_{A^*} to $A^* \times B^*$ and so on. Here the second term is zero, because $\widetilde{\ell}_X$ is a pullback from the first factor and $w \circ \widetilde{q}_{B^*}$ is a pullback from the second. Similarly the third term is zero. Next, the first term is the pullback of $a(X)(w \circ f) \circ q_{A^*}$ and the last term is the pullback of $-b^!(\eta)(w) \circ q_{B^*}$. We therefore have, on F^*,

$$a(X)(w \circ f) \circ \widetilde{q}_{A^*} = b^!(\eta)(w) \circ \widetilde{q}_{B^*}$$

and evaluating this on points of F^* gives the anchor condition.

Next take any $X, X' \in \Gamma A$ and write $F^!(X) = \eta$, $F^!(X') = \eta'$. At any point of $A^* \times B^*$,

$$\{\widetilde{\ell}_X - \widetilde{\ell}_\eta, \widetilde{\ell}_{X'} - \widetilde{\ell}_{\eta'}\} = \{\widetilde{\ell}_X, \widetilde{\ell}_{X'}\} - \{\widetilde{\ell}_X, \widetilde{\ell}_{\eta'}\} - \{\widetilde{\ell}_\eta, \widetilde{\ell}_{X'}\} + \{\widetilde{\ell}_\eta, \widetilde{\ell}_{\eta'}\}. \tag{28}$$

The first term on the RHS is $\widetilde{\ell}_{[X,X']}$, since both arguments are pullbacks from A^* and in $C^\infty(A^*)$ we have $\{\ell_X, \ell_{X'}\} = \ell_{[X,X']}$. For the second term, writing $\eta' = \sum u'_j \otimes Y'_j$, we have

$$\{\tilde{\ell}_X, \tilde{\ell}_{\eta'}\} = \sum \tilde{u}'_j \{\tilde{\ell}_X, \tilde{\ell}_{Y'_j}\} + \sum \tilde{\ell}_{Y'_j} \{\tilde{\ell}_X, \tilde{u}'_j\}.$$

Now $\{\tilde{\ell}_X, \tilde{\ell}_{Y'_j}\} = 0$ since the functions are pullbacks from different factors. In $\{\tilde{\ell}_X, \tilde{u}'_j\}$ the functions are both pullbacks from A^* and we know $\{\ell_X, u'_j \circ q_{A^*}\} = a(X)(u'_j) \circ q_{A^*}$, so

$$\{\tilde{\ell}_X, \tilde{u}'_j\} = \tilde{\ell}_{a(X)(u'_j)}.$$

Expanding out the final term on the RHS of (28), we have

$$\{\tilde{\ell}_\eta, \tilde{\ell}_{\eta'}\} = \sum \tilde{u}_i \tilde{u}'_j \{\tilde{\ell}_{Y_i}, \tilde{\ell}_{Y'_j}\} + \sum \tilde{\ell}_{Y_i} \tilde{\ell}_{Y'_j} \{\tilde{u}_i, \tilde{u}'_j\} +$$
$$\sum \tilde{u}_i \tilde{\ell}_{Y'_j} \{\tilde{\ell}_{Y_i}, \tilde{u}'_j\} + \sum \tilde{u}'_j \tilde{\ell}_{Y_i} \{\tilde{u}_i, \tilde{\ell}_{Y'_j}\}.$$

Here the first term on the RHS is $-\sum \tilde{u}_i \tilde{u}'_j \, \tilde{\ell}_{[Y_i, Y'_j]}$, the minus sign appearing because of the reversal of the Poisson structure on B^*. The second term is zero because both functions are pullbacks from A^*, and the functions on A^* are pullbacks from M. In the third and fourth terms the brackets are of functions pulled back from different factors, and hence are zero also.

Altogether we now have

$$\{\tilde{\ell}_X - \tilde{\ell}_\eta, \tilde{\ell}_{X'} - \tilde{\ell}_{\eta'}\} = \tilde{\ell}_{[X,X']} - \sum \tilde{\ell}_{a(X)(u'_j)} \tilde{\ell}_{Y'_j}$$
$$+ \sum \tilde{\ell}_{a(X')(u_i)} \tilde{\ell}_{Y_i} - \sum \tilde{u}_i \tilde{u}'_j \tilde{\ell}_{[Y_i, Y'_j]}. \quad (29)$$

From (27) and coisotropy it follows that the LHS of (28) vanishes on F^*. Writing the RHS of (29) as $\tilde{\ell}_{[X,X']} - \tilde{\ell}_\zeta$, where

$$\zeta = \sum a(X)(u'_j) \otimes Y'_j - \sum a(X')(u_i) \otimes Y_i + \sum u_i u'_j \otimes [Y_i, Y_j],$$

it follows from (27) again that

$$F^!([X,X']) = \sum a(X)(u'_j) \otimes Y'_j - \sum a(X')(u_i) \otimes Y_i + \sum u_i u'_j \otimes [Y_i, Y_j],$$

and so F is a morphism of Lie algebroids.

The converse is proved in the same way. $\qquad\square$

Proposition 10.4.10 *Consider Lie algebroids A on M and B on N. Let $f: M \to N$ be a smooth map and let $\gamma: f^! B \to A$ be a linear map. Denote by $F: A^* \to B^*$ the vector bundle map defined by*

$$\langle F(\varphi), Y \rangle = \langle \varphi, \gamma(m, Y) \rangle$$

for $\varphi \in A_m^*$. Then γ is a comorphism of Lie algebroids if and only if F is a Poisson map.

Proof This is straightforward once one observes that, for any $Y \in \Gamma B$ and $g \in C^\infty(N)$,

$$\ell_Y \circ F = \ell_{\overline{\gamma}(Y)}, \qquad (g \circ q_{B^*}) \circ F = (g \circ f) \circ q_{A^*}.$$

(The map $\overline{\gamma}: \Gamma B \to \Gamma A$ is defined in 4.3.16.) $\qquad\square$

Example 10.4.11 Suppose that $f: P \to Q$ is a Poisson map which maps a coisotropic submanifold $C \subseteq P$ to a coisotropic submanifold $D \subseteq Q$. Then $C(f): T^*Q \lessdot f \to T^*P$ (see the end of §10.1) is a Lie algebroid morphism and maps $(TC)^\circ \lessdot f$ into $(TD)^\circ$. It is clear from 10.4.2 that the restriction is a morphism of Lie algebroids. From 10.4.10 it now follows that the induced map of the normal bundles is Poisson with respect to the linearized structures. $\qquad\boxtimes$

The final result, which arises frequently, may be proved either by adapting the proof of 10.4.9, or by observing that C is the annihilator of the kernel and applying 10.4.5.

Proposition 10.4.12 Let $F: A \to B$ be a fibration of Lie algebroids over $f: M \to N$. Then $C \subseteq A^*$, the image of the embedding

$$f^! B^* \to A^*, \qquad (m, \psi) \mapsto F_m^* \psi,$$

is coisotropic in A^*.

10.5 Notes

§10.1. For thorough treatments of (finite–dimensional) Poisson geometry, see the books [Libermann and Marle, 1987] and [Vaisman, 1994]. The survey article [Weinstein, 1998] starts with the basics and covers a great deal of recent work. The monograph [Karasëv and Maslov, 1993] contains a great amount of material not readily available elsewhere. The book [Cushman and Bates, 1997] presents a great deal of the core theory of Poisson geometry in the context of classical integrable systems.

Poisson geometry may be done in terms of the Poisson 2–tensor, the Poisson bracket of functions, or the bracket of 1–forms. We emphasize the latter, but regard the Poisson anchor itself as primary.

Theorem 10.1.2 was found independently in the early 1980s by a number of different people. See [Huebschmann, 1990] for a detailed account and references.

A Poisson map $f: P \to P'$ is *complete* if the Hamiltonian vector field $X_{u' \circ f}$ on P of the pullback of a function $u' \in C^\infty(P')$ is complete whenever $X_{u'}$ on P' is. This is one of several notions of completeness used in Poisson geometry; see [Weinstein,

1998]. It would be reasonable to define an action of a Lie algebroid A on a smooth map $f \colon M' \to M$ to be *complete* if there is a set $\{X_\nu \mid \nu \in I\}$ of sections of A, which generates ΓA, and is such that each vector field $X_\nu^\dagger$ on M', for $\nu \in I$, is complete. It would then follow that a Poisson map is complete if and only if the Lie algebroid action (7) which it defines is complete.

§10.2. The notion of contravariant connection in Poisson geometry is due to Vaisman (see [Vaisman, 1994]) and has been extensively developed by [Fernandes, 1999, 2000].

Poisson cohomology was introduced by Lichnerowicz [1977] and developed by Koszul [1984] and Brylinski [1988]. For 10.2.3 in particular, see [Bhaskara and Viswanath, 1988b] and [Kosmann−Schwarzbach and Magri, 1990]. A thorough account of Poisson cohomology is given in [Vaisman, 1994]. Note that the spectral sequence of a regular Poisson manifold as treated in [Vaisman, 1994] may be obtained as a special case of the Lie algebroid spectral sequence in §7.4.

There is also a homology theory for Poisson manifolds. In the notation of §7.5, define $\partial_\pi = \mathfrak{L}_\pi \colon \Omega^\bullet(P) \to \Omega^{\bullet-1}(P)$; that is,

$$\partial_\pi = \iota_\pi \circ \delta - \delta \circ \iota_\pi.$$

This is the *Poisson codifferential*; it has square zero and is related to the bracket of differential forms by

$$[\varphi, \psi] = (-1)^i (\partial_\pi(\varphi \wedge \psi) - (\partial_\pi \varphi) \wedge \psi - (-1)^i \varphi \wedge \partial_\pi \psi) \qquad (30)$$

for $\varphi \in \Omega^i(P), \psi \in \Omega^\bullet(P)$. The homology of the complex $\Omega^\bullet(P)$ with ∂_π is the *canonical homology* of P, or the *Poisson homology* of P, and is denoted $H_\bullet^\pi(P)$. Equation (30) shows that $\Omega^\bullet(P)$ is a Batalin–Vilkovisky algebra.

The relations between Poisson cohomology and Poisson homology have been very much clarified by independent work of Evens, Lu, and Weinstein [1999] and Xu [1999].

§10.3. The linear Poisson structure on the dual of the Lie algebroid of a Lie groupoid is given in [Coste et al., 1987]. The equivalence between abstract Lie algebroid structures and linear Poisson structures on their duals, including 10.4.6, first appeared in [Courant, 1990]. The extension of this duality to arbitrary, base–changing morphisms and comorphisms is due to Higgins and Mackenzie [1993].

The term 'Poisson vector bundle' might seem too narrowly defined, since a linear Poisson structure on (E, q, M) does not induce a (non–zero) Poisson structure on M. This is, however, the concept which is obtained by requiring a vector bundle with a Poisson structure to be a Poisson groupoid (§11.4) with respect to its additive structure: in a Poisson groupoid the target and source maps are respectively Poisson and anti–Poisson, and if they coincide then the target manifold must have the zero Poisson structure.

Theorem 10.3.14 is Theorem 5.6 of [Mackenzie and Xu, 1994].

The semi–direct product Poisson structure of 10.3.8 was defined in [Marsden et al., 1984].

§10.4. For the importance of coisotropic submanifolds in Poisson geometry see [Weinstein, 1988], where 'coisotropic calculus' is developed as a systematic technique of general applicability. This paper introduced the Lie algebroid structure 10.4.2 on the conormal bundle of a coisotropic submanifold, and the characterization 10.4.4 of Poisson submersions.

The restriction we have made here to closed embedded submanifolds is merely in order to concentrate on the main ideas; see [Weinstein, 1988] for comments on more general cases.

Proposition 10.4.5 comes from [Xu, 1995].

The concept of Poisson submanifold is of less central importance for Poisson geometry than that of coisotropic submanifold; see [Marle, 2000] for work in which it is relevant.

The characterization 10.4.9 of Lie algebroid morphisms is stated in [Mackenzie and Xu, 1994, 6.1]; the base–preserving case was already noted in [Courant, 1990], and the general result was in the air before [Mackenzie and Xu, 1994]. The proof in [Mackenzie and Xu, 1994] proceeds via 10.4.5 and a characterization of Lie algebroid morphisms as those for which the graph is a Lie subalgebroid of the product. For the proof given here, one needs to know that if a submanifold of some M is defined as the zero set of finitely many functions $h_i \colon M \to \mathbb{R}$, then any smooth function which vanishes on the submanifold can be written as a linear combination in $C^\infty(M)$ of the h_i; see [Cushman and Bates, 1997, II§2].

The characterization 10.4.10 of comorphisms is from [Higgins and Mackenzie, 1993].

11

Poisson groupoids
and symplectic groupoids

Poisson groupoids provide a simultaneous generalization of Poisson Lie groups and symplectic groupoids. Both theories have considerable literatures of their own, and it is not the intention here to develop their individual features beyond the basic properties. The purpose of this chapter is to show how the general features of Poisson groupoids — and thus of the two main special cases — may be deduced by a systematic application of what one might call 'diagrammatic' methods to cotangent structures.

The way in which the Poisson structure and the group structure of a Poisson Lie group interact is fundamentally different to the relationship between a group structure and, say, the smooth structure in a Lie group. Immediate consequences of the definition of a Lie group are that right and left translations are smooth, and inversion is smooth, and there are many mixed algebraic–topological concepts which behave in a similar way. A Poisson Lie group is often defined as a Lie group with a Poisson structure for which the multiplication is a Poisson map, but despite the apparently standard nature of this compatibility condition, inversion in a Poisson Lie group is not Poisson but antiPoisson, and right and left translations are generally neither Poisson nor antiPoisson. In fact the compatibility condition is not as standard as it appears, since a Poisson structure is not, strictly speaking, defined on the manifold itself, but on the module of smooth functions, or 1–forms.

Here we define a Poisson Lie group to be a Lie group G equipped with a Poisson structure π such that $\pi^{\#} \colon T^*G \to TG$ is a Lie groupoid morphism from the cotangent groupoid $T^*G \rightrightarrows \mathfrak{g}^*$ to the tangent group(oid) $TG \rightrightarrows \{\cdot\}$. This is equivalent to the usual definition, but because it is formulated in terms of a morphism of Lie groupoids, all the apparatus of the Lie theory is available. In particular, we derive the

Lie algebra structure on $\mathfrak{g}^*$ as a quotient of the Lie algebroid structure on T^*G and show that, although inversion in the group is antiPoisson, inversion in the groupoid T^*G is a Lie algebroid isomorphism. By considering not the group and the Poisson structure on G, but the induced Lie groupoid and Lie algebroid structures on T^*G, the compatibility condition has become natural and functorial.

For a Poisson structure on a general Lie groupoid $G \rightrightarrows M$, the Poisson anchor has target the tangent groupoid $TG \rightrightarrows TM$ as described in 1.1.16. The structure on the domain is more complicated and we first consider the general concept of $\mathscr{VB}$–groupoid, of which TG and T^*G are instances. A $\mathscr{VB}$–groupoid is a Lie groupoid object in the category of vector bundles. Such structures may be dualized by an extension of the dualization process for double vector bundles dealt with in §9.2 (though the possibility of dualizing in alternating directions does not exist here). We present this duality in general in §11.2 and in §11.3 derive the groupoid structure of T^*G as a consequence.

§11.4 presents the basic properties of Poisson groupoids. We show that the base inherits a Poisson structure and that the dual of the Lie algebroid has itself a Lie algebroid structure. The relation between these two structures is embodied in the notion of Lie bialgebroid, and the crucial Theorem 11.4.5 is preparation for this. The results of this section rely only on the fact that the Poisson anchor is a Lie groupoid morphism and can be applied to more general situations involving compatible Lie groupoid and Lie algebroid structures; the specifics of Poisson geometry are only needed here to establish that the relevant Lie algebroids are induced by Poisson structures.

A symplectic groupoid is usually defined as a Lie groupoid Σ with a symplectic structure such that the graph of the groupoid multiplication is a Lagrangian submanifold of $\Sigma \times \overline{\Sigma} \times \overline{\Sigma}$; that is, multiplication is a canonical relation. In keeping with the spirit of this definition, a significant arsenal of symplectic techniques is used to obtain the Poisson structure on the base manifold P, and to prove that the Lie algebroid structure consequently induced on T^*P is isomorphic to the ordinary Lie algebroid $A\Sigma$. In the short first part of §11.5, we obtain these results by applying to the results of §11.4 the invertibility of the Poisson anchor. By treating symplectic groupoids as a special case of Poisson groupoids, we have bypassed all specifics of symplectic geometry. In the second part of the section we give the proof that $T^*G \rightrightarrows A^*G$ is a symplectic groupoid for any Lie groupoid $G \rightrightarrows M$. The proof given here, by a calculus of canonical isomorphisms, is more complex than that which works with the standard definition, but extends easily to more general situations.

11.1 Poisson–Lie groups

We begin with the standard definition of a Poisson–Lie group and immediately reformulate it in terms of a groupoid morphism condition.

Definition 11.1.1 Let G be a Lie group and let π be a Poisson structure on the manifold G. Then (G, π) is a *Poisson–Lie group* if the multiplication map $G \times G \to G$ is a Poisson map. $\qquad\square$

Theorem 11.1.2 *Let G be a Lie group and let π be a Poisson structure on the manifold G. Then (G, π) is a Poisson–Lie group if and only if the Poisson anchor $\pi^{\#} \colon T^{*}G \to TG$ is a groupoid morphism, from the cotangent groupoid $T^{*}G \rightrightarrows \mathfrak{g}^{*}$ of 1.1.17 to the usual tangent group $TG \rightrightarrows \{\cdot\}$.*

Proof By 10.4.8, multiplication is a Poisson map if and only if the graph C of the multiplication is coisotropic in $G \times \overline{G} \times \overline{G}$. Clearly,

$$TC = \{(Y \bullet X, Y, X) \mid Y \in T_h(G), \ X \in T_g(G), \ g, h \in G\}.$$

Take $\theta \in T^{*}_{hg}(G)$, $\psi \in T^{*}_{h}(G)$, $\varphi \in T^{*}_{g}(G)$. Then (θ, ψ, φ) is in the annihilator of TC if and only if

$$\langle \theta, Y \bullet X \rangle + \langle \psi, Y \rangle + \langle \varphi, X \rangle = 0$$

for all $Y \in T_h(G)$, $X \in T_g(G)$. Expanding out, this is

$$\langle \theta, T(R_g)(Y) \rangle + \langle \psi, Y \rangle = -\langle \theta, T(L_h)(X) \rangle - \langle \varphi, X \rangle,$$

and this can only be so if both sides are identically zero. We therefore have both $\theta = -\psi \circ T(R_{g^{-1}})$ and $\theta = -\varphi \circ T(L_{h^{-1}})$, whence in particular $\alpha(\psi) = \beta(\varphi)$ in $T^{*}G \rightrightarrows \mathfrak{g}^{*}$. Thus $(TC)^{\circ}$ is the set

$$\{(-\psi \bullet \varphi, \psi, \varphi) \mid \psi \in T^{*}_{h}(G), \ \varphi \in T^{*}_{g}(G), \ \alpha(\psi) = \beta(\varphi), \ h, g \in G\}.$$

Now, applying the Poisson anchor of $G \times \overline{G} \times \overline{G}$ to an element of $(TC)^{\circ}$, we obtain

$$(-\pi^{\#}(\psi \bullet \varphi), -\pi^{\#}(\psi), -\pi^{\#}(\varphi))$$

and the condition that this lies in TC is precisely the morphism condition. (Recall that vector negation $T^{*}G \to T^{*}G$ is an automorphism of the groupoid structure.) $\qquad\square$

The morphism condition may be written more explicitly as the requirment that for $\varphi \in T_g^*G$, $\psi \in T_h^*G$ with $\varphi \circ T(L_g) = \psi \circ T(R_h)$ we have

$$\pi^\#(\varphi \circ T(R_{h^{-1}})) = \pi^\#(\varphi) \bullet \pi^\#(\psi) = T(R_h)(\pi^\#(\varphi)) + T(L_g)(\pi^\#(\psi)). \tag{1}$$

Putting $g = 1$ we see that $\pi^\#(\varphi) = 0$ for all $\varphi \in \mathfrak{g}^*$. Thus π has rank zero at the identity.

Next we give the compatibility condition in terms of π itself. The right translation $T(R_h)(\pi(g))$ refers to the extension of $T(R_h)(X \wedge Y) = T(R_h)(X) \wedge T(R_h)(Y)$ and likewise with the left translation.

Theorem 11.1.3 *Let G be a Lie group and let π be a Poisson structure on the manifold G. Then (G, π) is a Poisson–Lie group if and only if, for all $g, h \in G$,*

$$\pi(gh) = T(R_h)(\pi(g)) + T(L_g)(\pi(h)). \tag{2}$$

Proof Assume that (G, π) is a Poisson–Lie group. Take any φ_{gh}, $\psi_{gh} \in T_{gh}^*(G)$, where the subscripts refer to the base points. We will calculate $\langle \varphi_{gh} \wedge \psi_{gh}, \pi(gh) \rangle$.

Define $\varphi_g = \varphi_{gh} \circ T(R_h) \in T_g^*(G)$ and $\varphi_h = \varphi_{gh} \circ T(L_g) \in T_h^*(G)$. Then $\tilde{\alpha}(\varphi_g) = \tilde{\beta}(\varphi_h)$ and

$$\varphi_{gh} = \varphi_g \bullet \varphi_h.$$

With the corresponding formula for ψ_{gh}, and using 11.1.2, this gives

$$\begin{aligned}
\langle \varphi_{gh} \wedge \psi_{gh}, \pi(gh) \rangle &= \langle \psi_{gh}, \pi^\#(\varphi_{gh}) \rangle \\
&= \langle \psi_g \bullet \psi_h, \pi^\#(\varphi_g) \bullet \pi^\#(\varphi_h) \rangle \\
&= \langle \psi_g, \pi^\#(\varphi_g) \rangle + \langle \psi_h, \pi^\#(\varphi_h) \rangle \\
&= \langle \varphi_g \wedge \psi_g, \pi(g) \rangle + \langle \varphi_h \wedge \psi_h, \pi(h) \rangle.
\end{aligned}$$

Now using $\varphi_g = \varphi_{gh} \circ T(R_h)$ and its like, we obtain the stated equation. The converse is a straightforward reversal. $\qquad\square$

Note that one cannot define a group structure on $TG \wedge TG$ which would make (2) a morphism condition.

Using the right translations in $TG \wedge TG$, define a map

$$\pi^R \colon G \to \mathfrak{g} \wedge \mathfrak{g}.$$

Then (2) becomes

$$\pi^R(gh) = \pi^R(g) + \mathrm{Ad}_g(\pi^R(h)). \tag{3}$$

This is precisely the condition that π^R be a cocycle with respect to the adjoint action of G on $\mathfrak{g} \wedge \mathfrak{g}$. Thus we have:

Corollary 11.1.4 *Let G be a Lie group with a Poisson structure π. Then (G, π) is a Poisson Lie group if and only if π^R is a cocycle with respect to the adjoint action of G on $\mathfrak{g} \wedge \mathfrak{g}$.*

The criterion in 11.1.4 is often taken as the definition. We now obtain the Lie algebra structure on $\mathfrak{g}^*$. There are several ways in which this may be done.

Given $\theta \in \Lambda^k(\mathfrak{g}^*)$, the *right–invariant* k–form $\overrightarrow{\theta}$ on G is defined as usual by

$$\overrightarrow{\theta}(g) = \theta \circ T(R_{g^{-1}}).$$

Observe that the notion of right–invariance can be extended to multivector fields on G in each of the usual ways. A multivector field $\xi \in \mathscr{X}^k(G)$ is *right–invariant* if $\xi(gh) = T(R_h)(\xi(g))$ for all $g, h \in G$; equivalently if $[\overrightarrow{Y}, \xi] = 0$ for all $Y \in \mathfrak{g}$; equivalently if $\langle \overrightarrow{\varphi}, \xi \rangle$ is constant over G for each $\varphi \in \Lambda^k(\mathfrak{g}^*)$.

Proposition 11.1.5 *Given $\theta_1, \theta_2 \in \mathfrak{g}^*$, the bracket $[\overrightarrow{\theta_1}, \overrightarrow{\theta_2}]$ is also right–invariant.*

Proof We prove that $\langle [\overrightarrow{\theta_1}, \overrightarrow{\theta_2}], \overrightarrow{X} \rangle$ is constant over G for each $X \in \mathfrak{g}$. Directly from the definition of d_π as a Lie algebroid coboundary operator it follows that

$$\langle [\overrightarrow{\theta_1}, \overrightarrow{\theta_2}], \overrightarrow{X} \rangle = -\langle \overrightarrow{\theta_1} \wedge \overrightarrow{\theta_2}, d_\pi \overrightarrow{X} \rangle$$

and so by 10.2.3 we have

$$\langle [\overrightarrow{\theta_1}, \overrightarrow{\theta_2}], \overrightarrow{X} \rangle = \langle \overrightarrow{\theta_1} \wedge \overrightarrow{\theta_2}, \mathcal{L}_{\overrightarrow{X}}(\pi) \rangle$$

and so it suffices to prove that $\mathcal{L}_{\overrightarrow{X}}(\pi)$ is right–invariant. Using the usual formula for the Lie derivative of a vector field, we have

$$(\mathcal{L}_{\overrightarrow{X}}(\pi))(g) = -\lim_{t \to 0} \frac{1}{t} \{T(L_{\exp tX})(\pi(\exp -tX\, g)) - \pi(g)\}.$$

Now $\pi(\exp -tX\, g) = T(L_{\exp -tX})(\pi(g)) + T(R_g)(\pi(\exp -tX))$ and so we obtain, since $\pi(1) = 0$,

$$-\lim_{t \to 0} \frac{1}{t} \{T(L_{\exp tX})(T(R_g)(\pi(\exp -tX)))\} = T(R_g)((\mathcal{L}_{\overrightarrow{X}}(\pi))(1)).$$

$\square$

We can therefore define a bracket on $\mathfrak{g}^*$ by $[\theta_1, \theta_2]_* = [\overrightarrow{\theta_1}, \overrightarrow{\theta_2}](1)$ and it follows that $\mathfrak{g}^*$ is a Lie algebra and that

$$[\overrightarrow{\theta_1}, \overrightarrow{\theta_2}] = \overrightarrow{[\theta_1, \theta_2]_*}.$$

This construction is an example of descent as in §4.4. The map $\widetilde{\beta} \colon T^*G \to \mathfrak{g}^*$, regarded as a morphism of vector bundles, has zero kernel in the usual sense and the right–invariant 1–form $\overrightarrow{\theta}$ corresponding to $\theta \in \mathfrak{g}^*$ is the unique section of the domain which projects to θ and is stable under the action of the pair groupoid $G \times G$ on T^*G. Of the conditions required in 4.4.2, only the first is not vacuous, and this is 11.1.5.

In particular, $\widetilde{\beta}$ is a morphism of Lie algebroids. Since $\widetilde{\beta}$ is a pullback of vector bundles, it follows that the Lie algebroid T^*G is isomorphic to the action Lie algebroid $\mathfrak{g}^* \ltimes G$ where $\mathfrak{g}^*$ acts on the manifold G by

$$\theta^\dagger = (\overrightarrow{\theta})^\#.$$

This is the *dressing transformation action* of $\mathfrak{g}^*$ on G.

The isomorphism of Lie algebroids $T^*G \to \mathfrak{g}^* \ltimes G$ induced by $\widetilde{\beta}$ is precisely the trivialization of T^*G by right–translations; denote it $\mathscr{R}^*$. This $\mathscr{R}^*$ is also an isomorphism of groupoids from T^*G to the action groupoid $\mathfrak{g}^* \rtimes G$ defined by the (right) coadjoint action $(\theta, g) \mapsto \theta \circ \mathrm{Ad}_g$, of G on $\mathfrak{g}^*$. Since $\pi^\#$ is a groupoid morphism by assumption, it follows that the composition, $\mathfrak{g}^* \rtimes G \to TG$, $(\theta, g) \mapsto \theta^\dagger(g)$, is a groupoid morphism; that is,

$$\theta^\dagger(hg) = \theta^\dagger(h) \bullet \varphi^\dagger(g), \tag{4}$$

where $\varphi = \theta \circ \mathrm{Ad}h$. Equivalently,

$$\theta^\dagger(hg) = T(R_g)(\theta^\dagger(h)) + T(L_h)((\mathrm{Ad}^*_{h^{-1}}\theta)^\dagger(g)). \tag{5}$$

This is the *twisted multiplicativity equation* for the dressing transformations.

It is important to notice that $\widetilde{\alpha}$ is also a morphism of Lie algebroids, with respect to the same structures. This follows immediately from the next result.

Proposition 11.1.6 *The groupoid inversion* $T^*G \to T^*G$ *is a morphism of Lie algebroids.*

Proof Denote the groupoid inversion $T^*G \to T^*G$ by i and denote the inversion $G \to G$ by i_0. That $\pi^\# \circ i = T(i_0) \circ \pi^\#$ follows from the fact that $\pi^\#$ is a groupoid morphism. Since the bracket in T^*G is determined by $\pi^\#$, the result should in principle follow directly. However the following intermediate steps are useful:

For $\varphi \in \Omega^1(G)$ denote by $i_*(\varphi)$ the transported section $(i_*(\varphi))(g) = i(\varphi(g^{-1}))$. Then

$$\pi(i_*(\varphi_1), i_*(\varphi_2)) = -\pi(\varphi_1, \varphi_2) \circ i_0. \qquad (6)$$

Using this, it follows that for any $\xi \in \mathcal{X}(G)$,

$$\mathcal{L}_{i_{0*}(\xi)}(i_*(\varphi)) = i_*(\mathcal{L}_\xi(\varphi)), \qquad (7)$$

where $i_{0*}(\xi)$ is the usual transport of the vector field. Finally, note that the pullback across i_0 of a 1–form φ and its inversion are related by

$$i_0^*(\varphi) = -i_*(\varphi). \qquad (8)$$

$\square$

Note that the inversion i_0 is antiPoisson. The reader may use the same techniques to prove the next result.

Proposition 11.1.7 *The groupoid multiplication $T^*G * T^*G \to T^*G$ is a Lie algebroid morphism, where the domain is the pullback Lie algebroid of $\widetilde{\alpha} \colon T^*G \to \mathfrak{g}^*$ over $\widetilde{\beta} \colon T^*G \to \mathfrak{g}^*$.*

Thus the two structures on T^*G are compatible in a very strong sense: the structure maps of the groupoid structure are all morphisms of Lie algebroids; that is, T^*G is a groupoid object in the category of Lie algebroids. Notice that the groupoid structure on $T^*G \rightrightarrows \mathfrak{g}^*$ arises entirely from the Lie group structure on G whereas the Lie algebroid structure on $T^*G \to G$ arises entirely from the Poisson structure.

Lastly, we know that $\pi^\#$ has rank zero at $1 \in G$, so the kernel of $\pi_1^\#$ is precisely $\mathfrak{g}^*$, with its Lie algebra structure obtained by linearization of π. As with any Lie algebroid at a point where the anchor is zero, $\mathfrak{g}^* \to \{\cdot\}$ is a Lie subalgebroid of $T^*G \to G$. Now the composite of this inclusion with $\widetilde{\beta}$ is the identity map $\mathfrak{g}^* \to \mathfrak{g}^*$ and since the inclusion and $\widetilde{\beta}$ are Lie algebroid morphisms, it follows that the identity map is also. Thus the Lie algebra structure on $\mathfrak{g}^*$ obtained by descent via $\widetilde{\beta}$ coincides with the linearized structure from π. Together with descent via $\widetilde{\alpha}$, this gives three different constructions of the one Lie algebra structure on $\mathfrak{g}^*$. There is a further construction, which is done in the more general setting of Poisson groupoids in §11.4.

11.2 𝒱ℬ–groupoids and their duals

In this section we use the notation Ω for a groupoid without implying that the groupoid is locally trivial.

Definition 11.2.1 A *𝒱ℬ–groupoid* $(\Omega; G, E; M)$ is a structure

$$
\begin{array}{ccc}
& \widetilde{q} & \\
\Omega & \longrightarrow & G \\[4pt]
\widetilde{\alpha}, \widetilde{\beta} \, \Big\Downarrow & & \Big\Downarrow \, \alpha, \beta \\[4pt]
E & \longrightarrow & M, \\
& q &
\end{array}
\qquad (9)
$$

in which Ω is a vector bundle over G, which is a Lie groupoid over M, and Ω is also a Lie groupoid over E, which is a vector bundle over M, subject to the condition that the structure maps of the groupoid structure (source, target, identity, multiplication, inversion) are vector bundle morphisms, and the 'double source map' $(\widetilde{q}, \widetilde{\alpha}) \colon \Omega \to G \times_M E$ is a surjective submersion.   □

Examples 11.2.2 In 1.1.16 we described the groupoid structure which any Lie groupoid $G \rightrightarrows M$ induces on its tangent bundle TG : namely the tangent prolongation groupoid structure $TG \rightrightarrows TM$ obtained by applying the tangent functor to the structure maps of $G \rightrightarrows M$. This makes $(TG; G, TM; M)$ a *𝒱ℬ–groupoid*, the *tangent 𝒱ℬ–groupoid of* $G \rightrightarrows M$.

Any double vector bundle may be regarded as a *𝒱ℬ–groupoid*.   ⊠

The compatibility conditions may be expanded out as in the case of double vector bundles. Thus we have

$$
\widetilde{q}(\eta\xi) = \widetilde{q}(\eta)\widetilde{q}(\xi), \qquad \widetilde{q}(\xi^{-1}) = \widetilde{q}(\xi)^{-1}, \qquad \widetilde{q}(\widetilde{1}_X) = 1_{qX},
$$
$$
\widetilde{\alpha}(\eta + \xi) = \widetilde{\alpha}(\eta) + \widetilde{\alpha}(\xi), \qquad \widetilde{\alpha}(t\xi) = t\widetilde{\alpha}(\xi), \qquad \widetilde{\alpha}(-\xi) = -\widetilde{\alpha}(\xi),
$$
$$
\widetilde{\alpha}(\widetilde{0}_g) = 0_{\alpha g}, \qquad \widetilde{0}_{hg} = \widetilde{0}_h\widetilde{0}_g, \qquad \widetilde{0}_{g^{-1}} = (\widetilde{0}_g)^{-1},
$$
$$
\widetilde{1}_{X+Y} = \widetilde{1}_X + \widetilde{1}_Y, \qquad \widetilde{1}_{tX} = t\widetilde{1}_X, \qquad \widetilde{1}_{-X} = -\widetilde{1}_X,
$$

for compatible $\eta, \xi \in \Omega$, $X, Y \in E$, $g, h \in G$. Corresponding equations hold for $\widetilde{\beta}$. We do not use a special notation for $\widetilde{1}_{0_m} = \widetilde{0}_{1_m}$. The interchange condition is

$$(\xi_1 + \xi_2)(\eta_1 + \eta_2) = \xi_1\eta_1 + \xi_2\eta_2$$

when the two sums on the left and the two products on the right are defined. Note that, for compatible ξ, η,

$$(-\eta)(-\xi) = -(\eta\xi), \qquad (\eta + \xi)^{-1} = \eta^{-1} + \xi^{-1}.$$

Define the *core* K *of* Ω to be

$$K = \{\xi \in \Omega \mid \widetilde{\alpha}(\xi) = 0_m, \ \widetilde{q}(\xi) = 1_m, \ \exists m \in M\}.$$

Thus K is the (vector bundle) pullback across $1\colon M \to G$ of the kernel of $\widetilde{\alpha}\colon \Omega \to E$. The restriction of $\widetilde{\beta}$ to K is a vector bundle morphism $K \to E$ over M which we denote by ∂_E. Note that for any core element k with $\partial_E k = Y$, the interchange law implies that

$$k^{-1} = \widetilde{1}_Y - k. \tag{10}$$

Proposition 11.2.3 *Let* $(\Omega; G, E; M)$ *be a* $\mathscr{VB}$-*groupoid with core* K. *Considering* K *as a groupoid, let it act on* $q_E\colon E \to M$ *by* $(k, X) \mapsto \partial_E(k) + X$, *where* $k \in K, X \in E$, $q_K(k) = q_E(X)$. *Then*

$$K \lessdot q_E \blacktriangleright\!\!\!-\!\!\!\longrightarrow \Omega \overset{\widetilde{q}}{-\!\!\!\longrightarrow\!\!\!\gg} G \tag{11}$$

is an exact sequence of Lie groupoids, where the injection is $(k, X) \mapsto k + \widetilde{1}_X$.
Likewise

$$\beta^!K \blacktriangleright\!\!\!-\!\!\!\longrightarrow \Omega \overset{\widetilde{\alpha}}{-\!\!\!\longrightarrow\!\!\!\gg} E \tag{12}$$

is an exact sequence of vector bundles, where the injection is $(k, g) \mapsto k\widetilde{0}_g$.

Proof Take $(k_1, X_1), (k_2, X_2) \in K \lessdot q_E$ with $X_1 = \partial_E(k_2) + X_2$. Multiplying the images in Ω, and using the interchange law, we have

$$\begin{aligned}
(k_1 + \widetilde{1}_{X_1})(k_2 + \widetilde{1}_{X_2}) &= (k_1 + \widetilde{1}_{X_2} + \widetilde{1}_{\partial_A(k_2)})(\widetilde{1}_{X_2} + k_2) \\
&= (k_1 + \widetilde{1}_{X_2})\widetilde{1}_{X_2} + \widetilde{1}_{\partial_E(k_2)}k_2 \\
&= k_1 + \widetilde{1}_{X_2} + k_2,
\end{aligned}$$

which is the image of $(k_1 + k_2, X_2)$ under the injection. So the injection is a groupoid morphism. Exactness is easily verified. The proof of the second sequence is similar. □

The concept of core section extends to this situation. Consider $k \in \Gamma K$. Define the *core section* $\overline{k} \in \Gamma_G \Omega$ by

$$\overline{k}(g) = k(\beta g)\widetilde{0}_g.$$

We will need a characterization of core sections in §11.4. Denote by $\overline{K}$ the kernel of $\widetilde{\alpha}$. Given any $\xi \in \Gamma\overline{K}$, there is a well-defined section $\xi * \widetilde{0}$ of the pullback vector bundle $\Omega * \Omega \to G * G$ given by $(\xi * \widetilde{0})(g, h) = (\xi(g), \widetilde{0}_h)$.

Lemma 11.2.4 *Let ξ be a section of $\overline{K}$. Then ξ is a core section if and only if $\xi * \widetilde{0} \overset{\kappa}{\sim} \xi$.*

Proof The condition $\xi * \widetilde{0} \overset{\kappa}{\sim} \xi$ is precisely that $\xi(g)\widetilde{0}_h = \xi(gh)$ for all compatible $g, h \in G$. If this holds, define $k \in \Gamma K$ by $k(m) = \xi(1_m)$; then $\xi = \overline{k}$. The converse is trivial.   □

We now consider the dual of $(\Omega; G, E; M)$ where Ω^* is the dual of Ω as a vector bundle over G. Define a groupoid structure on Ω^* with base K^* as follows. Take $\Phi \in \Omega_g^*$ where $g \in G_m^n$. Then the source and target of Φ in K_m^* and K_n^* respectively are

$$\langle \widetilde{\alpha}_*(\Phi), k \rangle = \langle \Phi, -\widetilde{0}_g k^{-1} \rangle, \; k \in K_m, \quad \langle \widetilde{\beta}_*(\Phi), k \rangle = \langle \Phi, k\widetilde{0}_g \rangle, \; k \in K_n. \tag{13}$$

The lack of symmetry is only superficial, and reflects the choice of α in the definition of the core.

For the composition, take $\Psi \in \Omega_h^*$ with $\widetilde{\alpha}_*(\Psi) = \widetilde{\beta}_*(\Phi)$. Any element of Ω_{hg} can be written as a product $\eta\xi$ where $\eta \in \Omega_h$ and $\xi \in \Omega_g$. Now the compatibility condition on Ψ and Φ ensures that

$$\langle \Psi\Phi, \eta\xi \rangle = \langle \Psi, \eta \rangle + \langle \Phi, \xi \rangle \tag{14}$$

is well defined. The identity element of Ω^* at $\theta \in K_m^*$ is $\widetilde{1}_\theta \in \Omega_{1_m}^*$ defined by

$$\langle \widetilde{1}_\theta, \widetilde{1}_X + k \rangle = \langle \theta, k \rangle, \tag{15}$$

where any element of Ω_{1_m} can be written as $\widetilde{1}_X + k$ for some $X \in E_m$ and $k \in K_m$.

It is straightforward to check the details of the following result.

Proposition 11.2.5 *Given a $\mathscr{VB}$-groupoid $(\Omega, G, E; M)$, the construction above yields a $\mathscr{VB}$-groupoid $(\Omega^*; G, K^*; M)$.*

This is the *dual $\mathcal{VB}$-groupoid to* Ω, shown in Figure 11.1(b). The core of Ω^* is the vector bundle $E^* \to M$, with the core element corresponding to $\varphi \in E_m^*$ being $\overline{\varphi} \in \Omega_{1_m}^*$ defined by

$$\langle \overline{\varphi}, \widetilde{1}_X + k \rangle = \langle \varphi, X + \partial_A(k) \rangle \tag{16}$$

for $X \in E_m$, $k \in K_m$.

Note that the dual of Ω^* identifies canonically with Ω as a $\mathcal{VB}$-groupoid, the signs in (13) cancelling. The ∂ map $E^* \to K^*$ for Ω^* is ∂_E^*.

$$
\begin{array}{ccc}
\Omega & \longrightarrow & G \\
\Downarrow & & \Downarrow \\
E & \longrightarrow & M
\end{array}
\qquad\qquad
\begin{array}{ccc}
\Omega^* & \longrightarrow & G \\
\Downarrow & & \Downarrow \\
K^* & \longrightarrow & M
\end{array}
$$

(a)   (b)

Fig. 11.1.

Now consider a morphism of $\mathcal{VB}$-groupoids which preserves the lower groupoids:

$$(F; \mathrm{id}_G, f; \mathrm{id}_M) \colon (\Omega; G, E, M) \to (\Omega'; G, E'; M)$$

and denote the core morphism $K \to K'$ by f_K. The proof of the following result is a simple check.

Proposition 11.2.6 *The dual morphism $F^* \colon \Omega'^* \to \Omega^*$ is a morphism of the dual $\mathcal{VB}$-groupoids, with base map $f_K^* \colon K'^* \to K^*$ and core morphism $f^* \colon E'^* \to E^*$.*

Example 11.2.7 The core of TG is precisely the vector bundle AG, and the map ∂ is the anchor $AG \to TM$. The two sequences (11) and (12) are

$$AG \vartriangleleft \!\!\! \frac{p_M}{} \!\!\!\! >\!\!\!-\!\!\!- \!\!\! > TG \xrightarrow{p_G} \!\!\!\! \gg G \quad \text{and} \quad \beta^! AG >\!\!\!-\!\!\!- \!\!\! \xrightarrow{\mathscr{R}} TG \xrightarrow{T(\alpha)} \!\!\!\! \gg TM,$$

where $\mathscr{R}$ is the right translation map $(X, g) \mapsto T(R_g)(X)$. The core section corresponding to $X \in \Gamma AG$ is the right–invariant vector field $\overrightarrow{X}$.

We deal with this example in more detail in the next section.   ⊠

11.3 The structure of T^*G

Consider a Lie groupoid $G \rightrightarrows M$. The dual construction of the preceding section, applied to the tangent $\mathscr{VB}$–groupoid $(TG; G, TM; M)$, yields a $\mathscr{VB}$–groupoid $(T^*G; G, A^*G; M)$, called the *cotangent groupoid of* G.

The source and target maps $T^*G \to A^*G$ are

$$\langle \widetilde{\alpha}(\Phi), X \rangle = \langle \Phi, T(L_g)(X - T(1)(aX)) \rangle, \quad \langle \widetilde{\beta}(\Phi), Y \rangle = \langle \Phi, T(R_g)(Y) \rangle, \tag{17}$$

where $\Phi \in T_g^*G$ and $Y \in A_{\beta g}G$, $X \in A_{\alpha g}G$. Here we have used $X - T(1)(aX) = -T(i)(X)$, from (10) or 1.4.14.

The multiplication in T^*G is given by

$$\langle \Phi \bullet \Psi, X \bullet Y \rangle = \langle \Phi, X \rangle + \langle \Psi, Y \rangle. \tag{18}$$

To define the identity $\widetilde{1}_\varphi \in T_{1_m}^*G$ corresponding to $\varphi \in A_m^*G$, note that every element ξ of $T_{1_m}G$ can be written uniquely in the form $T(1)(x) + X$ where $x = T(a)(X) \in T_mM$ and $X \in A_mG$. We can therefore define

$$\langle \widetilde{1}_\varphi, T(1)(x) + X \rangle = \langle \varphi, X \rangle \tag{19}$$

and it is straightforward to check that $\widetilde{1}_\varphi$ is indeed an identity for the multiplication and that this structure makes $T^*G \rightrightarrows A^*G$ a groupoid with inverses

$$\langle \Phi^{-1}, X^{-1} \rangle = -\langle \Phi, X \rangle. \tag{20}$$

For any 1–form on the base, $\omega \in T_m^*M$, define the core element $\overline{\omega} \in T_{1_m}^*G$ by

$$\langle \overline{\omega}, T(1)(x) + X \rangle = \langle \omega, x + aX \rangle. \tag{21}$$

Denote the pullback of ω to T_g^*G, where $\beta g = m$, by $\beta_g^*\omega$. Then

$$\beta_{1_m}^*\omega = \overline{\omega}, \qquad \beta_g^*\omega = \beta_{1_m}^*\omega \bullet \widetilde{0}_g = \overline{\omega} \bullet \widetilde{0}_g, \qquad (\alpha_g^*\omega)^{-1} = -\beta_{g^{-1}}^*\omega$$

where $\widetilde{0}_g$ is the zero of T^*G at g. Thus the core section defined by $\omega \in \Omega^1(M)$ is the pullback form $\beta^*\omega \in \Omega^1(G)$. Furthermore, for any $\omega \in T_m^*M$,

$$(\overline{\omega})^{-1} = \widetilde{1}_{a^*\omega} - \overline{\omega}. \tag{22}$$

Note that $a^*\omega$ always denotes the image of ω under $a^* \colon T^*M \to A^*G$, not the pullback of the 1–form ω to a 1–form on AG.

In the case of T^*G the exact sequences of 11.2.3 take the following forms:

$$T^*M \lhd q_* \!\!>\!\!-\!\!-\!\!>T^*G \xrightarrow{\tilde{q}}\!\!\gg G, \tag{23}$$

$$\beta^! T^*M \!\!>\!\!-\!\!-\!\!>T^*G \xrightarrow{\tilde{\alpha}}\!\!\gg A^*G. \tag{24}$$

Here (23) is an exact sequence of Lie groupoids with the vector bundle T^*M acting on $q_* \colon A^*G \to M$ as an additive groupoid by $(\omega, \varphi) \mapsto a^*\omega + \varphi$. The injection is $(\omega, \varphi) \mapsto \overline{\omega} + \widetilde{1}_\varphi$.

On the other hand, (24) is an exact sequence of vector bundles and the injection is $(\omega, g) \mapsto \beta_g^* \omega$.

Example 11.3.1 For a pair groupoid $G = M \times M$ the cotangent manifold is $T^*M \times T^*M$. However the source and target are given by

$$\tilde{\beta}(\psi, \varphi) = \psi, \qquad \tilde{\alpha}(\psi, \varphi) = -\varphi.$$

Thus it is not strictly the case that $T^*M \times T^*M \rightrightarrows T^*M$ is a pair groupoid.   ⊠

11.4 Poisson groupoids

Definition 11.4.1 Let $G \rightrightarrows P$ be a Lie groupoid with a Poisson structure π on the manifold G. Then (G, π) is a *Poisson groupoid* if the Poisson anchor $\pi^\# \colon T^*G \to TG$ is a morphism of groupoids over some map $A^*G \to TP$.   □

We denote the base map $A^*G \to TP$ by a_*; it is necessarily a vector bundle morphism. Since $\pi^\#$ is certainly a vector bundle morphism over G, it is a morphism of $\mathcal{VB}$–groupoids (with respect to the identity maps on G and P). Since $\pi^\#$ is skew–symmetric, by 11.2.6 the core morphism $T^*P \to AG$ is $-a_*^*$.

Proposition 11.4.2 *The inversion* $i \colon G \to G$ *is an antiPoisson map.*

Proof The equation (20) can be rewritten $\Phi^{-1} = -\Phi \circ T_{g^{-1}}(i)$ for $\Phi \in T_g^*G$. Hence

$$\pi^\#(i^*\Phi) = \pi^\#(\Phi \circ T(i)) = -\pi^\#(\Phi^{-1}).$$

Since $\pi^\#$ is a morphism, this is equal to

$$-\pi^\#(\Phi)^{-1} = -T(i)(\pi^\#(\Phi))$$

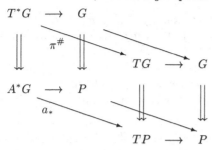

Fig. 11.2.

and since i is a diffeomorphism, this can be rewritten

$$T(i)(\pi^{\#}(i^*\Phi)) = -\pi^{\#}(\Phi),$$

which shows that i is antiPoisson. □

It now follows, as in 11.1.6, that inversion in the groupoid T^*G is a Lie algebroid automorphism.

We next show that the Poisson structure on G induces a Poisson structure on P. For any $\omega \in \Omega^1(P)$, the pullback form $\beta^*\omega$ is the core section of T^*G corresponding to ω. Since the core morphism of $\pi^{\#}$ is $-a_*^*$, we have

$$\pi^{\#}(\beta^*\omega) = -\overrightarrow{a_*^*\omega}, \tag{25}$$

recalling that the core section of TG corresponding to any $X \in \Gamma AG$ is $\overrightarrow{X}$. We therefore have

$$\langle \beta^*\omega_2, \pi^{\#}(\beta^*\omega_1) \rangle = -\langle \omega_2, (a \circ a_*^*)(\omega_1) \rangle \circ \beta,$$

using $T(\beta)(\overrightarrow{X}) = aX$. Define a 2–vector field π_P on P by

$$\pi_P(\omega_1, \omega_2) = -\langle \omega_2, (a \circ a_*^*)(\omega_1) \rangle.$$

We now have $\pi(\beta^*\omega_1, \beta^*\omega_2) = \pi_P(\omega_1, \omega_2) \circ \beta$ for all $\omega_1, \omega_2 \in \Omega^1(P)$ or, briefly,

$$\pi \overset{\beta}{\sim} \pi_P,$$

extending to multivector fields the usual notion of projectability of vector fields. So $[\pi, \pi] \overset{\beta}{\sim} [\pi_P, \pi_P]$ and since $[\pi, \pi] = 0$ and β is surjective, it follows that $[\pi_P, \pi_P] = 0$. In a similar fashion one proves that π_P is skew–symmetric. This completes the proof of the following:

Proposition 11.4.3 *The tensor π_P just defined is a Poisson structure on P with $\pi_P^\# = -a \circ a_*^* \colon T^*P \to TP$, and $\beta \colon G \to P$ is a Poisson map.*

Since β is a surjective submersion, π_P is the unique Poisson structure on P with respect to which β is Poisson. From 11.4.2 it further follows that α is antiPoisson.

Although π projects to $\pm \pi_P$ under the target and source projections, the values of π on identity elements are not determined by π_P alone. Using (25) and

$$\pi^\#(\widetilde{1}_\varphi) = T(1)(a_*\varphi), \tag{26}$$

for $\varphi \in A^*G$, which is the identity preservation part of the morphism condition on π, we have

$$\begin{aligned}
\pi(\overline{\omega} + \widetilde{1}_\varphi, \overline{\theta} + \widetilde{1}_\psi) &= \langle \theta, a_*\varphi \rangle - \langle \omega, a_*\psi \rangle - \langle a^*\theta, a_*^*\omega \rangle \\
&= \pi_P(\omega, \theta) + \langle \theta, a_*\varphi \rangle - \langle \omega, a_*\psi \rangle \quad (27)
\end{aligned}$$

for $\omega, \theta \in T^*P$ and $\varphi, \psi \in A^*G$.

Note that the skew–symmetry of π_P implies that

$$-a \circ a_*^* = a_* \circ a^*. \tag{28}$$

Another immediate consequence of (25) is that, for any $\omega, \theta \in \Omega^1(P)$,

$$\{\beta^*\omega, \alpha^*\theta\} = 0. \tag{29}$$

For, using (25), we have

$$\pi(\beta^*\omega, \alpha^*\theta) = \langle \alpha^*\theta, -\overrightarrow{a_*^*\omega} \rangle$$

and this is zero, since right–invariant vector fields are α–vertical.

Next, since β is a Poisson map, we have that $\beta^*[\omega_1, \omega_2] = [\beta^*\omega_1, \beta^*\omega_2]$ for $\omega_1, \omega_2 \in \Omega^1(P)$. Applying (25) we get that

$$a_*^*[\omega_1, \omega_2] = -[a_*^*\omega_1, a_*^*\omega_2]. \tag{30}$$

Together with $\pi_P^\# = a \circ (-a_*^*)$ this shows that a_*^* is an anti–morphism of Lie algebroids. Dually, $a_* \colon A^*G \to TP$ is an antiPoisson map.

Proposition 11.4.4 *The manifold 1_P of identity elements is coisotropic in G.*

Proof Take $\Phi \in T^*_{1_m}G$. Now every element $\xi \in T_{1_m}G$ can be written uniquely in the form $\xi = T(1)T(\alpha)(\xi) + X$ where $X \in A_mG$. So the condition that Φ be in $T(1_P)^\circ$ is that $\langle \Phi, \xi \rangle$ depends on Φ and X only, and so the condition is satisfied if and only if $\Phi = \tilde{1}_\varphi$ where $\varphi \in A_m^* G$ is the restriction of Φ to $A_m G$. Thus $T(1_P)^\circ$ is the base manifold $\{\tilde{1}_\varphi \mid \varphi \in A^* G\}$ of $T^* G \rightrightarrows A^* G$.

Now the morphism condition on $\pi^{\#}$ ensures that each $\tilde{1}_\varphi$ is mapped to $T(1)(a_*(\varphi))$, which is tangent to 1_P. □

It now follows that the conormal bundle of 1_P in G has a Lie algebroid structure. From the final subsection of §3.5 this conormal bundle is $A^* G$. The proof of 10.4.2 shows that the anchor regarded as $T(1_P)^\circ \to T(1)(TP)$ is $\tilde{1}_\varphi \mapsto T(1)(a_* \varphi)$, and the bracket of $\tilde{1}_\varphi$, $\tilde{1}_\psi$, where φ, $\psi \in \Gamma A^* G$, is defined by

$$[\tilde{1}_\varphi, \tilde{1}_\psi] = [\Phi, \Psi] \circ 1 \tag{31}$$

where Φ, $\Psi \in \Omega^1(G)$ are any forms for which $\Phi \circ 1 = \tilde{1}_\varphi$, $\Psi \circ 1 = \tilde{1}_\psi$. The coisotropy of 1_P ensues that $[\Phi, \Psi] \circ 1$ is of the form $\tilde{1}_\chi$ for some $\chi \in \Gamma A^* G$ and we write $\chi = [\varphi, \psi]_*$. In practice of course, we usually transfer the anchor and bracket to $A^* G$ itself. Then $A^* G$ has the *dual Lie algebroid structure induced by the Poisson structure on* G. The anchor is now precisely a_* and the bracket is given by

$$\tilde{1}_{[\varphi, \psi]_*} = [\tilde{1}_\varphi, \tilde{1}_\psi]. \tag{32}$$

We could alternatively write $[\varphi, \psi]_*(m) = \tilde{\alpha}([\Phi, \Psi](1_m))$, and equally with $\tilde{\beta}$.

Note that (32) and (26) show that $\tilde{1} \colon A^* G \to T^* G$ is a morphism of Lie algebroids over $1 \colon P \to G$. Observe also that the pullback Lie algebroid of $T^* G$ across 1 can be identified with $A^* G$ which may thus also be regarded as a Lie subalgebroid of $T^* G$.

The following crucial result is the key to the infinitesimal structure of Poisson groupoids.

Theorem 11.4.5 *Let* $G \rightrightarrows P$ *be a Poisson groupoid. Then for any* $g, h \in G$ *for which* gh *is defined,*

$$\pi(gh) = T(R_\tau)(\pi(g)) + T(L_\sigma)(\pi(h)) - T(R_\tau)T(L_\sigma)(\pi(1_y)) \tag{33}$$

where $y = \alpha g = \beta h$ *and* σ, τ *are (local) bisections of* G *with* $\sigma(y) = g$ *and* $\tau(\alpha h) = h$.

Proof Write $x = \alpha h$ and $z = \beta g$. First consider a single element $\Phi_{gh} \in T_{gh}^*(G)$; as in 11.1.3, the subscripts denote the base points. Define

$$\Phi_g = \Phi_{gh} \circ T_g(R_\tau) \in T_g^* G.$$

From the general formula (17) we have $\widetilde{\beta}(\Phi_g) = \Phi_{gh} \circ T_g(R_\tau) \circ T_{1_z}(R_g)$. Because $T_g(R_\tau)$ here is restricted to the image of $T_{1_z}(R_g)$, this can be written $\Phi_{gh} \circ T_g(R_h) \circ T_{1_z}(R_g)$ and so we have

$$\widetilde{\beta}(\Phi_g) = \widetilde{\beta}(\Phi_{gh}).$$

In a similar way, define

$$\Phi_h = \Phi_{gh} \circ T_g(L_\sigma) \in T_h^* G,$$

and we have $\widetilde{\alpha}(\Phi_h) = \widetilde{\alpha}(\Phi_{gh})$. We can therefore define

$$\Phi_{1_y} = \Phi_g^{-1} \bullet \Phi_{gh} \bullet \Phi_h^{-1} \in T_{1_y}^*(G).$$

We need to prove that

$$\Phi_{1_y} = (\Phi_{gh} \circ T(R_\tau) \circ T(L_\sigma))^{-1}. \tag{34}$$

Recall that any element Υ of $T_{1_y}^*(G)$ is of the form $\overline{\omega} + \widetilde{1}_\varphi$ with $\omega = \Upsilon \circ T(1)$ and $\varphi = \widetilde{\alpha}(\Upsilon)$. We first prove that, for any $v \in T_y(P)$,

$$\langle \Phi_g^{-1} \bullet \Phi_{gh} \bullet \Phi_h^{-1}, T(1)(v) \rangle = -\langle \Phi_{gh} \circ T(R_\tau) \circ T(L_\sigma), T(1)(v) \rangle. \tag{35}$$

Take any $\xi \in T_g G$ and $\eta \in T_h G$ with $T(\alpha)(\xi) = T(\beta)(\eta) = v$. Then since $T(1)(v) = \xi^{-1} \bullet (\xi \bullet \eta) \bullet \eta^{-1}$, we have

$$\begin{aligned}
\langle \Phi_g^{-1} \bullet \Phi_{gh} \bullet \Phi_h^{-1}, T(1)(v) \rangle &= -\langle \Phi_g, \xi \rangle + \langle \Phi_{gh}, \xi \bullet \eta \rangle - \langle \Phi_h, \eta \rangle \\
&= -\langle \Phi_g, \xi \rangle + \langle \Phi_{gh}, \xi \bullet \eta \rangle - \langle \Phi_h, \eta \rangle \\
&= -\langle \Phi_{gh}, T_g(R_\tau)(\xi) \rangle - \langle \Phi_{gh}, T_h(L_\sigma)(\eta) \rangle \\
&\quad + \langle \Phi_{gh}, T(R_\tau)(\xi) + T(L_\sigma)(\eta) - T(R_\tau)T(L_\sigma)(T(1)(v)) \rangle \\
&= -\langle \Phi_{gh}, T(R_\tau)T(L_\sigma)T(1)(v) \rangle.
\end{aligned}$$

This establishes (35). Write $\omega = \Phi_{1_y} \circ T(1) \in T_y^* P$.

Next consider the source φ of Φ_{1_y}. Clearly $\varphi = \widetilde{\beta}(\Phi_h)$. And, for any $Y \in A_y G$,

$$\langle \widetilde{\beta}(\Phi_h), Y \rangle = \langle \Phi_h, T(R_h)(Y) \rangle = \langle \Phi_{gh} \circ T(L_\sigma) \circ T(R_h), Y \rangle,$$

and since Y is α–vertical, we can replace $T(R_h)$ here by $T(R_\tau)$. Thus φ is the restriction to $A_y G$ of $\Phi_{gh} \circ T(L_\sigma) \circ T(R_\tau)$.

The source of $\Phi_{gh} \circ T(R_\tau) \circ T(L_\sigma)$, on the other hand, is given, for $Y \in A_y G$, by

$$\begin{aligned}
\langle \widetilde{\alpha}(\Phi_{gh} &\circ T(R_\tau) \circ T(L_\sigma)), Y \rangle \\
&= \langle \Phi_{gh} \circ T(R_\tau) \circ T(L_\sigma), Y \rangle - \langle \Phi_{gh} \circ T(R_\tau) \circ T(L_\sigma), T(1)(aY) \rangle
\end{aligned}$$

so we have $\tilde{\alpha}(\Phi_{gh} \circ T(R_\tau) \circ T(L_\sigma)) = \varphi + a^*\omega$. Altogether, therefore,

$$\Phi_{gh} \circ T(R_\tau) \circ T(L_\sigma) = -\overline{\omega} + \tilde{1}_{\varphi + a^*\omega}$$

and, using (22), this is the inverse of $\overline{\omega} + \tilde{1}_\varphi$. We have thus proved (34).

The remainder of the proof is similar to that of 11.1.3. We take another $\Psi_{gh} \in T^*_{gh}(G)$. We then have

$$\langle \Phi_{gh} \wedge \Psi_{gh}, \pi(gh) \rangle = \langle \Phi_g \bullet \Phi_{1_y} \bullet \Phi_h, \pi^\#(\Psi_g \bullet \Psi_{1_y} \bullet \Psi_h) \rangle$$
$$= \langle \Psi_g, \pi^\#(\Phi_g) \rangle + \langle \Psi_{1_y}, \pi^\#(\Phi_{1_y}) \rangle + \langle \Psi_h, \pi^\#(\Phi_h) \rangle,$$

using the morphism condition on $\pi^\#$. Now

$$\langle \Psi_g, \pi^\#(\Phi_g) \rangle = \langle \Phi_g \wedge \Psi_g, \pi(g) \rangle = \langle \Phi_{gh} \wedge \Psi_{gh}, T(R_\tau)(\pi(g)) \rangle$$

and likewise with the term at h. For the term at 1_y, again using the morphism condition, and (20),

$$\langle \Psi_{1_y}, \pi^\#(\Phi_{1_y}) \rangle = -\langle \Psi_{1_y}^{-1}, \pi^\#(\Phi_{1_y}^{-1}) \rangle = -\langle \Phi_{1_y}^{-1} \wedge \Psi_{1_y}^{-1}, \pi(1_y) \rangle$$
$$= -\langle \Phi_{gh} \wedge \Psi_{gh}, T(R_\tau)T(L_\sigma)\pi(gh) \rangle$$

which completes the proof. $\qquad\qquad\square$

This is the first occasion on which we have used the full morphism condition.

The key to the above proof is worth stating separately, since it provides an intrinsic description of the groupoid structure of T^*G.

Proposition 11.4.6 *Let $G \rightrightarrows M$ be any Lie groupoid. For any $g, h \in G$ such that gh is defined, and any $\Phi_{gh} \in T^*_{gh}(G)$,*

$$\Phi_g^{-1} \bullet \Phi_{gh} \bullet \Phi_h^{-1} = (\Phi_{gh} \circ T(R_\tau) \circ T(L_\sigma))^{-1}, \qquad (36)$$

where Φ_g, Φ_h are as defined in the proof of 11.4.5 and σ, τ are (local) bisections with $\sigma(\alpha g) = g$, $\tau(\alpha h) = h$.

Theorem 11.4.7 *Let $G \rightrightarrows P$ be a Poisson groupoid. Then for all $X \in \Gamma AG$,*

$$\mathcal{L}_{\overrightarrow{X}}(\pi) = -\overrightarrow{d_* X}. \qquad (37)$$

Proof Firstly note that because $\overrightarrow{X}$ projects under α to the zero vector field, and π projects to $-\pi_P$, we have

$$\mathcal{L}_{\overrightarrow{X}}(\pi) = [\overrightarrow{X}, \pi] \overset{\alpha}{\sim} [0, -\pi_P] = 0,$$

so $\mathcal{L}_{\overrightarrow{X}}(\pi)\colon G \to TG \wedge TG$ in fact takes values in $T^\alpha G \wedge T^\alpha G$.

We prove that $\mathcal{L}_{\overrightarrow{X}}(\pi)$ is right–invariant. Take $h \in G$, say $\beta h = y$. The flow of $\overrightarrow{X}$ is $L_{\mathrm{Exp}\, tX}$, so

$$(\mathcal{L}_{\overrightarrow{X}}(\pi))(h) = -\lim_{t\to 0} \tfrac{1}{t}\{T(L_{\mathrm{Exp}\, tX})(\pi(gh)) - \pi(h)\}$$

where $g = \mathrm{Exp}\, -tX\,(y)$. We apply (33) with $\sigma = \mathrm{Exp}\, -tX$ and τ any local bisection with $\tau(\alpha h) = h$. This leads to

$$(\mathcal{L}_{\overrightarrow{X}}(\pi))(h) = T(R_\tau)(-\lim_{t\to 0}\tfrac{1}{t}\{T(L_{\mathrm{Exp}\, tX})(\pi(L_{\mathrm{Exp}\, -tX}(1_y))) - \pi(1_y)\})$$

$$= T(R_\tau)(\mathcal{L}_{\overrightarrow{X}}(\pi)(1_y))$$

and since $\mathcal{L}_{\overrightarrow{X}}(\pi)(1_y)$ is known to be in $A_y G \wedge A_y G$, this can be written as $T(R_h)(\mathcal{L}_{\overrightarrow{X}}(\pi)(1_y))$.

We therefore have $\mathcal{L}_{\overrightarrow{X}}(\pi) = \overrightarrow{\rho}$ for some $\rho \in \Gamma\Lambda^2(AG)$ and it remains to prove that $\rho = -d_*X$. We apply 11.4.8 below to G with $\xi = \overrightarrow{X}$ and with Φ, Ψ such that $\Phi \circ 1 = \widetilde{1}_\varphi$, $\Psi \circ 1 = \widetilde{1}_\psi$.

First note that since the pairing is defined pointwise, we have

$$\langle \Phi \wedge \Psi, \mathcal{L}_\xi(\pi)\rangle \circ 1 = \langle (\Phi \circ 1) \wedge (\Psi \circ 1), (\mathcal{L}_{\overrightarrow{X}}(\pi)) \circ 1\rangle = \langle \varphi \wedge \psi, \rho\rangle,$$

where in the last step we identify $\widetilde{1}_\varphi$ with φ as usual. Likewise, in view of the definition of the bracket $[\,.\,]_*$, $\langle [\Phi, \Psi], \xi\rangle \circ 1$ is identified with $\langle [\varphi, \psi]_*, X\rangle$.

Lastly consider $\iota_{\pi^\#(\Psi)}\delta\langle \Phi, \xi\rangle$. Interpreting $\Psi \circ 1 = \widetilde{1}_\psi$ in terms of Figure 11.2, it is clear that $\pi^\#(\Psi) \circ 1 = T(1)(a_*\psi)$. Now

$$\langle \delta\langle \Phi, \overrightarrow{X}\rangle, \pi^\#(\Psi)\rangle \circ 1 = \langle (\delta\langle \Phi, \overrightarrow{X}\rangle) \circ 1, T(1)(a_*(\psi))\rangle = \langle \delta\langle \varphi, X\rangle, a_*\psi\rangle$$

where the first two δs are on G and the last is on P. Altogether, we have

$$-\langle \varphi \wedge \psi, \rho\rangle + \langle [\varphi, \psi]_*, X\rangle = a_*(\varphi)\langle \psi, X\rangle - a_*(\psi)\langle \varphi, X\rangle$$

so $\rho = -d_*X$ as required. $\qquad\square$

Lemma 11.4.8 *In any Poisson manifold,*

$$-\langle \Phi \wedge \Psi, \mathcal{L}_\xi(\pi)\rangle + \langle [\Phi, \Psi], \xi\rangle = \iota_{\pi^\#(\Phi)}\delta\langle \Psi, \xi\rangle - \iota_{\pi^\#(\Psi)}\delta\langle \Phi, \xi\rangle,$$

where ξ is a vector field and Φ, Ψ are 1–forms.

Proof Expand the LHS using the first line of (5) on p. 383, and then apply (36) on p. 307 repeatedly. $\qquad\square$

Corollary 11.4.9 *For any $\xi \in \Lambda^{\bullet}(AG)$,*

$$[\pi, \overrightarrow{\xi}] = \overrightarrow{d_* \xi}. \tag{38}$$

Proof This is a simple induction on the degree of ξ, using equations (22) and (23) on p. 304. $\square$

Another consequence of 11.4.7 is the following:

Proposition 11.4.10 *The source and target projections of T^*G are Lie algebroid morphisms.*

Proof We give the proof for $\widetilde{\beta} \colon T^*G \to A^*G$. Firstly, note that we have $a_* \circ \widetilde{\beta} = T(\beta) \circ \pi^{\#}$ because $\pi^{\#}$ is a groupoid morphism. Hence, by 4.3.8, it is sufficient to check that if $\Phi \in \Omega^1(G)$ and $\Psi \in \Omega^1(G)$ project under $\widetilde{\beta}$ to $\varphi \in \Gamma A^*G$ and $\psi \in \Gamma A^*G$, then $[\Phi, \Psi]$ projects to $[\varphi, \psi]$.

That Φ projects to φ is equivalent to

$$\langle \Phi, \overrightarrow{X} \rangle = \langle \varphi, X \rangle \circ \beta \tag{39}$$

for all $X \in \Gamma AG$. Take Φ projectable to φ and Ψ projectable to ψ, and any $X \in \Gamma AG$. First calculate

$$(d_{\pi}\overrightarrow{X})(\Phi, \Psi) = \pi^{\#}(\Phi)(\langle \overrightarrow{X}, \Psi \rangle) - \pi^{\#}(\Psi)(\langle \overrightarrow{X}, \Phi \rangle) - \langle \overrightarrow{X}, [\Phi, \Psi] \rangle).$$

Now $\pi^{\#}(\Phi)(f \circ \beta) = (a_*\varphi)(f) \circ \beta$ from the anchor condition, and so

$$(d_{\pi}\overrightarrow{X})(\Phi, \Psi) = (a_*\varphi)(\langle X, \psi \rangle) \circ \beta - (a_*\psi)(\langle X, \varphi \rangle) \circ \beta - \langle \overrightarrow{X}, [\Phi, \Psi] \rangle). \tag{40}$$

Next note that $d_{\pi}\overrightarrow{X} = -\mathcal{L}_{\overrightarrow{X}}(\pi) = \overrightarrow{d_*X}$. Also, (39) extends to 2–vector fields ξ to give

$$\langle \Phi \wedge \Psi, \overrightarrow{\xi} \rangle = \langle \varphi \wedge \psi, \xi \rangle \circ \beta.$$

We therefore have

$$\overrightarrow{d_*X}(\Phi, \Psi) = (d_*X(\varphi, \psi)) \circ \beta.$$

Expanding the RHS of this out, and comparing it with (40), it follows that $\langle [\Phi, \Psi], \overrightarrow{X} \rangle = \langle [\varphi, \psi], X \rangle \circ \beta$ as required. $\square$

This result may also be proved by the method of 10.3.6.

As in the case of Poisson Lie groups, 11.4.10 shows that the Lie algebroid structure on A^*G is a quotient of that on T^*G. The short exact sequence of vector bundles

$$a^! T^*P \rightarrowtail T^*G \overset{\widetilde{\beta}}{\twoheadrightarrow} A^*G \tag{41}$$

where the injection map is $(g, \omega) \mapsto \omega \circ T_g(\alpha)$, follows directly from the $\mathscr{VB}$-groupoid structure of T^*G. From the fact that α is antiPoisson we already know that $\ker \widetilde{\beta}$ is closed under the bracket; this is very much weaker than the result that $\widetilde{\alpha}$ is a Lie algebroid morphism. From

$$[\alpha^*\omega_1, \alpha^*\omega_2] = -\alpha^*[\omega_1, \omega_2]$$

it also follows that the Lie algebroid structure on $\ker \widetilde{\beta}$ is the action Lie algebroid $T^*P \lessdot (-\alpha)$ arising from the Poisson map $-\alpha$ as in §10.1.

Example 11.4.11 Given a Poisson manifold (P, π) denote by $P \times \overline{P}$ the manifold $G = P \times P$ with the Poisson structure $\pi_G = (\pi, -\pi)$. Now

$$\pi_G^\# = (\pi^\#, -\pi^\#)$$

and so, recalling from 11.3.1 the structure of T^*G, the pair groupoid G with π_G is a Poisson groupoid which we call, with some abuse of terminology, the *pair Poisson groupoid*. Note that the Lie algebroid structure on $A^*G = T^*P$ is that defined in 10.1.2.   ⊠

Definition 11.4.12 A *morphism of Poisson groupoids* $F \colon G \to H$ over $f \colon P \to Q$, where $G \rightrightarrows P$ and $H \rightrightarrows Q$ are Poisson groupoids, is a morphism of Lie groupoids which is also a Poisson map.   □

It follows that f is Poisson. As in §10.1, it also follows that the Lie algebroid T^*H acts on F and that T^*Q acts on f. Further, since $F(1_P) \subseteq 1_Q$, we can apply 10.4.11 to show that $AF \colon AG \to AH$ is Poisson and that the dual comorphism is the restriction of $C(F) \colon T^*H \lessdot F \to T^*G$ to $A^*H \lessdot f \to A^*G$, where A^*H acts on f by

$$\psi^\dagger = a_*(\ell_\psi \circ A(F))$$

for $\psi \in \Gamma A^*H$. To summarize:

Proposition 11.4.13 Let $F \colon G \to H$, $f \colon P \to Q$ be a morphism of Poisson groupoids. Then:
   (i) *the base map* $f \colon P \to Q$ *is Poisson, and*
   (ii) *the morphism of Lie algebroids* $AF \colon AG \to AH$ *is a Poisson map.*

From (i) it follows that if G and H are Poisson groupoids on the same base *manifold* and if there is a base–preserving Poisson morphism between them, then G and H induce the same Poisson structure on the base.

From (ii) it follows in particular that for any Poisson groupoid G on base P, the anchor $a\colon AG \to TP$ is a Poisson map; since β is Poisson and α is antiPoisson, the groupoid anchor $(\beta, \alpha)\colon G \to P \times \overline{P}$ is a morphism of Poisson groupoids. Hence by 10.4.6, $a^*\colon T^*P \to A^*G$ is a morphism of Lie algebroids.

We gather together for reference the properties of the various projection maps.

Theorem 11.4.14 *Let* $G \rightrightarrows P$ *be a Poisson groupoid. Then:*

(i) $\beta\colon G \to P$ *is a Poisson map, and* $\alpha\colon G \to P$ *is antiPoisson;*

(ii) $a\colon AG \to TP$ *is Poisson from the Poisson structure on* AG *dual to* A^*G, *to the tangent lift structure on* TP;

(iii) $a^*\colon T^*P \to A^*G$ *is a Lie algebroid morphism;*

(iv) $a_*\colon A^*G \to TP$ *is antiPoisson from the Poisson structure dual to the Lie algebroid* AG, *to the tangent lift structure on* TP;

(v) $-a_*^*\colon T^*P \to AG$ *is a Lie algebroid morphism;*

(vi) $\{\alpha^*\omega_1, \beta^*\omega_2\} = 0$ *for all* $\omega_1, \omega_2 \in \Omega^1(P)$;

(vii) 1_P *is a coisotropic submanifold of* G.

The bundle projections q, q_* are usually not Poisson maps.

11.5 Symplectic groupoids

Definition 11.5.1 Let $\Sigma \rightrightarrows P$ be a Lie groupoid with a symplectic structure $\omega \in \Omega^2(\Sigma)$. Then (Σ, ω) is a *symplectic groupoid* if Σ is a Poisson groupoid with respect to the associated Poisson structure. □

We will usually write π, or π_ω if clarity is needed, for the Poisson structure on a symplectic groupoid.

For a symplectic groupoid (Σ, ω) the fact that $\pi^\#\colon T^*\Sigma \to T\Sigma$ is an isomorphism of vector bundles implies immediately that $a_*\colon A^*\Sigma \to TP$ is an isomorphism of vector bundles. Now the rank of $A^*\Sigma$ is the dimension of the α–fibres of Σ, so we have

$$\dim \Sigma - \dim P = \dim P.$$

It is clear that this is an extremely strong constraint on the groupoid.

Definition 11.5.2 Consider a symplectic manifold (M, ω). A closed submanifold S of M is *Lagrangian* if it is coisotropic with respect to the Poisson structure and $\dim S = \frac{1}{2} \dim M$. □

It follows from 11.4.14(vii) that 1_P is a Lagrangian submanifold of Σ.

Next $a_* : A^*\Sigma \to TP$, being the anchor of $A^*\Sigma$, is a morphism of Lie algebroids; since it is an isomorphism of vector bundles, it is an isomorphism of Lie algebroids.

Similarly using 11.4.14(v), $-a_*^* : T^*P \to A\Sigma$ is an isomorphism of Lie algebroids. We will usually denote $-a_*^*$ by $\mathcal{W}$.

What is most remarkable about these isomorphisms is that T^*P has its Lie algebroid structure entirely from the Poisson structure on P, whilst $A\Sigma$ has its Lie algebroid structure entirely from the Lie groupoid structure on Σ. These isomorphisms show that the symplectic structure and the groupoid structure are very tightly related.

We summarize for reference:

Theorem 11.5.3 *Let* $\Sigma \rightrightarrows P$ *be a symplectic groupoid. Then:*

(i) $\beta : \Sigma \to P$ *is a Poisson map, and* $\alpha : \Sigma \to P$ *is antiPoisson;*

(ii) $\mathcal{W} = -a_*^*$ *is an isomorphism of Lie algebroids from the cotangent Lie algebroid* T^*P *to the Lie algebroid* $A\Sigma$ *of* Σ;

(iii) $a : A\Sigma \to TP$ *is a Poisson diffeomorphism, from the Poisson structure on* $A\Sigma$ *dual to* $A^*\Sigma$, *to the tangent lift structure on* TP;

(iv) $a^* : T^*P \to A^*\Sigma$ *is a Lie algebroid isomorphism;*

(v) $a_* : A^*\Sigma \to TP$ *is an antiPoisson diffeomorphism from the Poisson structure dual to the Lie algebroid* $A\Sigma$, *to the tangent lift structure on* TP;

(vi) $\{\alpha^*\omega_1, \beta^*\omega_2\} = 0$ *for all* $\omega_1, \omega_2 \in \Omega^1(P)$;

(vii) 1_P *is a Lagrangian submanifold of* Σ.

Notice that the Poisson structure on P will not itself be symplectic unless $A\Sigma$ is transitive, and in this case, providing P is connected, Σ will be a quotient of the fundamental groupoid of P.

Recall from 10.1.9 that for any Poisson manifold P, the transitivity orbits of T^*P are the symplectic leaves of P. When P is the base of a symplectic groupoid, this result is easily strengthened.

Proposition 11.5.4 *Let* $\Sigma \rightrightarrows P$ *be an* α–*connected symplectic groupoid. Then the transitivity orbits of* Σ *are the symplectic leaves of* Σ.

$$*\qquad*\qquad*\qquad*\qquad*$$

In the remainder of this section we prove that $T^*G \rightrightarrows A^*G$, for any Lie groupoid G, is a symplectic groupoid. This requires an extension of the canonical isomorphisms J, R, I and Θ from manifolds to Lie groupoids.

Consider a Lie groupoid $G \rightrightarrows M$. We begin by analyzing the canonical isomorphism J_G.

Proposition 11.5.5 *The canonical isomorphism* $J_G \colon T^2G \to T^2G$ *is a morphism of Lie groupoids over* $J_M \colon T^2M \to T^2M$.

Proof For any smooth map $\varphi \colon S_1 \to S_2$ of manifolds,

$$T^2(\varphi) \circ J_{S_1} = J_{S_2} \circ T^2(\varphi).$$

Applying this to α, to β, and to 1, we find that J_G commutes with the source, target, and identity maps of T^2G with respect to J_M. Denote the multiplication in G by $\kappa \colon G \times_M G \to G$. Then $T^2(\kappa) \circ J_{G \times_M G} = J_G \circ T^2(\kappa)$. Now, recalling that the tangent functor commutes with pullbacks, we can identify $T^2(G \times_M G)$ with $T^2G \times_{T^2M} T^2G$, and it follows that J_G preserves multiplication. $\qquad\qquad\square$

The following terminology will be needed shortly.

Definition 11.5.6 Given a $\mathscr{VB}$-groupoid $(\Omega; G, E; M)$, the *tangent prolongation* $\mathscr{VB}$-*groupoid* is $(T\Omega; TG, TE; TM)$ obtained by applying the tangent functor to the structure of Ω. $\qquad\qquad\square$

If $K \to M$ denotes the core of Ω, then the core of the tangent prolongation is $TK \to TM$.

Return now to the fixed Lie groupoid $G \rightrightarrows M$. It is possible to consider T^2G as a $\mathscr{VB}$-groupoid in two distinct ways, shown in Figure 11.3. In (a) we have the tangent $\mathscr{VB}$-groupoid of $TG \rightrightarrows TM$; its core is

$$
\begin{array}{ccc}
& p_{TG} & \\
T^2G & \longrightarrow & TG \\
\Big\Downarrow & & \Big\Downarrow \\
T^2M & \longrightarrow & TM \\
& p_{TM} &
\end{array}
\qquad\qquad
\begin{array}{ccc}
& T(p_G) & \\
T^2G & \longrightarrow & TG \\
\Big\Downarrow & & \Big\Downarrow \\
T^2M & \longrightarrow & TM \\
& T(p_M) &
\end{array}
$$

$$\text{(a)} \qquad\qquad\qquad \text{(b)}$$

$$\text{Fig. 11.3.}$$

ATG, the Lie algebroid of $TG \rightrightarrows TM$. In (b) we have the tangent prolongation of the tangent $\mathscr{VB}$–groupoid of $G \rightrightarrows M$ with core TAG. The proof of the proposition which follows is straightforward.

Proposition 11.5.7 J_G *and* J_M *together with* id_{TG} *and* id_{TM} *constitute a morphism of* $\mathscr{VB}$–*groupoids from the* $\mathscr{VB}$–*groupoid in Figure* 11.3(a) *to the* $\mathscr{VB}$–*groupoid in Figure* 11.3(b). *The core map is* $(j_G)^{-1} \colon ATG \to TAG$.

Now consider the dual $\mathscr{VB}$–groupoids of the structures in Figure 11.3. Figure 11.4(a) shows the dual of Figure 11.3(a); the upper horizontal

$$
\begin{array}{ccc}
T^*TG & \overset{c_{TG}}{\longrightarrow} & TG \\
\Downarrow & & \Downarrow \\
A^*TG & \longrightarrow & TM
\end{array}
\qquad\qquad
\begin{array}{ccc}
T^\bullet TG & \longrightarrow & TG \\
\Downarrow & & \Downarrow \\
T^\bullet AG & \longrightarrow & TM
\end{array}
$$

(a)   (b)

Fig. 11.4.

structure is the straightforward cotangent dual of a tangent bundle, and the lower is the Lie algebroid dual of the Lie algebroid of $TG \rightrightarrows TM$. The core is T^*TM. In Figure 11.4(b) the vector bundles are duals of prolonged structures and the core is $T^\bullet TM$.

We now take the dual over TG of the morphism of $\mathscr{VB}$–groupoids in 11.5.7. Since we have specified the domain and codomain structures, there is only one direction in which the dual can be taken, and we denote it J_G^*. Applying 11.2.6 we have:

Proposition 11.5.8 *The map* J_G^* *shown in Figure* 11.5(a) *and obtained by dualization of the morphism of* $\mathscr{VB}$–*groupoids in* 11.5.7 *is an isomorphism of* $\mathscr{VB}$–*groupoids from the* $\mathscr{VB}$–*groupoid in Figure* 11.4(b) *to the* $\mathscr{VB}$–*groupoid in Figure* 11.4(a) *over* id_{TG} *and the inverse of* $(j_G \,{}^{\star}TM)$. *The core map is* $J_M^* \colon T^\bullet TM \to T^*TM$.

In extending the maps I, Θ and R to the context of groupoids, there is no added work in considering a general $\mathscr{VB}$–groupoid, and the general case is in fact rather clearer. Until the end of the section, let $(\Omega; G, E; M)$ be a $\mathscr{VB}$–groupoid as in (9).

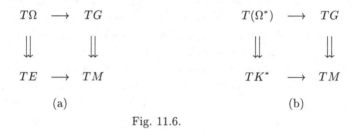

Fig. 11.5.

Applying the tangent functor to Ω and to its dual, we obtain two $\mathcal{VB}$-groupoids as in Figure 11.6 below.

$$
\begin{array}{ccc}
T\Omega & \longrightarrow & TG \\
\Downarrow & & \Downarrow \\
TE & \longrightarrow & TM
\end{array}
\qquad\qquad
\begin{array}{ccc}
T(\Omega^*) & \longrightarrow & TG \\
\Downarrow & & \Downarrow \\
TK^* & \longrightarrow & TM
\end{array}
$$

(a)   (b)

Fig. 11.6.

Proposition 11.5.9 *The map $I_\Omega\colon T\Omega^* \to T^\bullet\Omega$ which is shown in Figure 11.5(b) is an isomorphism of Lie groupoids over $I_K\colon TK^* \to T^\bullet K$. Furthermore, it is an isomorphism of $\mathcal{VB}$-groupoids, with respect to id_{TG} and id_{TM}, from $(T\Omega^*;TK^*,TG;TM)$ to $(T^\bullet\Omega;T^\bullet K,TG;TM)$ and as such has core morphism $I_E\colon TE^* \to T^\bullet E$.*

Proof The pairing of Ω^* and Ω over G is a pairing of Lie groupoids and so the tangent pairing of $T\Omega^*$ and $T\Omega$ over TG is a pairing of the tangent Lie groupoids. □

Corollary 11.5.10 *The map $I_{TG}\colon TT^*G \to T^\bullet TG$ is an isomorphism of Lie groupoids over $I_{AG}\colon TA^*G \to T^\bullet AG$. Furthermore, it is an isomorphism of $\mathcal{VB}$-groupoids with respect to id_{TG} and id_{TM} and as such has core morphism $I_{TM}\colon TT^*M \to T^\bullet TM$.*

Together the previous two results yield the following:

Theorem 11.5.11 *The Tulczyjew isomorphism Θ_G is an isomorphism of Lie groupoids*

$$
\begin{array}{ccc}
& \Theta_G & \\
TT^*G & \longrightarrow & T^*TG \\[4pt]
\Updownarrow & & \Updownarrow \\[4pt]
TA^*G & \longrightarrow & A^*TG \\
& \vartheta_G &
\end{array}
$$

where the base map $(j_G \,{\ast}\, TM)^{-1} \circ I_{AG}$ is denoted ϑ_G. Furthermore, it is an isomorphism of $\mathcal{VB}$–groupoids over id_{TG} and id_{TM} and as such has core morphism $\Theta_M \colon TT^*M \to T^*TM$.

We call ϑ_G the *groupoid Tulczyjew map for G*. From the properties of j_G and I_{AG} it follows that ϑ_G is an isomorphism of double vector bundles over A^*G and TM and is a Poisson map from the tangent prolongation of the Poisson structure on A^*G, to the Poisson structure on A^*TG dual to the Lie algebroid structure from $TG \rightrightarrows TM$.

Before proceeding to consider R, we need another internalization map.

Again consider a general $\mathcal{VB}$–groupoid Ω. Because the Lie functor sends groupoid morphisms to vector bundle morphisms (in fact to Lie algebroid morphisms) and preserves pullbacks, the Lie algebroid of Ω is a double vector bundle as in Figure 11.7(a) below. Likewise we can

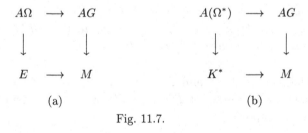

$$
\begin{array}{ccc}
A\Omega & \longrightarrow & AG \\
\downarrow & & \downarrow \\
E & \longrightarrow & M \\
& \text{(a)} &
\end{array}
\qquad\qquad
\begin{array}{ccc}
A(\Omega^*) & \longrightarrow & AG \\
\downarrow & & \downarrow \\
K^* & \longrightarrow & M \\
& \text{(b)} &
\end{array}
$$

Fig. 11.7.

apply the Lie functor to the dual of Ω and obtain the double vector bundle in Figure 11.7(b).

Now the canonical pairing of the vector bundle Ω with Ω^* is a map

$$
\langle\,,\,\rangle \colon \Omega \times_G \Omega^* \to \mathbb{R}
$$

and, as noted already, this is a Lie groupoid morphism: the domain is the pullback in the category of Lie groupoids of the projections $\Omega \to G$ and $\Omega^* \to G$, and the definition (14) of the multiplication in Ω^* may be read precisely as the statement that the canonical pairing is a groupoid morphism into the additive group(oid) $\mathbb{R}$. We can therefore apply the Lie functor and get a pairing

$$\langle\!\langle \, , \, \rangle\!\rangle = A(\langle \, , \, \rangle) \colon A\Omega \times_{AG} A\Omega^* \to \mathbb{R} \tag{42}$$

of $A\Omega$ with $A\Omega^*$ over AG.

Theorem 11.5.12 *The pairing* (42) *is non–degenerate and is a pairing of the double vector bundles in Figure* 11.7 *in the sense of* 9.2.3.

Proof That (42) satisfies (iv) and (v) of 9.2.3 follows because $\langle\!\langle \, , \, \rangle\!\rangle$ is itself a vector bundle morphism. To prove (i) of 9.2.3, take $X \in E_m$ and $\varphi \in E_m^*$ and write

$$\widetilde{0}_X = \frac{d}{dt}\widetilde{1}_X\bigg|_0 \, , \qquad \overline{\varphi} = \frac{d}{dt}(t\varphi)\bigg|_0$$

where the $t\varphi$ is itself a core element in Ω^*. Then

$$\langle\!\langle \widetilde{0}_X, \overline{\varphi} \rangle\!\rangle = \frac{d}{dt}\langle \widetilde{1}_X, t\varphi \rangle\bigg|_0 = \langle \varphi, X \rangle$$

which proves (i). Likewise, $\langle\!\langle \overline{k}, \widetilde{0}_\kappa^{(*)} \rangle\!\rangle = \langle \kappa, k \rangle$ for $k \in K_m, \kappa \in K_m^*$. The proof of (iii) is similar.

To show that $\langle\!\langle \, , \, \rangle\!\rangle$ is nondegenerate, take $(\Xi; X, x; m)$ in $A\Omega$ and assume that $\langle\!\langle \mathscr{X}, \Xi \rangle\!\rangle = 0$ for all $(\mathscr{X}; \kappa, x; m)$ in $A(\Omega^*)$. We first prove that $X = 0$. Take any $\varphi \in E_m^*$ and define

$$\mathscr{X} = \frac{d}{dt}\Big(\widetilde{0}_{g_t}^{(*)}(t\overline{\varphi})^{-1}\Big)\bigg|_0$$

where $x = \frac{d}{dt}g_t\big|_0$ and $\widetilde{0}_{g_t}^{(*)}$ is the zero of Ω^* above g_t. Using (14) and (15) we get

$$\langle \widetilde{0}_{g_t}^{(*)}(t\overline{\varphi})^{-1}, \xi_t \widetilde{1}_X \rangle = \langle \widetilde{0}_{g_t}^{(*)}, \xi_t \rangle + \langle (t\overline{\varphi})^{-1}, \widetilde{1}_X \rangle = 0 - t\langle \varphi, X \rangle$$

where $\Xi = \frac{d}{dt}\xi_t\big|_0$. Hence $\langle\!\langle \mathscr{X}, \Xi \rangle\!\rangle = -\langle \varphi, X \rangle$. Since φ was arbitrary, we have $X = 0$ and so $\Xi = A(\widetilde{0})(x) \underset{E}{+} \overline{k}$ for some $k \in K_m$. Now take any $\kappa \in K_m^*$ and some element $(\mathscr{X}; \kappa, x; m)$ in $A(\Omega^*)$. Using (iv) of 9.2.3 we expand $\langle\!\langle A(\widetilde{0})(x) \underset{E}{+} \overline{k}, \mathscr{X} \underset{K^*}{+} \widetilde{0}_\kappa \rangle\!\rangle$ and get $\langle\!\langle \mathscr{X}, \Xi \rangle\!\rangle = \langle \kappa, k \rangle$ and so $k = 0$. Thus Ξ is the zero of $A\Omega \to AG$ over x. $\qquad\square$

Accordingly there is an isomorphism of double vector bundles, defined for $\mathscr{X} \in A(\Omega^*)$, $\Xi \in A\Omega$ by

$$\mathfrak{I}_\Omega \colon A(\Omega^*) \to A^\bullet\Omega, \qquad \langle \mathfrak{I}_\Omega(\mathscr{X}), \Xi \rangle_{AG} = \langle\!\langle \mathscr{X}, \Xi \rangle\!\rangle, \qquad (43)$$

which preserves the side bundles AG and K^* and the core E^*. This is the $\mathscr{VB}$-*groupoid internalization map*.

In the case $\Omega = TG$ we obtain an isomorphism, the *groupoid internalization map*

$$i_G = \mathfrak{I}_{TG} \colon A(T^*G) \to A^\bullet TG, \qquad \langle i_G(\mathscr{X}), \Xi \rangle_{AG} = \langle\!\langle \mathscr{X}, \Xi \rangle\!\rangle. \qquad (44)$$

$$* \quad * \quad * \quad * \quad *$$

We turn now to the reversal isomorphism. Again, we start by working with a general $\mathscr{VB}$-groupoid $(\Omega; G, E; M)$. We proceed as in §9.5 with D now the double vector bundle $(A\Omega; AG, E; M)$, and its two duals as in Figure 11.8.

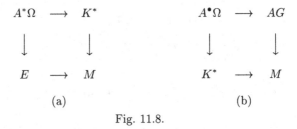

(a) (b)

Fig. 11.8.

Applying 9.2.2 to $D = A\Omega$ we get the pairing

$$\lfloor \varphi, \mathfrak{f} \rfloor = \langle \varphi, \xi \rangle_E - \langle \mathfrak{f}, \xi \rangle_{AG} \qquad (45)$$

for $\varphi \in A^*\Omega$, $\mathfrak{f} \in A^\bullet\Omega$ and suitable $\xi \in A\Omega$. Composing the isomorphism Z_E from 9.2.4 with the dual over K^* of the isomorphism $\mathfrak{I}_\Omega$, we get an isomorphism of double vector bundles

$$(\mathfrak{I}_\Omega \mathbin{\overline{*}} K^*) \circ Z_E \colon A^*\Omega \to A^*(\Omega^*).$$

Still proceeding as in §9.5, we obtain:

Theorem 11.5.13 *The map* $\mathfrak{R}_\Omega \colon A^*\Omega^* \to A^*\Omega$ *defined by*

$$\langle \mathfrak{R}_\Omega(\Phi), \xi \rangle_E = \langle\!\langle \mathscr{X}, \xi \rangle\!\rangle_{AG} - \langle \Phi, \mathscr{X} \rangle_{K^*} \qquad (46)$$

for $\xi \in A\Omega$, $\mathscr{X} \in A\Omega^*$, $\Phi \in A^*\Omega^*$ *such that* ξ *and* $\mathscr{X}$ *have the same* *projection into AG, $\mathscr{X}$ and Φ have the same projection into K^*, and Φ and ξ have the same projection into E, is an isomorphism of double vector bundles, preserving the side bundles E and K^*, and inducing $-\mathrm{id}\colon A^*G \to A^*G$ on the cores.*

We call $\mathfrak{R}_\Omega$ the $\mathscr{VB}$-groupoid reversal isomorphism. For $\Omega = TG$, we call

$$\mathfrak{r}_G = \mathfrak{R}_{TG}\colon A^*T^*G \to A^*TG \tag{47}$$

the *groupoid reversal isomorphism*.

Theorem 11.5.14 *The map*

$$
\begin{array}{ccc}
 & R_\Omega & \\
T^*\Omega^* & \longrightarrow & T^*\Omega \\
\Downarrow & & \Downarrow \\
A^*\Omega^* & \longrightarrow & A^*\Omega \\
 & \mathfrak{R}_\Omega &
\end{array}
$$

is an isomorphism of Lie groupoids.

Proof We drop the subscripts Ω on the maps. We first show that the targets are preserved. Take $\Phi \in T^*\Omega^*$. Both $\mathfrak{R}(\widetilde{\beta}(\Phi))$ and $\widetilde{\beta}(R(\Phi))$ are elements of $A^*\Omega$. To see that each has the same projections, it is best to consider the cubes in Figure 11.9; each projection of either of these elements can be seen as a projection from a projection into Ω or Ω^*, which we know that R preserves.

Denote the projection of Φ into Ω by ω. We need to show that

$$\langle \mathfrak{R}(\widetilde{\beta}(\Phi)), \xi \rangle_E = \langle \widetilde{\beta}(R(\Phi)), \xi \rangle_E$$

for all $\xi \in A\Omega|_{\beta\omega}$. By (46), the LHS is

$$\langle \mathfrak{R}(\widetilde{\beta}(\Phi)), \xi \rangle_E = \langle\!\langle \mathscr{X}, \xi \rangle\!\rangle_{AG} - \langle \widetilde{\beta}(\Phi), \mathscr{X} \rangle_{K^*}$$
$$= \langle\!\langle \mathscr{X}, \xi \rangle\!\rangle_{AG} - \langle \Phi, T(R_\varphi)(\mathscr{X}) \rangle_{\Omega^*}$$

where φ is the projection of Φ into Ω^* and $\mathscr{X}$ is a suitable element of $A\Omega^*|_{\beta\varphi}$. Denote by X the common projection of ξ and $\mathscr{X}$ into AG. Note that the R_φ is right translation.

Likewise the RHS is, for a suitable $\mathscr{Y} \in T\Omega^*$,

$$\langle R(\Phi), T(R_\omega)(\xi) \rangle_\Omega = \langle\!\langle \mathscr{Y}, T(R_\omega)(\xi) \rangle\!\rangle_{TG} - \langle \Phi, \mathscr{Y} \rangle_{\Omega^*}.$$

We can choose $\mathscr{Y} = T(R_\varphi)(\mathscr{X})$ and we then have

$$\langle\!\langle T(R_\varphi)(\mathscr{X}), T(R_\omega)(\xi)\rangle\!\rangle_{TG} - \langle \Phi, T(R_\varphi)(\mathscr{X})\rangle_{\Omega^*}$$

and the result follows from the lemma below. The proof for the source projections is similar.

We now need to prove that R preserves the groupoid multiplication. For $i = 1, 2$, take $\Phi_i \in T^*\Omega^*$ with projections ω_i into Ω and φ_i into Ω^*, and such that $\Phi_1 \bullet \Phi_2$ is defined. We have to show that

$$\langle R(\Phi_1 \bullet \Phi_2), \Xi\rangle_\Omega = \langle R(\Phi_1) \bullet R(\Phi_2), \Xi\rangle_\Omega$$

for all $\Xi \in T_{\omega_1\omega_2}\Omega$. Now any such Ξ can be written as $\Xi_1 \bullet \Xi_2$ for $\Xi_i \in T_{\omega_i}\Omega$. The RHS is then

$$\langle R(\Phi_1), \Xi_1\rangle_\Omega + \langle R(\Phi_2), \Xi_2\rangle_\Omega.$$

We now choose $\mathfrak{X}_i \in T_{\varphi_i}\Omega^*$ with the same projection into AG as Ξ_i. The result then follows from use of (37) on p. 361, from the morphism properties of the pairings, and from using Figure 11.9 to keep track of the various projections. $\qquad\square$

Lemma 11.5.15 *Let* $\mathscr{X} \in A\Omega^*|_{\beta\varphi}$ *and* $\xi \in A\Omega|_{\beta\omega}$ *have the same projection into* AG. *Then*

$$\langle\!\langle T(R_\varphi)(\mathscr{X}), T(R_\omega)(\xi)\rangle\!\rangle_{TG} = \langle\!\langle \mathscr{X}, \xi\rangle\!\rangle_{AG},$$

Proof Written in terms of the tangent groupoids $T\Omega^*$ and $T\Omega$ the LHS is $\langle\!\langle \mathscr{X} \bullet \widetilde{0}_\varphi, \xi \bullet \widetilde{0}_\omega\rangle\!\rangle_{TG}$. Now the pairing is a morphism of the tangent groupoids, since it is the tangent of the pairing of Ω^* and Ω. So we have

$$\langle\!\langle \mathscr{X} \bullet \widetilde{0}_\varphi, \xi \bullet \widetilde{0}_\omega\rangle\!\rangle_{TG} = \langle\!\langle \mathscr{X}, \xi\rangle\!\rangle_{TG} + \langle\!\langle \widetilde{0}_\varphi, \widetilde{0}_\omega\rangle\!\rangle_{TG}$$

and the last term is zero. The final equality holds because $\mathscr{X} \in A\Omega^*$ and $\xi \in A\Omega$. $\qquad\square$

Corollary 11.5.16 *The reversal map* $\mathfrak{R}_\Omega\colon A^*\Omega^* \to A^*\Omega$ *is an antiPoisson diffeomorphism.*

Proof The map $R_\Omega\colon T^*\Omega^* \to T^*\Omega$ is an antisymplectomorphism and the target projections of both groupoid structures are Poisson submersions. $\qquad\square$

Corollary 11.5.17 *The reversal isomorphism* $R_{TG}\colon T^*T^*G \to T^*TG$ *is an isomorphism of Lie groupoids over* $\mathfrak{r}_G\colon A^*T^*G \to A^*TG$.

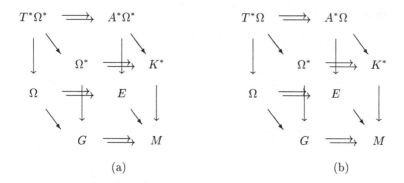

Fig. 11.9.

From 9.6.7 we have $\Theta_G \circ (\delta \nu_G)^\# = R_{TG}$ and so it follows that $(\delta \nu_G)^\#$ is an isomorphism of Lie groupoids over

$$a_* = \vartheta_G^{-1} \circ \mathfrak{r}_G : A^* T^* G \to T A^* G. \tag{48}$$

Hence $T^* G \rightrightarrows A^* G$ is a symplectic groupoid. From 11.5.3 we know that $\mathscr{W} = -a_*^*$ is an isomorphism of Lie algebroids. To summarize:

Theorem 11.5.18 *For any Lie groupoid $G \rightrightarrows M$, the Lie groupoid $T^* G \rightrightarrows A^* G$ is a symplectic groupoid with respect to the canonical symplectic structure on $T^* G$. The induced Poisson structure on the base is the Poisson structure dual to the Lie algebroid structure on AG.*

For clarity we will usually denote $\mathscr{W} = \mathscr{W}_{T^* G}$ by $\mathfrak{w}_G$ or merely $\mathfrak{w}$. The map a_* above will then be $-\mathfrak{w}^*$. The minus sign here refers to the vector bundle structure over $A^* G$. The map $\mathfrak{w}$ is a morphism of double vector bundles over AG and $A^* G$ which on the cores $T^* M$ is $-\mathrm{id}$.

11.6 Table of canonical isomorphisms

This section summarizes the various canonical isomorphisms which have been introduced in this and previous chapters.

For any manifold M:

- The canonical involution

$$J_M \colon T^2 M \to T^2 M$$

is an isomorphism of double vector bundles over TM and TM and induces id: $TM \to TM$ on the cores. 9.6.1

- The Tulczyjew isomorphism

$$\Theta_M \colon TT^* M \to T^* TM$$

is an isomorphism of double vector bundles over $T^* M$ and TM and induces id: $T^* M \to T^* M$ on the cores. It is a symplectomorphism from the tangent prolongation of the canonical structure on $T^* M$ to the canonical structure on $T^* TM$. 9.6.2

- The map associated to the canonical symplectic structure $\delta \nu$ on $T^* M$,

$$(\delta \nu_M)^{\#} \colon T^* T^* M \to TT^* M$$

is an isomorphism of double vector bundles over $T^* M$ and TM and induces $-$id: $T^* M \to T^* M$ on the cores. 10.1.8

For any vector bundle $E \to M$:

- The reversal isomorphism

$$R_E \colon T^* E^* \to T^* E$$

is an isomorphism of double vector bundles over E and E^* and induces $-$id: $T^* M \to T^* M$ on the cores. It is an antisymplectomorphism of the canonical structures. §9.5

- The internalization map

$$I_E \colon TE^* \to T^{\bullet} E$$

is an isomorphism of double vector bundles over E and TM and induces id: $E \to E$ on the cores. 9.3.2

For any Lie groupoid $G \rightrightarrows M$:

- The canonical isomorphism

$$j_G \colon TAG \to ATG$$

is an isomorphism of double vector bundles over TM and induces the identity on the cores AG. It is an isomorphism of Lie algebroids over TM.   9.7.5

- The groupoid internalization map

$$i_G \colon AT^*G \to A^\bullet TG$$

is an isomorphism of double vector bundles over AG and A^*G and is the identity on the cores T^*M.   (44)

- The groupoid Tulczyjew isomorphism

$$\vartheta_G \colon TA^*G \to A^*TG$$

is an isomorphism of double vector bundles over A^*G and TM with core map id_{A^*G}. It is a Poisson map from the tangent prolongation of the Poisson structure on A^*G, to the Poisson structure on A^*TG dual to the Lie algebroid structure from $TG \rightrightarrows TM$.   11.5.11

- The groupoid reversal map

$$\mathfrak{r}_G \colon A^*T^*G \to A^*TG$$

is an isomorphism of double vector bundles over TM and A^*G, inducing $-\mathrm{id} \colon A^*G \to A^*G$ on the cores.   (47)

For any $\mathcal{VB}$-groupoid $(\Omega; G, E; M)$:

- The $\mathcal{VB}$-groupoid internalization map

$$\mathfrak{I}_\Omega \colon A\Omega^* \to A^\bullet\Omega$$

is an isomorphism of double vector bundles over AG and K^* and is the identity on the cores E^*.   (43)

- The $\mathcal{VB}$-groupoid reversal map

$$\mathfrak{R}_\Omega \colon A^*\Omega^* \to A^*\Omega$$

is an isomorphism of double vector bundles over E and K^*, inducing $-\mathrm{id} \colon A^*G \to A^*G$ on the cores.   11.5.13

*   *   *   *   *

For any manifold M, the diagram in Figure 11.10 commutes. The lower–right triangle is Definition 9.6.2 and the upper–left triangle is Theorem 9.6.7.

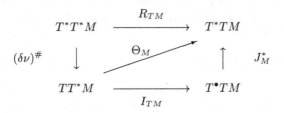

Fig. 11.10.

For any Lie groupoid $G \Rightarrow M$, the diagram in Figure 11.11 commutes. The lower–right triangle is from 11.5.11 and the upper–left triangle is (48).

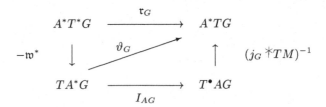

Fig. 11.11.

11.7 Notes

§11.1 There are several books which incorporate accounts of Poisson–Lie groups; notably [Chari and Pressley, 1994] and [Majid, 1995]. See also the detailed set of notes [Kosmann − Schwarzbach, 2004] which contains a great deal of material not readily available elsewhere.

The approach given in this section, that the various 'twisted' identities satisfied by the actions in a Poisson Lie group arise naturally from the associated Lie groupoid and Lie algebroid structures, comes from [Mackenzie, 1992]. This was itself a development of the idea, introduced in various forms by Kosmann − Schwarzbach and Magri [1988],

Lu and Weinstein [1990] and Majid [1990], that these equations could be obtained naturally from the structure which was called, respectively, a 'twilled extension' or (the infinitesimal form of) a 'double Lie group' or a 'matched pair'.

The compatibility between the Lie groupoid and the Lie algebroid structures on T^*G, for G a Poisson Lie group, makes T^*G an $\mathscr{LA}$-*groupoid* [Mackenzie, 1992, 4.12]. In general an $\mathscr{LA}$-groupoid consists of a quadruple $(\Omega; G, A; M)$ in which Ω is both a Lie groupoid over A, which is a Lie algebroid over M, and a Lie algebroid over G, which is a Lie groupoid over M; in the groupoid structure on Ω, the source, target, multiplication and identity maps are to be morphisms of Lie algebroids. This notion is central to the author's approach to general notions of double; see the discusssion in the Appendix.

Some of the details of the presentation in this section have been taken from [Mackenzie, 2000].

§11.2, §11.3 The construction of the general cotangent groupoid $T^*G \rightrightarrows A^*G$ was given by Coste, Dazord, and Weinstein [1987]. Immediately thereafter, Pradines [1988] exhibited it as an instance of a general duality for the notion of $\mathscr{VB}$-groupoid introduced there. The approach to $\mathscr{VB}$-groupoids which we give here differs from Pradines' in using the kernel of the source projection; this imposes a lack of symmetry (apparent in (13) and (17)), but this is only superficial. This approach is consistent with the construction of the Lie algebroid of a Lie groupoid which we use, whereas Pradines' conventions extend the symmetric construction given at the end of §3.5. Our approach to $\mathscr{VB}$-groupoids was first given in [Mackenzie, 1992, §4, §5]. See also [Mackenzie, 2000b, §2]. The conventions for $T^*G \rightrightarrows A^*G$ which we follow here are precisely those of [Mackenzie and Xu, 1994].

The pairing of AT^*G with ATG was introduced in [Mackenzie and Xu, 1994, §7]. Theorem 11.5.12 comes from [Mackenzie, 1999].

A recent account of the tangent and cotangent groupoids has been given by Landsman [2002].

Symplectic groupoids. The notion of symplectic groupoid was introduced independently by Karasëv [1989, 1986], by Weinstein [1987], Coste, Dazord, and Weinstein [1987], and by Zakrzewski [1990a,b]. Since the earliest work on Poisson structures — indeed since the work of Lie — one approach to the study of a Poisson manifold P has been to represent P as a quotient of a symplectic manifold S under a surjective submersive Poisson map $S \to P$; this is known as the *(full) realizability problem*: see [Weinstein, 1982]. If one has a symplectic realization of P it is reasonable to think that one may be able to lift problems from P to S, solve them on S using techniques specific to symplectic structures, and then seek to quotient the solution back down to P. The notion of symplectic groupoid emerged from observing that many realizations — most notably T^*G as a realization of $\mathfrak{g}^*$ — had additional structure, and that this structure was that of a symplectic groupoid.

At about the same time — the early–mid 1980s — a large number of people independently observed the bracket structure on the 1–forms of a Poisson manifold, and that this gives T^*P the structure of a Lie algebroid. (See [Huebschmann, 1990] for a detailed history.) This observation could of itself have given a lead into the use of Lie groupoids in Poisson geometry.

The work on symplectic groupoids done by its originators made heavy use of the special properties of symplectic structures: for example, the existence of local Lagrangian bisections and the method of characteristics. The approach which we have given, deducing the basic properties of symplectic groupoids from the general Poisson case, shows that this reliance is unnecessary; one may work in an entirely diagrammatic fashion, using the basic results of the Lie theory of Lie groupoids and Lie algebroids.

We have not gone into further developments with symplectic groupoids, since — apart from issues of length — this subject has been dramatically revitalized very recently by the work of Cattaneo and Felder [2001] and subsequently of Crainic and

Fernandes [arXiv:0210152]. In the broader sense of [Cattaneo and Felder, 2001], every Poisson manifold may be said to integrate to a symplectic groupoid with singularities; this is a matter in which the full ramifications remain to be seen.

In particular, the strict concept of symplectic groupoid with which we have dealt here admits few broad classes of examples, and a detailed exposition would lead us far afield. The reader may consult [Mikami, 1991] and [Weinstein, 1989]; further references are given in [Mackenzie, 1995].

For Lie groupoids with a compatible contact structure, see [Libermann, 1995].

Poisson groupoids. The notion of Poisson groupoid was introduced by Weinstein [1988], together with an analysis of the principal examples and the nature of duality in this setting. The same paper introduced the coisotropic calculus and used it to construct the Lie algebroid structure on the Lie algebroid dual (see 11.4.4 and the discussion which follows it). The coisotropic calculus allows one to work directly on the manifolds concerned, rather than, as we do here, on the cotangent and tangent bundles of the groupoids and Lie algebroids; it is also closer to standard methods in symplectic geometry. The definition we use here, 11.4.1, may be proved to be equivalent to that of [Weinstein, 1988] by the method of 11.1.2.

The concept of Poisson groupoid was further developed by Xu in the important paper [Xu, 1995], which gave significantly different approaches to some of the constructions of [Mackenzie and Xu, 1994].

The account in §11.4 draws on that of the author and Ping Xu [Mackenzie and Xu, 1994, 1998, 2000], and that of Xu [1995] but the account given here makes greater use of the concept of $\mathcal{VB}$–groupoid, which enables the nature of many arguments to be seen diagrammatically. The crucial Theorem 11.4.5, due to Xu [1995], is here proved directly from the structure of T^*G, rather than in terms of the Poisson structure itself.

We have not devoted much attention to examples of Poisson groupoids, except for those belonging to fairly general families, or arising from the basic canonical constructions. This reflects our concern with Poisson groupoids as themselves examples of double structures, though the full development of this point of view must take place elsewhere (see [Mackenzie, 1998] for an overview).

Beyond the constructions of Poisson groupoids studied in the text, we mention those associated with classical dynamical Yang–Baxter equations [Etingof and Varchenko, 1998], [Bangoura and Kosmann–Schwarzbach, 1998], [Li and Parmentier, arXiv:0209212] and those for which the Poisson structure is exact [Liu and Xu, 1996].

In [Mackenzie, 1999] the author showed that every double Lie groupoid gives rise to a pair of Poisson groupoids in duality, by a construction which extends to the double setting the canonical examples $T^*G \rightrightarrows A^*G$ of symplectic groupoids.

Affine 2–vector fields. In the terminology of Xu [1995] and Weinstein [1990], condition (33) shows that the Poisson tensor on a Poisson groupoid is an *affine* 2–vector field.

As with the case of 1–vector fields (see §9.8), there is no requirement in (33) that $\pi(1_y)$ be tangent to 1_M. We already observed in the text that in a Poisson groupoid $G \rightrightarrows P$, the values of π on 1_P are not determined by π_P alone. We want to emphasize that it is this fact which makes π affine, and not the absence of a groupoid structure on $TG \wedge TG$. In [Xu, 1995] it is proved that a Poisson tensor π on a Lie groupoid G makes (G, π) a Poisson groupoid if it satisfies (33) and if in addition, 1_P is coisotropic, $\pi^\#(\beta^*\omega)$ is right–invariant for all $\omega \in \Omega^1(P)$, and pullbacks under source and target always Poisson commute (as in (29)).

Affine Poisson structures have been studied in [Dazord and Sondaz, 1991] as well as in [Weinstein, 1990].

Duality. For a Poisson Lie group (G, π), the Lie algebra dual $\mathfrak{g}^*$ always integrates to a connected and simply–connected Lie group G^*, the *dual group of* (G, π). The relationship between G and G^* was the study of work by Lu and Weinstein [1990] and Majid [1990], among others; the most general result so far was obtained in [Lu and Weinstein, 1989], which showed that G and G^* form the side group(oid)s of a symplectic double groupoid.

For general Poisson groupoids one may say [Weinstein, 1988] that $G \rightrightarrows P$ and $H \rightrightarrows P$ are in *duality* if $AG \cong A^*H$ and $AH \cong A^*G$ as Lie algebroids; thus in particular a symplectic groupoid $\Sigma \rightrightarrows P$ and the pair groupoid $P \times \overline{P} \rightrightarrows P$ are always in duality. In [Mackenzie, 1999] the author showed that the two side groupoids of a symplectic double groupoid are Poisson groupoids in duality, and one may thus seek to extend the result of [Lu and Weinstein, 1989] to general Poisson groupoids. These matters are still under development.

Canonical isomorphisms of $\mathscr{VB}$–groupoids. The material from 11.5.5 onward to the end of the section is an expansion of part of [Mackenzie, 1999, §3]; Theorem 11.5.14 was stated in [Mackenzie, 1999] without proof.

The proof of 11.5.18 given here uses some of the same techniques as the proof in [Mackenzie and Xu, 1994], but differs significantly: in place of the commutativity relations in Figure 11.11, [Mackenzie and Xu, 1994] defines a map $\theta_G = (j_G \, {}^*\!AG) \circ i_G$ and shows that $\theta_G = R_{AG} \circ s$ where $s = \mathscr{W}$ (see Figure 12.2). The point of the approach used here is that it applies more generally.

As the reader has undoubtedly surmised, most of the canonical isomorphisms in this section are actually isomorphisms of triple structures. A fuller development of this aspect will be given elsewhere.

12

Lie bialgebroids

Lie bialgebroids are the infinitesimal form of Poisson groupoids. Applying the Lie functor to a Poisson groupoid (G, π_G) results in the usual Lie algebroid AG together with a Lie algebroid structure on the vector bundle dual A^*G. This may be obtained as the conormal Lie algebroid of the base manifold of G, as a quotient of the cotangent Lie algebroid T^*G, or as the dual of a Poisson structure on AG obtained by prolongation of π_G. The pair AG and A^*G constitute the Lie bialgebroid of (G, π_G). As usual, there is much more work required to give an abstract formulation of the compatibility condition between two Lie algebroid structures on a pair of dual bundles A and A^* which models the case of AG and A^*G.

Taking (G, π_G) to be first a Poisson–Lie group and, second, a Poisson pair groupoid $P \times \overline{P}$, shows that Lie bialgebras and (TP, T^*P), for P a Poisson manifold, are Lie bialgebroids. The usual abstract definition of a Lie bialgebra depends upon the fact that the bracket of a Lie algebra is a (bi)linear map, and can therefore be dualized; this is not the case with general Lie algebroid brackets. The 'algebraic' definition of a Lie bialgebroid which we give in §12.1 uses the Lie algebroid coboundary as a replacement for the dual of the bracket. This first section uses the Schouten calculus of §7.5 to establish the self–duality of the definition and to develop the most basic properties.

The definition of abstract Lie bialgebroid is not obviously related to that of Poisson groupoid. We clarify this in §12.2 by showing that a pair of Lie algebroid structures on a vector bundle A and its dual A^* form a Lie bialgebroid if and only if the Poisson anchor $T^*A \to TA$ corresponding to the Poisson structure on A, which is dual to the Lie algebroid structure on A^*, gives a morphism of Lie algebroids from T^*A^*

to TA when combined with the canonical isomorphism $R\colon T^*A^* \to T^*A$ of §9.5.

In §12.3 we complete the circle of descriptions by showing that, for a Poisson groupoid (G, π_G), the Poisson structure induced on AG by the Lie algebroid structure on A^*G, is the linearization of the Poisson structure on G. This result again depends on the use of canonical isomorphisms.

The final section, §12.4, is a very brief introduction to moment maps in terms of Lie groupoids and Lie algebroids, and shows, in particular, that the existence of moment maps can be deduced from the integrability of Lie algebroid morphisms.

12.1 Lie bialgebroids

A Lie bialgebroid consists of Lie algebroid structures on a vector bundle A and on its dual A^*, subject to a compatibility condition. Before stating this condition, we set up the notation.

We consider a Lie algebroid (A, q, M) with anchor a and bracket $[\ ,\ ]$. The coboundary operator in $\Gamma\Lambda^\bullet(A^*)$ is denoted as usual by d. The two Lie derivatives, of cochains and of elements of $\Gamma\Lambda^\bullet(A)$, are denoted as usual by $\mathcal{L}_X$. The insert operators are denoted as usual by ι_ξ for $\xi \in \Gamma\Lambda^\bullet(A^*)$.

The Lie algebroid structure on the dual bundle (A^*, q_*, M) has anchor denoted a_* and bracket $[\ ,\ ]_*$. The coboundary operator in $\Gamma\Lambda^\bullet(A)$ is denoted d_*. The two Lie derivatives are denoted $\mathcal{L}^*_\varphi$ for $\varphi \in \Gamma A^*$. The insert operators are denoted by ι_φ for $\varphi \in \Gamma\Lambda^\bullet(A)$, with no asterisk.

Definition 12.1.1 Let (A, q, M) be a vector bundle equipped with Lie algebroid structures both on A itself and on the dual bundle (A^*, q_*, M). Then these two structures constitute a *Lie bialgebroid* structure on A if, for all $X, Y \in \Gamma A$,

$$d_*[X, Y] = [d_*X, Y] + [X, d_*Y]. \tag{1}$$

□

We will usually denote a Lie bialgebroid simply by (A, A^*). Before proceeding we note the most fundamental examples.

Example 12.1.2 For a Poisson manifold (P, π) consider (TP, T^*P)

where TP is the standard tangent bundle and T^*P is the cotangent Lie algebroid. Then (1) becomes

$$d_\pi[X, Y] = [d_\pi X, Y] + [X, d_\pi Y].$$

Since $d_\pi = [\pi, -]$ by 10.2.3, this is an easy consequence of the graded Jacobi identity, and so (TP, T^*P) is a Lie bialgebroid, the *standard Lie bialgebroid of* (P, π). ⊠

Example 12.1.3 Consider a Poisson groupoid $(G \rightrightarrows P, \pi)$. By 11.4.7 we have that, for all $X \in \Gamma AG$,

$$\mathcal{L}_{\overrightarrow{X}}(\pi) = -\overrightarrow{d_*X}.$$

So, with $Y \in \Gamma AG$ also, applying graded Jacobi gives

$$\overrightarrow{d_*[X,Y]} = [\pi, \overrightarrow{[X,Y]}] = -[[\overrightarrow{X}, \pi], \overrightarrow{Y}] - [[\pi, \overrightarrow{Y}], \overrightarrow{X}] = [\overrightarrow{d_*X}, \overrightarrow{Y}] + [\overrightarrow{X}, \overrightarrow{d_*Y}]$$

and (1) follows. Thus (AG, A^*G) is a Lie bialgebroid, the *Lie bialgebroid of* $(G \rightrightarrows P, \pi)$. ⊠

Our aim now is to show that the definition is self–dual (Theorem 12.1.9 below) and to do this we need a considerable number of working equations.

Lemma 12.1.4 *Consider any Lie algebroid A on base M. For $\xi \in \Gamma \Lambda^i(A)$ and $f \in C^\infty(M)$,*

$$[\xi, f] = (-1)^{i-1} \iota_{df}(\xi).$$

Proof For $i = 1$ we have $[X, f] = a(X)(f) = \langle df, X \rangle = \iota_{df}(X)$. Assume the result for a given i. Then, for $\xi \in \Gamma \Lambda^i(A)$ and $X \in \Gamma A$, using first the graded Leibniz identity (equation (22) on p. 304) and then the hypothesis leads to

$$
\begin{aligned}
[\xi \wedge X, f] &= -[\xi, f] \wedge X + \xi \wedge [X, f] \\
&= (-1)^i \iota_{df}(\xi) \wedge X + \xi \wedge \iota_{df}(X).
\end{aligned}
$$

By (30) on p. 306, this is equal to $(-1)^i \iota_{df}(\xi \wedge X)$ and extending by additivity completes the induction. □

From now on until 12.1.15, we are considering a Lie bialgebroid as just defined.

In (1) replace Y by fY and expand out. By $d_*(fZ) = d_*f \wedge Z + f d_*Z$ the LHS becomes

$$d_*f \wedge [X,Y] + f d_*[X,Y] + d_*(a(X)(f)) \wedge Y + a(X)(f) d_*Y.$$

On the RHS we first have, by the graded Leibniz identity,

$$\begin{aligned}
[d_*X, fY] &= f[d_*X, Y] + [d_*X, f] \wedge Y \\
&= f[d_*X, Y] - \iota_{df}(d_*X),
\end{aligned}$$

using 12.1.4 in the second line. The second term on the RHS is

$$\begin{aligned}
[X, d_*(fY)] &= [X, d_*f \wedge Y] + [X, f d_*Y] \\
&= [X, d_*f] \wedge Y + d_*f \wedge [X,Y] + a(X)(f) d_*Y + f[X, d_*Y].
\end{aligned}$$

Equating the LHS and RHS and cancelling, we have

$$([X, d_*f] - d_*(a(X)(f)) - \iota_{df}(d_*X)) \wedge Y = 0$$

for all Y. This completes the proof of:

Lemma 12.1.5 *For all $X \in \Gamma A$ and $f \in C^\infty(M)$,*

$$[X, d_*f] = d_*(a(X)(f)) + \iota_{df}(d_*X).$$

This equation has various formulations. Firstly, since $a(X)(f) = \iota_{df}(X)$, and $\mathcal{L}_\varphi^* = d_* \circ \iota_\varphi + \iota_\varphi \circ d_*$, it can be written as

$$\mathcal{L}_{d_*f}(X) + \mathcal{L}_{df}^*(X) = 0. \tag{2}$$

Secondly, writing $a(X)(f) = [X, f]$ and $\iota_{df}(d_*X) = -[d_*X, f]$, we obtain

$$d_*[X, f] = [d_*X, f] + [X, d_*f]. \tag{3}$$

The next step is to carry out the same device on (3). Replacing X by gX and expanding out, the LHS becomes

$$g d_*[X, f] + [X, f] d_*g.$$

On the RHS we have, firstly,

$$\begin{aligned}
[d_*(gX), f] &= [d_*g \wedge X, f] + [g d_*X, f] \\
&= -\iota_{df}(d_*g \wedge X) + g[d_*X, f] \\
&= -\langle df, d_*g \rangle X + [X, f] d_*g + g[d_*X, f],
\end{aligned}$$

and, secondly,

$$[gX, d_*f] = g[X, d_*f] - a(d_*f)(g)(X).$$

Equating and cancelling, this gives

$$0 = -\langle df, d_*g \rangle X - a(d_*f)(g)(X)$$

for all $X \in \Gamma A$, or

$$\langle df, d_*g \rangle = -a(d_*f)(g). \tag{4}$$

Rewriting $a(d_*f)(g) = \langle \delta g, a(a_*^*(\delta f)) \rangle = \langle a^*(\delta g), a_*^*(\delta f) \rangle = \langle dg, d_*f \rangle$, this becomes

$$\langle df, d_*g \rangle = -\langle dg, d_*f \rangle. \tag{5}$$

Since $\langle df, d_*g \rangle = [d_*g, f]$, (5) can also be written as

$$d_*[f, g] = [d_*f, g] + [f, d_*g].$$

Together with (3) and (1) this proves that

$$d_*[\xi, \eta] = [d_*\xi, \eta] + [\xi, d_*\eta] \tag{6}$$

holds for all $\xi, \eta \in \Gamma \Lambda^\bullet(A)$ where both are of degree $\leqslant 1$. The proof of the following now follows by induction.

Theorem 12.1.6 *Equation* (6) *holds for all* $\xi, \eta \in \Gamma \Lambda^\bullet(A)$.

For the next proof, note that (5) can also be written as

$$\mathcal{L}_{d_*f}(g) + \mathcal{L}_{df}^*(g) = 0. \tag{7}$$

Lemma 12.1.7 *For all* $\varphi \in \Gamma A^*$ *and* $f \in C^\infty(M)$,

$$[\varphi, df]_* = d(\iota_{d_*f}\varphi) + \iota_{d_*f}(d\varphi).$$

Proof First observe that the RHS is $\mathcal{L}_{d_*f}(\varphi)$. Take any $X \in \Gamma A$. Then, by (36) on p. 307,

$$
\begin{aligned}
\langle \mathcal{L}_{d_*f}(\varphi), X \rangle &= \mathcal{L}_{d_*f}\langle \varphi, X \rangle - \langle \varphi, [d_*f, X] \rangle \\
&= -\mathcal{L}_{df}^*\langle \varphi, X \rangle + \langle \varphi, d_*(a(X)(f)) \rangle + \langle \varphi, \iota_{df}(d_*X) \rangle,
\end{aligned}
$$

where we used (7) and 12.1.5. Now the last two terms combine to give $\langle \varphi, \mathcal{L}_{df}^*(X) \rangle$ which is equal to

$$\mathcal{L}_{df}^*\langle \varphi, X \rangle - \langle \mathcal{L}_{df}^*(\varphi), X \rangle.$$

Since $\mathcal{L}_{df}^*(\varphi) = [df, \varphi]_*$, this completes the proof. $\qquad \square$

As with the equation in 12.1.5, this result can be rewritten

$$\mathcal{L}^*_{df}(\varphi) = -\mathcal{L}_{d_*f}(\varphi). \tag{8}$$

Furthermore, it can be rewritten as

$$d[\varphi, f]_* = [d\varphi, f]_* + [\varphi, df]_*. \tag{9}$$

For this it is necessary to use 12.1.4, which is valid for any Lie algebroid and so may be applied to A^* to give

$$[\varphi, f]_* = (-1)^{i-1} \iota_{d_*f}(\varphi), \tag{10}$$

for any $\varphi \in \Gamma\Lambda^i(A^*)$ and $f \in C^\infty(M)$.

Lemma 12.1.8 *For $X \in \Gamma A$, $\varphi \in \Gamma A^*$ and $f \in C^\infty(M)$,*

$$[\mathcal{L}_X, \mathcal{L}^*_\varphi](f) - \mathcal{L}^*_{\mathcal{L}_X(\varphi)}(f) + \mathcal{L}_{\mathcal{L}^*_\varphi(X)}(f) = \mathcal{L}_{d_*\langle\varphi, X\rangle}(f). \tag{11}$$

Proof Firstly, note that, merely using $\mathcal{L}_Z(f) = \langle df, Z \rangle$ and (36) on p. 307,

$$\mathcal{L}_{\mathcal{L}^*_\varphi(X)}(f) = \mathcal{L}^*_\varphi \mathcal{L}_X(f) - \langle [\varphi, df]_*, X \rangle.$$

Likewise,

$$\mathcal{L}^*_{\mathcal{L}_X(\varphi)}(f) = \mathcal{L}_X \mathcal{L}^*_\varphi(f) - \langle \varphi, [X, d_*f] \rangle.$$

Subtracting these, we obtain

$$[\mathcal{L}_X, \mathcal{L}^*_\varphi](f) - \mathcal{L}^*_{\mathcal{L}_X(\varphi)}(f) + \mathcal{L}_{\mathcal{L}^*_\varphi(X)}(f) = \langle \varphi, [X, d_*f] \rangle - \langle [\varphi, df]_*, X \rangle.$$

Now considering the RHS of (11), we have

$$\mathcal{L}_{d_*\langle\varphi, X\rangle}(f) = \langle d_*\langle\varphi, X\rangle, df \rangle = -\langle d_*f, d\langle\varphi, X\rangle\rangle = -\mathcal{L}_{d_*f}\langle\varphi, X\rangle$$

using (5). Expanding this out we get $\langle \varphi, \mathcal{L}_{d_*f}(X) \rangle = \langle \varphi, [d_*f, X] \rangle$ and

$$-\langle \mathcal{L}_{d_*f}(\varphi), X \rangle = \langle \mathcal{L}^*_{df}(\varphi), X \rangle = \langle [df, \varphi]_*, X \rangle,$$

using (8), and this completes the proof. $\qquad\qquad\square$

We finally arrive at the dual form of the defining equation.

Theorem 12.1.9 *For all $\varphi, \psi \in \Gamma A^*$,*

$$d[\varphi, \psi]_* = [d\varphi, \psi]_* + [\varphi, d\psi]_*. \tag{12}$$

Proof Rewrite (12) as $\mathcal{L}^*_\varphi(d\psi) - \mathcal{L}^*_\psi(d\varphi) - d[\varphi,\psi]_* = 0$. We will prove that, for all $X, Y \in \Gamma A$, the expression

$$K = (\mathcal{L}_X(d_*Y) - \mathcal{L}_Y(d_*X) - d_*[X,Y])(\varphi,\psi)$$
$$- (\mathcal{L}^*\varphi(d\psi) - \mathcal{L}^*_\psi(d\varphi) - d[\varphi,\psi]_*)(X,Y)$$

is zero. Expanding out the first term $(\mathcal{L}_X(d_*Y))(\psi,\psi)$, we obtain

$$\mathcal{L}_X\mathcal{L}^*_\varphi\langle Y,\psi\rangle - \mathcal{L}_X\mathcal{L}^*_\psi\langle Y,\varphi\rangle - \mathcal{L}_X\langle Y,[\varphi,\psi]_*\rangle$$
$$- \mathcal{L}^*_{\mathcal{L}_X(\varphi)}\langle Y,\psi\rangle + \mathcal{L}^*_\psi\langle Y,\mathcal{L}_X(\varphi)\rangle + \langle Y,[\mathcal{L}_X(\varphi),\psi]_*\rangle$$
$$- \mathcal{L}^*_\varphi\langle Y,\mathcal{L}_X(\psi)\rangle + \mathcal{L}^*_{\mathcal{L}_X(\psi)}\langle Y,\varphi\rangle + \langle Y,[\varphi,\mathcal{L}_X(\psi)]\rangle.$$

Treating each term of K in this way yields 42 terms. Of these, 10 immediately cancel in pairs. Using 12.1.8, a further 16 give

$$\mathcal{L}_{d_*\langle\varphi,X\rangle}\langle Y,\psi\rangle, \quad -\mathcal{L}_{d_*\langle\psi,X\rangle}\langle Y,\varphi\rangle, \quad -\mathcal{L}_{d_*\langle\varphi,Y\rangle}\langle X,\psi\rangle, \quad \mathcal{L}_{d_*\langle\psi,Y\rangle}\langle X,\varphi\rangle,$$

and these cancel in pairs due to (5). The remaining 16 terms form groups of four, which are zero due to identities of the form

$$\mathcal{L}^*_\psi\langle Y,\mathcal{L}_X(\varphi)\rangle = \mathcal{L}_X\langle\mathcal{L}^*_\psi(Y),\varphi\rangle - \langle[X,\mathcal{L}^*_\psi(Y)],\varphi\rangle + \langle Y,[\psi,\mathcal{L}_X(\varphi)]_*\rangle.$$

$\square$

Theorem 12.1.9 shows that (A^*, A) is also a Lie bialgebroid; that is, the concept of Lie bialgebroid is self–dual. That d is a derivation of the Schouten algebra $\Gamma\Lambda^\bullet(A^*)$ now follows from 12.1.6:

Corollary 12.1.10 *Equation (12) holds for all* $\varphi, \psi \in \Gamma\Lambda^\bullet(A^*)$.

We now need to set out the data associated to a Lie bialgebroid. To show that a Lie bialgebroid structure induces a Poisson structure on the base manifold, we need the following lemma.

Lemma 12.1.11 *For all* $f, g \in C^\infty(M)$,

$$d\{f,g\} = [df, dg]_*, \qquad d_*\{f,g\} = -[d_*f, d_*g].$$

Proof In (12) put $\varphi = f$ and $\psi = dg$. Then $d[f, dg]_* = [df, dg]_*$. On the other hand, $[f, dg]_* = -\langle dg, d_*f\rangle$ from (10). The companion equation is similar. $\square$

Proposition 12.1.12 *The bracket on $C^\infty(M)$ defined by*

$$\{f, g\} = \langle df, d_* g \rangle$$

is a Poisson structure on M.

Proof That $\{\ ,\ \}$ is skew–symmetric follows from (5), and that it is a derivation in each entry is immediate. To verify Jacobi, define $X_f = \{f, -\}$ as in (2) on p. 382. Then $X_f = \mathcal{L}^*_{df}$ and so, using 12.1.11,

$$X_{\{f,g\}} = \mathcal{L}^*_{d\{f,g\}} = \mathcal{L}^*_{[df,dg]_*} = [\mathcal{L}^*_{df}, \mathcal{L}^*_{dg}] = [X_f, X_g],$$

which is equivalent to Jacobi. □

We declare this to be the *Poisson structure on M induced by* (A, A^*) and denote it π_M. Thus:

$$\pi_M^{\#} = a_* \circ a^* = -a \circ a_*^*. \tag{13}$$

Although the notion of Lie bialgebroid is symmetric between A and A^*, the choice of a particular Poisson structure on the base imposes a distinction; the Lie bialgebroid (A^*, A) induces the opposite Poisson structure. In practice, the notation (A, A^*) already places a distinction on the two Lie algebroids.

The following results are easily proved from (13) and 12.1.11.

Proposition 12.1.13 (i) *The map $a^* \colon T^*M \to A^*$ is a morphism of Lie algebroids, and $a \colon A \to TM$ is a Poisson map from the linear Poisson structure on A corresponding to the Lie algebroid structure on A^* to the tangent lift of the Poisson structure on M.*

(ii) *The map $a_*^* \colon T^*M \to A$ is an antimorphism of Lie algebroids, and $a_* \colon A^* \to TM$ is an anti–Poisson map from the linear Poisson structure on A^* corresponding to the Lie algebroid structure on A to the tangent lift of the Poisson structure on M.*

By an antimorphism of Lie algebroids we mean a map $\varphi \colon A \to B$ such that $-\varphi$ is a morphism of Lie algebroids.

Note that the bundle projections q, q_* are only Poisson if the induced structure on M is null.

The next result will be needed in §12.2.

Proposition 12.1.14 *For $X \in \Gamma A$, $\varphi \in \Gamma A^*$ and $\omega \in \Omega^1(M)$,*

$$\mathcal{L}^*_{a^*\omega}(X) = [X, a_*^*\omega] - a_*^*(\iota_{aX}\delta\omega),$$
$$\mathcal{L}_{a_*^*\omega}(\varphi) = [\varphi, a^*\omega]_* - a^*(\iota_{a_*\varphi}\delta\omega).$$

Proof We prove the first identity. It suffices to consider $\omega = g\delta f$ for $f, g \in C^\infty(M)$. Expanding the LHS by (1) on p. 262, we have

$$g\mathcal{L}^*_{df}(X) + d_*g \wedge \iota_{df}(X).$$

Now $\mathcal{L}^*_{df}(X) = -\mathcal{L}_{d_*f}(X)$ by (2), so this becomes

$$-g[d_*f, X] + a(X)(f)d_*g$$

and a simple expansion of the RHS brings it to the same form. The second equation is proved in the same way. □

Definition 12.1.15 Let (A, A^*) and (B, B^*) be Lie bialgebroids on P and Q respectively. Then a *morphism of Lie bialgebroids from* (A, A^*) *to* (B, B^*) is a morphism of Lie algebroids $\mathscr{F} \colon A \to B$ which is also a Poisson map. □

It follows that f is also a Poisson map. There is an associated action of B^* on f defined by

$$\psi^\dagger = a_*(\ell_\psi \circ \mathscr{F})$$

for $\psi \in \Gamma B^*$. From 10.4.10 it follows that the dual of $\mathscr{F}$ is a comorphism of Lie algebroids $B^* \triangleleft f \to A^*$.

This notion of morphism is clearly not symmetric; it does however suffice for most purposes. Notice that when $P = Q$ as manifolds, the existence of a morphism of Lie bialgebroids implies that the Poisson structures must be identical.

Example 12.1.16 For any Lie bialgebroid (A, A^*) on base manifold P, the anchor map $a \colon A \to TP$ is a morphism of Lie bialgebroids from (A, A^*) to (TP, T^*P). ⊠

We close the section with a handful of simple examples.

Example 12.1.17 Suppose that P is a point so that $A = \mathfrak{g}$ is a Lie algebra with a Lie algebra structure on its dual $\mathfrak{g}^*$. Then, for $X \in \mathfrak{g}$ and $\varphi, \psi \in \mathfrak{g}^*$, we have that $\langle \varphi \wedge \psi, d_*X \rangle = -\langle [\varphi, \psi]_*, X \rangle$; since d_* and the bracket $\mathfrak{b}_* \colon \mathfrak{g}^* \otimes \mathfrak{g}^* \to \mathfrak{g}^*$ are linear we can write this as $d_* = -\mathfrak{b}^*_*$. Equation (1) can be written as

$$d_*[X, Y] = \mathrm{ad}_X(d_*Y) - \mathrm{ad}_Y(d_*X)$$

where $\mathrm{ad}_X \colon \Lambda^2(\mathfrak{g}) \to \Lambda^2(\mathfrak{g})$ is defined, in accordance with (22) on p. 304, by $\mathrm{ad}_X(Y \wedge Z) = \mathrm{ad}_X(Y) \wedge Z + Y \wedge \mathrm{ad}_X(Z)$. ⊠

Example 12.1.18 For any Lie algebroid A on base M, the zero Lie algebroid structure on A^* makes (A, A^*) a Lie bialgebroid.   ⊠

Example 12.1.19 Consider a Lie algebroid A on base M and suppose given $\Lambda \in \Gamma\Lambda^2(A)$ such that $[\Lambda, \Lambda] = 0$. Then, imitating the basic constructions on a Poisson manifold, define $\Lambda^\#: A^* \to A$ by $\langle \psi, \Lambda^\#(\varphi) \rangle = \Lambda(\varphi, \psi)$. Then with $a_* = a \circ \Lambda^\#$ and bracket

$$[\varphi, \psi]_* = \mathscr{L}_{\Lambda^\#\varphi}(\psi) - \mathscr{L}_{\Lambda^\#\psi}(\varphi) - d(\Lambda(\varphi.\psi)),$$

A^* is a Lie algebroid and $\Lambda^\#$ is a morphism of Lie algebroids. Furthermore, $d_* X = [\Lambda, X]$ for $X \in \Gamma A$ and it follows readily, as in 12.1.2, that (A, A^*) is a Lie bialgebroid. Instances of this construction are called *triangular Lie bialgebroids*.   ⊠

12.2 The morphism criterion for Lie bialgebroids

Theorem 12.2.1 gives a criterion for Lie algebroid structures on a vector bundle and its dual to constitute a Lie bialgebroid. This is an infinitesimal form of the definition of a Poisson groupoid. The proof will take us to the end of the section.

Theorem 12.2.1 *Suppose that A is a Lie algebroid with base manifold P such that its dual vector bundle A^* also has a Lie algebroid structure. Then (A, A^*) is a Lie bialgebroid if and only if*

$$
\begin{array}{ccc}
T^*(A^*) & \xrightarrow{\Pi} & TA \\
\downarrow & & \downarrow \\
A^* & \xrightarrow[a_*]{} & TP
\end{array}
\tag{14}
$$

is a Lie algebroid morphism, where the domain $T^*(A^*) \to A^*$ is the cotangent Lie algebroid induced by the Poisson structure on A^*, the target $TA \to TP$ is the tangent prolongation of A, and Π is the composition of $R: T^*A^* \to T^*A$ with $\pi_A^\#: T^*A \to TA$.

Note that the domain and codomain Lie algebroid structures are derived from the Lie algebroid A and the map from the Lie algebroid A^*. Throughout the section we assume that A and A^* satisfy the hypotheses of the Theorem. As usual, denote the projections of A and A^* by q and q_* and the anchors by a and a_*.

We use the characterization 10.4.9 of Lie algebroid morphisms in terms of their dual relations. Much of the proof consists of describing the algebra of functions that vanish on the graph of the relation dual to Π and calculating the bracket relations for them.

The dual relation to Π is a subset of $TA^* \times T^\bullet A$, but we will identify $T^\bullet A$ with TA^* via I. Denote the set corresponding to Π^* by $\mathscr{C}$. Thus $\mathscr{C}$ consists of all elements $(\mathfrak{X}, \mathfrak{Y}) \in T_\varphi A^* \times T_\psi A^*$ such that $a_*(\varphi) = T(q_*)(\mathfrak{Y})$ and $\langle \mathfrak{X}, \mathfrak{F} \rangle = \langle\!\langle \mathfrak{Y}, \Pi(\mathfrak{F}) \rangle\!\rangle$ for all $\mathfrak{F} \in T_\varphi^* A^*$. We begin by giving another description of $\mathscr{C}$.

For $X \in \Gamma A$, choose a function $f_{d_* X} \colon A^* \times A^* \to \mathbb{R}$ which extends $d_* X \colon A^* \times_P A^* \to \mathbb{R}$ and define F_X on $TA^* \times TA^*$ by

$$F_X(\mathfrak{X}, \mathfrak{Y}) = \mathfrak{X}(\ell_X) - \mathfrak{Y}(\ell_X) + f_{d_* X}(\varphi, \psi).$$

Since $A^* \times_P A^*$ is coisotropic in $A^* \times \overline{A^*}$, the values on $A^* \times_P A^*$ of any bracket $\{G, f_{d_* X}\}$, for $G \in C^\infty(A^* \times A^*)$, do not depend on the choice of extension.

Proposition 12.2.2 *Let* $(\mathfrak{X}, \mathfrak{Y}) \in T_{\varphi_m} A^* \times T_{\psi_m} A^*$. *Then* $(\mathfrak{X}, \mathfrak{Y}) \in \mathscr{C}$ *if and only if the three conditions*

$$a_*(\varphi_m) = T(q_*)(\mathfrak{Y}), \quad a_*(\psi_m) = T(q_*)(\mathfrak{X}), \quad F_X(\mathfrak{X}, \mathfrak{Y}) = 0, \quad (15)$$

hold, the last for all $X \in \Gamma A$.

Proof From 9.4.1 we know that $T_{\varphi_m}^* A^*$ is spanned by the covectors of the form $(\delta \ell_X)(\varphi_m)$, for $X \in \Gamma A$, and those of the form $(q_*^* \omega)(\varphi_m)$, for $\omega \in T_m^* P$. So $(\mathfrak{X}, \mathfrak{Y}) \in \mathscr{C}$ if and only if the conditions

$$a_*(\varphi_m) = T(q_*)(\mathfrak{Y}), \qquad \langle \mathfrak{X}, (\delta \ell_X)(\varphi_m) \rangle = \langle\!\langle \mathfrak{Y}, \Pi(\delta \ell_X(\varphi_m)) \rangle\!\rangle,$$
$$\langle \mathfrak{X}, (q_*^* \omega)(\varphi_m) \rangle = \langle\!\langle \mathfrak{Y}, \Pi((q_*^* \omega)(\varphi_m)) \rangle\!\rangle, \qquad (16)$$

hold for all $X \in \Gamma A$, $\omega \in T_m^* (P)$.

The first of these is common to both (15) and (16), so we can assume it throughout. For the second we have

$$\langle\!\langle \mathfrak{Y}, \Pi(\delta \ell_X(\varphi_m)) \rangle\!\rangle = \mathfrak{Y}(\ell_X) + \Pi(\delta \ell_X(\varphi_m))(\ell_\psi) - T(q_*)(\mathfrak{Y})(\langle \psi, X \rangle)$$

by 3.4.6, where $\psi \in \Gamma A^*$ is a section passing through ψ_m. For the second term on the RHS, apply 9.5.3 to $\delta \ell_\psi(X_m)$ and 10.3.9 to A^*, and we obtain

$$\pi_A(R(\delta \ell_X(\varphi_m)), R(\delta \ell_X(\psi_m))) + \pi_A(R(\delta \ell_X(\varphi_m)), (q^* \delta \langle \psi, X \rangle)(X_m))$$
$$= -d_* X(\varphi_m, \psi_m) + a_*(\varphi_m)(\mathfrak{Y})\langle \psi, X \rangle.$$

Using $a_*(\varphi_m) = T(q_*)(\mathfrak{Y})$ again, the second condition in (16) is equivalent to $F_X(\mathfrak{X}, \mathfrak{Y}) = 0$ for all $X \in \Gamma A$.

Similarly, again using 10.3.9, $\langle\!\langle \mathfrak{Y}, \Pi((q_*^*\omega)(\varphi_m)) \rangle\!\rangle$ becomes

$$
\begin{aligned}
\mathfrak{Y}(\ell_0) &+ \Pi((q_*^*\omega)(\varphi_m))(\ell_\psi) - T(q_*)(\mathfrak{Y})(\langle\psi, 0\rangle) \\
&= \pi_A(R((q_*^*\omega)(\varphi_m), \delta\ell_\psi(0_m) + \overline{\omega}) - \pi_A(R((q_*^*\omega)(\varphi_m)), \overline{\omega}) \\
&= \pi_A(R((q_*^*\omega)(\varphi_m)), R((q_*^*\omega)(\varphi_m)) \\
&\qquad -\pi_A(R((q_*^*\omega)(\varphi_m)), R((q_*^*\omega)(\varphi_m)) \\
&= \langle \omega, a_*(\psi_m) - a_*(\varphi_m)\rangle - \langle\omega, a_*(0_m) - a_*(\varphi_m)\rangle \\
&= \langle\omega, a_*(\psi_m)\rangle,
\end{aligned}
$$

and so the third condition of (16) is equivalent to $a_*(\psi_m) = T(q_*)(\mathfrak{X})$. $\qquad\square$

Corollary 12.2.3 *The projection of $T^*A^* \times T^*A^*$ onto $A^* \times A^*$ maps $\mathscr{C}$ onto $A^* \times_P A^*$.*

For any $\omega \in \Omega^1(P)$, define functions G_ω and H_ω on $TA^* \times TA^*$ by:

$$
G_\omega(\mathfrak{X}, \mathfrak{Y}) = \langle\omega, a_*(\varphi) - T(q_*)\mathfrak{Y}\rangle, \qquad H_\omega(\mathfrak{X}, \mathfrak{Y}) = \langle\omega, a_*(\psi) - T(q_*)\mathfrak{X}\rangle,
$$

where $\mathfrak{X} \in T_\varphi(A^*)$, $\mathfrak{Y} \in T_\psi(A^*)$. Proposition 12.2.2 shows that $\mathscr{C}$ is the set of common zeros of the three families of real–valued functions F_X, G_ω, H_ω, for $X \in \Gamma A$, $\omega \in \Omega^1(P)$. The next step is to calculate the Poisson brackets of these functions on $TA^* \times T\overline{A^*}$.

Theorem 12.2.4 *Fix $X, Y \in \Gamma A$. Then for $(\mathfrak{X}, \mathfrak{Y}) \in T_\varphi A^* \times T_\psi \overline{A^*}$,*

$$
\{F_X, F_Y\}(\mathfrak{X}, \mathfrak{Y}) - F_{[X,Y]}(\mathfrak{X}, \mathfrak{Y}) = (\mathfrak{L}_X d_* Y - \mathfrak{L}_Y d_* X - d_*[X, Y])(\varphi, \psi).
$$

Again we start with a lemma. Let $\xi \in \Gamma\Lambda^2(A)$ be any 2–section of A. Then, regarding ξ as a function $A^* \times_P A^* \to \mathbb{R}$, let f_ξ be any extension of ξ to $A^* \times A^*$.

Lemma 12.2.5 *For any $X \in \Gamma A$, let ℓ_X^1 and ℓ_X^2 denote the linear functions on $A^* \times A^*$ defined by $\ell_X^1(\varphi, \psi) = \ell_X(\varphi)$, and $\ell_X^2(\varphi, \psi) = \ell_X(\psi)$, for any $(\varphi, \psi) \in A^* \times A^*$, respectively. Then, for $(\varphi, \psi) \in A^* \times_P A^*$,*

$$
\{\ell_X^1, f_\xi\}(\varphi, \psi) - \{\ell_X^2, f_\xi\}(\varphi, \psi) = (\mathfrak{L}_X\xi)(\varphi, \psi). \tag{17}
$$

Proof This is essentially a product rule. For a 1–section $Y \in \Gamma A$, we have $\ell_{[X,Y]} = X(\ell_Y)$ and this is easily extended to 2–sections ξ. It is thus sufficient to treat $\mathcal{L}_X \xi$ as the Lie derivative of a function of two variables. The minus sign on the LHS arises from the reversal of the Poisson structure on the second factor. $\square$

Proof of Theorem 12.2.4 Expanding out $\{F_X, F_Y\}(\mathfrak{X}, \mathfrak{Y})$ we get nine terms, of four different types. Firstly consider $\{\mathfrak{X}(\ell_X), \mathfrak{X}(\ell_Y)\}$. The function $\mathfrak{X} \mapsto \mathfrak{X}(\ell_X)$ is, in the notation of (20) on p. 395, $\ell_{\delta \ell_X}$; that is, it is the lift to $TA^* \to \mathbb{R}$ of $\ell_X \colon A^* \to \mathbb{R}$. Therefore

$$\{\mathfrak{X}(\ell_X), \mathfrak{X}(\ell_Y)\} = \mathfrak{X}(\ell_{[X,Y]}).$$

Next, $\{\mathfrak{Y}(\ell_X), \mathfrak{X}(\ell_Y)\}$ is zero, since the first argument depends only on the second TA^* component, and the second depends only on the first.

Now consider $\{\mathfrak{X}(\ell_X), f_{d_*Y}(\varphi, \psi)\}$. Since the first argument depends only on the first component, we may regard the second argument as the pullback to TA^* of $f_{d_*Y}(-, \psi)$. Using (20) on p. 395 again, and the fact that $\pi_{A^*}^\#(\delta \ell_X) = H_X$, the bracket is the pullback to $TA^* \times TA^*$ of $\{\ell_X, f_{d_*Y}(-, \psi)\}(\varphi)$, that is, $\{\ell_X^1, f_{d_*Y}(\varphi, \psi)\}$.

Lastly, $\{f_{d_*X}(\varphi, \psi), f_{d_*Y}(\varphi, \psi)\}$ is zero, since both functions are pullbacks. Altogether we have

$$\mathfrak{X}(\ell_{[X,Y]}) - \mathfrak{Y}(\ell_{[X,Y]}) + \{\ell_X^1, f_{d_*Y}\}(\varphi, \psi)$$
$$- \{\ell_X^2, f_{d_*Y}\}(\varphi, \psi) - \{\ell_Y^1, f_{d_*X}\}(\varphi, \psi) + \{\ell_Y^2, f_{d_*X}\}(\varphi, \psi).$$

The result now follows immediately from 12.2.5. $\square$

Theorem 12.2.6 *For any* ω, $\theta \in \Omega^1(P)$, *and any* $\mathfrak{X} \in T_\varphi(A^*)$, $\mathfrak{Y} \in T_\psi(A^*)$,

$$\{G_\omega, G_\theta\}(\mathfrak{X}, \mathfrak{Y}) = 0, \qquad \{H_\omega, H_\theta\} = 0,$$
$$\{G_\omega, H_\theta\}(\mathfrak{X}, \mathfrak{Y}) = \langle a^*\theta, a_*^*\omega \rangle (q_*\varphi) + \langle a^*\omega, a_*^*\theta \rangle (q_*\psi).$$

If (A, A^*) *is a Lie bialgebroid, then the third expression is zero for all* $(\varphi, \psi) \in A^* \times_P A^*$.

Proof Expanding out $\{G_\omega, G_\theta\}(\mathfrak{X}, \mathfrak{Y})$. we have

$$\{\langle a_*(\varphi), \omega \rangle, \langle a_*(\varphi), \theta \rangle\} - \{\langle a_*(\varphi), \omega \rangle, \langle T(q_*)\mathfrak{Y}, \theta \rangle\}$$
$$- \{\langle T(q_*)\mathfrak{Y}, \omega \rangle, \langle a_*(\varphi), \theta \rangle\} + \{\langle T(q_*)\mathfrak{Y}, \omega \rangle, \langle T(q_*)\mathfrak{Y}, \theta \rangle.\}$$

Writing $\langle a_*\varphi, \theta \rangle = \langle \varphi, a_*^*\theta \rangle$, this function is $\ell_{a_*^*\theta}$, where $a_*^*\theta \in \Gamma A$. As a function of $\mathfrak{X}$, it is $\ell_{a_*^*\theta} \circ p_{A^*}$. Similarly $\langle T(q_*)(\mathfrak{X}), \omega \rangle = \ell_{q_*^*\omega}(\mathfrak{X})$ where $q_*^*\omega \in \Omega^1(A^*)$.

Thus the first term is the bracket of two pullbacks, and so zero. In $\{\langle a_*(\varphi), \omega \rangle, \langle T(q_*)\mathfrak{Y}, \theta \rangle\}$, the two pairings are functions of $\mathfrak{X}$ alone, and $\mathfrak{Y}$ alone, respectively, and so the bracket is zero also.

Lastly, $\{\langle T(q_*)\mathfrak{Y}, \omega \rangle, \langle T(q_*)\mathfrak{Y}, \theta \rangle\}$ is $\{\ell_{q_*^*\omega}, \ell_{q_*^*\theta}\} = \ell_{[q_*^*\omega, q_*^*\theta]}$. Here $[q_*^*\omega, q_*^*\theta]$ is the Lie algebroid bracket of 1–forms on A^* and since both forms are pullbacks, the bracket is zero.

The proof that $\{H_\omega, H_\theta\} = 0$ is similar. For the third identity, expanding out $\{G_\omega, H_\theta\}(\mathfrak{X}, \mathfrak{Y})$ we have

$$\{\langle a_*(\varphi), \omega \rangle, \langle a_*(\psi), \theta \rangle\} - \{\langle a_*(\varphi), \omega \rangle, \langle T(q_*)\mathfrak{X}, \theta \rangle\}$$
$$- \{\langle T(q_*)\mathfrak{Y}, \omega \rangle, \langle a_*(\psi), \theta \rangle\} + \{\langle T(q_*)\mathfrak{Y}, \omega \rangle, \langle T(q_*)\mathfrak{X}, \theta \rangle\}.$$

The two functions in the first term depend on independent arguments, so the bracket is zero; likewise with the fourth term. For the second term, we have

$$\{\langle T(q_*)(\mathfrak{X}), \theta \rangle, \langle a_*\varphi, \omega \rangle\} = \{\ell_{q_*^*\theta}, \ell_{a_*^*\omega} \circ p_{A^*}\}(\mathfrak{X}) = \pi_{A^*}^\#(q_*^*\theta)(\ell_{a_*^*\omega})(\varphi)$$

by (20) on p. 395. Now $\pi_{A^*}^\#(q_*^*\theta) = -(a^*\theta)^\uparrow$ by (15) on p. 391 so, using (21) on p. 112, we have

$$\{\langle T(q_*)(\mathfrak{X}), \theta \rangle, \langle a_*\varphi, \omega \rangle\} = \langle a^*\theta, a_*^*\omega \rangle(q_*\varphi). \tag{18}$$

The third term is similar, but takes the opposite sign, since it depends entirely on the second factor.

For the final statement, use (13). $\qquad\qquad\square$

Theorem 12.2.7 *Suppose that (A, A^*) is a Lie bialgebroid. For any $X \in \Gamma A$ and $\omega \in \Omega^1(P)$,*

$$\{F_X, G_\omega\} = G_\tau \qquad \text{and} \qquad \{F_X, H_\omega\} = H_\tau,$$

where $\tau = \delta\langle \omega, a(X) \rangle + \iota_{a(X)}\delta\omega$.

Proof Expanding out $\{F_X, G_\omega\}(\mathfrak{X}, \mathfrak{Y})$ and dropping terms which are zero for reasons by now familiar, we have

$$\{\mathfrak{X}(\ell_X), \langle a_*(\varphi), \omega \rangle\} + \{\mathfrak{Y}(\ell_X), \langle T(q_*)\mathfrak{Y}, \omega \rangle\} - \{f_{d_*X}(\varphi, \psi), \langle T(q_*)\mathfrak{Y}, \omega \rangle\}.$$

The first term is $\{\ell_{\delta\ell_X}, \ell_{a_*^*\omega} \circ q_{A^*}\}(\mathfrak{X})$. This is

$$\pi_{A^*}^\#(\delta\ell_X)(\ell_{a_*^*\omega}(\varphi)) = H_X(\ell_{a_*^*\omega})(\varphi) = \ell_{[X, a_*^*\omega]}(\varphi).$$

Using 12.1.14, this expands to $\langle \mathfrak{L}^*_{a^*\omega}(X), \varphi \rangle + \langle a^*_*(\iota_{aX}\delta\omega), \varphi \rangle$ and standard exterior calculus in A^* gives

$$\langle a_*(\varphi), \tau \rangle + d_*X(a^*\omega, \varphi).$$

Next, $\{\mathfrak{Y}(\ell_X), \langle T(q_*)\mathfrak{Y}, \omega \rangle\} = \{\ell_{\delta\ell_X}, \ell_{q_*^*\omega}\}(\mathfrak{Y}) = \langle [\delta\ell_X, q_*^*\omega], \mathfrak{Y} \rangle$. We expand this by (5) on p. 383, getting

$$-\iota_{(\delta\ell_X)^\#}(\delta q_*^*\omega) + 0 - \delta(\pi_{A^*}(\delta\ell_X, q_*^*\omega)),$$

where the signs are reversed since this is a function of the second argument alone. Evaluating at $\mathfrak{Y}$, the first of these terms is

$$-(q_*^*\delta\omega)(H_X, \mathfrak{Y}) = -\delta\omega(aX, T(q_*)(\mathfrak{Y})) = -\langle \iota_{aX}(\delta\omega), T(q_*)(\mathfrak{Y}) \rangle$$

and the last is

$$-\langle \delta(\pi_{A^*}(\delta\ell_X, q_*^*\omega)), \mathfrak{Y} \rangle = -\langle T(q_*)(\mathfrak{Y}), \delta\langle aX, \omega \rangle \rangle$$

since $\pi^\#_{A^*}(\delta\ell_X) = H_X$ and H_X projects under q_* to aX. Altogether we have $-\langle T(q_*)\mathfrak{Y}, \tau \rangle$.

Finally consider $-\{f_{d_*X}(\varphi, \psi), \langle T(q_*)\mathfrak{Y}, \omega \rangle\}$. In the by now familiar way, and using $\pi^\#_{A^*}(q_*^*\omega) = -(a^*\omega)^\uparrow$, this is equal to $d_*X(\varphi, a^*\omega)$. This completes the proof that $\{F_X, G_\omega\} = G_\tau$. The second identity is proved similarly. $\square$

Proof of Theorem 12.2.1 Suppose that (A, A^*) is a Lie bialgebroid. Then 12.2.7 holds and, together with 12.2.6 and 12.2.4, this shows that the set of all F_X, G_ω, H_θ is closed under the Poisson bracket. From 12.2.2 it follows that the set of functions vanishing on $\mathscr{C}$ is closed, and so $\mathscr{C}$ is coisotropic. By 10.4.9, Π is a morphism of Lie algebroids.

Conversely assume that Π is a Lie algebroid morphism. Then $\mathscr{C}$ is a coisotropic submanifold in $TA^* \times T\overline{A^*}$, by 10.4.9, and so 12.2.4 implies that $(d_*[X, Y] - \mathfrak{L}_X d_* Y + \mathfrak{L}_Y d_* X)(\mathfrak{X}, \mathfrak{Y}) = 0$, for all $(\mathfrak{X}, \mathfrak{Y}) \in \mathscr{C}$. From 12.2.3 it therefore follows that

$$d_*[X, Y] = \mathfrak{L}_X d_* Y - \mathfrak{L}_Y d_* X$$

identically on $A^* \times_P A^*$, for all $X, Y \in \Gamma A$. $\square$

12.3 Further Poisson groupoids

In the definition of the Lie bialgebroid structure associated to a Poisson groupoid G, the Lie algebroid structure on the dual A^*G was deduced from the coisotropy of the base manifold. In this section we show that the dual Poisson structure on AG may be obtained as a 'prolongation' to AG of the Poisson structure on G. This is, for many purposes, an easier description to work with. We first need an alternative version of the triangles of canonical isomorphisms given in §11.5 for an arbitrary Lie groupoid.

Let $G \rightrightarrows M$ be a Lie groupoid. At the end of §11.5 we showed that $(\delta\nu)^\#\colon T^*T^*G \to TT^*G$ is an isomorphism of Lie groupoids over $-\mathfrak{w}^*\colon A^*T^*G \to TA^*G$. We now need to consider it as a morphism of $\mathcal{VB}$-groupoids, with domain as shown in Figure 12.1(a) and codomain

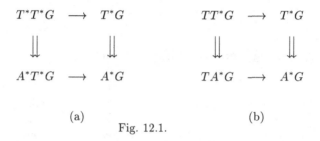

$$
\begin{array}{ccc}
T^*T^*G & \longrightarrow & T^*G \\
\Downarrow & & \Downarrow \\
A^*T^*G & \longrightarrow & A^*G
\end{array}
\qquad\qquad
\begin{array}{ccc}
TT^*G & \longrightarrow & T^*G \\
\Downarrow & & \Downarrow \\
TA^*G & \longrightarrow & A^*G
\end{array}
$$

(a) (b)

Fig. 12.1.

as shown in Figure 12.1(b). Here (a) is the cotangent $\mathcal{VB}$-groupoid of $T^*G \rightrightarrows A^*G$, and (b) is the tangent $\mathcal{VB}$-groupoid of $T^*G \rightrightarrows A^*G$. The cores are therefore, respectively, T^*A^*G and AT^*G.

Proposition 12.3.1 *The map* $(\delta\nu)^\#\colon T^*T^*G \to TT^*G$ *is an isomorphism of the* $\mathcal{VB}$*-groupoids just described, over* $-\mathfrak{w}^*$ *and the identity of* T^*G. *The core map is* $\mathfrak{w}\colon T^*A^*G \to AT^*G$.

Proof The first statement is easily checked. The second follows from the skew–symmetry of $(\delta\nu)^\#$ and 9.2.1. □

In Figure 12.2, the lower triangle is the definition of the map θ_G. It is an isomorphism of double vector bundles over AG and A^*G.

Theorem 12.3.2 *The upper triangle in Figure* 12.2 *commutes.*

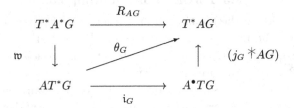

Fig. 12.2.

Proof First consider $R: T^*T^*G \to T^*TG$ applied to a core element $\overline{\Phi}$ of the domain. Giving Φ the form

$$
\begin{array}{ccc}
\Phi & \longmapsto & X \\
\downarrow & & \downarrow \\
\varphi & \longmapsto & m
\end{array}
\qquad
\begin{array}{ccc}
T^*A^*G & \longrightarrow & AG \\
\downarrow & & \downarrow \\
A^*G & \longrightarrow & M,
\end{array}
\tag{19}
$$

the corresponding core element has the form

$$
\begin{array}{ccc}
\overline{\Phi} & \longmapsto & \overline{X} \\
\downarrow & & \downarrow \\
\widetilde{1}_\varphi & \longmapsto & 1_m
\end{array}
\qquad
\begin{array}{ccc}
T^*T^*G & \longrightarrow & TG \\
\downarrow & & \downarrow \\
T^*G & \longrightarrow & G.
\end{array}
$$

Applying R to $\overline{\Phi}$, we have

$$
\langle R(\overline{\Phi}), \xi \rangle = \langle\!\langle \mathfrak{X}, \xi \rangle\!\rangle - \langle \overline{\Phi}, \mathfrak{X} \rangle
\tag{20}
$$

where $\xi \in T^2G$ and $\mathfrak{X} \in TT^*G$ must be compatible with $\overline{\Phi}$ and each other in the usual way.

Now applying R_{AG} to $\Phi \in T^*A^*G$, we similarly have

$$
\langle R_{AG}(\Phi), \eta \rangle = \langle\!\langle \mathfrak{Y}, \eta \rangle\!\rangle - \langle \Phi, \mathfrak{Y} \rangle
$$

for compatible $\eta \in TAG$ and $\mathfrak{Y} \in TA^*G$. Now $TAG \subseteq T^2G$ so in (20) we may take $\xi = \eta$. Furthermore, TA^*G may be mapped into TT^*G by $T(\widetilde{1})$, which is a morphism of double vector bundles over $\widetilde{1}$, $T(1)$ and 1, so in (20) we may take $\mathfrak{X} = T(\widetilde{1})(\mathfrak{Y})$. Since $\overline{\Phi}$ is the pullback of Φ across $T(\widetilde{\beta})$, and since of course $\widetilde{\beta} \circ \widetilde{1} = \text{id}$, it follows that the restriction map $T^*TG \to T^*AG$ sends $R(\overline{\Phi})$ to $R_{AG}(\Phi)$.

Next consider $\Theta: TT^*G \to T^*TG$. For compatible $\mathfrak{X} \in TT^*G$ and

$\xi \in T^2G$ we have, by 9.6.4, that

$$\langle\!\langle \mathfrak{X}, \xi \rangle\!\rangle = \langle \Theta(\mathfrak{X}), J(\xi) \rangle.$$

Now take $\mathfrak{X} \in AT^*G$ of the form $(\mathfrak{X}; \varphi, X; m)$, so that the corresponding core element has the form

$$
\begin{array}{ccc}
\overline{\mathfrak{X}} & \longmapsto & \overline{X} \\
\downarrow & & \downarrow \\
\widetilde{1}_\varphi & \longmapsto & 1_m
\end{array}
\qquad
\begin{array}{ccc}
TT^*G & \longrightarrow & TG \\
\downarrow & & \downarrow \\
T^*G & \longrightarrow & G.
\end{array}
\qquad (21)
$$

Then, restricting $\Theta(\overline{\mathfrak{X}}) \in T^*TG$ to $AG \subseteq TG$, we have that, for $\xi \in ATG$,

$$\langle \Theta(\overline{\mathfrak{X}}), j_G^{-1}(\xi) \rangle = \langle\!\langle \mathfrak{X}, \xi \rangle\!\rangle,$$

where the pairing on the right is the pairing of AT^*G with ATG over AG. Thus the restriction of Θ to $AT^*G \to T^*TG$, composed with the restriction map $T^*TG \to T^*AG$, is θ_G.

The commutativity is now an immediate consequence of 9.6.7. $\qquad \square$

It is possible to regard Figure 12.2 as the diagram of core maps for Figure 11.10 (with M replaced by G). However, this requires more work than the proof above.

Consider a Poisson groupoid $G \rightrightarrows P$. By definition, $\pi_G^\#$ is a Lie groupoid morphism over $a_* \colon A^*G \to TP$. Apply the Lie functor. We obtain a morphism of double vector bundles $A(\pi_G^\#) \colon AT^*G \to ATG$. On the side structures this is $a_* \colon A^*G \to TP$ and $\mathrm{id} \colon AG \to AG$; the core map is $-a_*^*$. Now define $\pi_{AG}^\# \colon T^*AG \to TAG$ by

$$
\begin{array}{ccc}
AT^*G & \xrightarrow{\ A(\pi_G^\#)\ } & ATG \\[4pt]
\theta_G \downarrow & & \downarrow j_G^{-1} \\[4pt]
T^*AG & \xrightarrow[\ \pi_{AG}^\#\ ]{} & TAG.
\end{array}
\qquad (22)
$$

Now, by 9.7.5, j_G is an isomorphism of Lie algebroids over TP. Writing

$$\pi_{AG}^\# \circ \theta_G = \pi_{AG}^\# \circ R_{AG} \circ \mathfrak{w}^{-1} = \Pi \circ \mathfrak{w}^{-1},$$

it follows from the fact that $\mathfrak{w}$ is an isomorphism of Lie algebroids over A^*G that $\pi_{AG}^\#$ satisfies the criterion of Theorem 12.2.1. This will provide an alternative proof that (AG, A^*G) is a Lie bialgebroid once we establish that this Poisson structure is the dual to the Lie algebroid structure defined on A^*G in §11.4. This requires a mechanism for relating forms on G to forms on AG.

$$* \qquad * \qquad * \qquad * \qquad *$$

Consider for the time being any Lie groupoid G on M.

Definition 12.3.3 A *multiplicative 1–form* on G is a pair (Φ, φ) with $\Phi \in \Omega^1(G)$ and $\varphi \in \Gamma A^*G$ such that

$$
\begin{array}{ccc}
T^*G & \Longrightarrow & A^*G \\[1em]
\Phi \uparrow & & \uparrow \varphi \\[1em]
G & \Longrightarrow & M
\end{array}
\tag{23}
$$

is a morphism of Lie groupoids.

A *star 1–form* on G is similarly a pair (Φ, φ) where $\Phi \in \Omega^1(G)$ and $\varphi \in \Gamma A^*G$ such that $\tilde{\alpha} \circ \Phi = \varphi \circ \alpha$ and $\Phi \circ 1 = \tilde{1} \circ \varphi$. $\qquad \square$

Now let (Φ, φ) be a multiplicative 1–form and let (ξ, x) be a multiplicative vector field on G. Writing $F = \langle \Phi, \xi \rangle \colon G \to \mathbb{R}$, we have, for $h, g \in G$ compatible,

$$F(hg) = \langle \Phi(h) \bullet \Phi(g), \xi(h) \bullet \xi(g) \rangle;$$

from the definition of the multiplication $\bullet$ in T^*G it follows that this is equal to $\langle \Phi(h), \xi(h) \rangle + \langle \Phi(g), \xi(g) \rangle$. Thus F is a multiplicative function.

Now consider a star 1–form (Φ, φ) on G. Applying the Lie functor gives a vector bundle morphism $A(\Phi) \colon AG \to AT^*G$. Define

$$\widetilde{\Phi} = \theta_G \circ A(\Phi); \tag{24}$$

since θ_G is an isomorphism of double vector bundles, $(\widetilde{\Phi}, \varphi)$ is a linear 1–form on AG in the obvious sense.

Proposition 12.3.4 Let (Φ, φ) be a star 1–form and let $X \in \Gamma AG$ and let (ξ, x) a star vector field on G. Then

$$\langle \widetilde{\Phi}, X^{\uparrow} \rangle = \langle \varphi, X \rangle \circ q, \qquad \langle \widetilde{\Phi}, \widetilde{\xi} \rangle = \widetilde{\langle \Phi, \xi \rangle}, \tag{25}$$

where for any function $F \colon G \to \mathbb{R}$, we denote $A(F) \colon AG \to \mathbb{R}$ by $\widetilde{F}$.

Proof The first equation follows from the fact that $r \circ \widetilde{\Phi} = \varphi \circ q$, as in (32) on p. 357. For the second, since $\langle\!\langle \, , \, \rangle\!\rangle = A(\langle \, , \, \rangle)$, we have

$$\langle \widetilde{\Phi}, \widetilde{\xi} \rangle = \langle j'_G \circ A(\Phi), j_G^{-1} \circ A(\xi) \rangle = \langle\!\langle A(\Phi), A(\xi) \rangle\!\rangle = A(\langle \Phi, \xi \rangle) = \widetilde{\langle \Phi, \xi \rangle}.$$

$\square$

These two equations describe the behaviour of the forms $(\widetilde{\Phi}, \varphi)$ on $\mathfrak{X}(AG)$. To complete the description of the 1–forms on AG we need to include the pullbacks of forms on the base manifold.

Given $\omega \in \Omega^1(M)$ there is the 1–form $q^*\omega \in \Omega^1(AG)$. Here it is immediate that

$$\langle q^*\omega, \widetilde{\xi} \rangle = \langle \omega, x \rangle \circ q, \qquad \langle q^*\omega, X^\uparrow \rangle = 0, \qquad (26)$$

for star vector fields (ξ, x) on G and $X \in \Gamma AG$, since $\widetilde{\xi}$ projects to x under q and $X^\uparrow$ projects to 0.

Lemma 12.3.5 *Given any $\varphi \in \Gamma A^*G$, there is a star 1–form (Φ, φ) on G.*

Proof From 11.4.10 we know that $\widetilde{\alpha} \colon T^*G \to A^*G$ is a fibration of Lie algebroids over α. Hence, given $\varphi \in \Gamma A^*G$, there is a 1–form Φ on G such that $\widetilde{\alpha} \circ \Phi = T(\alpha) \circ \alpha$. Evaluating this equation at an identity element shows that $\Phi \circ 1 = \widetilde{1} \circ \varphi$. $\square$

Proposition 12.3.6 *The 1–forms $\widetilde{\Phi}$, where (Φ, φ) is a star 1–form on G, together with the pullbacks $q^*\omega$, where $\omega \in \Omega^1(M)$, generate $\Omega^1(AG)$.*

Proof Take $\Upsilon \in T^*_X AG$ with $r(\Upsilon) = \varphi(m)$. Extend $\varphi(m)$ to a section φ of ΓA^*G. By 12.3.5, there is a star 1–form (Φ, φ) on G. So at X we have $r(\widetilde{\Phi}(X)) = \varphi(m)$ and hence

$$\Upsilon = \widetilde{\Phi}(X) + q^*\omega$$

for some $\omega \in \Omega^1(M)$. $\square$

*   *   *   *   *

We now return to consider a Poisson groupoid G on base P.

Theorem 12.3.7 *Let (Φ, φ) and (Ψ, ψ) be star 1–forms on G. Then*

$$[\widetilde{\Phi}, \widetilde{\Psi}] = \widetilde{[\Phi, \Psi]}. \tag{27}$$

Proof We verify this by evaluating both sides on all $\widetilde{\zeta}$, where (ζ, z) is a star vector field on G, and all $Z^{\uparrow}$, where $Z \in \Gamma AG$. By Proposition 9.8.14, this will suffice.

For any star vector field (ζ, z), we have

$$\langle [\widetilde{\Phi}, \widetilde{\Psi}], \widetilde{\zeta} \rangle = \widetilde{\xi}(\langle \widetilde{\Psi}, \widetilde{\zeta} \rangle) - \widetilde{\eta}(\langle \widetilde{\Phi}, \widetilde{\zeta} \rangle) + \langle \widetilde{\Phi}, [\widetilde{\eta}, \widetilde{\zeta}] \rangle - \langle \widetilde{\Psi}, [\widetilde{\xi}, \widetilde{\zeta}] \rangle - \widetilde{\zeta}(\langle \widetilde{\Phi}, \widetilde{\xi} \rangle),$$

where $\xi = \pi^{\#}(\Phi)$, $\eta = \pi^{\#}(\Psi)$. Using (25) and 9.8.13, it is straightforward to verify that this is the tilde of $\langle [\Phi, \Psi], \zeta \rangle$, and is therefore equal to $\langle \widetilde{[\Phi, \Psi]}, \widetilde{\zeta} \rangle$.

For any $Z \in \Gamma AG$, (25) gives $\langle \widetilde{[\Phi, \Psi]}, Z^{\uparrow} \rangle = \langle [\varphi, \psi], Z \rangle \circ q$. Similarly expanding out the LHS, the equation reduces to

$$\langle [\varphi, \psi], Z \rangle = a_{*}(\varphi)(\langle \psi, Z \rangle) - a_{*}(\psi)(\langle \varphi, Z \rangle)$$
$$+ \langle \varphi, D_{\eta}(Z) \rangle - \langle \psi, D_{\xi}(Z) \rangle - Z(\langle \Psi, \xi \rangle). \tag{28}$$

Now the equality of these two functions on P follows from similarly expanding out $\langle [\widetilde{\Phi}, \widetilde{\Psi}], \overrightarrow{Z} \rangle$ on G and restricting the result to the identity elements of G. $\qquad\square$

The bracket on the LHS of (27) is the bracket of 1–forms on AG induced, via θ, from the prolongation of the Poisson structure on G. On the other hand, the bracket on the RHS is the bracket on $\Gamma A^{*}G$ obtained in 11.4.10 by descent from $\Omega^{1}(G)$; this refers to the conormal Lie algebroid structure. We thus have the following result:

Theorem 12.3.8 *For a Poisson groupoid G, the Poisson structure on AG obtained by prolongation of the structure on G coincides with the dual of the Lie algebroid structure on $A^{*}G$ defined in §11.4.*

The construction of the Poisson structure on AG by prolongation produces the Lie algebroid structure on $A^{*}G$ and, simultaneously, the fact that $(AG, A^{*}G)$ is a Lie bialgebroid; it also applies in more general situations involving compatible Lie groupoid and Lie algebroid structures. On the other hand, the construction of the Lie algebroid structure on $A^{*}G$ as the conormal bundle of the coisotropic submanifold 1_P applies to any coisotropic submanifold, and makes clear the role of the Schouten calculus, a technique which is needed at some stage in either approach.

12.4 Poisson actions and moment maps

This section is a very brief introduction to the subjects of the title. We start with the standard definition of a Poisson Lie group action.

Definition 12.4.1 Let G be a Poisson Lie group and let $\sigma\colon G\times P \to P$ be a smooth action on a Poisson manifold P. The action is a *Poisson action* if σ is a Poisson map. □

Denote by σ_g the diffeomorphism $P \to P$ induced by $g \in G$ and by $\sigma_{,u}$ the evaluation map $G \to P$ at $u \in P$. As with any smooth action, σ induces a smooth action of the group TG on TP by

$$X \bullet \xi = T(\sigma)(X,\xi) = T(\sigma_g)(\xi) + T(\sigma_{,u})(X)$$

for $X \in T_g G$, $\xi \in T_u P$. The following criterion, or a rearrangement of it, is sometimes taken as the definition.

Proposition 12.4.2 *Let σ be a smooth action of a Poisson Lie group G on a Poisson manifold P. Then σ is a Poisson action if and only if, for all $\Phi, \Psi \in T^*_{gu}P$,*

$$\langle \Psi, \pi^\#_P(\Phi)\rangle = \langle \Psi \circ T(\sigma_g), \pi^\#_P(\Phi \circ T(\sigma_g)))\rangle + \langle \Psi \circ T(\sigma_{,u}), \pi^\#_G(\Phi \circ T(\sigma_{,u})))\rangle.$$

Proof The proof follows the same method as that of 11.1.2. Denote the graph $\{(g,u,gu) \mid g \in G,\ u \in P\}$ of the group multiplication by C. Thus

$$TC = \{(X,\xi,X \bullet \xi) \mid X \in TG,\ \xi \in TP\}$$

and so (φ, Φ, Ψ), where $\varphi \in T^*_g G$, $\Phi \in T^*_u P$, $\Psi \in T^*_{gu}P$, is in $(TC)^\circ$ if and only if

$$\varphi = -\Phi \circ T(\sigma_{,u}) \qquad \text{and} \qquad \Phi = -\Psi \circ T(\sigma_g). \tag{29}$$

Eliminating Ψ, and noting that $\sigma_{g^{-1}} \circ \sigma_{,u} \circ L_g = \sigma_{,u}$, we have

$$\widetilde{\alpha}(\varphi) = \varphi \circ T(L_g) = \mathfrak{p}(\Phi). \tag{30}$$

This, together with the second equation in (29), characterizes $(TC)^\circ$. The coisotropy condition now asserts that for such φ, Φ, Ψ we have

$(\pi_G^\#(\varphi), \pi_P^\#(\Phi), -\pi_P^\#(\Psi)) \in TC$; that is,

$$-\pi_P^\#(\Psi) = \pi_G^\#(\varphi) \bullet \pi_P^\#(\Phi). \tag{31}$$

Using the second equation in (29), this is

$$\pi_P^\#(\Phi \circ T(\sigma_{g^{-1}})) = T(\sigma_g)(\pi_P^\#(\Phi)) + T(\sigma_{,u})(\pi_G^\#(\varphi)).$$

Using (30) to eliminate φ, and renaming $\Phi \circ T(\sigma_{g^{-1}})$ as Φ, this becomes

$$\pi_P^\#(\Phi) = T(\sigma_g)(\pi_P^\#(\Phi \circ T(\sigma_g))) + T(\sigma_{,u})(\pi_G^\#(\mathfrak{p}(\Phi \circ T(\sigma_g)) \circ T(L_{g^{-1}}))).$$

Now unravel the last term and pair the equation with any $\Psi \in T_{gu}^* P$.
□

The proof of the next criterion follows that of 11.1.3 very closely.

Proposition 12.4.3 *Let σ be a smooth action of a Poisson Lie group G on a Poisson manifold P. Then σ is a Poisson action if and only if, for all $g \in G$, $u \in P$,*

$$\pi_P(gu) = T_g(\sigma_{,u})(\pi_G(g)) + T_u(\sigma_g)(\pi_P(u)).$$

Any smooth action $G \times P \to P$ induces an action of $T^*G \rightrightarrows \mathfrak{g}^*$ on the pith $\mathfrak{p} \colon T^*P \to \mathfrak{g}^*$; see 1.6.9.

Proposition 12.4.4 *Let σ be a smooth action of a Poisson Lie group G on a Poisson manifold P. Then σ is a Poisson action if and only if*

$$(\pi_G^\#, \pi_P^\#) \colon T^*G \lessdot T^*P \to TG \lessdot TP$$

is a morphism of Lie groupoids over $\pi_P^\#$.

Proof The source projections always commute. The condition that the target projections commute is that

$$\pi_G^\#(\varphi) \bullet \pi_P^\#(\Phi) = \pi_P^\#(\varphi \bullet \Phi) \tag{32}$$

for $\varphi \in T_g^*G$, $\Phi \in T_u^*P$ with $\widetilde{\alpha}(\varphi) = \mathfrak{p}(\Phi)$. From the proof of 12.4.2, this is equivalent to σ being a Poisson action. (Note that (31), together with the second equation in (29), is equivalent to (32).)

It is a simple general result that when the source and target projections commute, a map such as $(\pi_G^\#, \pi_P^\#)$ is a morphism.
□

The next result is an infinitesimal criterion: it is stated here as a property of Poisson Lie groups, but the converse may easily be proved when the group is connected.

Theorem 12.4.5 *Let σ be a Poisson action of a Poisson Lie group G on a Poisson manifold P. Then $\mathfrak{p} \colon T^*P \to \mathfrak{g}^*$ is a morphism of Lie bialgebroids.*

Proof That $\mathfrak{p}$ is a Poisson map, for any smooth action, follows from the fact that it is the dual of a morphism of Lie algebroids. More directly, for any $X \in \mathfrak{g}$, note that

$$\ell_X \circ \mathfrak{p} = \ell_{X^\dagger} \colon T^*P \to \mathbb{R}.$$

Hence $\{\ell_X \circ \mathfrak{p}, \ell_Y \circ \mathfrak{p}\}_{T^*P} = \{\ell_{X^\dagger}, \ell_{Y^\dagger}\}_{T^*P} = \ell_{[X,Y]^\dagger} = \{\ell_X, \ell_Y\}_{\mathfrak{g}^*} \circ \mathfrak{p}$.

To show that $\mathfrak{p}$ is a morphism of Lie algebroids, consider the corresponding map into the pullback Lie algebroid:

$$T^*P \to TP \oplus (P \times \mathfrak{g}^*), \qquad \varphi \mapsto \pi_P^\#(\varphi) \oplus \mathfrak{p}(\varphi),$$

where φ here is a 1–form and $\mathfrak{p}(\varphi)$ is $P \to \mathfrak{g}^*$. We must show that, for $\varphi, \psi \in \Omega^1(P)$,

$$\pi_P^\#(\varphi)(\mathfrak{p}(\psi)) - \pi_P^\#(\psi)(\mathfrak{p}(\varphi)) - \mathfrak{p}([\varphi, \psi]) = [\mathfrak{p}(\varphi), \mathfrak{p}(\psi)].$$

It is sufficient to show that this holds when ℓ_X, for any $X \in \mathfrak{g}$, is applied to both sides. Since ℓ_X is linear, this becomes:

$$\pi_P^\#(\varphi)(\langle \psi, X^\dagger \rangle) - \pi_P^\#(\psi)(\langle \varphi, X^\dagger \rangle) - \langle [\varphi, \psi], X^\dagger \rangle = \ell_X \circ [\mathfrak{p}(\varphi), \mathfrak{p}(\psi)].$$

The LHS is $\langle \varphi \wedge \psi, d_{\pi_P}(X^\dagger) \rangle$ and from 10.2.3, we have

$$d_{\pi_P}(X^\dagger) = [\pi_P, X^\dagger] = -\mathfrak{L}_{X^\dagger}(\pi_P).$$

Now, by an argument similar to that in the proof of 11.1.5, we have that

$$(\mathfrak{L}_{X^\dagger}(\pi_P))(u) = -T(\sigma_{,u})(\mathfrak{L}_{\overrightarrow{X}}(\pi_G)(1)).$$

Applying this to $\varphi, \psi \in \Omega^1(P)$, we obtain

$$\langle \varphi \wedge \psi, \mathfrak{L}_{X^\dagger}(\pi_P) \rangle = -\langle \mathfrak{p}(\varphi) \wedge \mathfrak{p}(\psi), \mathfrak{L}_{\overrightarrow{X}}(\pi_G)(1) \rangle = -\langle X, [\mathfrak{p}(\varphi), \mathfrak{p}(\psi)] \rangle(u)$$

and this completes the proof. $\qquad\square$

Consider a Poisson action σ. The manifold underlying $T^*G \lessdot T^*P$ is the pullback of $\tilde{\alpha}$ and $\mathfrak{p}$ and since these are both morphisms of Lie algebroids, $T^*G \lessdot T^*P$ has the pullback Lie algebroid structure with base $G \times P$. The map $(\pi_G^\#, \pi_P^\#)$ is the anchor of this structure.

Example 12.4.6 Consider the case in which G has the zero Poisson structure and P is symplectic and simply–connected. If the action is Poisson, then it is an action by symplectomorphisms.

Since P is symplectic, $T^*P \cong TP$ as Lie algebroids, and since P is simply–connected, the monodromy groupoid is $P \times P$. The Lie algebroid morphism $\mathfrak{p}$ therefore integrates to a morphism $\mathfrak{P} \colon P \times P \to \mathfrak{g}^*$, where the codomain is the additive group. Now $\mathfrak{p}$ is G–equivariant and it is easy to deduce from the uniqueness of $\mathfrak{P}$ that it also is equivariant.

Any groupoid morphism $\mathfrak{P} \colon P \times P \to \mathfrak{g}^*$ is of the form

$$\mathfrak{P}(v, u) = \mu(v) - \mu(u)$$

where $\mu(w) = \mathfrak{P}(w, u_0)$ for a chosen reference point $u_0 \in P$. It is easy to see that such a $\mu \colon P \to \mathfrak{g}^*$ is a moment map in the standard sense. If P has a fixed point u_0, then the moment map obtained from u_0 is equivariant. $\boxtimes$

12.5 Notes

§12.1 Lie bialgebroids were defined by the author and Ping Xu in [Mackenzie and Xu, 1994]. A much clearer formulation of the basic Schouten and exterior calculus associated with them was immediately given by Kosmann–Schwarzbach [1995]; in particular 12.1.6 (and 12.1.10) come from [Kosmann–Schwarzbach, 1995]. The account given here uses elements of both formulations. Work on the structure of general Lie bialgebroids is still at an early stage; see [Liu and Xu, 2002] for one result.

The proof given for 12.1.3, that (AG, A^*G) for G a Poisson groupoid is a Lie bialgebroid, is due to Xu [1995]. For the proof given in [Mackenzie and Xu, 1994], see §12.3.

In view of the essential symmetry of the notion of Lie bialgebroid it is desirable for the concept of morphism, and everything which depends upon it, to be developed in a symmetric fashion. This may be obtained as a byproduct of the theory of Dirac structures; see [Liu, Weinstein, and Xu, 1998].

It was shown in 7.5.2 that Lie algebroid structures on a vector bundle A are in bijective correspondence with brackets on $\mathscr{A} = \Gamma \Lambda^\bullet(A)$ which, together with the exterior algebra structure, make $\mathscr{A}$ a Schouten algebra. Under the same correspondence, Lie bialgebroid structures on A correspond to strong differential Gerstenhaber algebra structures; see [Kosmann–Schwarzbach, 1995], [Xu, 1999].

The concept of Lie bialgebroid may also be extended to a purely algebraic setting, in the same way that Lie pseudoalgebras are a purely algebraic form of Lie algebroids.

The triangular Lie bialgebroids of 12.1.19 were introduced in [Mackenzie and Xu, 1994], developing a construction of Kosmann–Schwarzbach and Magri [1990]. It is possible to extend to general triangular Lie bialgebroids the relations which hold between the two Schouten algebras attached to a Poisson manifold; see [Kosmann–Schwarzbach, 2000].

As with Poisson groupoids, we have not given detailed treatments of examples of Lie bialgebroids. For Poisson groupoids arising from the CDYBE, the corresponding Lie bialgebroids have been constructed by Bangoura and Kosmann–Schwarzbach [1998]; see also the references in Chapter 11. For Lie bialgebroids associated to Poisson–Nijenhuis structures see [Kosmann–Schwarzbach, 1996b]. Several authors have studied Jacobi structures compatible with Lie groupoids, and have called the infinitesimal concept a *generalized Lie bialgebroid*; see [Iglesias–Ponte and Marrero, 2003] and references given there.

The largest single body of work on Lie bialgebroids concerns the concept of a double. For a Lie bialgebroid (A, A^*), Liu, Weinstein, and Xu [1998] defined a bracket on $A \oplus A^*$ which enables many of the properties of the double of a Lie bialgebra to be extended to Lie bialgebroids. Abstracting the properties of this bracket, Liu et al. [1998] introduced the concept of Courant algebroid; this simultaneously provides a means of extending the classical Dirac bracket on a submanifold of a Poisson manifold to this more general setting.

For any Lie algebroid A the Poisson structure on A^* induces a Lie algebroid structure on the cotangent $T^*A^* \to A^*$. If A^* is also a Lie algebroid then $T^*A \to A$ is likewise a Lie algebroid. Using the isomorphism of double vector bundles $R: T^*A^* \to T^*A$ from 9.5.1, these may be regarded as two Lie algebroid structures on, say, T^*A. In terms of the author's concept of double Lie algebroid [Mackenzie, 1998], [Mackenzie, arXiv:0011212], it then follows that A and A^* constitute a Lie bialgebroid if and only if T^*A, with these structures, is a double Lie algebroid. This result extends the Manin triple characterization of the double of a Lie bialgebra. For (A, A^*) a Lie bialgebroid, call the double Lie algebroid T^*A the *cotangent double* of (A, A^*). The general notion of double Lie algebroid overlaps with that of Courant algebroid only in the case of Lie bialgebroids; see the Appendix for some further discusssion.

Notions of double of a Lie bialgebroid involving the cotangents have also been introduced in the context of super geometry; see [Roytenberg, arXiv:9910078], [Voronov, 2002].

§12.2 This section has been taken from §6 of [Mackenzie and Xu, 1994]. In that paper it was used to establish that the Lie algebroid of a Poisson groupoid, together with its Poisson structure, form a Lie bialgebroid (see §12.3). In [Mackenzie and Xu, 2000] this process was reversed to show that a Lie bialgebroid structure on (AG, A^*G), for G an α–connected and α–simply connected Lie groupoid, integrates to a Poisson groupoid structure on G.

§12.3 The material of this section is from [Mackenzie and Xu, 1998], which gives a fuller treatment of the calculus of 1–forms on Poisson groupoids.

§12.4 Most accounts of Poisson Lie groups include a treatment of the basics of Poisson actions. Theorem 12.4.5 is due to Xu [1995] and has been generalized to Poisson groupoid actions in [He et al., 2002], [Zhong and He, 2003].

The view taken of moment maps in this section is based loosely on [Mackenzie, 2000], though 12.4.4 has not appeared in this form before. Proposition 12.4.4 is part of the proof, given in [Mackenzie, 2000], that for a Poisson action, the structure $(T^*G \lhd T^*P; G \lhd P, T^*P; P)$ is an $\mathscr{LA}$–groupoid.

The viewpoint of 12.4.6 can be applied very widely. Mikami and Weinstein [1988] noted that a Hamiltonian action $G \times P \to P$ with equivariant moment map $\mu: P \to \mathfrak{g}^*$ induces an action of the symplectic groupoid $T^*G \rightrightarrows \mathfrak{g}^*$ on $\mu: P \to \mathfrak{g}^*$; for this reason maps on which a Lie groupoid acts are sometimes called moment maps. The action of T^*G on μ is *symplectic* in the sense that its graph is a Lagrangian submanifold of $T^*G \times P \times \overline{P}$. This paper was the first to consider a case of Poisson actions in this way.

More generally, Mikami and Weinstein [1988] consider any symplectic action of a symplectic groupoid: given a symplectic groupoid $\Sigma \rightrightarrows M$ and a symplectic manifold P, provided with a smooth map $f: P \to M$, an action of Σ on P is a *symplectic action* if the graph is a Lagrangian submanifold as above. As in the group case, take the infinitesimal action $\Gamma A\Sigma \to \mathfrak{X}(P)$ and dualize it to $\mathfrak{p}: T^*P \to A^*\Sigma$. Again it may be proved [Xu, 1995] that this is a Lie algebroid morphism, with base map $f: P \to M$. Composing with the canonical isomorphisms $T^*P \cong TP$ and

$A^*\Sigma \cong TM$ we can in fact identify $\mathfrak{p}$ with $T(f)\colon TP \to TM$. This integrates (without any simple connectivity hypothesis) to $f \times f\colon P \times P \to M \times M$ which may in turn be identified with f, which [Mikami and Weinstein, 1988] shows may be regarded as the moment(um) map of the action.

Consider further a Poisson action $G \times P \to P$ of a Poisson Lie group on a Poisson manifold. If P is no longer assumed to be symplectic, but integrates to an α–simply connected symplectic groupoid $\Pi \rightrightarrows P$ and if G^* is a dual group, integrating the Lie algebra $\mathfrak{g}^*$, then $\mathfrak{p}$ may be integrated to a Lie groupoid morphism $\mathfrak{P}\colon \Pi \to G^*$. If a moment map $\mu\colon P \to G^*$ in the sense of Lu [1991] exists, then $\mathfrak{P}$ is the composite of the anchor $\Pi \to P \times P$ with the map $P \times P \to G^*$, $(u_2, u_1) \mapsto \mu(u_2)\mu(u_1)^{-1}$, derived from μ; compare [Xu, 1995, §6].

In fact all these cases are special cases of results for actions of Poisson groupoids or Lie bialgebroids; this most general construction is the subject of ongoing work.

Appendix

This Appendix describes recent and ongoing work on three topics which are closely related to the main text and which, in an ideal world, the author would have covered in full. Some important recent developments are not covered here.

Foliations

The most important single topic which has not been covered is that of the relationship between groupoid theory and foliation theory. An involutive distribution Δ on a manifold M is a Lie subalgebroid of $TM = A\Pi(M)$. When $\Delta = T^f M$ is the vertical bundle of a surjective submersion $f \colon M \to N$, the Lie subgroupoid $R(f) \subseteq M \times M$ of 1.1.8 has Lie algebroid $T^f M$, but in general an involutive distribution is not the Lie algebroid of a Lie subgroupoid of $M \times M$ or $\Pi(M)$.

The usual holonomy groups of a foliation $\mathscr{F}$ are the vertex groups of the holonomy groupoid of the foliation, introduced in the 1950s and one of the key examples considered by Pradines [1966]. (For historical background, see [Haefliger, 1980].) This groupoid is best understood by first considering the monodromy groupoid of $\mathscr{F}$. Algebraically, the *monodromy groupoid* $\mathscr{M}(\mathscr{F})$ is the disjoint union of the fundamental groupoids of the leaves of $\mathscr{F}$ and the *holonomy groupoid* $\mathscr{H}(\mathscr{F})$ is the quotient of $\mathscr{M}(\mathscr{F})$ over the totally intransitive subgroupoid which consists of the elements represented by loops with trivial holonomy.

The monodromy groupoid has a natural smooth structure which is not necessarily Hausdorff but in all other respects satisfies the conditions of a Lie groupoid, and is such that the fundamental Lie groupoid of each leaf is a Lie subgroupoid. This structure may then be quotiented to give the holonomy groupoid a smooth structure which, again, need not be Hausdorff but in all other respects satisfies the conditions of a Lie groupoid. In view of these examples, Pradines [1966] in his definition of a differentiable groupoid allowed a non–Hausdorff arrow space, but required the base manifold and the α–fibres to be Hausdorff. In this Appendix we use his terminology *differentiable groupoid* for this more general concept. Thus $\mathscr{M}(\mathscr{F})$ and $\mathscr{H}(\mathscr{F})$ are differentiable groupoids. For $\mathscr{M}(\mathscr{F})$ this was stated in [Pradines, 1966] and proved in [Winkelnkemper, 1983]; an efficient proof for both monodromy and holonomy is given in [Moerdijk and Mrčun, 2003, §5.2].

The following result, due to Crainic and Moerdijk [2001], describes the situation comprehensively.

Theorem A *Let $\mathscr{F}$ be a (smooth, regular) foliation on M with distribution Δ. For any α–connected differentiable groupoid G with $AG = \Delta$, the natural projection $\mathscr{M}(\mathscr{F}) \to \mathscr{H}(\mathscr{F})$ factorizes into base–preserving smooth morphisms*

$$\mathscr{M}(\mathscr{F}) \longrightarrow G \longrightarrow \mathscr{H}(\mathscr{F})$$

both of which are surjective and étale.

This analysis may now be used to resolve the problem of integrating general Lie subalgebroids. Consider a Lie groupoid G on base M and a wide Lie subalgebroid $A' \leqslant AG$. As in the transitive case (§6.2), translate A' over G to give an involutive distribution Δ on G with corresponding foliation $\mathscr{F}$. Then Moerdijk and Mrčun [2002] show that $\mathscr{M}(\mathscr{F})$ may be quotiented over the induced action of G to give a Lie groupoid G' on M and a base–preserving morphism $G' \to G$ which is an immersion and induces the inclusion $A' \to AG$. In general G' cannot be injected into G; the case of distributions already shows this. The relationship between G' and G is related in a deep way to the transverse structure of $\mathscr{F}$; see [Moerdijk and Mrčun, arXiv:0406558].

Again as in the transitive case, the integrability of morphisms may be deduced from the integrability of Lie subalgebroids. The following result is proved in [Mackenzie and Xu, 2000] and in [Moerdijk and Mrčun, 2002].

Theorem B *Let $G \rightrightarrows M$ and $H \rightrightarrows N$ be Lie groupoids and suppose that $\varphi\colon AG \to AH$, $f\colon M \to N$ is a morphism of Lie algebroids. Then if G is α–simply connected, there is a unique morphism of Lie groupoids $F\colon G \to H$ over f such that $A(F) = \varphi$.*

Finally, Theorem A may be applied to the integrability problem for actions of Lie algebroids:

Theorem C *Let $G \rightrightarrows M$ be an α–simply connected Lie groupoid and suppose that AG acts on a proper map $f\colon M' \to M$. Then the action integrates to a global action of G on f.*

Theorem C is from [Moerdijk and Mrčun, 2002]. This paper also establishes general integrability results for derivative representations of one Lie algebroid on another, as in 4.5.2.

In his original paper, Pradines [1966] outlined a theory of *microdifferentiable groupoids* for the solution of the problems related to holonomy and monodromy. Very roughly speaking, given a foliation $\mathscr{F}$ on M, let $H = H(\mathscr{F})$ denote the equivalence relation defined by the partition of M into leaves; thus H is a set–subgroupoid of $M \times M$. The foliation charts of M may be used to define a smooth structure on a subset of H which contains the identity elements and which generates H in the algebraic sense. A germ equivalence class of such a smooth structure is a microdifferentiable groupoid structure on H. The theory of holonomy and monodromy groupoids described in [Pradines, 1966] was designed to apply in this situation. This approach was pursued in a series of papers by Brown and coauthors (see, for example, [Aof and Brown, 1992, Brown and Mucuk, 1995]), which was geared particularly to the case of topological groupoids.

There remain a considerable number of difficult questions concerning the relationship between foliations and differentiable groupoids. It is often useful to regard a groupoid as a desingularization of its transitivity foliation. However this point of view cuts across the usual perception of difficulty associated with foliations: a regular foliation may have a non–Hausdorff holonomy groupoid whereas a foliation defined by a group action, though usually singular, has the corresponding action groupoid, always Hausdorff, as a desingularization. An answer to the question of which singular foliations are transitivity foliations of differentiable groupoids would be very interesting: for regular symplectic foliations — that is, for regular Poisson manifolds — this is the realizability problem.

The General Integrability Obstructions of
Crainic and Fernandes

Unexpectedly simple and general conditions for the integrability of an arbitrary Lie algebroid were given by Crainic and Fernandes [2003] using an approach to integrability developed by Cattaneo and Felder [2001]. Until these papers, there seemed no reason to believe that straightforward criteria for the integrability of arbitrary Lie algebroids would exist.

Prior to [Crainic and Fernandes, 2003], the most general known integrability criterion was perhaps the cohomological obstruction for transitive Lie algebroids, due to the author [Mackenzie, 1987a] and treated here in Chapter 8. There were also a number of results for more specialized classes of Lie algebroids; an account of these is given in [Mackenzie, 1995] and will not be repeated here.

Firstly consider a Lie groupoid $G \rightrightarrows M$. An α–path $\nu \colon I \to G$ as in §6.1 can be reformulated as the morphism $\nu^{\div} \colon I \times I \to G$, $\nu^{\div}(s,t) = \nu(s)\nu(t)^{-1}$ from the pair groupoid $I \times I$ to G, and this therefore induces a morphism of Lie algebroids $A(\nu^{\div}) \colon TI \to AG$ called an A–path in AG. In a similar way α–homotopies give rise to A–homotopies of A–paths in AG. Now the concepts of A–path and A–homotopy may be formulated for any abstract Lie algebroid A and, by a construction which is analogous to that of the monodromy groupoid of a Lie groupoid, but involves new difficulties, Crainic and Fernandes [2003] construct a groupoid $\mathscr{G}(A)$ from A–paths in A which they name the *Weinstein groupoid* of A. As a set, $\mathscr{G}(A)$ is a groupoid on M, and has a natural topology making it a topological groupoid. If A is integrable, then $\mathscr{G}(A)$ has a smooth structure with respect to which it is the α–simply connected Lie groupoid integrating A.

The problem of integrability of A is thus reduced to the problem of establishing a suitable smooth structure on $\mathscr{G}(A)$. For each $x \in M$, Crainic and Fernandes [2003] define a subgroup $N_x(A)$ of the centre of the Lie algebra $L_x = \ker(a_x \colon A_x \to T_x M)$. The following theorem is their main result.

Theorem D *Let A be a Lie algebroid on M. With the above notation, A is integrable if and only if:*

(i) *$N_x(A)$ is a discrete subgroup of the centre of L_x, for all $x \in M$;*

(ii) *for all $x \in M$, $\liminf_{y \to x} r(y) > 0$ where $r(y)$ is the distance from the origin of L_y to the set of nonzero elements of $N_y(A)$, measured in terms of any norm on A.*

The construction of the Weinstein groupoid of an abstract Lie algebroid is an extension of the approach developed for the integrability of Poisson manifolds by Cattaneo and Felder [2001]; see Cattaneo [2004]. For the use of morphisms of Lie algebroids as σ–models see [Schaller and Strobl, 1994].

Although the results of [Crainic and Fernandes, 2003] provide a complete answer to the integrability problem on the theoretical level, work still remains to be done to relate their obstructions to some earlier results on integrability and obstruction, and to obtain descriptions of the Weinstein groupoid tailored to various specific classes of Lie algebroid.

Double Lie groupoids and double Lie algebroids

As remarked in the Prologue, one of the crucial strengths of the groupoid concept is its capacity to be 'doubled': a groupoid object in the category of groupoids is a genuinely new and rich object, unlike a group object in the category of groups. Multiple structures have been used to provide algebraic models of homotopy types since at least the 1970s; in particular, double groupoids of a particularly special type [Brown and Spencer, 1976a] are equivalent to crossed modules over groupoids and thus provide algebraic models of homotopy 2–types [Brown and Higgins, 1978].

In the Lie theory of ordinary groupoids it is useful to think of groupoid elements as elements of length, and the Lie functor as a linearization process which replaces them with infinitesimal line elements. Ordinary Lie groupoids model a very large variety of phenomena in differential geometry, with their Lie algebroids embodying the corresponding first–order infinitesimal invariants. (This view is developed at length in [Mackenzie, 1995].) In the same way elements of a double groupoid may be thought of as elements of area, and the Lie functor in this case is accordingly a second–order process. Application of the Lie functor to a double Lie groupoid yields an $\mathscr{LA}$–groupoid— that is, a Lie groupoid object in the category of Lie algebroids — and taking the Lie algebroid of its groupoid structure yields the double Lie algebroid. (For these two steps see, respectively, [Mackenzie, 1992] and [Mackenzie, 2000b].)

The concept of $\mathscr{LA}$–groupoid is of interest in its own right for, given any Poisson groupoid G on base P, the cotangent T^*G is an $\mathscr{LA}$–groupoid, with respect to the Lie algebroid structure induced by the Poisson structure and the groupoid structure induced by the groupoid structure on G. The double Lie algebroid in this case is the cotangent double Lie algebroid as described briefly in the Notes to Chapter 11 and, in the opposite direction, one may seek a symplectic double Lie groupoid Σ with $\mathscr{LA}$–groupoid isomorphic to T^*G; in the case of Poisson groups the existence of Σ was established by Lu and Weinstein [1989].

Most general constructions involving Poisson groupoids may be regarded as constructions for the corresponding $\mathscr{LA}$–groupoids. The Lie theory of double groupoids thus embodies a very large number of differentiation processes and integrability problems arising from Poisson groupoid theory.

The construction of the double Lie algebroid of a double Lie groupoid, though technically involved, follows a direct and natural path. It gives no indication, however, of how to define an abstract concept of double Lie algebroid. Whereas the notions of double vector bundle and of double Lie groupoid can be defined diagrammatically in terms of the categories respectively of vector bundles and of manifolds, a definition of Lie algebroid requires consideration of modules of sections, and these lie outside the categories to which the structure maps belong. The solution to this difficulty lies in a systematic application of the duality between Lie algebroids and Poisson vector bundles, in the setting of the duality of $\mathscr{VB}$–groupoids. An overview with references is given in Mackenzie [1998].

Bibliography

R. Abraham and J. Marsden. *Foundations of Mechanics*. Addison-Wesley, second edition, 1985.

R. Almeida and A. Kumpera. Structure produit dans la catégorie des algèbroids de Lie. *An. Acad. Brasil Ciênc*, 53:247–250, 1981.

R. Almeida and P. Molino. Suites d'Atiyah et feuilletages transversalement complets. *C. R. Acad. Sci. Paris Sér. I Math.*, 300:13–15, 1985.

I. Androulidakis. Connections on Lie algebroids and the Weil–Kostant theorem. In *Proceedings of the 4th Panhellenic Conference on Geometry (Patras, 1999)*, volume 44, pages 51–57, 2000.

I. Androulidakis. *Extensions, cohomology and classification for Lie algebroids and Lie groupoids*. PhD thesis, University of Sheffield, 2001.

I. Androulidakis. Connections and holonomy for extensions of Lie groupoids. arXiv:math.DG/0307282, .

M. E. S. A. F. Aof and R. Brown. The holonomy groupoid of a locally topological groupoid. *Topology Appl.*, 47:97–113, 1992.

M. F. Atiyah. Complex analytic connections in fibre bundles. *Trans. Amer. Math. Soc.*, 85:181–207, 1957.

M. Bangoura and Y. Kosmann–Schwarzbach. Équation de Yang–Baxter dynamique classique et algébroïdes de Lie. *C. R. Acad. Sci. Paris Sér. I Math.*, 327(6):541–546, 1998.

A. A. Beilinson and V. V. Schechtmann. Determinant bundles and Virasoro algebras. *Comm. Math. Phys.*, 118:651–701, 1988.

A. L. Besse. *Manifolds all of whose Geodesics are Closed*, volume 93 of *Ergebnisse des Mathematik und ihrer Grenzgebiete*. Springer–Verlag, 1978.

K. H. Bhaskara and K. Viswanath. Calculus on Poisson manifolds. *Bull. London Math. Soc.*, 20:68–72, 1988a.

K. H. Bhaskara and K. Viswanath. *Poisson Algebras and Poisson Manifolds*, volume 174 of *Pitman Research Notes in Mathematics Series*. Longman, Harlow, Essex, 1988b.

R. L. Bishop and R. J. Crittenden. *Geometry of Manifolds*, volume 15 of *Pure and Applied Mathematics*. Academic Press, 1964.

R. Bkouche. Structures (K, A)–linéaires. *C. R. Acad. Sci. Paris, Série A*, 262:373–376, 1966.

R. Bott and L. W. Tu. *Differential Forms in Algebraic Topology*. Springer–Verlag, New York, 1982.

N. Bourbaki. *Variétés Différentielles et Analytiques. Fasicule de Résultats*. Éléments de Mathématique. Diffusion C.C.L.S., 1982. Nouveau tirage.

R. A. Bowshell. Abstract velocity functors. *Cahiers Topologie Géom. Différentielle*, 12:57–91, 1971.

R. Brown. Fibrations of groupoids. *J. Algebra*, 15:103–132, 1970.

R. Brown. Groupoids as coefficents. *Proc. London Math. Soc.* (3), 25: 413–426, 1972.

R. Brown. From groups to groupoids: a brief survey. *Bull. London Math. Soc.*, 19:113–134, 1987.

R. Brown. *Topology: a Geometric Account of General Topology, Homotopy Types and the Fundamental Groupoid*. Ellis Horwood, Chichester, revised, updated and expanded edition, 1988.

R. Brown, G. Danesh–Naruie, and J. P. L. Hardy. Topological groupoids: II. Covering morphisms and G–spaces. *Math. Nachr.*, 74:143–156, 1976.

R. Brown and J. P. L. Hardy. Topological groupoids: I. Universal constructions. *Math. Nachr.*, 71:273–286, 1976.

R. Brown and P. J. Higgins. On the connection between the second relative homotopy groups of some related spaces. *Proc. London Math. Soc.* (3), 36:193–212, 1978.

R. Brown and K. C. H. Mackenzie. Determination of a double Lie groupoid by its core diagram. *J. Pure Appl. Algebra*, 80(3):237–272, 1992.

R. Brown and O. Mucuk. The monodromy groupoid of a Lie groupoid. *Cahiers Topologie Géom. Différentielle Catégoriques*, 36(4):345–369, 1995.

R. Brown and C. B. Spencer. Double groupoids and crossed modules. *Cahiers Topologie Géom. Différentielle*, 17:343–362, 1976a.

R. Brown and C. B. Spencer. $\mathscr{G}$–groupoids, crossed modules and the fundamental groupoid of a topological group. *Nederl. Akad. Wetensch. Proc. Ser. A* **79**=*Indag. Math.*, 38(4):296–302, 1976b.

J.-L. Brylinski. A differential complex for Poisson manifolds. *J. Differential Geom.*, 28:93–114, 1988.

H. Bursztyn, M. Crainic, A. Weinstein, and C. Zhu. Integration of twisted Dirac brackets. arXiv:math.DG/0303180.

H. Cartan and S. Eilenberg. *Homological Algebra*. Princeton University Press, Princeton, 1956.

P. Cartier. Some fundamental techniques in the theory of integrable systems. In *Lectures on Integrable Systems (Sophia–Antipolis, 1991)*, pages 1–41. World Sci. Publishing, River Edge, NJ, 1994.

A. S. Cattaneo. On the integration of Poisson manifolds, Lie algebroids, and coisotropic submanifolds. *Lett. Math. Phys.*, 67(1):33–48, 2004.

A. S. Cattaneo and G. Felder. Poisson sigma models and symplectic groupoids. In *Quantization of Singular Symplectic Quotients*, volume 198 of *Progr. Math.*, pages 61–93. Birkhäuser, Basel, 2001.

V. Chari and A. Pressley. *A Guide to Quantum Groups*. Cambridge University Press, Cambridge, 1994. ISBN 0-521-43305-3.

P. M. Cohn. *Lie Groups*. Cambridge University Press, 1957.

A. Connes. *Noncommutative Geometry*. Academic Press, 1994.

A. Coste, P. Dazord, and A. Weinstein. Groupoïdes symplectiques. In *Publications du Département de Mathématiques de l'Université de Lyon, I*, number 2/A-1987, pages 1–65, 1987.

T. J. Courant. Dirac manifolds. *Trans. Amer. Math. Soc.*, 319:631–661, 1990.

M. Crainic and R. L. Fernandes. Integrability of Lie brackets. *Ann. of Math. (2)*, 157(2):575–620, 2003.

M. Crainic and R. L. Fernandes. Integrability of Poisson brackets, . arXiv:math.DG/0210152.

M. Crainic and I. Moerdijk. A homology theory for étale groupoids. *J. Reine Angew. Math.*, 521:25–46, 2000.

M. Crainic and I. Moerdijk. Foliation groupoids and their cyclic homology. *Adv. Math.*, 157(2):177–197, 2001.

M. Crainic and I. Moerdijk. Deformations of Lie brackets: cohomological aspects, . arXiv:math.DG/0403434.

R. H. Cushman and L. M. Bates. *Global Aspects of Classical Integrable Systems*. Birkhäuser Verlag, Basel, 1997. ISBN 3-7643-5485-2.

A. C. da Silva and A. Weinstein. *Geometric Models for Noncommutative Algebras*, volume 10 of *Berkeley Mathematics Lecture Notes*. American Mathematical Society, 1999.

M. K. Dakin and A. K. Seda. G–spaces and topological groupoids. *Glasnik Matematički*, 12:191–198, 1977.

P. Dazord and D. Sondaz. Groupes de Poisson affines. In Dazord and Weinstein [1991], pages 99–128.

P. Dazord and A. Weinstein, editors. *Symplectic Geometry, Groupoids and Integrable Systems*, Séminaire Sud Rhodanien de Géométrie (1989), 1991. Springer–Verlag, MSRI Publications, 20.

C. M. de Barros. Espaces infinitésimaux. *Cahiers Topologie Géom. Différentielle*, 7, 1964. xi + 96 pages.

J. Dieudonné. *Treatise on Analysis*, volume III. Academic Press, 1972. Translated by I. G. Macdonald.

M. P. do Carmo. *Differential Geometry of Curves and Surfaces*. Prentice–Hall, 1976.

A. Douady and M. Lazard. Espaces fibrés en algèbres de Lie et en groupes. *Invent. Math.*, 1:133–151, 1966.

V. G. Drinfel'd. Hamiltonian structures on Lie groups, Lie bialgebras and the geometric meaning of the classical Yang–Baxter equation. *Soviet. Math. Dokl.*, 27:68–71, 1983.

A. C. Ehresmann, editor. *Charles Ehresmann: Œuvres Complètes et Commentées*. Seven volumes. Imprimerie Evrard, Amiens, 1984.

C. Ehresmann. Les connexions infinitésimales dans un espace fibré différentiable. In *Colloque de Topologie (Espaces Fibrés), Bruxelles, 1950*, pages 29–55. Georges Thone, Liège, 1951. Contained in [Ehresmann, 1984, Parties I–1 et I–2].

C. Ehresmann. Sur les connexions d'ordre supérieur. In *Dagli Atti del V Congresso dell'Unione Matematica Italiana, Pavia-Torino*, pages 344–346, 1956. Contained in [Ehresmann, 1984, Parties I–1 et I–2].

C. Ehresmann. Gattungen von lokalen strukturen. *Jahres. d. Deutschen Math.*, 60-2:49–77, 1957. Contained in [Ehresmann, 1984, Partie II–1].

C. Ehresmann. Catégories topologiques et catégories différentiables. In *Colloque de Géométrie Différentielle Globale*, pages 137–150. Centre Belge de Recherches Mathématiques, Bruxelles, 1959. Contained in [Ehresmann, 1984, Parties I–1 et I–2].

C. Ehresmann. *Catégories et Structures*. Dunod, 1965.

C. Ehresmann. Sur les catégories différentiables. In *Atti. Conv. Int. Géom. Diff. Bologna*, pages 31–40, 1967. Contained in [Ehresmann, 1984, Partie I].

P. Etingof and A. Varchenko. Geometry and classification of solutions of the classical dynamical Yang–Baxter equation. *Comm. Math. Phys.*, 192(1):77–120, 1998.

S. Evens, J.-H. Lu, and A. Weinstein. Transverse measures, the modular class and a cohomology pairing for Lie algebroids. *Quart. J. Math. Oxford Ser. (2)*, 50(200):417–436, 1999.

G. L. Fel'dman. Global dimension of rings of differential operators. *Trans. Moscow Math. Soc.*, 41(1):123–147, 1982.

R. L. Fernandes. Contravariant connections on Poisson manifolds. In *Summer School on Differential Geometry (Coimbra, 1999)*, pages 99–108. Univ. Coimbra, Coimbra, 1999.

R. L. Fernandes. Connections in Poisson geometry. I. Holonomy and invariants. *J. Differential Geom.*, 54(2):303–365, 2000.

M. Gerstenhaber. The cohomology structure of an associative ring. *Ann. of Math.* (2), 78:267–288, 1963.

H. Goldschmidt. The integrability problem for Lie equations. *Bull. Amer. Math. Soc. (N.S.)*, 84:531–546, 1978.

M. Golubitsky and V. Guillemin. *Stable Mappings and their Singularities*, volume 14 of *Graduate Texts in Mathematics*. Springer, 1973.

J. Grabowski. Quasi–derivations and QD–algebroids. *Rep. Math. Phys.*, 52(3):445–451, 2003.

W. Greub, S. Halperin, and R. Vanstone. *Connections, Curvature and Cohomology*, volume 1. Academic Press, 1972.

W. Greub, S. Halperin, and R. Vanstone. *Connections, Curvature and Cohomology*, volume 2. Academic Press, 1973.

W. Greub, S. Halperin, and R. Vanstone. *Connections, Curvature and Cohomology*, volume 3. Academic Press, 1976.

W. Greub and H. R. Petry. On the lifting of structure groups. In K. Bleuler, H. R. Petry, and A. Reetz, editors, *Differential Geometrical Methods in Mathematical Physics II, 1977*. Springer–Verlag, 1978. Lecture Notes in Mathematics 676.

V. Guillemin and S. Sternberg. *Geometric Asymptotics*, volume 14 of *Mathematical Surveys and Monographs*. American Mathematical Society, revised edition, 1990.

V. Guillemin and S. Sternberg. *Supersymmetry and Equivariant de Rham Theory*. Springer, 1999.

A. Haefliger. Sur l'extension du groupe structural d'une espace fibré. *C. R. Acad. Sci. Paris, Série A*, 243:558–560, 1956.

A. Haefliger. Ehresmann: Un Géométre. *Gazette des Mathématiciens*, 13:27–35, 1980. Reprinted in [Ehresmann, 1984], Partie I–1 et I–2.

L.-g. He, Z.-J. Liu, and D.-S. Zhong. Poisson actions and Lie bialgebroid morphisms. In *Quantization, Poisson Brackets and Beyond (Manchester, 2001)*, volume 315 of *Contemp. Math.*, pages 235–244. Amer. Math. Soc., Providence, RI, 2002.

R. Hermann. Analytic continuation of group representations, IV. *Comm. Math. Phys.*, 5:131–156, 1967.

J.-C. Herz. Pseudo–algèbres de Lie. *C. R. Acad. Sci. Paris, Série A*, 236:1935–1937, 1953a.

J.-C. Herz. Pseudo–algèbres de Lie, II. *C. R. Acad. Sci. Paris, Série A*, 236:2289–2291, 1953b.

P. J. Higgins. *Notes on Categories and Groupoids*. van Nostrand Reinhold, 1971.

P. J. Higgins and K. C. H. Mackenzie. Algebraic constructions in the category of Lie algebroids. *J. Algebra*, 129:194–230, 1990a.

P. J. Higgins and K. C. H. Mackenzie. Fibrations and quotients of differentiable groupoids. *J. London Math. Soc.* (2), 42:101–110, 1990b.

P. J. Higgins and K. C. H. Mackenzie. Duality for base–changing morphisms of vector bundles, modules, Lie algebroids and Poisson bundles. *Math. Proc. Cambridge Philos. Soc.*, 114(3):471–488, 1993.

G. Hochschild. Group extensions of Lie groups. *Ann. Math.*, 54:96–109, 1951.

G. Hochschild. Cohomology classes of finite type, and finite–dimensional kernels for Lie algebras. *Amer. J. Math.*, 76:763–778, 1954a.

G. Hochschild. Lie algebra kernels and cohomology. *Amer. J. Math.*, 76:698–716, 1954b.

G. Hochschild. Simple algebras with purely inseparable splitting fields of exponent 1. *Trans. Amer. Math. Soc.*, 79:477–489, 1955.

G. Hochschild and G. D. Mostow. Cohomology of Lie groups. *Illinois J. Math.*, 6:367–401, 1962.

G. Hochschild and J. Serre. Cohomology of Lie algebras. *Ann. Math.*, 57:591–603, 1953.

S. Hu. *Homotopy Theory*. Academic Press, 1959.

J. Huebschmann. Poisson cohomology and quantization. *J. Reine Angew. Math.*, 408:57–113, 1990.

J. Huebschmann. Lie–Rinehart algebras, Gerstenhaber algebras and Batalin–Vilkovisky algebras. *Ann. Inst. Fourier (Grenoble)*, 48(2): 425–440, 1998.

J. Huebschmann. Duality for Lie–Rinehart algebras and the modular class. *J. Reine Angew. Math.*, 510:103–159, 1999a.

J. Huebschmann. Extensions of Lie–Rinehart algebras and the Chern–Weil construction. In *Higher Homotopy Structures in Topology and Mathematical Physics (Poughkeepsie, NY, 1996)*, pages 145–176. Amer. Math. Soc., Providence, RI, 1999b.

D. Iglesias–Ponte and J. C. Marrero. Jacobi groupoids and generalized Lie bialgebroids. *J. Geom. Phys.*, 48(2-3):385–425, 2003. ISSN 0393-0440.

L. Illusie. *Complexe Cotangent et Déformations, II*. Springer–Verlag Lecture Notes in Mathematics, number 283, 1972.

N. Jacobson. On pseudo–linear transformations. *Proc. Nat. Acad. Sci.*, 21:667–670, 1935.

N. Jacobson. Pseudo–linear transformations. *Ann. Math.*, 38:485–506, 1937.

N. Jacobson. An extension of Galois theory to non–normal and non–separable fields. *Amer. J. Math.*, 66:1–29, 1944.

A. Jadczyk and D. Kastler. Graded Lie–Cartan pairs, I. *Rep. Math. Phys.*, 25:1–51, 1987a.

A. Jadczyk and D. Kastler. Graded Lie–Cartan pairs, II. *Ann. Physics*, 179:169–200, 1987b.

F. W. Kamber and P. Tondeur. *Invariant Differential Operators and the Cohomology of Lie Algebra Sheaves*. Number 113 in Mem. Amer. Math. Soc. American Mathematical Society, Providence, R.I., 1971.

M. V. Karasëv. Analogues of objects of the theory of Lie groups for nonlinear Poisson brackets. *Izv. Akad. Nauk SSSR Ser. Mat.*, 50: 508–538, 638, 1986. English translation: Math. USSR–Izv. 28 (1987), no. 3, 497–527.

M. V. Karasëv. The Maslov quantization conditions in higher cohomology and analogs of notions developed in Lie theory for canonical fibre bundles of symplectic manifolds. I, II. *Selecta Math. Soviet.*, 8(3):213–234, 235–258, 1989. Translated from the Russian by Pavel Buzytsky. Preprint, Moscov. Inst. Electron Mashinostroeniya, 1981, deposited at VINITI, 1982.

M. V. Karasëv and V. P. Maslov. *Nonlinear Poisson Brackets*. American Mathematical Society, Providence, RI, 1993. Geometry and quantization, Translated from the Russian by A. Sossinsky [A. B. Sosinskiĭ] and M. Shishkova.

D. Kastler and R. Stora. Lie–Cartan pairs. *J. Geom. Phys.*, 2:1–31, 1985.

S. Kobayashi and K. Nomizu. *Foundations of Differential Geometry*, volume 1. Interscience, 1963.

S. Kobayashi and K. Nomizu. *Foundations of Differential Geometry*, volume 2. Interscience, 1969.

I. Kolář, P. W. Michor, and J. Slovák. *Natural Operations in Differential Geometry*. Springer–Verlag, 1993.

K. Konieczna and P. Urbański. Double vector bundles and duality. *Arch. Math. (Brno)*, 35(1):59–95, 1999.

Y. Kosmann. Groupes de transformations et covariance des opérateurs différentiels. *C. R. Acad. Sci. Paris Sér. A-B*, 275:A1235–A1237, 1972.

Y. Kosmann. On Lie transformation groups and the covariance of differential operators. In *Differential geometry and relativity*, pages 75–89. Mathematical Phys. and Appl. Math., Vol. 3. Reidel, Dordrecht, 1976.

Y. Kosmann – Schwarzbach. Lie bialgebras, Poisson Lie groups and dressing transformations. In *Integrability of Nonlinear Systems*, volume 638 of *Lecture Notes in Phys.*, pages 107–173. Springer, Berlin, 2004.

Y. Kosmann – Schwarzbach and F. Magri. Poisson–Lie groups and complete integrability. I. Drinfel'd bialgebras, dual extensions and their canonical representations. *Ann. Inst. H. Poincaré Phys. Théor.*, 49 (4):433–460, 1988.

Y. Kosmann–Schwarzbach. Dérivées de Lie des morphismes de fibrés. In *Differential geometry (Paris, 1976/1977)*, pages 55–71. Univ. Paris VII, Paris, 1978.

Y. Kosmann–Schwarzbach. Vector fields and generalized vector fields on fibered manifolds. In *Geometry and Differential Geometry (Proc. Conf., Univ. Haifa, Haifa, 1979)*, volume 792 of *Lecture Notes in Mathematics*, pages 307–355. Springer, Berlin, 1980.

Y. Kosmann–Schwarzbach. Exact Gerstenhaber algebras and Lie bialgebroids. *Acta Appl. Math.*, 41:153–165, 1995.

Y. Kosmann–Schwarzbach. From Poisson algebras to Gerstenhaber algebras. *Ann. Inst. Fourier (Grenoble)*, 46(5):1243–1274, 1996a.

Y. Kosmann–Schwarzbach. The Lie bialgebroid of a Poisson–Nijenhuis manifold. *Lett. Math. Phys.*, 38:421–428, 1996b.

Y. Kosmann–Schwarzbach. Modular vector fields and Batalin–Vilkovisky algebras. In *Poisson geometry (Warsaw, 1998)*, pages 109–129. Polish Acad. Sci., Warsaw, 2000.

Y. Kosmann–Schwarzbach and K. C. H. Mackenzie. Differential operators and actions of Lie algebroids. In *Quantization, Poisson brackets and Beyond (Manchester, 2001)*, volume 315 of *Contemp. Math.*, pages 213–233. Amer. Math. Soc., Providence, RI, 2002.

Y. Kosmann–Schwarzbach and F. Magri. Poisson–Nijenhuis structures. *Ann. Inst. H. Poincaré Phys. Théor.*, 53(1):35–81, 1990.

B. Kostant. Quantization and unitary representations. In C. T. Taam, editor, *Lectures in Modern Analysis and Applications, III*, pages 87–208. Springer–Verlag Lecture Notes in Mathematics, number 170, 1970.

B. Kostant and S. Sternberg. Anti–Poisson algebras and current algebras. Preprint, 13pp., 1990.

J.-L. Koszul. Crochet de Schouten–Nijenhuis et cohomologie. In *Elie Cartan et les mathématiques d'aujourd'hui*, Astérisque, numéro hors série, pages 257–271, Lyon, juin 1984. Société Mathématique de France, 1985.

J. Kubarski. Lie algebroid of a principal fibre bundle. *Publ. Dép. Math. Nouvelle Sér. A, 89–1, Univ. Claude–Bernard, Lyon*, pages 1–66, 1989.

A. Kumpera. An introduction to Lie groupoids. Duplicated notes, Núcleo de Estudos e Pesquisas Científicas, Rio de Janeiro, 1971.

A. Kumpera. Invariants différentiels d'un pseudogroupe de Lie. I. *J. Differential Geom.*, 10:289–345, 1975.

A. Kumpera and D. C. Spencer. *Lie Equations. Volume I: General Theory*. Princeton University Press, 1972.

N. P. Landsman. *Mathematical Topics between Classical and Quantum Mechanics*. Springer–Verlag, New York, 1998.

N. P. Landsman. Quantization as a functor. In *Quantization, Poisson Brackets and Beyond (Manchester, 2001)*, volume 315 of *Contemp. Math.*, pages 9–24. Amer. Math. Soc., Providence, RI, 2002.

H. B. Lawson, Jr. Foliations. *Bull. Amer. Math. Soc.*, 80:369–418, 1974.

L.-C. Li and S. Parmentier. On dynamical Poisson groupoids I. Institut G. Desargues (UMR 5028) Preprint series. August 2002/07. arXiv:math.DG/0209212.

P. Libermann. Sur la géométrie des prolongements des espaces fibrés vectoriels. *Ann. Inst. Fourier (Grenoble)*, 14:145–172, 1964.

P. Libermann. *Sur les prolongements des fibrés principaux et des groupoïdes différentiables banachiques*, volume 42, Analyse Globale of *Seminaire de Mathématiques Supérieures*, pages 7–108. Les Presses de l'Université de Montréal, 1971.

P. Libermann. Sur les groupoïdes différentiables et le "presque parallélisme". *Sympos. Math.*, 10:59–93, 1972.

P. Libermann. Parallélismes. *J. Differential Geom.*, 8:511–539, 1973.

P. Libermann. On "fibre parallelism" and locally reductive spaces. In *Global analysis and its applications*, Lectures, International Seminar Course, ICTP, Trieste 1972. Volume III, pages 13–23. International Atomic Energy Agency, Vienna, 1974.

P. Libermann. On contact groupoids and their symplectification. In *Analysis and Geometry in Foliated Manifolds (Santiago de Compostela, 1994)*, pages 153–176. World Sci. Publishing, River Edge, NJ, 1995.

P. Libermann and C. Marle. *Symplectic Geometry and Analytical Mechanics*. Mathematics and its applications. D. Reidel, 1987. Translated by B. E. Schwarzbach.

A. Lichnerowicz. Les variétés de Poisson et leurs algèbres de Lie associées. *J. Differential Geom.*, 12:253–300, 1977.

Z.-J. Liu, A. Weinstein, and P. Xu. Dirac structures and Poisson homogeneous spaces. *Comm. Math. Phys.*, 192:121–144, 1998.

Z.-J. Liu and P. Xu. Exact Lie bialgebroids and Poisson groupoids. *Geom. Funct. Anal.*, 6:138–145, 1996.

Z.-J. Liu and P. Xu. The local structure of Lie bialgebroids. *Lett. Math. Phys.*, 61(1):15–28, 2002.

J.-H. Lu. Momentum mappings and reduction of Poisson actions. In Dazord and Weinstein [1991], pages 209–226.

J.-H. Lu and A. Weinstein. Groupoïdes symplectiques doubles des groupes de Lie–Poisson. *C. R. Acad. Sci. Paris Sér. I Math.*, 309: 951–954, 1989.

J.-H. Lu and A. Weinstein. Poisson Lie groups, dressing transformations, and Bruhat decompositions. *J. Differential Geom.*, 31:501–526, 1990.

S. Mac Lane. *Homology*. Classics in Mathematics. Springer-Verlag, Berlin, 1995. Reprint of the 1975 edition.

K. Mackenzie. Rigid cohomology of topological groupoids. *J. Austral. Math. Soc. (Series A)*, 26:277–301, 1978.

K. Mackenzie. Infinitesimal theory of principal bundles. Talk given to 50th ANZAAS Conference, Adelaide, 8 pp, 1980.

K. Mackenzie. *Lie Groupoids and Lie Algebroids in Differential Geometry*. London Mathematical Society Lecture Note Series, no. 124. Cambridge University Press, Cambridge, 1987a.

K. C. H. Mackenzie. Integrability obstructions for extensions of Lie algebroids. *Cahiers Topologie Géom. Différentielle Catégoriques*, 28: 29–52, 1987b.

K. C. H. Mackenzie. A note on Lie algebroids which arise from groupoid actions. *Cahiers Topologie Géom. Différentielle Catégoriques*, 28:283–302, 1987c.

K. C. H. Mackenzie. Infinitesimal characterization of homogeneous bundles. *Proc. Amer. Math. Soc.*, 103:1271–1277, 1988a.

K. C. H. Mackenzie. On extensions of principal bundles. *Ann. Global Anal. Geom.*, 6(2):141–163, 1988b.

K. C. H. Mackenzie. Double Lie algebroids and second–order geometry, I. *Adv. Math.*, 94(2):180–239, 1992.

K. C. H. Mackenzie. Lie algebroids and Lie pseudoalgebras. *Bull. London Math. Soc.*, 27(2):97–147, 1995.

K. C. H. Mackenzie. Drinfel'd doubles and Ehresmann doubles for Lie algebroids and Lie bialgebroids. *Electron. Res. Announc. Amer. Math. Soc.*, 4:74–87, 1998.

K. C. H. Mackenzie. On symplectic double groupoids and the duality of Poisson groupoids. *Internat. J. Math.*, 10:435–456, 1999.

K. C. H. Mackenzie. Affinoid structures and connections. In J. Grabowski and P. Urbański, editors, *Poisson Geometry: Stanisław Zakrzewski In Memoriam*, number 51 in Banach Center Publications, pages 175–186, Warsaw, 2000a.

K. C. H. Mackenzie. Double Lie algebroids and second–order geometry, II. *Adv. Math.*, 154:46–75, 2000b.

K. C. H. Mackenzie. Notions of double for Lie algebroids, . arXiv:math.DG/0011212.

K. C. H. Mackenzie. A unified approach to Poisson reduction. *Lett. Math. Phys.*, 53:215–232, 2000.

K. C. H. Mackenzie and P. Xu. Lie bialgebroids and Poisson groupoids. *Duke Math. J.*, 73(2):415–452, 1994.

K. C. H. Mackenzie and P. Xu. Classical lifting processes and multiplicative vector fields. *Quarterly J. Math. Oxford (2)*, 49:59–85, 1998.

K. C. H. Mackenzie and P. Xu. Integrability of Lie bialgebroids. *Topology*, 39:445–467, 2000.

S. Majid. Matched pairs of Lie groups associated to solutions of the Yang–Baxter equations. *Pacific J. Math.*, 141:311–332, 1990.

S. Majid. *Foundations of Quantum Group Theory*. Cambridge University Press, 1995.

M.-P. Malliavin. Algèbre homologique et opérateurs différentiels. In J. L. Bueso, P. Jara, and B. Torrecillas, editors, *Ring Theory*, pages 173–186. Springer–Verlag Lecture Notes in Mathematics, number 1328, 1988.

P. Malliavin. *Géométrie Différentielle Intrinsèque.* Hermann, Paris, 1972.

Y. I. Manin. Neveu–Schwarz sheaves and differential equations for Mumford superforms. *J. Geom. Phys.*, 5:161–181, 1988.

C. Marle. On submanifolds and quotients of Poisson and Jacobi manifolds. In *Poisson Geometry (Warsaw, 1998)*, volume 51 of *Banach Center Publ.*, pages 197–209. Polish Acad. Sci., Warsaw, 2000.

J. E. Marsden, T. Raţiu, and A. Weinstein. Semidirect products and reduction in mechanics. *Trans. Amer. Math. Soc.*, 281:147–177, 1984.

K. Mikami. Symplectic double groupoids over Poisson $(ax + b)$–groups. *Trans. Amer. Math. Soc.*, 324(1):447–463, 1991.

K. Mikami and A. Weinstein. Moments and reduction for symplectic groupoids. *Publ. Res. Inst. Math. Sci.*, 24:121–140, 1988.

J. Milnor. On the existence of a connection with curvature zero. *Comm. Math. Helv.*, 32:215–223, 1958.

J. Milnor. *Morse Theory.* Annals of Mathematics Studies. Princeton University Press, 1963.

I. Moerdijk. Lie groupoids, gerbes, and non–abelian cohomology. *K–Theory*, 28(3):207–258, 2003.

I. Moerdijk and J. Mrčun. On integrability of infinitesimal actions. *Amer. J. Math.*, 124(3):567–593, 2002.

I. Moerdijk and J. Mrčun. *Introduction to Foliations and Lie Groupoids*, volume 91 of *Cambridge Studies in Advanced Mathematics.* Cambridge University Press, Cambridge, 2003.

I. Moerdijk and J. Mrčun. On the integrability of subalgebroids, . arXiv:math.DG/0406558.

P. Molino. *Riemannian Foliations.* Birkhäuser, 1988.

G. Morandi, C. Ferrario, G. Lo Vecchio, G. Marmo, and C. Rubano. The inverse problem in the calculus of variations and the geometry of the tangent bundle. *Phys. Rep.*, 188(3–4):147–284, 1990.

M. Mori. On the three–dimensional cohomology group of Lie algebras. *J. Math. Soc. Japan*, 5:171–183, 1953.

P. Muhly, J. Renault, and D. Williams. Equivalence and isomorphism for groupoid C^*–algebras. *J. Operator Theory*, 17:3–22, 1987.

Y. Ne'eman. Commutateurs de courants locaux et termes à gradient dans une algèbre de Lie. Conférence donneé à Strasbourg au cours de la Douzième Rencontre entre Physiciens Théoriciens et Mathématiciens dans le cadre de la R.C.P. n° 25 du Centre National de la Recherche Scientifique du 1er au 3 juin 1971, unpublished.

E. Nelson. *Tensor Analysis*. Princeton University Press, Princeton, N.J., 1967.

A. Nijenhuis. Jacobi–type identities for bilinear differential concomitants of certain tensor fields. I, II. *Nederl. Akad. Wetensch. Proc. Ser. A.* **58** = *Indag. Math.*, 17:390–397, 398–403, 1955.

V. Nistor. Groupoids and the integration of Lie algebroids. *J. Math. Soc. Japan*, 52(4):847–868, 2000.

R. S. Palais. The cohomology of Lie rings. In *Proc. Sympos. Pure Math.*, *Vol. III*, pages 130–137. American Mathematical Society, Providence, R.I., 1961.

R. S. Palais. Differential operators on vector bundles. In R. S. Palais, editor, *Seminar on the Atiyah–Singer Index Theorem*, pages 51–93. Princeton University Press, 1965. Chapter IV.

R. S. Palais. *Foundations of Global Non–Linear Analysis*. W. A. Benjamin, Inc., New York–Amsterdam, 1968.

A. L. T. Paterson. *Groupoids, Inverse Semigroups, and their Operator Algebras*, volume 170 of *Progress in Mathematics*. Birkhäuser Boston Inc., Boston, MA, 1999.

J. F. Pommaret. Troisième théorème fondamental pour les pseudogroupes de Lie transitifs. *C. R. Acad. Sci. Paris, Série A*, 284: 429–432, 1977.

J. Pradines. Théorie de Lie pour les groupoïdes différentiables. Relations entre propriétés locales et globales. *C. R. Acad. Sci. Paris, Série A*, 263:907–910, 1966.

J. Pradines. Géométrie différentielle au–dessus d'un groupoïde. *C. R. Acad. Sci. Paris, Série A*, 266:1194–1196, 1967a.

J. Pradines. Théorie de Lie pour les groupoïdes différentiables. Calcul différentiel dans la catégorie des groupoïdes infinitésimaux. *C. R. Acad. Sci. Paris, Série A*, 264:245–248, 1967b.

J. Pradines. Troisième théorème de Lie pour les groupoïdes différentiables. *C. R. Acad. Sci. Paris, Série A*, 267:21–23, 1968.

J. Pradines. Fibrés vectoriels doubles et calcul des jets non holonomes. Notes polycopiées, Amiens, 1974a.

J. Pradines. Représentation des jets non holonomes par des morphismes vectoriels doubles soudés. *C. R. Acad. Sci. Paris, Série A*, 278:1523–1526, 1974b.

J. Pradines. Suites exactes vectorielles doubles et connexions. *C. R. Acad. Sci. Paris, Série A*, 278:1587–1590, 1974c.

J. Pradines. How to define the graph of a singular foliation. *Cahiers Topologie Géom. Différentielle*, 26:339–380, 1986a.

J. Pradines. Quotients de groupoïdes différentiables. *C. R. Acad. Sci. Paris Sér. I Math.*, 303:817–820, 1986b.

J. Pradines. Remarque sur le groupoïde cotangent de Weinstein–Dazord. *C. R. Acad. Sci. Paris Sér. I Math.*, 306:557–560, 1988.

J. Pradines. Private communication. 1989.

N. V. Quê. Du prolongement des espaces fibrés et des structures infinitésimales. *Ann. Inst. Fourier (Grenoble)*, 17:157–223, 1967.

N. V. Quê. Sur l'espace de prolongement différentiable. *J. Differential Geom.*, 2:33–40, 1968.

N. V. Quê. Nonabelian Spencer cohomology and deformation theory. *J. Differential Geom.*, 3:165–211, 1969.

N. V. Quê and A. A. M. Rodrigues. Troisième théorème fondamental de réalization de Cartan. *Ann. Inst. Fourier (Grenoble)*, 25(1):251–282, 1975.

G. S. Rinehart. Differential forms on general commutative algebras. *Trans. Amer. Math. Soc.*, 108:195–222, 1963.

D. J. S. Robinson. *A Course in the Theory of Groups*, volume 80 of *Graduate Texts in Mathematics*. Springer–Verlag, New York, 1982.

A. M. Rodrigues. The first and second fundamental theorems of Lie for Lie pseudo groups. *Amer. J. Math.*, 84:265–282, 1962.

D. Roytenberg. Courant algebroids, derived brackets and even symplectic supermanifolds. Thesis, Univ. California, Berkeley, 1999. arXiv:math.DG/9910078.

S. Runciman. *Byzantine Art and Civilization*. Folio Society, London, 2004.

P. Schaller and T. Strobl. Poisson structure induced (topological) field theories. *Modern Phys. Lett. A*, 9(33):3129–3136, 1994.

J. A. Schouten. On the differential operators of first order in tensor calculus. In *Convegno Internazionale di Geometria Differenziale, Italia, 1953*, pages 1–7. Edizioni Cremonese, Roma, 1954.

J. Serre. *Lie Algebras and Lie Groups*, volume 1500 of *Lecture Notes in Mathematics*. Springer–Verlag, second edition, 1992.

U. Shukla. A cohomology for Lie algebras. *J. Math. Soc. Japan*, 18:275–289, 1966.

I. M. Singer and J. A. Thorpe. *Lecture Notes on Elementary Topology and Geometry*. Scott, Foresman and Company, Glenview, 1967.

P. A. Smith. Some topological notions connected with a set of generators. In *Proceedings of the International Congress of Mathematicians, Cambridge, Mass., 1950, vol. 2*, pages 436–441, Providence, R. I., 1952. Amer. Math. Soc.

M. Spivak. *A Comprehensive Introduction to Differential Geometry.* Volumes 1–5. Publish or Perish, Berkeley, second edition, 1979.

J. D. Stasheff. Continuous cohomology of groups and classifying spaces. *Bull. Amer. Math. Soc.*, 84:513–530, 1978.

P. Stefan. Accessible sets, orbits, and foliations with singularities. *Proc. London Math. Soc. (3)*, 29:699–713, 1974.

P. Stefan. Integrability of systems of vector fields. *J. London Math. Soc. (2)*, 21(3):544–556, 1980. ISSN 0024-6107.

H. J. Sussmann. Orbits of families of vector fields and integrability of distributions. *Trans. Amer. Math. Soc.*, 180:171–188, 1973.

M. Sweedler and M. Takeuchi. From differential geometry to differential algebra. IBM Thomas J. Watson Research Centre preprint, 198 pp., 1986.

N. Teleman. A characteristic ring of a Lie algebra extension. *Atti. Accad. Naz. Lincei. Rend. Cl. Sci. Fis. Mat. Natur. (8)*, 52:498–506, 708–711, 1972.

W. M. Tulczyjew. *Geometric Formulation of Physical Theories*, volume 11 of *Monographs and Textbooks in Physical Science*. Bibliopolis, Naples, 1989.

A. Y. Vaĭntrob. Lie algebroids and homological vector fields. *Uspekhi Mat. Nauk*, 52(2(314)):161–162, 1997.

I. Vaisman. *Lectures on the Geometry of Poisson Manifolds*, volume 118 of *Progress in Mathematics*. Birkhäuser Verlag, Basel, 1994.

W. T. van Est. Group cohomology and Lie algebra cohomology in Lie groups I, II. *Nederl. Akad. Wetensch. Proc. Ser. A.*, 56:484–504, 1953.

W. T. van Est. On the algebraic cohomology concepts in Lie groups I, II. *Nederl. Akad. Wetensch. Proc. Ser. A.*, 58:225–233, 286–294, 1955a.

W. T. van Est. Une application d'une méthode de Cartan-Leray. *Nederl. Akad. Wetensch. Proc. Ser. A.*, 58:542–544, 1955b.

W. T. van Est. Local and global groups I, II. *Nederl. Akad. Wetensch. Proc. Ser. A.*, 65:391–425, 1962.

V. S. Varadarajan. *Lie Groups, Lie Algebras, and their Representations.* Prentice–Hall, Inc., Englewood Cliffs, N. J., 1974.

P. ver Eecke. Calculus of jets and higher–order connections. Mathematics research report, University of Melbourne, 1981. Translated, edited and revised by J. J. Cross and F. R. Smith. Second Edition.

J. Virsik. On the holonomy of higher–order connections. *Cahiers Topologie Géom. Différentielle*, 12:197–212, 1971.

T. Voronov. Graded manifolds and Drinfeld doubles for Lie bialgebroids. In *Quantization, Poisson Brackets and Beyond (Manchester, 2001)*, volume 315 of *Contemp. Math.*, pages 131–168. Amer. Math. Soc., Providence, RI, 2002.

A. Weil. *Variétés Kählériennes*. Hermann, Paris, 1958.

A. Weinstein. The symplectic "category". In H.-D. Doebner, S. I. Andersson, and H. R. Petry, editors, *Differential Geometric Methods in Mathematical Physics, Clausthal 1980*, pages 45–51. Springer–Verlag Lecture Notes in Mathematics, number 905, 1982.

A. Weinstein. Symplectic groupoids and Poisson manifolds. *Bull. Amer. Math. Soc. (N.S.)*, 16:101–104, 1987.

A. Weinstein. Coisotropic calculus and Poisson groupoids. *J. Math. Soc. Japan*, 40:705–727, 1988.

A. Weinstein. Blowing up realizations of Heisenberg–Poisson manifolds. *Bull. Sci. Math.* (2), 113:381–406, 1989.

A. Weinstein. Affine Poisson structures. *Internat. J. Math.*, 1:343–360, 1990.

A. Weinstein. Groupoids: Unifying internal and external symmetry. *Notices Amer. Math. Soc.*, 43:744–752, 1996.

A. Weinstein. Poisson geometry. *Differential Geom. Appl.*, 9(1-2):213–238, 1998.

A. Weinstein. Linearization of regular proper groupoids. *J. Inst. Math. Jussieu*, 1(3):493–511, 2002.

H. E. Winkelnkemper. The graph of a foliation. *Ann. Global Anal. Geom.*, 1:51–75, 1983.

N. M. J. Woodhouse. *Geometric Quantization*. Clarendon Press, Oxford, second edition, 1992.

P. Xu. Morita equivalence of Poisson manifolds. *Comm. Math. Phys.*, 142:493–509, 1991a.

P. Xu. Morita equivalent symplectic groupoids. In Dazord and Weinstein [1991], pages 291–311.

P. Xu. Symplectic groupoids of reduced Poisson spaces. *C. R. Acad. Sci. Paris Sér. I Math.*, 314:457–461, 1992.

P. Xu. On Poisson groupoids. *Internat. J. Math.*, 6(1):101–124, 1995.

P. Xu. Gerstenhaber algebras and BV–algebras in Poisson geometry. *Comm. Math. Phys.*, 200(3):545–560, 1999.

K. Yano and S. Ishihara. *Tangent and Cotangent Bundles*, volume 16 of *Pure and Applied Mathematics*. Marcel Dekker, Inc., 1973.

S. Zakrzewski. Quantum and classical pseudogroups. Part I: Union pseudogroups and their quantization. *Comm. Math. Phys.*, 134:347–370, 1990a.

S. Zakrzewski. Quantum and classical pseudogroups. Part II: Differential and symplectic pseudogroups. *Comm. Math. Phys.*, 134:371–395, 1990b.

D.-s. Zhong and L.-g. He. On actions of groupoids and morphisms of Lie bialgebroids. *Adv. Math. (China)*, 32(3):311–318, 2003.

N. T. Zung. Proper Groupoids and Momentum Maps: Linearization, Affinity and Convexity. arXiv:math.SG/0407208.

Index

Printed in the United States
By Bookmasters